Engineering Materials

The "Engineering Materials" series provides topical information on innovative, structural and functional materials and composites with applications in optical, electronical, mechanical, civil, aeronautical, medical, bio and nano engineering. The individual volumes are complete, comprehensive monographs covering the structure, properties, manufacturing process and applications of these materials. This multidisciplinary series is devoted to professionals, students and all those interested in the latest developments in the Materials Science field.

More information about this series at http://www.springer.com/series/4288

Pietro Pedeferri

Corrosion Science and Engineering

Edited by Luciano Lazzari and MariaPia Pedeferri

In Cooperation with Marco Ormellese, Andrea Brenna, Silvia Beretta, Fabio Bolzoni, Maria Vittoria Diamanti

 Springer

Pietro Pedeferri (Deceased)
Politecnico di Milano
Milan, Italy

ISSN 1612-1317 ISSN 1868-1212 (electronic)
Engineering Materials
ISBN 978-3-030-07380-0 ISBN 978-3-319-97625-9 (eBook)
https://doi.org/10.1007/978-3-319-97625-9

This Springer imprint is published by the registered company Springer Nature Switzerland AG
The registered company address is: Gewerbestrasse 11, 6330 Cham, Switzerland

Preface

This Pedeferri's *Corrosion Science and Engineering* textbook is the English edition of Pietro Pedeferri's *Corrosione e Protezione dei Materiali*, Polipress, Milano (2007), with many integrations made by his collaborators of the PoliLaPP, the Laboratory of Corrosion of Materials that Pedeferri founded. The main goal while translating and integrating the original Italian book, so far very appreciated in Italy with about 2000 copies printed, is to give a modern and updated handbook on corrosion and corrosion prevention for a twofold use: as a teaching textbook and a modern, technical support for industrial applications. This textbook stands as an ideal learning resource for students of corrosion courses in chemical, mechanical, energy and materials engineering at graduate and advanced undergraduate levels, as well as a valuable reference for engineers.

This English edition, integrated and updated, contains 30 chapters, dealing with corrosion theory (9 chapters), forms of corrosion (7), corrosion control and prevention methods (3), applications in different environments as waters, air, soil, concrete (4), and industrial applications as petrochemical plants, refinery and high temperature (2) as well as corrosion of implants in the human body. Four chapters are dedicated to design, corrosion monitoring, laboratory tests and the statistical processing of corrosion data. Chapters dedicated to the on-field applications propose an overview of the most used metals and relevant case histories. Emphasis has been devoted to cathodic protection and corrosion of reinforced concrete to give merit to the pioneering works carried out by Pietro Pedeferri. Each chapter is enriched by pictures of corrosion case studies analysed by PoliLaPP; most of the samples are actually available at the "Corrosion Museum", where Pietro Pedeferri and his school have collected the most significant corrosion case studies.

The book offers the reader and the user many case histories and an important number of questions and exercises to help check the acquired knowledge. Questions and exercises included in each chapter represent the experience gathered by Pedeferri and his school over the last 50 years as a fruit of teaching, research, consultancy on material selection, failure analysis and corrosion engineering. Answers and solutions of exercises for readers will be available on PoliLaPP website (http://polilapp.chem.polimi.it).

Finally, a warm thank to all collaborators Andrea Brenna Silvia Beretta, Fabio Bolzoni, Maria Vittoria Diamanti for their hard, precious and tenacious work in contributing to the translation, integration and revision of the chapters and the effort spent on collecting more than 300 exercises. Special mention to Marco Ormellese for the unparalleled contribution. Thanks to Roberto Chiesa for reviewing the chapter related to corrosion in the human body, Giorgio Re for the suggestions on chapters dedicated to environmental-assisted cracking, Eleonora Faccioli for the drawing of figures and tables and Davide Prando for the collection of the original pictures.

Milan, Italy   Luciano Lazzari
June 2018   MariaPia Pedeferri

Contents

About the Author

Pietro Pedeferri (1938–2008)

Pietro Pedeferri was a Full Professor in *Corrosion and Protection of Materials* at the School of Engineering at Politecnico di Milano, Italy.

He graduated in chemical engineering (cum laude) at Politecnico di Milano as Montecatini gold medal holder and won the De Nora Award with a thesis on electrochemistry under the supervision of Professor Roberto Piontelli. His career started and continued at Politecnico di Milano, as an Assistant Professor first and then Full Professor in electrochemistry and later in corrosion and protection of materials. In 1968, he was appointed as lecturer of the first ever course on corrosion and protection at an Italian university. He was a Visiting Professor at the University of Cambridge, UK, and the University of Connecticut, USA. From 1993 to 1999, he was Head of the Department of Applied Physical Chemistry at the Politecnico di Milano.

His first academic activity was electrochemistry research; then, in the 1963, he moved on to the corrosion field focusing on industrial and engineering aspects. His topics of study in electrochemistry were overvoltage in sulphamic solutions, anodic effects in Al production cells, anodic oxidation of Ti and so-called valve metals and relevant chromatic effects. His research in corrosion started with cold-worked stainless steels and continued with implanted metals in simulated physiological solutions, corrosion of bronze artefacts and cathodic protection. Since 1985, he dealt with corrosion of steel reinforcements in concrete, indicating factors and conditions for initiation and propagation. In 1991, he invented and proposed a new technique called *cathodic prevention* for concrete structures destined to be chloride contaminated, nowadays included in operative international standards. From the study of the corrosion behaviour of stainless steel reinforcements, he proposed a potential-to-chloride diagram for interpretation: this diagram is now called the *Pedeferri Diagram*.

Meanwhile, he continued his studies on Ti colouring, winning an award in 1988 in Paris, within the international event *Science pour l'art*, and displaying his work in the *Fondazione Corrente Gallery* in Milan, Italy. He revisited the publications of Alessandro Volta and Leopoldo Nobili and then rewrote several chapters of the history of electrochemistry. Some of the Pedeferri's findings on Volta priorities in corrosion are reported in this book.

He published 388 papers and 34 books, and took out 8 patents.

Symbols and Abbreviations

$a_{M^{z+}}$	Activity (or concentration) of ions of metal M in a solution (mol/L)
α	Coefficient (adimensional)
b	Tafel slope (module) (V/decade)
b_a	Tafel slope of the anodic curve (module) (V/decade)
b_c	Tafel slope of the cathodic curve (module) (V/decade)
b_{Fe}	Tafel slope of iron dissolution reaction (V/decade)
b_{H_2}	Tafel slope of hydrogen evolution reaction (V/decade)
b_{O_2}	Tafel slope of oxygen reduction reaction (V/decade)
C	Concentration (mol/L)
C_{rate}	Corrosion rate (mm/y)
$C_{rate,m}$	Mass loss rate (mdd)
CCGS	Critical crevice gap size (μm)
CCT	Critical crevice temperature (°C)
CIPP	Close interval potential profile
CP	Cathodic protection
CPrev	Cathodic prevention
CPCC	Critical pitting chloride concentration
CPT	Critical pitting temperature (°C)
CSE	Saturated copper sulphate electrode (+0.32 V SHE)
d	Distance (m)
d_{eq}	Diameter of the coating equivalent defect (m)
D	Diffusion coefficient (m^2/s)
DL	Design life
δ	Diffusion layer thickness (m)
e^-	Electron
E	Electrode potential (V)
E^{XY}	Potential difference between electrode X and Y (V)
E^0	Standard potential (V)
E_a	Anodic potential (V)
E_c	Cathodic potential (V)

E_{corr}	Free corrosion potential (V)
E_{eq}	Equilibrium potential given by Nernst equation (V)
$E_{IR\text{-}free}$	Potential free of the ohmic drop in CP applications (V)
E_{off}	Off-potential in CP applications (V)
E_{on}	On-potential in CP applications (V)
E_p	Passivation potential (V)
E_{pit}	Pitting potential or passivity breakdown potential (V)
E_{pp}	Primary passivation potential (V)
E_{prot}	Protection potential (V)
E_{rp}	Repassivation potential (V)
E_{tr}	Transpassive potential (V)
EMF	Electromotive force (V)
ΔE	Driving voltage or potential difference (V)
ε	Efficiency (unitary fraction)
F	Faraday constant (96,485 C)
FEM	Finite element method
ϕ	Diameter (m)
G	Gibbs free energy (J/mol)
GACP	Galvanic anode cathodic protection
γ	Mass density (g/cm^3)
ΔG°	Standard Gibbs free energy variation (J/mol)
H	Activation energy (J/mol)
HE	Hydrogen embrittlement
HIC	Hydrogen-induced cracking
HID	Hydrogen-induced damage
η	Overvoltage (with respect to the equilibrium potential) (V)
η_a	Anodic overvoltage (V)
η_{act, O_2}	Activation overvoltage of oxygen reduction (V)
η_c	Cathodic overvoltage (V)
η_{conc, O_2}	Concentration overvoltage of oxygen reduction (V)
η_{H_2}	Activation overvoltage of hydrogen evolution reaction (V)
η_M	Activation overvoltage of metal dissolution reaction (V)
η_{O_2}	Overvoltage of oxygen reduction (V)
i	Current density (mA/m^2)
i_a	Anodic current density (mA/m^2)
i_c	Cathodic current density (mA/m^2)
i_{corr}	Corrosion current density (mA/m^2)
i_{cp}	Critical passivation current density (mA/m^2)
i_{GC}	Current density in galvanic coupling (mA/m^2)
i_L	Oxygen limiting current density (mA/m^2)
i_0	Exchange current density (mA/m^2)
i_{0, H_2}	Exchange current density of hydrogen evolution (mA/m^2)
$i_{0,M}$	Exchange current density of metal M (mA/m^2)

i_{0, O_2}	Exchange current density of oxygen (mA/m^2)
i_p	Passivity current density (mA/m^2)
i_{prot}	Protection current density (mA/m^2)
I	Current (A)
I_a	Anodic current (A)
I_c	Cathodic current (A)
I_e	External current (A)
I_{el}	Current in the electrolyte (A)
I_{interf}	Interference current (A)
I_{prot}	Protection current (A)
ICCP	Impressed current cathodic protection
k	Constant (generic)
κ	Conductivity of an electrolyte (S/m)
K_s	Complex stability constant
L	Length (m)
L_{max}	Throwing power (m)
LSI	Langelier saturation index
m	Mass (g)
M	Generic metal, less noble metal in a coupling
M^{z+}	Oxidised metal species
MIC	Microbiologically influenced corrosion
MOB	Manganese oxidising bacteria
MMO	Mixed metal oxides (of noble metals Ir, Rh, Ru)
MW	Atomic or molecular weight (g/mol)
N	More noble metal in a coupling
N_a	Anode number
p	Porosity of a scale (unitary fraction)
p_{CO_2}	Partial pressure of CO_2 (bar)
p_{H_2S}	Partial pressure of H_2S (bar)
P	Pressure of a gas (bar)
PREN	Pitting resistance equivalent number
Q	Flux of electrical charges (C)
R	Generic ohmic resistance (Ω)
R	Gas constant (1.987 cal/mol K = 8.314 J/mol K)
R_0	Coating insulation resistance (Ω m^2)
R_a	Anode resistance (Ω)
R_c	Cathode resistance (Ω)
R_{cable}	Resistance of feeding cables (Ω)
R_{tot}	Total resistance (Ω)
RH	Relative humidity
RSI	Ryznar saturation index
ρ	Resistivity (Ω m)
ρ_{el}	Electrolyte resistivity (Ω m)
ρ_{met}	Metal resistivity (Ω m)

s	Thickness (m)
S	Surface (m^2)
S_a	Anodic surface (m^2)
S_c	Cathodic surface (m^2)
S_M	Surface of the less noble metal in a coupling (m^2)
S_N	Surface of the more noble metal in a coupling (m^2)
SHE	Standard hydrogen electrode
SCC	Stress corrosion cracking
SCE	Saturated calomel electrode (+0.24 V SHE)
SOHIC	Stress-oriented hydrogen-induced cracking
SRB	Sulphate-reducing bacteria
SSC	Silver/silver chloride reference electrode (+0.25 V SHE)
SSC	Sulphide stress cracking
σ	Conductivity (S/m)
t	Time (s)
T	Temperature (°C; K)
T/R	Transformer/rectifier
TDS	Total dissolved solids or salinity (g/L or mg/L)
v	Velocity (m/s)
V	Voltage or feeding voltage (V)
ΔV	Voltage drop or ohmic drop (V)
ξ	Coating efficiency (unitary fraction)
w	Anode consumption (kg/A y)
ψ	Polarisation or potential shift from the free corrosion potential (V)
ψ^*	Thermodynamic and kinetic contribution of electrode reactions (V)
z	Valence, number of electrons in an electrodic reaction (adimensional)
ZN	Zinc/sea water reference electrode (−0.8 V SHE)

Units

A	Ampere
cal	Calorie
C	Coulomb
°C	Degree centigrade
h	Hour
J	Joule
K	Degree Kelvin
L	Litre
m	Metre
M	Molar
mol	Mole
Ω	Ohm
s	Second

S	Siemens
V	Volt
W	Watt

Chapter 1
General Principles of Corrosion

You are dust and to dust you shall return.

Genesis, 3.19

Abstract This introductory chapter presents the general aspects of corrosion: its origins, the main forms it can take, the general mechanism and corrosion rate involved, and the impact it has on society, with particular reference to the economic aspects. All technical information here briefly mentioned will be addressed in more details in specific chapters.

Fig. 1.1 Case study at the PoliLaPP Corrosion Museum of Politecnico di Milano

© Springer Nature Switzerland AG 2018
P. Pedeferri, *Corrosion Science and Engineering*, Engineering Materials,
https://doi.org/10.1007/978-3-319-97625-9_1

1.1   Corrosion as Metallurgy in Reverse

Materials exposed to aggressive environments may undergo chemical and physical degradation (Fig. 1.1). This degradation is called *"corrosion"* when the material concerned is a metal. The term corrosion derives from Medieval Latin [*corrosionis*, from the verb *corrodere*], Middle English [*corosioun*] and Old French [*corrosion*].

Corrosion is often defined as *"destruction or degradation of a material caused by a reaction to its environment"* and also *"the spontaneous tendency of a metallic component to return to its original state as found in nature"* [quoted from Fontana]. For this reason, corrosion is also called *metallurgy in reverse*, because the corrosion process returns metals to their more thermodynamically stable natural state as oxides or sulphides or other compounds, from which metallurgy transforms to metal by supplying energy. Figure 1.2 shows the whole corrosion related life cycle of iron.

> **Definitions of Corrosion**
> The breaking down or destruction of a material, especially a metal, through chemical reactions. The most common form of corrosion is rusting which occurs when iron combines with oxygen and water [The American Heritage® Science Dictionary].
>
> Corrosion is the deterioration of a material due to interaction with its environment. It is the process by which metallic atoms leave the metal or form compounds in the presence of water and gases. Metal atoms are removed from a structural element until it fails or oxides build up inside a pipe until it is plugged [DOE Fundamentals Handbook Material Science, Volume 1 of 2].
>
> Corrosion can be defined as the deterioration of material by reaction to its environment. Corrosion occurs because of the natural tendency of most

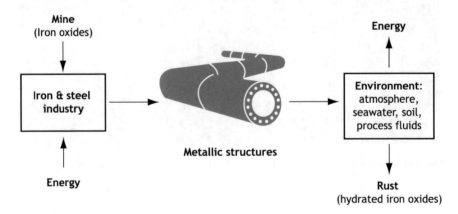

Fig. 1.2 The corrosion process as metallurgy in reverse (adapted from M. Fontana)

metals to return to their natural state; e.g., in the presence of moist air iron will revert to its natural state—iron oxide. Metals can be corroded by the direct reaction of metal to a chemical; e.g. zinc will react with dilute sulfuric acid and magnesium will react with alcohols [NASA, Corrosion Control And Treatment Manual-TM-584C Rev-C.].

Physicochemical interaction between a metal and its environment which results in changes in the properties of the metal and which may often lead to impairment of the function of the metal, the environment or the technical system of which these form a part [ISO 8044].

EFC Definition of Corrosion
In simple terms, corrosion processes may be considered metal's reactions to species in the environment to form chemical compounds. Note that in the definition of corrosion given by the ISO 8044-1986 international standard the term 'corrosion' applies to the process, not to the result, the latter being 'corrosion damage', deterioration or effect. Implicit in the concept of corrosion as a process is the corrosion reaction rate; implicit in the damage caused is the extent and nature of the damage in relation to the function of the systems concerned.

A broader, but widely accepted alternative definition, from the International Union of Pure and Applied Chemistry (IUPAC), encompasses the degradation of non-metals as well as metallic materials, as follows: "*Corrosion is an irreversible interfacial reaction of a material (metal, ceramic, polymer) with its environment which results in consumption of the material or in dissolution into the material of a component of the environment. Often, but not necessarily, corrosion results in effects which are detrimental to the usage of the material considered. Exclusively physical or mechanical processes such as melting or evaporation, abrasion or mechanical fracture are not included in the term corrosion*" [EFC Working Party 7: Corrosion Education].

1.2 The Economic Impact of Corrosion

Corrosion involves industry, infrastructure and cultural heritage concurrently. No sector is left out: energy, transport, chemistry, food and beverages, oil and gas, pharmaceutics, machinery, civil engineering. Corrosion hits metallic and reinforced concrete structures, pipelines transporting hydrocarbons and water, aerial, terrestrial and naval transport infrastructure, bridges, piers, offshore structures, chemical plants and nuclear reactors, power plants, electronic devices, body implants, cultural heritage, artefacts and many more.

The entity of corrosion damage is impressive. It is well summarized by a sign displayed at an NACE conference in capital letters:

While You Are Reading This Message, 10 Tons of Iron Is Corroding Around the World

Estimates by the UK and Japanese Ministries of Industry, the National Bureau of Standards on behalf of the US Congress and the US National Institute of Science and Technology have confirmed that the so-called *cost of corrosion* accounts for around 3–4% of the GNP of industrialized countries. A recent study by NACE (IMPACT 2016) estimates the global cost of corrosion to be US$2.5 trillion, roughly 3.4% of the global Gross Domestic Product (GDP) of a generic country.

The cost of corrosion is given by the sum of *direct* (for example, the costs of damaged components and associated substitution costs,[1] prevention method costs— e.g. protective coatings, cathodic protection, corrosion allowance or redundant solutions, corrosion resistant materials[2]) and *indirect costs* (for example, production loss costs, pollution related costs, loss of image and so on) which also encompass unquantifiable costs such as the loss of human life when catastrophic failures occur.[3] Indirect costs are often difficult to assess and can exceed direct costs.

Corrosion cannot be halted but only reduced to a much greater extent than is normally achieved. In 1971, Hoar suggested, based on various sources, that the costs of corrosion could be reduced by 15–35% simply by applying basic knowledge of corrosion principles, adopting the most familiar techniques such as cathodic protection, corrosion inhibitor injection or selecting a resistant material and improving design. The NACE IMPACT report (2016) confirmed that implementing corrosion prevention best practices could result in global savings of between 15 and 35% of the cost of damage or US$375–875 billion. In a later chapter, the cost of corrosion will be assessed in the context of economic appraisal in material selection.

It must be emphasized that corrosion does not always mean damage but simply an industrial cost, which also has human, social and cultural aspects. Corrosion prevention (i.e., associated corrosion costs) greatly contributes to making industrial processes possible, reducing energy and raw material consumption, making energy conservation, making plants safer and more reliable, preserving cultural heritage and much more.

There is also 'constructive' metal corrosion such as, for example, chemical attack used to highlight its microstructure or to make it rough or glossy, build up a protective or attractive layer, carve its surface, perform selective removal of material, produce specific corrosion components or hydrogen and more. Some

---

[1]It has been estimated that about 40% of steel produced is made to replace corroded steel.

[2]Cost is calculated as that in excess of the carbon steel solution cost, taken as reference.

[3]Eminent historians (Mommsen among them) have suggested that the decline and fall of the Roman Empire was caused by the corrosion of the lead employed for food and beverage pots which poisoned people.

applications lead to artistic or creative corrosion, as was common in the Middle Ages, to decorate weapons, armour and other objects by chemical etching and produce *acqueforti* from steel and copper plates etched with nitric acid. Anodic oxidation of titanium allowed an artist—one of the many Pedeferri's facets—to recreate boreal dawns or magical soap bubbles on its surface. All of these are corrosion processes, which open a window into the world of art.

1.3   Corrosion Forms

Corrosion damage presents two main morphologies with regard to the environmentally exposed surface: on the whole surface, so-called *generalized corrosion*, only on a small portion of it, so-called *localized corrosion*. Moreover, in specific conditions and in the presence of a tensile load, attacks can cause cracks perpendicular to the tensile stress to form, so-called *stress corrosion cracking*. In general, corrosion processes bring all the constituents of the material into solution even if in some cases only one constituent is dissolved (*selective corrosion*) or only the grain border is attacked producing *intergranular corrosion*. Figure 1.3 shows the morphology of the various forms of corrosion schematically.

1.3.1   Uniform or Generalized Corrosion

This is a form of corrosion which affects the whole surface of the metallic material in contact with the corrosive environment. If the attack is spread evenly, it is referred to as *uniform corrosion*, otherwise corrosion is *unevenly generalized*. Material thinning, called also thickness loss, is generated at a typically predictable rate provided that the environmental conditions are known. For example, the corrosion of carbon steel exposed to the atmosphere takes place at a rate varying from a few tens to a few hundred µm/year depending on humidity, temperature, the presence of chlorides and other pollutants. When zinc coatings are used in the same environment, the corrosion rate drops to a value about 10–30 times lower.

Is Corrosion Really a Cost?
The 'corrosion is a cost' approach is misleading. Corrosion has an economic impact in the sense that it implies a net cost that someone has to pay. However, the corrosion cost is inevitable like the corrosion process as thermodynamics states. The logical consequence of this is that corrosion costs should be considered the best option cost, once design targets are achieved, for instance, transporting fluids in total safety for people and the environment.

If this new approach is accepted, it means that the cost of using a corrosion resistant material or using a corrosion prevention method should not be listed

as a direct cost because it is most likely necessary in order to ensure reliability
and safety level targets as should be specified at the design stage. In other
words, the minimum inevitable cost is not the mild steel cost, as is assumed in
an economic appraisal, but rather that which fits the quality requirements
specified in the design documents.

Finally, corrosion prevention costs are not pure costs which could theoretically be saved but rather necessary costs (to be reduced as far as possible,
of course) to minimize or even eliminate the risk of material failure.

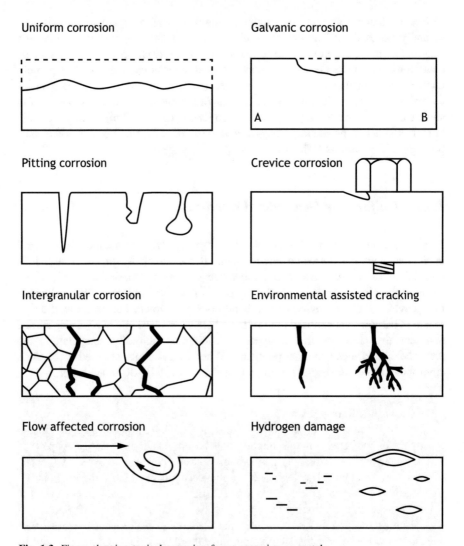

Fig. 1.3 Figure showing typical corrosion forms occurring on metals

1.3.2   Localized Corrosion

Localized corrosion takes place on specific sections of an exposed surface because of two general conditions: a non-homogeneous condition for the material or the environment and a specific localized attack due to the presence of aggressive species.

Non-homogeneous conditions lead to *galvanic corrosion* (also called *bimetallic corrosion*) when different materials are in contact (i.e., a material coupling in which a noble material is present, as in the case of aluminium and copper) or to *differential aeration corrosion* caused by non-homogeneous oxygen availability on a metal surface or *under deposit corrosion* when oxygen diffusion is impeded by a deposit (scales or even former corrosion products) or by so-called *interstitial corrosion* (*crevice corrosion*) when a gap is present on the material (a typical example is under a gasket in flanged joints) or *stray current corrosion* when an electrical field is present in the environment (typically in soil).

Localized attacks can occur because of environmental flows or micro motions between two metals: the former leads to *erosion-corrosion* (due to the turbulence of the aggressive solution) or *impingement corrosion* (when a liquid hits the material) or *cavitation corrosion* as in case of pump impellers, the latter to *fretting corrosion*.

Localized attacks also take place in homogeneous environments when aggressive species locally destroy the passive film present on a metal surface as in the case of stainless steels and many other passive alloys. The morphology of an attack depends on diameter-to-depth ratio and varies from large cavities (craters) to small pinholes, called *pitting*, often showing penetration rates up to more than 1 mm/year.

1.3.3   Stress Corrosion Cracking

For specific material-to-environment couplings and under an enduring tensile load, corrosion attack can take the form of cracks penetrating materials in a direction which is perpendicular to its tensile stress.

This is called *stress corrosion cracking* (SCC) or *corrosion fatigue* depending on whether tensile stress is constant or varies cyclically. It is particularly dangerous because it can jeopardize the reliability of a plant where it occurs.

If crack growth is due to the action of atomic hydrogen, produced on a metal surface for various reasons and then diffusing in the metal, cracking is called *hydrogen embrittlement* (HE). Furthermore, atomic hydrogen in metal, in addition to cracks, can cause other types of damage such as blister or bulge formation or small internal cracks (*blistering* and *stepwise cracking*).

1.4   Corrosion Rates

Regardless of the morphology of an attack, corrosion leads to a loss in mass. In this event, corrosion rates can be expressed as mass loss per time per unit area as is often used in laboratories. In engineering, penetration rate (or thickness loss rate) is preferred. For some forms of corrosion, such as SCC or corrosion-fatigue giving rise to crack formation, mass loss and reductions in thickness are of no practical use since time to failure or crack propagation rate (crack growth) is more useful.

1.4.1   Uniform Corrosion

Uniform corrosion, i.e., when corrosion attacks are uniformly distributed over the surface of a material, can be measured by the mass loss rate per unit area, $C_{rate,m}$, exposed to the aggressive environment and calculated by the equation:

$$C_{rate,m} = \frac{\Delta m}{S \cdot t} \tag{1.1}$$

where Δm is mass loss in time, t, and S is exposed surface area.

The most frequently used units of measurement of mass loss rate are **mg/dm²day** (*mdd*) and **mg/m²hour** (*mmh*). These units are of interest when the amount of dissolved metal is of concern, for example in pollution related matters, as in the case of the contamination of tin in tomato cans. It is most frequently used in laboratory testing.

In industrial applications, thinning rates (thickness loss rate or penetration rate) are preferred to mass loss as a corrosion measurement and given as, C_{rate} in the equation:

$$C_{rate} = \frac{\Delta m}{\gamma \cdot S \cdot t} = \frac{C_{rate,m}}{\gamma} \tag{1.2}$$

where γ is mass metal density. The most frequently used units of measurement of thickness loss rates are *mm/y* and *mpy* (mils per year) (1 mpy = 0.025 mm/year or 25.4 µm/year).

The relationship between the mass loss rate, $C_{rate,m}$, and the penetration rate, C_{rate}, for most industry-used metals such as iron, copper and zinc, which have a mass density in the range of 7–8 Mg/m³ (tons per cubic meter) is:

$$1\,\text{mdd} \cong 5\,\mu\text{m/year} \quad 1\,\text{mmh} \cong 1\,\mu\text{m/year} \tag{1.3}$$

Table 1.1 shows an industry accepted corrosion rate classification for metallic materials first proposed by Fontana's classic *Corrosion Engineering* textbook. The values reported are valid for the oil and gas and chemical industries (although they

Table 1.1 Classification of industry accepted corrosion rates (adapted from Fontana 1986)

Corrosion behaviour	Corrosion rate				
	mpy (mils/year)	μm/year	nm/h	pm/s	mm/year
Excellent	<1	<25	<2	<1	<0.02
Very good	1–5	25–100	2–10	1–5	0.02–0.01
Good	5–20	100–500	10–50	5–20	0.1–0.5
Average	20–50	500–1000	50–150	20–50	0.5–1
Poor	50–200	1000–5000	150–500	50–200	1–5
Inacceptable	>200	>5000	>500	>200	>5

Units as nm/h and pm/s are reported for comparison purposes only

seem quite high) but not for other industrial sectors such as energy, nuclear, food and beverage, pharmaceutical, biomedical and construction. For example, the corrosion of steel reinforcements in concrete structures is considered negligible when it is below 1.5–2 μm/year or, according to European standards, nickel release rates from objects in contact with human skin must be less than 2 μg/cm^2 week which is equivalent to 0.12 μm/year (nickel density 8.9 Mg/m^3).

1.4.2 Localized Corrosion

When corrosion is localized a distinction has to be made between mass loss rate, which represents an average value related to the entire exposed surface, and the maximum penetration rate which is the value of concern. The equivalence considered above for uniform corrosion attacks is obviously not applicable.

1.5 Corrosion Mechanisms

Metal corrosion follows two different mechanisms: *high temperature corrosion* which is typical of metals exposed to hot gas, for instance in boilers and turbines, and *aqueous corrosion* which takes place on metals exposed to waters, soil, chloride contaminated or carbonated concrete and many process fluids, in a word to an electrolyte. However, there are environments, such as melted salts or melted metals and non-aqueous solutions, whose corrosion attacks do not correspond to one alone of the above mechanisms but show features of both.

The two distinct types of corrosion imply two different mechanisms. The former is an electrochemical process which is the result of two simultaneous and complementary reactions, one anodic and one cathodic, in which electrons play a key role. Wet corrosion follows thermodynamic laws and electrochemistry kinetics.

The second mechanism relates to chemical reactions, which obey the thermodynamic laws and chemistry kinetics of heterogeneous reactions. Since hot corrosion involves the formation of protective layers (typically oxides) the kinetics of the corrosion process are generally more complicated and depend on a range of factors such as adhesion, consistence, porosity, type of conduction (ionic or electronic) and film conductivity.

1.6 Questions and Exercises

1.1 A zinc plate with a surface area of 1 m^2 is exposed to an aggressive solution then suffers a uniform mass loss of 2 g/day. Calculate the corrosion rate expressed in: mdd (mg/dm^2day), mmh (mg/m^2h) and in $\mu m/year$ (Zn density = 7.14 Mg/m^3).

1.2 A carbon steel alloy suffers a uniform mass loss of 1 mdd. How much is it the corrosion rate in $\mu m/y$ (Fe density = 7.85 Mg/m^3)?

1.3 A home heating system is made of carbon steel (MW_{Fe} = 55.85 g/mol). Total volume is 1 m^3. After the circuit is filled with water, calculate mass loss and thickness loss assuming that the oxygen content in water is 8 mg/L (MW_O = 16 g/mol) and the total exposed internal surface area is 3.6 m^2. The corrosion reaction is: $Fe + \frac{1}{2} O_2 + H_2O \rightarrow Fe(OH)_2$.

1.4 Define the mathematical equation to determine the thickness loss of the piping system in a home heating plant as a function of pipe diameter and oxygen content.

1.5 The corrosion rate of a tin plate is 2.6 mdd ($Sn \rightarrow Sn^{2+} + 2e^-$). Which is the corresponding corrosion current density (mA/m^2) and penetration rate ($\mu m/y$)? (MW_{Sn} = 118.7 g/mol, Sn density = 7.3 Mg/m^3).

1.6 Calculate the maximum allowed corrosion rate of the tin layer of a tomato containing can (can size 7.6 cm in diameter and 10 cm high) if max tin concentration after 2 years should not exceed 50 mg/kg (the European Standard reports 200 mg/kg as the maximum permitted value—CE 1881/2006—19.12.06). Tin density is 7.28 Mg/m^3. Tomato sauce density is 1.11 kg/dm^3.

1.7 Calculate the expiration time of a tin layer of a tomato containing can (can size 7.6 cm in diameter and 10 cm high) if max tin concentration should not exceed 100 mg/kg. Tin average corrosion rate in declared working condition is 0.05 $\mu m/year$.

1.8 The maximum allowable corrosion rate of a carbon steel rebar in concrete is 2 $\mu m/year$. Assuming a bar diameter of 10 mm, estimate the cross section reduction after 20 years.

1.9 According to European standards, nickel release rates from objects in contact with human skin must be less than 2 $\mu g/cm^2$ week. How much is it the corrosion rate in $\mu m/year$? (Ni density = 8.9 Mg/m^3).

1.10   Why is corrosion considered "metallurgy in reverse"?
1.11   Which are the main differences between uniform and localized corrosion?

Great Steps Forward in the Study of Corrosion Phenomena
Right up to the beginning of the 1950s, corrosion experts, «*corrosionists*» were a sort of old fashioned local medical officer who worked in the field and in all fields «visiting» corroded plants, perforated pipelines, shut-down boilers, burst reactors and «sick» structures and then making their diagnoses, perhaps followed by an appropriate remedy on the basis of their limited knowledge of processes of deterioration and prevention, common sense and professional experience. The «corrosionist's» business was said to be an art not a science. Indeed, even at that time knowledge was not behind the times and not even solely empirical. U. R. Evans, who pointed out the electrochemical mechanism of corrosion phenomenon in the 1920s, published important books (in 1926, 1937 and 1948, others were to appear later) covering the phenomena of passivation, corrosion by differential aeration, by galvanic coupling and by stray currents and provided a great deal of information on other forms of attack, on techniques of prevention and on the corrosion related properties of various materials. In 1948, H. H. Uhlig edited a manual on corrosion with the intention of collating data that was scattered through the already plentiful scientific and technical literature. In 1951, E. Rabald published his «*Corrosion Guide*» describing the behaviour of thousands of metal/environment couples of interest to industry. Already before the Second World War, Vernon made clear the effect of the principal factors of atmospheric corrosion such as relative humidity or pollutants. In the case of buried structures, the knowledge of the day went back to experiments that the National Bureau of Standards had initiated in 1910, exactly like those of today (in particular, Romanoff had already published results obtained by exposing no fewer than thirty-seven thousand samples in 97 different types of soil and on timeframes that varied from a few to 17 years!) to these conditions. As far as stainless steel in particular is concerned, intergranular corrosion, and the method used to block it, had been known since the 1930s to the extent that stabilised steels with a low carbon content were already available for use in welded structures. Similarly well-known were the conditions that could foster pitting or stress corrosion, the influence of the main environmental factors (chlorides, pH, temperature), those related to composition (e.g. the effect of molybdenum) and to the structure of the steel. In short, a great deal of knowledge, above all empirical, already existed. Not everyone who dealt with corrosion, however, was aware of it.

Those Fabulous Sixties. The end of the fifties saw the conditions that enabled «corrosionists» to make a great leap forward develop. These, five in number, were as follows. The first three concern what may be termed the «software» of corrosion, that is: (1) the availability of electrochemical models, originally introduced by Evans and then gradually developed and adapted to explain the different forms of corrosion and control; (2, 3) potential/pH diagrams, conceived and developed by Pourbaix, permitting the evaluation of driving forces for different corrosion processes and specifying the ranges of pH and potential in which conditions of immunity, passivity or activity are established; finally, (4, 5) potential/current curves permitting the identification and definition of the conditions for the functioning of corrosive systems. The last two conditions, on the other hand, concern the «hardware», the availability of new equipment facilitating electrochemical measurements and the study of surfaces.

This included the potentiostat, an instrument that simplified, and increased the accuracy of, the tracing of potential/current curves for the most diverse metals and in the most diverse environmental conditions, and new instruments for the study of surfaces (optical and electronic microscopes, X-ray) made possible by post-war development in this sector. Starting from these five assets, research in the sixties developed extensively and clarified many aspects of corrosion phenomena that had up to then been unknown. The obvious consequence was the development of methods of corrosion prevention and control: starting with corrosion-resistant materials (from stainless steel to super-alloys to plastics) and continuing through inhibitors, treatment and surface coatings to the control of the environment and cathodic protection. All this contributed to the formation of a body of knowledge that was based on electrochemistry but had links to metallurgy and crossed the border into applied chemistry, electrics and mechanics rising in rank to become a genuine scientific discipline which could include corrosion phenomena and methods for controlling them and took the name «Corrosion Science».

The Beatles Years. Over the five years from 1963 to 1968, an impressive series of texts were published on the foundations of this discipline. These were the years of the Beatles. In 1963, the Liverpool four had just recorded their first single, *Love me Do*, and were working on *She Loves You* and *I Want to Hold Your Hand* when U. R. Evans published—«*An Introduction to Metallic Corrosion*», L. L. Sheir—«*Corrosion*»; H. H. Uhlig—«*Corrosion and Corrosion Control*», F. L. La Que and H. R. Copson—«*Corrosion Resistance of Metals and Alloys*». By 1964–65, the Beatles were a success and drew crowds with *A Hard Day's Night, Yesterday, We Can Work it Out* and, on the corrosion front, J. Benard sent «*L'oxydation des Métaux*» and K. Hauffe «*Oxidation of Metals*» for printing. We're now in 1966–67. The Liverpool four were singing *Yellow Submarine, Penny Lane, Strawberry Fields for Ever* when M. Pourbaix, N. Thomashov, J. M. West, J. C. Scully and H. Kaeshe, respectively, published their *«Atlas of Electrochemical Equilibria in Aqueous Solutions»*, *«Theory of the Corrosion and Protection of Metals»*, *«Electro-deposition and Corrosion Processes»*, *«The Fundamentals of Corrosion»* and *«Die Korrosion der Metallen»*. Finally, in 1967–68, the time of *All You Need is Love, Lady Madonna*, and *Hey Jude*, *«Corrosion Engineering»* by M. G. Fontana and N. D. Green and *«Corrosione e Protezione dei Metalli»* by G. Bianchi and F. Mazza were appearing in the bookshops. As 1969 neared, the golden age of the Beatles was coming to an end: they had yet to write *Something* and *Come Together* and little else and then they split up. And on the corrosion front also, the era of assembly-line publication was over. Those books went everywhere, even if not quite like the Beatles' records, and enabled the «corrosionists» of the new generation to base their professional training on solid theoretical foundations. These books are still today the most widely read texts on corrosion, true «evergreens» just like many of the songs of the «fabulous four».

No Longer «A Devoted Subject of Empiricism». In the course of those years, things changed to such an extent that Professor Roberto Piontelli who was still defining the world of corrosion as «a devoted subject of empiricism» in 1961 referred to «corrosionists» in these terms in 1968:

«Corrosionists must concern themselves above all with correlating the properties of composition, structure, and surface condition of metals with their behaviour; with establishing the boundaries of compatibility, foreseeing the onset of corrosion phenomena, their probable type (nature and distribution), the course they follow in time, with diagnosing the causes of the phenomena that have occurred, with suggesting expedients that can prevent or limit them. Within an ambit of industrial activity in which, in addition to other risks, pace imposes an exceptional economic burden on any shut-down, they must prevent the catastrophic forms in which corrosion phenomena may occur. They must therefore, know how to arrange things so that any possible deterioration of their materials will be negligible, or gradual, so as to permit an adequately precise estimate of their working lives in safe working conditions. To face up to this extremely exacting burden of tasks and duties, they seek assistance from thermodynamics, in order to know the conditions for the possible onset of such dreaded phenomena in advance, investigating structure (internal or surface) no longer simply by using metallography or X-rays but exploiting all the most modern resources (electronic microprobes, Mossbauer Effect, neutron diffraction, slow electrons). They mobilise the most complex equipment for kinetic electrochemical investigation and analysis. With the help of all these means, they patiently create their atlas of «pathological anatomy» of metals exposed to the most diverse corrosive

environments, build up a 'corpus' of diagnostics, develop an increasingly efficient anti-corrosion pharmacology». (Incidentally, perhaps Piontelli let things get a little out of hand in mentioning the Mossbauser Effect or neutron diffraction, which very few «corrosionists» know the applications of.)

Anticorrosion Engineering. At the end of the sixties it was recognised that, in an industrialised country, corrosion generates extremely serious losses—in terms of wasted resources, reduced service life of consumer goods, cost of preventative measures—and it was evident that certain kinds of technical progress—those which condition the future of mankind itself, involving petroleum, the nuclear industry, water, the conquest of the depths of the ocean —are blocked and precisely by these very problems of corrosion. This gave rise to a number of initiatives to channel the knowledge gained by the newly born science of corrosion into the fight against corrosion, first of all by spreading this new knowledge among technicians. In this context corrosion engineering, or as some prefer to term it, anti-corrosion engineering, was born. The new discipline shifted attention from the metal that corroded, to the system, i.e. to the structure, the equipment, the plant, the manufactured article, in which the phenomenon took place, and from the mechanism and the generic laws that govern the phenomenon, to the means and procedures necessary for the system to be able to function in conditions of deterioration that are acceptable. The change modified the approach to corrosion and introduced the concepts of reliability and service life and drew the attention of corrosion engineers to the design and construction of the structure and on the programmes for the inspection, monitoring and maintenance to which it must be subject.

The Seventies Also Dawn. The propulsive thrust imparted by the turning point of the sixties, though weakened, continued into the following decade, on both scientific and corrosion engineering fronts. In 1972, Pourbaix defined the conditions of perfect and imperfect passivity, opening the way to the cathodic protection of materials with an active-passive type of behaviour, like stainless steel. Parkins demonstrated the importance of conditions of slow deformation, the «slow strain rate», in causing the growth of cracks due to a combination of stress and corrosion. Meanwhile, fear that failures in off-shore platform protection and consequent collapses due to corrosion-fatigue in the early years of North Sea oil-wells could be repeated resulted in the development of research in the sector which led to enormous progress in the field of cathodic protection and huge increases in knowledge of fatigue phenomena in sea water. Moreover, the need to exploit deeper oil wells richer in carbon dioxide and hydrogen sulphide required the development of petrochemical industry specific materials. At the end of the seventies another sector opened up: ever more frequently occurring damage in reinforced concrete structures that, up to that time, had been considered everlasting, or almost, brought corrosion engineers into the world of building construction. Much progress was made possible by the development of surface analysis techniques some of which had just recently become available. These included photo-electronic spectroscopy (XPS or ESCA) and AUGER electronic spectroscopy (AES) which are sensitive to all the elements that involve corrosion, with the exception of hydrogen, and hence of great utility in the study of surface films.

Meanwhile, and to an even greater extent in the two decades which followed, trade and professional associations, cultural societies, public authorities and, above all, standards institutes began, more and more frequently, to issue guidelines, recommendations and standards concerning corrosion covering a wide range of applications. I remember in particular the ASTM, NACE, ISO and CEN standards. It was precisely on the basis of the information contained in these directives, and the know-how within the various companies concerned with plant design and running, that the guide-lines that all companies use in the choice of material were developed and, at the same time, expert systems and intelligent data banks were set up. Soon the importance of this «structured» knowledge was recognized since it contained both the theoretical laws derived from science and from corrosion engineering and the rules deriving from experience. It is a fine reward for the corrosion engineers of the writer's generation who for years thought that there could be no more room for empirical knowledge in this sector.

Pietro Pedeferri, Pianeta inossidabili/XLIII, 2002

Bibliography

Bardal E (2004) Corrosion and protection. Springer-Verlag London Limited, UK

Bianchi G, Mazza F (1989) Corrosione e protezione dei metalli, 3rd edn. Masson Italia Editori, Milano (in Italian)

Evans UR (1948) An introduction to metallic corrosion. Edward Arnold, London, UK

Fontana M (1986) Corrosion engineering, 3rd edn. McGraw-Hill, New York

Hoar TP (1971) Report of the committee on corrosion and protection, Department of Trade and Industry, H.M.S.O., London, UK

Jacobson G (2016) IMPACT report international measure of prevention, application and economics of corrosion technologies study. NACE International, Houston

NBS publication 511-1-2-3 (1978) Economic effects of metallic corrosion in the United States, Report to Congress by the National Bureau of Standards, Washington D.C

Piontelli R (1961) Elementi di teoria della corrosione a umido dei materiali metallici. Longanesi, Milano (in Italian)

Pourbaix M (1973) Lectures on electrochemical corrosion. Plenum Press, New York

Roberge PR (1999) Handbook of corrosion engineering. McGraw-Hill, London

Shreir LL, Jarman RA, Burstein GT (1994) Corrosion. Butterworth-Heinemann, London
Speller FN (1926) Corrosion. Causes and prevention. McGraw-Hill, London
Tomashov N (1966) Theory of corrosion and protection of metals: the science of corrosion.
   McMillan, New York
Winston Revie R (2000) Uhlig's corrosion handbook, 2nd edn. Wiley, London

Chapter 2
Electrochemical Mechanism

Possibly it is really the strangeness of corrosion reactions which causes the orthodox physical chemist to regard the whole subject of corrosion with suspicion.

U. R. Evans

Abstract Wet corrosion is based on an electrochemical mechanism in which two reactions sum up to give the overall corrosion process; a cathodic reaction that consumes electrons and an anodic one, where electrons are released by the metal oxidation. In this chapter, the electrochemical mechanism is examined in details, and the most important anodic and cathodic processes are described. From these basic principles stoichiometric considerations will be drawn, leading to the correlation between corrosion rate and current density by the use of Faraday law.

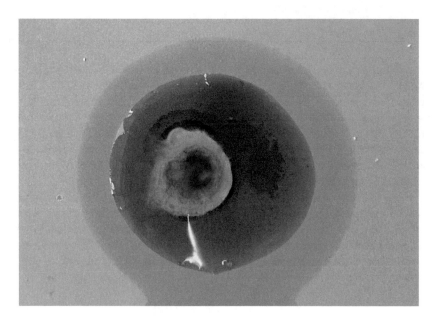

Fig. 2.1 Evans drop test at the PoliLaPP Corrosion Museum of Politecnico di Milano

© Springer Nature Switzerland AG 2018
P. Pedeferri, *Corrosion Science and Engineering*, Engineering Materials,
https://doi.org/10.1007/978-3-319-97625-9_2

2.1   Electrochemical Processes

The corrosion of a metal, M, can be expressed by the following general reaction:

$$M + aggressive\ environment \rightarrow corrosion\ products \tag{2.1}$$

If the environment is an electrolyte (*aqueous* or *wet corrosion*) the corrosion reaction is the sum of two *electrochemical reactions*, as highlighted experimentally in Fig. 2.1:

- an *anodic process* which consists of the oxidation of the metal
- a *cathodic process* which is a reduction reaction, typically oxygen reduction or hydrogen ion reduction.

In the case of iron, the anodic reaction is:

$$2Fe \rightarrow 2Fe^{2+} + 4e^- \tag{2.2}$$

where electrons are made available; these electrons are taken by the cathodic reaction, for instance by oxygen, with production of alkalinity:

$$O_2 + 2H_2O + 4e^- \rightarrow 4OH^- \tag{2.3}$$

or by hydrogen ions in the case of an acidic solution, with consumption of acidity (i.e., production of alkalinity):

$$4H^+ + 4e^- \rightarrow 4H_2O + 2H_2 \tag{2.4}$$

In addition to the above reactions, a corrosion process implies two further processes, namely: (1) an electron flow within the metal from the anodic area, where electrons are released, to the cathodic zone, where electrons are consumed; the electron flow direction is opposite to the conventional current direction (since electron charge is negative). (2) a current flow (circulation) within the electrolyte by ion transportation, from the anode to the cathode zone; positive ions move in the same direction of the current and negative ions in the opposite one. In short, a corrosion process consists of four processes in series, as depicted in Fig. 2.2.

These four processes occur at the same rate. In fact, (a) the number of electrons released by the anodic reaction, i.e., the anodic current, I_a, exchanged on the metal surface, (b) the number of electrons consumed by the cathodic reaction, i.e., the cathodic current, I_c, (c) the current flowing within the metal from the cathode to the anodic zone, I_m, and (d) finally the current circulating within the electrolyte, I_{el}, must be the same:

$$I_a = I_c = I_m = I_{el} = I_{corr} \tag{2.5}$$

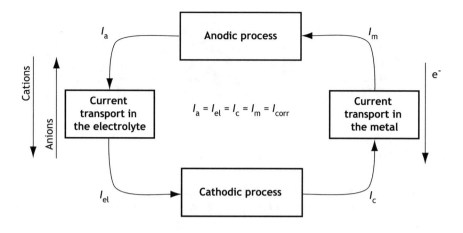

Anodic process
Iron → Oxidation products + Electrons (in metallic phase)

Cathodic process
Oxygen + Water + Electrons (in metallic phase) → Alkalinity

Ionic current in the electrolyte
Cations move as current direction, anions in opposite direction

Electronic current in the metal
Electrons move from anode, where they are released, to cathode
(i.e., in opposite direction of conventional current)

Fig. 2.2 Electrochemical mechanism of a corrosion process

This common current flow, I_{corr}, measures, in electrochemical units, the corrosion process rate.

Electrochemical Mechanism of Corrosion
An *electrochemical reaction* implies the participation of chemical species (neutral molecules or ions) and *electrons*. They are oxidation reactions, called *anodic reactions*, which make available free electrons in the metal, and reduction reactions, called *cathodic reactions*, which take those electrons.

As far as *anode* and *cathode* terms are concerned, an anode is an electrode hosting an oxidation and the electric current leaves the anode toward the electrolyte; a cathode is an electrode hosting a reduction and the electric current enters the cathode from the electrolyte.

In a galvanic or bi-electrode system, the *positive* pole is the one at higher (more noble) potential which is connected to: (1) an *anode*, when current flows due to an external source (a DC current feeder); (2) a *cathode*, when the system is a voltaic pile as in the corrosion process.

2.2   Historical Notes

It is generally quoted that the Swiss scientist August De La Rive, around 1830, first advanced the hypothesis that corrosion is produced by an electrochemical mechanism, even though important observations on the matter were made previously by the Florentine Lorenzo Fabbroni in 1792 and the Italian Alessandro Voltà shortly after his invention of the battery in 1800. Also, the English Humphrey Davy in 1824 showed that it was possible to protect the copper sheets which at that time covered the hulls of wooden ships, through a connection with blocks of iron or zinc.

It might be of some surprise to note that despite this promising start—shortly followed by the brief and successful incursion into the matter of Shömbein and Faraday, who dealt with the passivity of metals in the late 1830s—the interest in corrosion faded for the whole XIX century.

At that time, electrochemists addressed their attention to problems aroused by Volta's invention, the solution of which was also important for building a solid scientific basis for the corrosion phenomena; namely, the laws between chemical effects and electrical charge (Faraday 1835), the conductivity of the solutions (Arrhenius 1880) and electrochemistry related energy (Nernst and Ostwald 1890).

2.2.1   Evans's Experiences

Starting in 1923, Evans developed a series of ingenious and simple laboratory experiences, that have become historical, to prove the corrosion theory. He used mild steel strips, an aerated neutral solution containing potassium chloride, KCl (3%) and two indicators: potassium ferricyanide, which turns blue when iron ions, Fe^{2+}, released by the corrosion reaction are complexed by ferricyanide ions, and phenolphthalein, which turns pink for pH greater than 9. In order to measure the circulating current, Evans used an ammeter. The two most well-known experiences are illustrated below.

Evans's first experience. As shown in Fig. 2.3, a droplet of the above solution is laid on a mild steel strip. Soon, small blue dots and pink spots form randomly (Fig. 2.3a). The blue areas indicate the presence of Fe^{2+} ions, identifying the points at which the oxidation of iron occurs; while the pink zones identify where pH increased, that is, where oxygen is reduced.

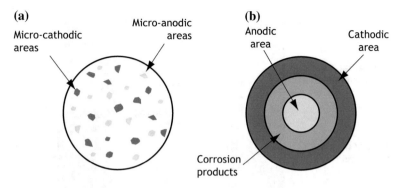

Fig. 2.3 Evans experience (or Evans drop test): **a** initial condition; **b** steady condition

The distribution of the coloured areas changes over time and, within a few hours or a few days, the centre of the droplet becomes blue, while its border becomes pink (Fig. 2.1). Meanwhile, as the Fe^{2+} ions diffuse towards the oxygen-rich border, they are oxidized to Fe^{3+} ions, which precipitate as $Fe(OH)_3$ in an intermediate region between the centre and the edge of the droplet (Fig. 2.3b). At the end of the experiment, the metal surface in correspondence of the droplet border is not corroded, while in the centre a corrosion crater is formed.

This experience proves that even in neutral aerated solution the corrosion attack of the iron is produced through an electrochemical mechanism implying two electrochemical processes: the oxidation of iron (anodic process) and oxygen reduction (cathodic process). The two processes take place on separated areas of the metal surface, which act as anodic and cathodic zones, respectively. Since the oxidation (anodic) process releases electrons, while the reduction (cathodic) process consumes them, it can be concluded that within the metal a current circulates from cathode to anode and within the electrolyte in the opposite direction as shown in Fig. 2.4.

This experience of Evans's also shows that the electrochemical mechanism is "self-organizing:" at the beginning, corrosion seems to be randomly distributed over the entire surface of the steel strip; then, once a steady state condition is reached, corrosion localizes and proceeds at the centre of the droplet where the oxygen diffusion slows down. The organizing criterion is governed by the

Fig. 2.4 Current flow in Evans drop test

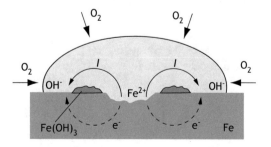

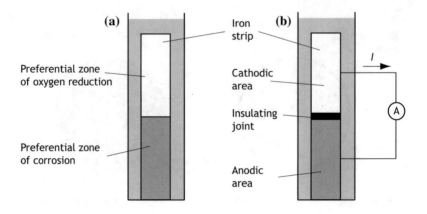

Fig. 2.5 Evans second experience

non-uniform distribution of oxygen inside the droplet, which eventually determines the morphology of the attack. Instead, one could expect a casual morphology determined by the initial attack, which is intrinsically random because of the presence of a continuous, oxygen-containing film.

Evans's second experience. In a second experiment, Evans highlighted the current flow between the cathodic and anodic zones established on the metal surface. Figure 2.5 illustrates the set up that consists of a strip of iron (or zinc) immersed in a cylinder containing the solution above described. Evans noted that corrosion attack occurred preferentially on the bottom, which is far from the solution free surface (Fig. 2.5a), where the cathodic process of oxygen reduction took place. As in the case of the droplet, the attack tends to localize in areas where the oxygen initially present is consumed and not replaced. Then, Evans proceeded by cutting the strip along the separation line between the corroded areas (at the bottom) and those not corroded (at the top). He then connected the two pieces by means of an insulating joint, then restoring the electrical continuity through an external metallic circuit. An ammeter was inserted in the circuit for measuring the current, as shown in Fig. 2.5b. The ammeter showed a current flow from the cathodic upper region toward the corroding lower anodic zone: the electrochemical nature of the process was then brilliantly demonstrated.

Corrosion Mechanism
August De La Rive proposed an electrochemical mechanism based on the observation that impure zinc in acidic solutions corroded faster than pure zinc, which he attributed to an electrical effect between matrix and impurities.

Towards the end of the century, corrosion experienced a new awakening as soon as it was considered a normal chemical reaction between metals and acids; even rust was believed to be the result of a reaction with the carbonic acid present in the atmosphere. Curiously, no one recognized the enormous importance of the role played by oxygen, even the Swedish Palmaer who rediscovered the electrochemical mechanism at the beginning of the Twentieth century. He reconsidered De La Rive's observation about the influence of impurities on the corrosion of zinc in acidic solutions, and pointed out how this was related to the action of micro cells made of impurities and the surrounding metal matrix. This suggested to the scientist the wrong theory—endorsed by the scientific community at the time—that a perfectly pure metal, assuming it would be manufactured, cannot corrode because of the lack of impurities, i.e., the lack of the tiny local electrodes to set up the corrosion microcell.

Only twenty years later, **U. R. Evans** and his School at the University of Cambridge showed that metals corrode even in the absence of impurities, in environments of any pH, often because of the presence of dissolved oxygen in the solution; he also contributed to give experimental and quantitative support to the electrochemical theory of corrosion.

2.3 Local Cell Theory

Evans's experience is based on a separation of anodic and cathodic regions (at micro scale the former, and at macro scale the latter). Indeed, this separation is confirmed by direct visual check with a microscope. This evidence suggested the so-called *local cell theory*, which can be summarized by the two following statements:

- The mechanism of corrosion processes is electrochemical
- Different zones of the metal surface could assume a behaviour either anodic or cathodic

According to this theory, corrosion consists of electrodes in short circuit like the short-circuited battery shown in Fig. 2.6. M and N are the regions exhibiting the anodic and cathodic behaviour, respectively, I^{MN} is current of the local cell flowing from cathode to anode within the metal, and ε is the electrolyte (aggressive environment).

However, in most corrosion case studies, local cells cannot be found even at micro scale, although laboratory experiments proved that affecting parameters and their relationships are the same. This evidence suggested the presence of heterogeneities or impurities or metallurgical defects at a "submicroscale" acting as anodic and cathodic zones. On this basis, the local cell theory was considered valid also for those case studies.

Fig. 2.6 Electrochemical cell
model of a corrosion process
(M anodic surface, N cathodic
surface; short-circuited)

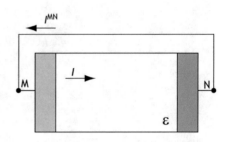

2.3.1   Mixed Potential Theory

Truthfully, some doubts still remained. There are homogeneous systems, such as
the amalgams, in which the local cell at "sub-microscale" cannot exist; in this case
study, the local cell theory seems to fail. To eliminate this contradiction, in 1938
Wagner and Traud proposed that on homogeneous materials exposed to homogeneous environments both anodic and cathodic reactions could take place simultaneously and alternatively on the same place. This theory is called *mixed potential* or
Wagner and Traud theory, which states as follows:

- The corrosion process follows an electrochemical mechanism (even in the
  absence of any kind of heterogeneity), and proceeds through electrochemical
  anodic and cathodic reactions over the entire metal surface
- The rate of each anodic or cathodic process depends on potential only,
  regardless any other processes
- Anodic and cathodic processes are complementary. In other words, the sum of
  all anodic process rates equals the sum of all cathodic process rates. This is the
  principle of conservation of charge: the number of electrons released at the
  anode is the same of those used at the cathode.

Wagner and Traud came to the formulation of their theory after studying, in acidic
solutions, the rate of hydrogen evolution on mercury and homogeneous amalgams.
The two scientists noted that if pure mercury was maintained at the potential to
which amalgam of zinc freely corrodes, the rate of hydrogen evolution was the
same. This proved that hydrogen evolution is determined by the potential regardless
if the process is spontaneous (as it was for the amalgam) or imposed (as it was for
the pure mercury).

After this initial assertion, Wagner and Traud extended their theory to heterogeneous conditions either for the metal surface or the environment, assuming that a
zone could be *predominantly* anodic or cathodic and not merely anodic or cathodic,
as in the local cell theory. In this light, the local cell theory can be regarded as a
special case of the more general mixed potential theory.

2.4 Corrosion Reactions

The corrosion process involves various aspects, which deal with anodic and cathodic reactions, electrophoretic migration of ionic species within the electrolyte, secondary chemical reactions between metal and anodic or cathodic products, as illustrated hereafter.

2.4.1 Anodic Process

A generic *anodic process* of a metal is an oxidation reaction, which can produce a metallic ion, releasing electrons:

$$M = M^{z+} + ze^- \qquad (2.6)$$

where z^- is the number of electrons, M is a generic metal and M^{z+} is a metal ion which passes into solution. Or, in specific ranges of pH or in the presence of particular species, metal ions are separated as insoluble compounds (oxides, hydroxides, salts), as for example:

$$M + zH_2O = M(OH)_z + zH^+ + ze^- \qquad (2.7)$$

where $M(OH)_z$ is an insoluble product, in this reaction example a hydroxide.

Examples of anodic reactions, with metal dissolution and formation of metal ions, complexes or precipitates, are:

$$Ag \rightarrow Ag^+ + e^-$$
$$Fe \rightarrow Fe^{2+} + 2e^-$$
$$Al \rightarrow Al^{3+} + 3e^-$$
$$Al + 2H_2O \rightarrow AlO_2^- + 4H^+ + 3e^-$$
$$Mg + 2OH^- \rightarrow Mg(OH)_2 + 2e^-$$
$$Ag + Cl^- \rightarrow AgCl + e^-$$
$$Fe + CO_3^{2-} \rightarrow FeCO_3 + 2e^-$$

In anodic processes the species electron is produced by the reaction; therefore, it appears in the second member of the reaction.

Corrosion Science and Corrosion Engineering
Starting from the third decade of last century, research in the corrosion field headed to electrochemistry, and in half a century, scientist like Evans first and then Vernon, Pourbaix, Piontelli, Uhlig, Hoar, Tomashov, Stern and others created a body of knowledge framed by electrochemistry, with links to metallurgy, applied chemistry, electronics, and mechanics. In the '70s of the

XX century this knowledge rose to the rank of a scientific discipline that took the name of *Corrosion Science*, which enables to rationalize corrosion phenomena and their control methods.

In those years, it was realized that corrosion produces very high losses, especially in industrialized countries. This appeared as a clutch for technological developments in strategic industry related activities for the future of humanity, such as oil and gas, nuclear, deep water and many others. Consequently, within the fight against corrosion, a series of actions developed to steer the knowledge of the newborn corrosion science, primarily by teaching and training. In this context, corrosion scientists/engineers—such as Fontana, Green, Bianchi, Parkins, Staehle and many others—gave rise to *Corrosion Engineering*.

This new discipline shifted the focus from the corroding metal itself, to the system in which corrosion occurs (i.e., structures, equipment, plants, components) and to the laws that govern the phenomenon, its mechanism and control methods with the aim to make corrosion acceptable from an industry viewpoint.

This change altered the approach to corrosion introducing the concepts of risk, reliability, service life. It also brought to the attention of corrosion engineers not only the choice of materials, but also design, construction, inspection, monitoring and maintenance programs to guarantee the design life. In the early 1990s, **Giuseppe Bianchi** (1919–1996) indicated "corrosion informatics" as the new branch of corrosion.

2.4.2   Cathodic Processes

The *cathodic process* takes the released electrons from the metal to reduce the chemical species present in the environment. The generic reaction is:

$$OX^{y+} + ze^- = RD^{w+} \tag{2.8}$$

where $z = y - w$, OX^{y+} is oxidized species and RD^{w+} is reduced species.

Cathodic reactions of practical interest for corrosion are limited in number. In the case of corrosion in an acidic solution, the cathodic process is the reduction of hydrogen ions to produce either atomic or molecular hydrogen, according to the reactions:

$$H^+ + e^- = H \quad \text{or} \quad 2H^+ + 2e^- = H_2 \tag{2.9}$$

In neutral or alkaline environments and in natural environments (atmosphere, soil and waters), the most important cathodic reaction is the oxygen reduction reaction:

$$O_2 + 2H_2O + 4e^- = 4OH^- \tag{2.10}$$

In acidic solution, the corresponding reaction is:

$$O_2 + 4H^+ + 4e^- = 2H_2O \tag{2.10'}$$

The reactant oxygen that appears in the above reactions is the molecular oxygen dissolved in water, the concentration of which varies from 0 to 12 milligrams per kilogram of water (ppm).

In addition, water can give a cathodic reaction in a specific potential range, as follows:

$$2H_2O + 2e^- \rightarrow 2OH^- + H_2 \tag{2.10''}$$

Another relevant cathodic reaction is the reduction of chlorine to chloride:

$$Cl_2 + 2e^- \rightarrow 2Cl^- \tag{2.10'''}$$

2.4.3 Other Cathodic Processes

There are other possible cathodic processes that occur in peculiar conditions:

1. *reduction of metal ions* to a lower valence:

$$Fe^{3+} + e^- \rightarrow Fe^{2+}$$
$$Cu^{2+} + e^- \rightarrow Cu^+$$
$$Hg^{2+} + e^- \rightarrow Hg^+$$

2. *reduction of anions*:

$$2ClO^- + 2H^+ + 2e^- \rightarrow 2Cl^- + H_2O$$
$$NO_2^- + 2H^+ + e^- \rightarrow NO + H_2O$$
$$Cr_2O_7^{2-} + 14H^+ + 6e^- \rightarrow 2Cr^{3+} + 7H_2O$$

2.4.4 Complementary Processes

Since the circulation of current and subsequent chemical modifications do not alter the electrical state of the system, anodic and cathodic reactions must be complementary from the viewpoint of the electrical charge balance, and must be simultaneous from the viewpoint of the reaction rate. This implies that the stoichiometric coefficients of the electron, usually denoted by z, must be identical: this is called *equivalence of the chemical reaction* (see examples in box).

2.5 Stoichiometry (Faraday Law)

The stoichiometry of electrochemical reactions is governed by the Faraday law, which reads as follows. The mass, $\Delta m(g)$, of the chemical species and the charge, Q (Coulomb), exchanged in the electrode process are linked through the electrochemical equivalent, according to the relation:

$$\Delta m = e_{ech} \cdot Q = \frac{e_{chem}}{F} \cdot Q = \frac{MW}{z \cdot F} \cdot I \cdot t \qquad (2.11)$$

where e_{ech} is the electrochemical equivalent, e_{chem} is the chemical equivalent, F is Faraday's constant (96,485 C or 26.8 A h), I is current (A), t is time (s) and MW is atomic or molecular mass (g/mol). Electrochemical equivalent, e_{ech}, and chemical equivalent, e_{chem}, represent the mass relative to a charge of 1 C or to 1 F, respectively.

2.5.1 Corrosion Current Density

The corrosion rate can be obtained through Eq. 2.11 as follow:

$$C_{rate} = \frac{C_{rate,m}}{\gamma} = \frac{\Delta m}{\gamma \cdot S \cdot t} = \frac{e_{ech} \cdot Q}{\gamma \cdot S \cdot t} = \frac{1}{\gamma} \cdot e_{ech} \cdot i_a \qquad (2.12)$$

where i_a is the anodic current density, also called *corrosion current density*.

In electrochemical units, the corrosion rate is measured in mA/m^2, or in $\mu A/cm^2$ ($1\,\mu A/cm^2 = 10\,mA/m^2$). The constant e_{ech}/γ in Eq. 2.12 depends on metal as reported in Table 2.1. As rule of thumb, penetration rate, C_{rate}, expressed in $\mu m/year$ corresponds approximately, in electrochemical units, to mA/m^2, for many metals such as iron, copper, nickel, aluminium.

2.6 Change of the Environment

The environment composition varies as corrosion proceeds. First, there may be significant variations in pH. In particular, in cathodic zones, alkalinity increases due to both oxygen reduction and hydrogen evolution; whereas, in anodic regions,

Table 2.1 Equivalence of the corrosion rate (mm/year) for an anodic current density of 1 mA/m^2

Metal	Valence	Density (g/cm^3)	Equivalent mass (g/eq)	Corrosion rate (μm/year)
Iron	Fe^{2+}	7.87	27.92	1.17
Nickel	Ni^{2+}	8.90	29.36	1.09
Copper	Cu^{2+}	8.96	31.77	1.17
Aluminium	Al^{3+}	2.70	8.99	1.09
Lead	Pb^{2+}	11.34	103.59	2.84
Zinc	Zn^{2+}	7.13	2.68	1.50
Tin	Sn^{2+}	7.30	59.34	2.67
Titanium	Ti^{2+}	4.51	23.95	1.75
Zirconium	Zr^{4+}	6.50	22.80	1.91
AISI 304	Fe^{2+}, Cr^{3+}, Ni^{2+}	7.90	25.12	1.04
AISI 316	Fe^{2+}, Cr^{3+}, Ni^{2+}, Mo^{3+}	8.00	24.62	1.04

acidity increases due to the hydrolysis of corrosion products. Other changes in composition also take place. For example, metal ions accumulate in the anodic regions when, in suitable conditions, current is mainly transported by anions; or in the cathodic regions, the dissolved oxygen content decreases.

Examples of Complementary Processes

1. The corrosion of iron in an acidic solution takes place as a result of two reactions:

$$Fe \rightarrow Fe^{2+} + 2e^-$$
$$2H^+ + 2e^- \rightarrow H_2$$

Therefore, the global reaction is:

$$Fe + 2H^+ \rightarrow Fe^{2+} + H_2$$

The equivalence of this reaction is then 2. Recalling that the atomic mass of iron is 55.8, and 2 g/mol for hydrogen, the oxidation of 55.8 g of iron results in the development of 2 g of hydrogen gas.

2. The corrosion of iron in an aerated solution takes place as a result of the following two reactions:

$$2Fe \rightarrow 2Fe^{2+} + 4e^-$$
$$O_2 + 2H_2O + 4e^- \rightarrow 4OH^-$$

Therefore, the global reaction is:

$$2Fe + 2H_2O + O_2 \rightarrow 2Fe^{2+} + 4OH^-$$

In this case, the equivalence of the reaction is 4. The oxidation of 111.6 g (= 55.8 g × 2) of Fe to Fe^{2+}, requires the reduction of 32 g of O_2 (since 32 g/mol is the molecular mass of oxygen), and the consumption of 36 g of water, being 18 g/mol the molecular mass of the water.

3. If the corrosion of iron takes place according to the following two reactions:

$$2Fe \rightarrow 2Fe^{3+} + 6e^-$$
$$3/2O_2 + 3H_2O + 6e^- \rightarrow 6OH^-$$

the global reaction is:

$$2Fe + 3H_2O + 3/2\,O_2 \rightarrow 2Fe^{3+} + 6OH^-$$

and the equivalence of the reaction is 6. The oxidation of 111.6 g (= 55.8 g × 2) of Fe to Fe^{3+}, requires the reduction of 48 g of O_2 (1.5 × 32 g), and the consumption of 54 g of water (= 3 × 18 g).

How to Calculate Chemical (MW/z) and Electrochemical (MW/zF) Equivalent

MW is the molar mass and z the valence to form an ion M^{z+}. For example, in the case of the reaction $Fe = Fe^{2+} + 2e^-$, the mass of iron which is oxidized by a charge of 1 F or 1 C is equal to respectively 55.8 g / 2 = 27.6 g and 55.8 g / (2 × 96,485) = 1.036 × 10^{-5} g.

MW is the molar mass of a species partially oxidized or reduced by a valence change of z. For example, in the case of the reaction $Fe^{2+} = Fe^{3+} + e^-$, the mass of ferrous ion which oxidizes by a charge of 1 F or 1 C is, respectively, 55.8 g and (55.8 g / 96,485) = 2.072 × 10^{-5} g.

(MW/F)·(z/Z) or MW·(z/Z), where MW is the molecular mass of a generic neutral species participant to the electrode process with stoichiometric coefficient z, and Z is the stoichiometric coefficient of the electron. For example, in the case of the reaction $O_2 + H_2O + 4e^- \rightarrow 4OH^-$, the mass of oxygen consumed by a charge of 1 F or 1 C is, respectively, 32 g / 4 = 8 g and 32 / 4 F, that is 8 / 96,485 = 8.29 × 10^{-5} g.

In the case of alloys, the electrochemical equivalent is approximately the weighted average of that of each element of the alloy composition. For example, the electrochemical equivalent of a stainless steel of composition 19% Cr, 9.25% Ni and 71.75% Fe (all other elements with content less than 1% are neglected), is 25.12, bearing in mind that the valence of the three elements is 3, 2 and 2, and their molecular weight is 52, 59 and 56, respectively.

The specific enunciation of the Faraday law is not necessary if in the formulation of the electrochemical reaction the electron is considered as any other chemical species. Then, the symbol e^- represents, from the stoichiometry viewpoint, a mole of electrons, i.e., a number of electrons equal to Avogadro's number (6.022×10^{23}), and then with charge 1 F, or 96,485 C or 26.8 A h (obtained by multiplying the elementary charge of the electron, equal to 1.602×10^{-19} C, to Avogadro's number). Since the 'chemical species' electron is monovalent, one mole of electrons also corresponds to one gram-equivalent of electrons; precisely engaging, in an electrochemical reaction, a gram-equivalent of substance.

Therefore, for example:

- A mole of hydrogen (2 g), according to the reaction $2H^+ + 2e^- = H_2$, needs two moles of electrons, i.e., 2F and then $2 \times 96,485$ C, or 2×26.8 A h to oxidize one mole of iron (55.8 g) to ferrous ions; according to the reaction $Fe = Fe^{3+} + 3e^-$, 3 mol of electrons are required, i.e., 3F and then $3 \times 96,485$ C, or 3×26.8 A h to oxidize one mole of iron (55.8 g) to ferric ions

- To reduce one mole of oxygen (32 g), according to the reaction $O_2 + 2H_2O + 4e^- = 4OH^-$, 4 mol of electrons are needed that is, 4F and then $4 \times 96,485$ C, or 4×26.8 A h.

It is worthwhile recalling that an anodic current of 1 A brings into solution a metal mass of $e_{chem}/96.485 = 0.00001036 \cdot e_{chem}$ (g/s), or $e_{chem}/26.8 = 0.000373 \cdot e_{chem}$ (g/h), and also $326 \cdot e_{chem}$ (g/year). For example, 1 m^2 iron plate which corrodes according to the anodic reaction $Fe = Fe^{2+} + 2e^-$, for a total current of 1 A, the amount of iron which goes into solution in a year is equal to $C_{rate,m} = 326 \cdot e_{chem}$ (g/year) $= 326 \cdot (55.8/2) = 9.1$ kg/year. Since the density of iron, γ, is 7.85 Mg/m^3, the penetration rate, C_{rate}, is equal to $\Delta W/(S\gamma) = 9.1$ kg/(1 m$^2 \cdot 7.85$ Mg/m^3) $= 1.17$ mm/year. In conclusion, an anodic current of 1 A/m^2 produces a corrosion attack of about 1 mm/year. Then, a current density of 1 mA/m^2 causes an attack of about 1 μm/y.

The magnitude of the variation of composition strongly depends on electrode reactions and on electrophoretic transport, diffusion and convection within the solution. The latter become particularly important in those cases—which will be studied more in detail later—in which particular geometries (created by the presence of cracks, dead spaces, corrosion products or deposits) lead to a split of the electrolyte composition between anode and cathode, with the formation of occluded cells, where aggressive species can concentrate.

Furthermore, reduced or oxidized species formed by cathodic and anodic processes can react with the separation of basic salts, oxides, hydroxides often able to produce layers on the metal surface which play a key role in determining the corrosion behaviour of the metal in that environment.

2.7 Questions and Exercises

2.1 In a corrosion cell, which is the positive and the negative electrode? Which reactions occur at the positive and negative electrode? Give an electrical explanation.

2.2 In natural environments, such as seawater, fresh water, soil and condensed water (dew) in the atmosphere, which is the dominating cathodic reaction in the corrosion of mild steel? Write the reaction equation. Can rust formation affect the corrosion rate?

2.3 Which property of the corrosion medium (the aqueous solution) is the most important prerequisite for electrochemical corrosion?

2.4 What can be said about the relationship between the anodic current, I_a, and the cathodic current, I_c, in a corrosion process? And what about current densities, respectively?

2.5 Estimate carbon steel corrosion rate corresponding to 1 mA/m^2 anodic current density.

2.6 A steel plate has corroded on both sides in seawater. After 10 years, a thickness reduction of 3 mm is measured. Calculate the average corrosion current density. Take into consideration that the dissolution reaction is mainly $Fe = Fe^{2+} + 2e^-$, and that the density and the atomic mass of iron are 7.8 Mg/m^3 and 56 g/mol, respectively.

2.7 Suppose and reproduce Evans's first experience on a copper plate with a seawater drop. Write the corrosion reaction and indicate inside the seawater drop where presumably the corrosion product forms, namely the copper oxy-chloride (approximately $Cu(OH)Cl$).

2.8 Evans's second experience does not work when using copper. Try to indicate possible causes.

2.9 Write the cathodic reaction occurring by copper ion displacement and determine the value of constant e_{ech}/γ of Eq. 2.12. [Hint: refer to Sect. 7.2.5]

2.10 Write the corrosion reaction of silver exposed in a solution containing H_2S. By means of Faraday Law, calculate the corrosion rate ($\mu m/y$), if corrosion current is 0.01 A, and if the silver sample has a surface area of 0.1 m^2.

2.11 How much oxygen is necessary to completely corrode a square carbon steel plate (10 cm in side, 10 mm thick).

2.12 The corrosion current density on a galvanized steel plate (steel coated with zinc coating) is 1.25×10^{-7} A/cm^2. Zinc coating thickness is 0.02 mm. After how many years the coating will be completely corroded? ($MW_{Zn} = 65.3$ g/mol, $\gamma_{Zn} = 7.14$ Mg/m^3).

Alessandro Volta Priorities

Pietro has reviewed and read almost completely the writings and letters of Alessandro Volta, highlighting some interesting discoveries obtained by the scientist. This book includes four of them, related to the first law of Faraday (hereafter), the Ohm's law (Chap. 5), the definition of the potential ranking and driving force (Chap. 10) and the principle of cathodic protection (Chap. 19).

Alessandro Volta and the First Faraday Law

From the letter that Volta sent to Van Marum (June 1802 [1]), it is clear how Alessandro Volta knew not only to identify the physical elements related to the pile but also to specify the quantitative relationship that links them together. In this letter, Volta anticipates many of the chemical effects produced by the current circulation, that Faraday will obtain thirty years later.

Volta asks Van Marum to carried out a series of tests, as he is not able to perform because he did not have powerful generators. In the letter he wrote *"An important thing is to try to obtain the hydrogen gas evolution, and the oxidation of the two metal wires immersed in water [...] produced by the continuous functioning of the pile, to obtain the effect, I say, with many charges [of capacitor banks] reiterated by shooting from the current of a good pile. Things can be easily arranged in such a way that such charges and alternate discharges occur with the interval of half a second or less. But I would like it even more if you succeed in another way that I have already proposed to you: with the direct electric current using your big generator. This copious current, perhaps like that of a good pile (you believe it even more abundant, but I doubt it a lot), forced by a convenient arrangement to move from one wire through the water to another in free communication with the wet soil or, even better, with the bearings of the machine in action, it should make almost the same amount of hydrogen gas appearing around a wire and oxygen gas or metal oxide around the other, as with the pile. Yes, the same quantity and in the same way and with the same appearances, if really your great and prodigious generator is able to provide and to allow to flow through endless conductors as much electric fluid in every moment or in a given time, as the pile supplies. It will therefore be the success of the experiment that decides which of the two devices is able to supply more. For other generators that are not as big and excellent as your, it is already proven that they provide much less current than even a small pile"*

---

[1] The letter is part of Volta's correspondence with the Dutch physics Van Marum, who had a powerful scrubbing electric machine in Rotterdam. Unfortunately, Van Marum, unlike what he did with previous letters, did not make it public. The letter was only disclosed in 1905 when J. Bosscha published the correspondence Volta-Van Marum.

Volta makes some fundamental statements. He hypothesises the identity, as regards the chemical effects produced, of the electric fluid regardless of whether it is generated by a battery, a capacitors bank or a scrubbing machine. Anticipate the first Faraday law of electrochemical stoichiometry. He proposes to compare the quantity of electric fluid produced by two different generators on the basis of the extent of the chemical effects resulting from the current circulation. Moreover, he states that a similar evaluation has already been made (evidently by him and in the manner just indicated) to compare the amount of electric fluid supplied by the pile and by generators less powerful than that of Van Marum.

The text shows how Volta in 1802 and Faraday in 1832–33 follow the same scientific path. Both start from the problem of verifying the identity of the electric fluid produced by different generators,[2] then arriving at the law that links the mass formed or transformed at the electrodes to the circulated charge, finally they both propose to apply the law to the measurement of the exchanged charge.[3]

We can now compare the law indicated by Volta in the case of hydrogen evolution, oxygen or oxide formation and the general law enunciated thirty years later by Faraday.

Volta in 1802 wrote: *"The same quantity of hydrogen is produced at the cathode and the oxygen or oxide are formed at the anode [...]—the same quantity, in the same way, with the same appearances—if really the electric scrubbing machine is able to provide and to allow circulating [...] as much electric fluid in every moment or in a given time as it provides and passes the pile."*

Faraday, thirty years later, wrote: *"Electricity, whatever may be its source, is identical in its nature. [...] For a constant quantity of electricity the amount of electro-chemical action is also constant."*

---

[2]Faraday checked the chemical effects not only produced by the electricity obtained from the pile, from the friction machine or from the batteries of condensers—which called "voltaic electricity" or "common electricity"—but also the effect of the "animal electricity", of the "thermoelectric" (Seebek effect) and of the "magneto-electric" one (by induction).

[3]Pietro Pedeferri wondered a lot of time why Faraday proposed in 1833 to call Volta-electrometer and then in 1838 Voltameter, the instrument he developed to determine the charge circulated through the measurement of the chemical effects produced, since Volta was at that time was accused (as, moreover, happens today: read for example what the former president of the Senate Pera wrote on Volta in his book "The Ambiguous Frog", Princeton University Press) to have always disinterested in the correlation between the circulated charge and the chemical effects, indeed the chemical effects "tout court". In light of the priority just mentioned, it is clear that Faraday could not find a more suitable name, even if he could not know it. In fact, the English scientist could not know the content of the letter that Volta had written to Van Marum that will be disclosed only in 1905. "And I do not even think—Pietro Pedeferri said—that when, in 1812 in Milan, Faraday, still very young and unknown, took part with Davy at the meeting with the almost seventy, very famous and acclaimed Volta, the inventor of the pile may have talked about this with him.

Bibliography

Evans UR (1948) An introduction to metallic corrosion. Edward Arnold, London

Fontana M (1986) Corrosion engineering, 3rd edn. McGraw-Hill, New York

Piontelli R (1961) Elementi di teoria della corrosione a umido dei materiali metallici. Longanesi, Milano (in Italian)

Shreir LL, Jarman RA, Burstein GT (1994) Corrosion. Butterworth-Heinemann, London

Tomashov N (1966) Theory of corrosion and protection of metals: the science of corrosion. McMillan, New York

Winston Revie R (2000) Uhlig's corrosion handbook, 2nd edn. Wiley, London

Chapter 3
Thermodynamics of Aqueous Corrosion

There is nothing more practical than a good theory.
W. Nernst

Abstract This chapter addresses the thermodynamic aspects of corrosion, starting from the concept of free energy: indeed, corrosion can take place only if the free energy variation associated with the reaction is negative, i.e., the reaction is thermodynamically favoured. This translates in terms of variation of potential, outlined as driving voltage, or electromotive force, for the reaction. Standard potentials and equilibrium potentials of anodic and cathodic reactions are defined, together with conditions for corrosion and for immunity. Reference electrodes are presented, which allow to measure the potential as difference between a given electrode and a well defined reference electrode that has the property of maintaining its potential constant. Finally, electrochemical cells, as Daniell or concentration cells, are introduced.

Fig. 3.1 Case study at the PoliLaPP Corrosion Museum of Politecnico di Milano

© Springer Nature Switzerland AG 2018
P. Pedeferri, *Corrosion Science and Engineering*, Engineering Materials,
https://doi.org/10.1007/978-3-319-97625-9_3

3.1 Driving Voltage and Free Energy

A chemical reaction, such as a corrosion process, exemplified in Fig. 3.1 can be described by a reaction of the following type:

$$aA + bB + \cdots \rightarrow cC + dD + \cdots \tag{3.1}$$

To this reaction, it is possible to associate the variation of a *state function*, called *free energy*, G, which decreases as the reaction proceeds (i.e., there is a *driving voltage* available for the occurrence of this process). This reduction, denoted ΔG, is also named *reaction affinity*.

A positive driving voltage $(-\Delta G > 0$, then $\Delta G < 0)$ is a necessary condition for a reaction to occur. Conversely, the fading of the driving voltage $(\Delta G = 0)$ or the presence of a negative driving voltage (i.e., $\Delta G > 0$) is a sufficient condition to exclude the possibility that the reaction will take place. In the first case $(\Delta G = 0)$, the reaction is in a condition of equilibrium, whilst in the second case $(\Delta G > 0)$, the system tends to evolve in the opposite direction to that indicated, unless there is the intervention of an external energy.

These thermodynamic concepts are more easily understood when considering a mechanical analogy, in which the system is a body placed on an inclined plane (Fig. 3.2a). The direction of the spontaneous movement of the body can only be the one that corresponds to a reduction of the potential energy, that is, toward lower heights (i.e., there is a positive driving voltage given by the gravity force). The equilibrium condition is reached when the plane is horizontal, which corresponds to the zeroing of this work (Fig. 3.2b). Finally, the movement of the body in the direction that corresponds to an increase of height cannot occur spontaneously: instead, it is obtained only through the intervention of external forces (Fig. 3.2c).

It is worth remembering that the assessment of the free energy change, ΔG, associated with any reaction, implies the knowledge of all chemical species involved and their thermodynamic levels. These levels are: in the case of dissolved species, their *activities* (i.e., their concentration modified by an appropriate correction coefficient, called *activity coefficient*, which takes into account environmental effects); in the case of gaseous species, the *fugacity* (i.e., their partial pressure modified by an appropriate correction coefficient, called *fugacity coefficient*, which takes into account environmental effects).

Fig. 3.2 The mechanical analogy of a corrosion process: **a** spontaneous; **b** equilibrium; **c** non-spontaneous

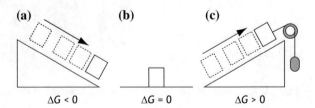

The free energy variation, ΔG, of Eq. 3.1 can be expressed as:

$$\Delta G = \Delta G^0 + RT \ln \frac{a_C^c \cdot a_D^d \cdots}{a_A^a \cdot a_B^b \cdots} \tag{3.2}$$

where ΔG^0 is the standard Gibbs free energy change involved in the reaction (i.e., when occurring at standard conditions, with unitary activity for species in liquid and solid phase, and fugacity 1 atm for gaseous species), R is the universal gas constant equal to 8.314 J/mol K, T is the absolute temperature, and, finally, a_I^i are the activities of species, I, elevated to their stoichiometric coefficient.

3.2 Corrosion and Immunity Condition

A corrosion process is represented by the following reactions:

$$M + A \rightarrow M^{z+} + (z/a)A^{a-} \text{ (global reaction)} \tag{3.3}$$

$$M \rightarrow M^{z+} + ze^- \text{ (anodic reaction)} \tag{3.4}$$

$$A + ze^- \rightarrow (z/a)A^{a-} \text{ (cathodic reaction)} \tag{3.5}$$

Since a corrosion process is of electrochemical nature, the free energy variation, ΔG, can be expressed through the electrical energy variation:

$$\Delta G = -z \cdot F \cdot \Delta E \tag{3.6}$$

where ΔE is the electromotive force (EMF) of the reaction, z and F have the known meaning. In the following, ΔE is called *driving voltage* or *potential difference*.

The thermodynamic condition for a spontaneous process ($\Delta G < 0$) becomes:

$$\Delta G < 0 \Rightarrow \Delta E > 0 \tag{3.7}$$

that is, a positive driving voltage ($\Delta E > 0$). From the expression of free energy Eq. 3.2 by introducing the potential, the well known Nernst[1] equation is obtained (for its general formulation see Eq. 3.17).

For the anodic reaction:

$$E^{M^{z+}/M} = E^0 + \frac{RT}{zF} \ln \frac{[M^{z+}]}{[M]}, \quad \text{then } E_a = E_a^0 + \frac{RT}{zF} \ln[M^{z+}] \tag{3.8}$$

---

[1]**Walther Herman Nernst (1864–1941)** was a German chemist. He received the Nobel Prize for Chemistry in 1920.

For the cathodic reaction:

$$E^{A/A^{a-}} = E^0 - \frac{RT}{zF}\ln\frac{[A^{a-}]}{[A]}, \quad \text{then } E_c = E_c^0 - \frac{RT}{zF}\ln[A^{a-}] \tag{3.9}$$

The free energy variation, ΔG, is expressed by the potential as follows:

$$\Delta E = E_c - E_a \tag{3.10}$$

hence, the condition for a spontaneous corrosion process is:

$$\Delta E > 0 \Rightarrow E_c - E_a > 0 \quad E_c > E_a \tag{3.11}$$

that is, the equilibrium potential of the cathodic reaction must be more positive (*more noble*) than the anodic process (metal oxidation).

Corrosion as an electrochemical process can be represented by the cell depicted in Fig. 3.3, where the anodic reaction occurs on M, while the complementary cathodic process takes place on N, in the electrolyte, ε (the electrochemical cell is indicated as: M/ε/N). The potential difference between M and N $(E^{MN}{}_{eq})$ can be derived as the difference between potentials measured against a same remote reference electrode, R:

$$E_{eq}^{MN} = E_{eq}^{MR} + E_{eq}^{RN} = E_{eq}^{MR} - E_{eq}^{NR} \tag{3.12}$$

where $E_{eq}^{MR}\left(=E_{eq,a}\right)$ is the anodic equilibrium potential and $E_{eq}^{NR}\left(=E_{eq,c}\right)$ is the cathodic equilibrium potential, both measured against the same reference, R. It results:

$$E_{eq}^{MN} = \Delta E^{MN} = E_{eq,c} - E_{eq,a} \tag{3.13}$$

as already above obtained.

Fig. 3.3 Equilibrium potential $(E_{eq}{}^{MN})$ of the corrosion system *MN* as sum of anodic and cathodic equilibrium potentials

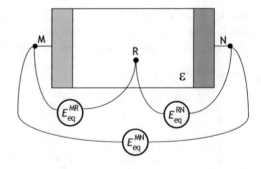

In conclusion:

- The necessary condition for the occurrence of a corrosion process is that the driving force is positive. This occurs if the equilibrium potential of the cathodic process, $E_{eq,c}$, is greater than that of the anodic process, $E_{eq,a}$
- The sufficient condition to prevent any corrosion process is that the driving force is zero or negative. This occurs if the equilibrium potential of the cathodic process, $E_{eq,c}$, is less than or equal to that of the anodic process, $E_{eq,a}$
- To evaluate the driving voltage it is necessary to know the equilibrium potentials of the individual electrode reactions taking place on the surface of materials.

3.3 Standard Potential

Let's consider the galvanic cell, $M/\varepsilon_1//\varepsilon_2/H$, at 25 °C obtained by coupling a metal, M, in equilibrium with a solution, ε_1, of unitary concentration of metal ions, with a reference electrode consisting of a platinised platinum wire[2] immersed in a solution, ε_2, of unitary acid concentration (pH = 0), bubbling hydrogen gas at a pressure of 1 atm. This electrode is called *Standard Hydrogen Electrode* (SHE). A salt bridge, consisting of a glass tube filled with agar-agar gel saturated with KCl, electrically connects the two solutions, ε_1 and ε_2, keeping them physically separated (Fig. 3.4). The equilibrium reaction on the surface of metal M is:

$$M \rightarrow M^{z+} + ze^- \tag{3.14}$$

and the one on the platinum surface is:

$$2H^+ + 2e^- \rightarrow H_2 \tag{3.15}$$

These two reactions, when the circuit is open, are in equilibrium.

The potential difference between terminals M and H, respectively in contact with metal, M, and platinum, is the equilibrium potential of metal M measured toward the standard hydrogen electrode, SHE, whose potential is taken conventionally equal to zero at all temperatures. The *standard potential* of metal M at the temperature considered is E^0, i.e., the equilibrium potential at unitary concentration (1 mol/L).

The list of standard potentials, E^0, of the various elements, sorted by increasing potential, is the so-called *series of standard potentials* (Table 3.1). Potentials are all referred to the reduction reactions. It is conventional to refer to the term *nobility*

---

[2]A platinum wire is subjected to anodic and cathodic cycles, which form a deposit of black platinum powder on the wire surface to increase the effective surface. The platinum so treated is called *platinised platinum*.

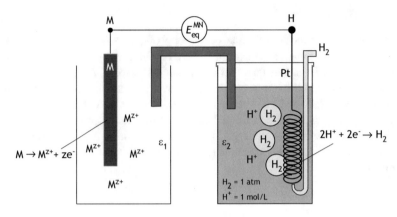

Fig. 3.4 Galvanic cell used for the measurement of the equilibrium potential of metal M versus the standard hydrogen electrode (SHE)

Table 3.1 Standard potentials series

Electrode reactions	E (V SHE)	Electrode reactions	E (V SHE)
$F_2 + 2H^+ + e^- \rightarrow 2HF$	+3.030	$Cu^{2+} + e^- \rightarrow Cu^+$	+0.158
$O_3 + 2H^+ + e^- \rightarrow O_2 + H_2O$	+2.070	$Sn^{4+} + 2e^- \rightarrow Sn^{2+}$	+0.150
$Co^{3+} + 3e^- \rightarrow Co$	+1.842	$2H^+ + 2e^- \rightarrow H_2$	0
$Au^+ + e^- \rightarrow Au$	+1.680	$2D^+ + 2e^- \rightarrow D_2$	−0.003
$Mn^{3+} + e^- \rightarrow Mn^{2+}$	+1.510	$Fe^{3+} + 3e^- \rightarrow Fe$	−0.036
$Au^{3+} + 3e^- \rightarrow Au$	+1.500	$Pb^{2+} + 2e^- \rightarrow Pb$	−0.126
$MnO_4^- + 8H^+ + 5e^- \rightarrow Mn^{2+} + 4H_2O$	+1.491	$Sn^{2+} + 2e^- \rightarrow Sn$	−0.136
$PbO_2 + 4H^+ + 2e^- \rightarrow Pb^{2+} + 2H_2O$	+1.467	$Ge^{4+} + 4e^- \rightarrow Ge$	−0.150
$Cl_2 + 2e^- \rightarrow 2Cl^-$	+1.358	$Mo^{3+} + 3e^- \rightarrow Mo$	−0.200
$Cr_2O_7^{2-} + 14H^+ + 6e^- \rightarrow 2Cr^{3+} + 7H_2O$	+1.330	$Ni^{2+} + 2e^- \rightarrow Ni$	−0.250
$O_2 + 4H^+ + 4e^- \rightarrow 2H_2O$	**+1.230**	$Co^{2+} + 2e^- \rightarrow Co$	−0.280
$CrO_4^{2-} + 8H^+ + 3e^- \rightarrow Cr^{3+} + 4H_2O$	+1.195	$Mn^{3+} + 3e^- \rightarrow Mn$	−0.283
$Pt^{2+} + 3e^- \rightarrow Pt$	+1.190	$In^{3+} + 3e^- \rightarrow In$	−0.342
$Br_2 + 2e^- \rightarrow 2Br^-$	+1.087	$Cd^{2+} + 2e^- \rightarrow Cd$	−0.400
$HNO_3 + 3H^+ + 3e^- \rightarrow NO + 2H_2O$	+0.960	$Cr^{3+} + e^- \rightarrow Cr^{2+}$	−0.410
$2Hg^{2+} + 2e^- \rightarrow Hg_2^{2+}$	+0.920	$Fe^{2+} + 2e^- \rightarrow Fe$	−0.440
$Hg^{2+} + 2e^- \rightarrow 2Hg$	+0.851	$Cr^{3+} + 3e^- \rightarrow Cr$	−0.740
$Ag^+ + e^- \rightarrow Ag$	+0.800	$Zn^{2+} + 2e^- \rightarrow Zn$	−0.760
$Hg_2^{2+} + 2e^- \rightarrow 2Hg$	+0.796	$V^{3+} + 3e^- \rightarrow V$	−0.876
$Fe^{3+} + e^- \rightarrow Fe^{2+}$	+0.770	$Cr^{2+} + 2e^- \rightarrow Cr$	−0.913
$O_2 + 2H^+ + 2e^- \rightarrow H_2O$	+0.682	$Nb^{3+} + 3e^- \rightarrow Nb$	−1.100
$Hg_2SO_4 + 2e^- \rightarrow 2Hg + SO_4^{2-}$	+0.620	$Mn^{2+} + 2e^- \rightarrow Mn$	−1.180
$MnO_4^- + 2 H_2O + 3e^- \rightarrow MnO_2 + 4 OH^-$	+0.588	$V^{2+} + 2e^- \rightarrow V$	−1.180

(continued)

Table 3.1   (continued)

Electrode reactions	E (V SHE)	Electrode reactions	E (V SHE)
$I_2 + 2e^- \rightarrow 2I^-$	+0.534	$Ti^{3+} + 3e^- \rightarrow Ti$	−1.210
$Cu^+ + e^- \rightarrow Cu$	+0.522	$Zr^{4+} + 4e^- \rightarrow Zr$	−1.530
$2NO_2^- + 4H_2O + 6e^- \rightarrow N_2 + 8OH^-$	+0.420	$Ti^{2+} + 2e^- \rightarrow Ti$	−1.630
$SO_4^{2-} + 6e^- + 8H^+ \rightarrow S + 4H_2O$	+0.360	$Al^{3+} + 3e^- \rightarrow Al$	−1.660
$Cu^{2+} + 2e^- = Cu$	+0.340	$Mg^{2+} + 2e^- \rightarrow Mg$	−2.360
$2NO_3^- + 6H_2O + 10e^- \rightarrow N_2 + 12OH^-$	+0.250	$Na^+ + e^- \rightarrow Na$	−2.710
$AgCl + e^- \rightarrow Ag + Cl^-$	+0.220	$Ca^{2+} + 2e^- \rightarrow Ca$	−2.860
$SO_4^{2-} + 2e^- + 2H^+ \rightarrow SO_3^{2-} + H_2O$	+0.170	$Li^+ + e^- \rightarrow Li$	−3.050

depending on the position in the series: the more positive the potential, the higher the nobility. It is worth noting that on the top of the scale there are the first metals produced by man (in order gold, silver, then copper and iron) which are more noble, while those on the bottom have been obtained only in recent times (aluminium, titanium, magnesium, sodium).

3.4   Potential of an Electrochemical Reaction

To calculate the potential of an electrochemical reaction at any condition, it is necessary to know the chemical species involved, as well as their thermodynamic levels in terms of activity for dissolved species and fugacity for gas. For corrosion systems, activities and fugacities are approximated to concentrations and pressures, respectively, with the exception for concentrated solutions or when complexes or high pressure systems are present.

For any electrochemical reaction as the one below, written as anodic:

$$aA + bB + \cdots \rightarrow cC + dD + \cdots + ze^- \tag{3.16}$$

the equilibrium potential is given by the Nernst equation derived from Eq. 3.2:

$$E_{eq} = E^0 + \frac{RT}{zF} \ln \frac{a_C^c \cdot a_D^d}{a_A^a \cdot a_B^b} \tag{3.17}$$

E^0 is the standard potential of the reaction at standard conditions (i.e., unitary activity for dissolved species and fugacity 1 atm for gas), z is the number of electrons involved and F is the Faraday constant.

The Nernst equation shorts to:

$$E_{eq} = E^0 + \frac{0.059}{z} \log \frac{a_C^c \cdot a_D^d}{a_A^a \cdot a_B^b} \tag{3.18}$$

where 0.059 V is the term 2.3 RT/F at 25 °C (2.3 is the conversion coefficient from natural to decimal-base logarithm).

When H^+ (or OH^-) ions participate in the reaction, it is useful to highlight the effect of pH. Therefore, the above Eq. 3.14 becomes:

$$aA + bB \rightarrow cC + hH^+ + ze^- \tag{3.19}$$

and the equilibrium potential is:

$$E_{eq} = E^0 - 0.059 \frac{h}{z} pH + \frac{0.059}{z} \log \frac{a_C^c}{a_A^a} \tag{3.20}$$

From the above equation, it results that in an E-pH plot the equilibrium potential is a straight line with slope 0.059 h/z. This is when the reaction involves H^+ or OH^- ions, while it is worth to notice that it is a horizontal line when these ions do not participate, according to Eq. 3.18.

3.5 Potential of Metal Dissolution Reaction

For a metal dissolution reaction, $M \rightarrow M^{z+} + ze^-$, the Nernst equation becomes:

$$E_{eq} = E^0 + \frac{2.3RT}{zF} \log \frac{a_{M^{z+}}}{a_M} \tag{3.21}$$

where E^0 is the standard potential of metal, M, a_M^{z+} is the ion concentration and a_M is the metal concentration, where for pure metals $a_M = 1$.

Metal ion concentration a_M^{z+} is assumed 10^{-6} mol/L, as suggested by Pourbaix, for electrolytes not containing metal ions, as in the case of metals exposed to waters or buried in soil. The equilibrium potential shorts to:

$$E_{eq} = E^0 - \frac{0.354}{z} \tag{3.22}$$

3.5.1 Corrosion and Immunity Conditions

For a general anodic reaction of metal dissolution as Eq. 3.4, the thermodynamic condition to proceed toward anodic direction (*corrosion*) becomes $E > E_{eq}$, which corresponds to the condition $\Delta G < 0$, or $\Delta E = E - E_{eq} > 0$. Conversely, if $E < E_{eq}$ (or $\Delta G > 0$), the anodic reaction does not take place, instead it occurs in the opposite direction. This condition is called *immunity*.

3.6 Potential of Cathodic Processes

Among possible cathodic processes, the rank of occurrence is established by the equilibrium potential: the most noble process is the first, sequentially followed by less noble ones, as in the following sequence (standard potential, V vs SHE, ordered downward):

- Chlorine reduction to give chloride (+1.36 V)
- Oxygen reduction (+1.23 V)
- Copper ions reduction (+0.34 V)
- Hydrogen ion reduction (0 V).

In most of the corrosive environments, two cathodic processes occur, that is, oxygen reduction and hydrogen ion reduction. Hence, in general, to assess the corrosion of metals, it is useful to distinguish between acidic solutions and neutral or alkaline solutions. Oxygen reduction always takes place, if present. Whereas, hydrogen evolution occurs in acidic environments, at least for ferrous alloys. For instance, for carbon steel exposed to natural environments such as water, soil, atmosphere and concrete, the cathodic process is oxygen reduction, only.

Example 1
Let's consider a solution containing ferrous ions with a concentration 10^{-3} mol/L of Fe^{2+} ions at 25 °C. If an iron specimen (for instance a strip), immersed in that solution, has a potential of −500 mV or −600 mV SHE, respectively, what would you expect? Corrosion or deposition of iron?

Answer. Under standard conditions at 25 °C iron has an equilibrium potential equal to −0.44 V SHE; therefore, in a solution with a concentration 10^{-3} mol/L of Fe^{2+} ions, the equilibrium potential is:

$$E_{eq} = E^0 + \frac{RT}{zF} \ln a_{Fe^{2+}} = -0.44 + \frac{0.059}{2}(-3) = -0.527 \text{ V SHE}$$

When the measured potential is -500 mV SHE, since -500 mV > -0.527 V iron tends to pass into solution; instead, when the potential is -600 mV SHE, iron ions tend to be deposited.

Example 2

A storage tank for fresh water is made of iron (carbon steel). Can you decide whether corrosion takes place or not according to the potential measured?

Answer. If the measured potential is E_M, and the equilibrium potential is E_{eq}, there is corrosion if $E_M > E_{eq}$, or immunity if $E_M < E_{eq}$.

The equilibrium potential is calculated with the Nernst equation introducing the concentration of 10^{-6} mol/L of iron ions in the solution (corresponding in the case of iron to 0.056 ppm). The equilibrium potential of reaction $Fe = Fe^{2+} + 2e^-$ is

$$E_{eq} = -0.44 - 0.059/2 \log 10^{-6} = -0.62 \text{ V SHE.}$$

It appears evident that the greater the difference between measured and equilibrium potential, the greater the driving force available for the occurrence of the corrosion process.

3.6.1 Potential of Hydrogen Evolution Reaction

The hydrogen evolution process in acidic solutions is given by the reaction:

$$2H_3O^+ + 2e^- \rightarrow H_2 + 2H_2O \qquad (3.23)$$

It can also occur in neutral or alkaline solutions when the metal is electronegative, like aluminium or magnesium, by the following reaction:

$$2H_2O + 2e^- \rightarrow 2OH^- + H_2 \qquad (3.24)$$

The two reactions above are energetically equivalent and therefore characterized by the same equilibrium potential.[3] Assuming $pH_2 = 1$ bar and $a_{H_2O} = 1$, according to the Nernst equation, the equilibrium potential depends on pH as follows:

---

[3]The combination of the equation of the hydrogen evolution reaction ($H_2O + e^- = \frac{1}{2}H_2 + OH^-$, reaction a) with the water dissociation reaction ($H^+ + OH^- = H_2O$, reaction c) gives $H^+ + e^- = \frac{1}{2}H_2$ (reaction b). Therefore, $\Delta G_b = \Delta G_a + \Delta G_c$. At equilibrium conditions $\Delta G_c = 0$, therefore $\Delta G_a = \Delta G_b$, as well as the associated potentials with respect to the same reference

Fig. 3.5 The equilibrium potential of most common cathodic processes in function of pH: oxygen reduction and hydrogen evolution

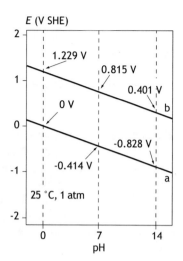

$$E_{eq} = E^0 + \frac{0.059}{z} \log a_{H^+} = 0.059 \log a_{H^+} = -0.828 - 0.059 \log a_{OH^-} = -0.059 \, pH$$

$$(3.25)$$

In a potential - pH diagram, called Pourbaix diagram (Fig. 3.5), the equilibrium potential of hydrogen evolution is given by the *straight line a*, having a slope of -0.059 at 25 °C.

3.6.2   Potential of Oxygen Reduction Reaction

In aerated acidic solutions, the oxygen reduction reaction is the following:

$$O_2 + 4H^+ + 4e^- \rightarrow 2H_2O \qquad (3.26)$$

and in neutral or alkaline aerated solutions it is:

$$O_2 + 2H_2O + 4e^- \rightarrow 4OH^- \qquad (3.27)$$

---

electrode. From an energy viewpoint, the processes (a) and (b) are equivalent. The potentials of the two processes are mutually correlated through the ionic dissociation constant of water (at 25 °C $K_w = a_{H^+} \cdot a_{OH^-} = 10^{-14}$). For example, at pH 14, in conditions where both $a_{H_2O} = 1$ and $P_{H2} = 1$ bar, depending on which process a) or b) reference is made, the following is obtained:
$E_b = E_b^0 + \frac{2.3RT}{F} \log a_{H^+} = 0 + 0.059 \cdot (-14) = -0.828 \, V \, SHE$

The two reactions are equivalent and therefore characterized by the same equilibrium potential.[4] With reference to the standard conditions, $p_{O2} = 1$ bar, $a_{H_2O} = 1$ and 25 °C, the equilibrium potential is given by the Nernst equation:

$$E_{eq} = 0.401 - 0.059 \log a_{OH^-} = 1.229 + 0.059 \log a_{H^+} = 1.229 - 0.059\,pH$$

$$(3.28)$$

The equilibrium potential of the oxygen reduction reaction in the E-pH diagram (Fig. 3.5) follows *line b*, which is parallel to *line a* of the hydrogen evolution reaction. The equilibrium potential changes as the oxygen content—which determines oxygen partial pressure—varies. On the basis of Henry's law, 1 ppm variation of the oxygen content determines a variation of equilibrium potential of 25 mV. Often, in practice, for 1 ppm oxygen content change, about 50 mV shift is considered, which also includes some activation overvoltage contributions, as discussed in Chap.5.

3.6.3 Applications of Thermodynamic Criteria

Thermodynamic criteria help determine whether a corrosion reaction can occur, once equilibrium potentials of both cathodic and anodic processes are known.

De-aerated solutions. In *acidic de-aerated solutions*, i.e., in the absence of dissolved oxygen, the only possible cathodic process is the reduction of hydrogen ion to give hydrogen evolution which takes place at the potential $E_{eq} = -0.059 \cdot pH$ (often taken as about 0). All metals with equilibrium potential $E_{eq,M}$ more negative than $-0.059 \cdot pH$ can corrode; the more negative (less noble) the equilibrium potential of the metal, the greater is the driving voltage. Examples of metals that corrode in acidic de-aerated solutions (listed in descending order of nobility) are: lead, tin, nickel, cobalt, thallium, cadmium, iron, chromium, zinc, aluminium, magnesium. Instead, examples of metals that do not corrode in acidic de-aerated solutions (listed in order of increasing nobility) are: copper, mercury, silver. In *neutral de-aerated solutions*, in particular water for which the equilibrium potential of hydrogen evolution is -0.414 V SHE, metals with equilibrium potential lower than that—such as iron, chromium, zinc, aluminium, magnesium—can corrode. Finally, in *alkaline de-aerated solutions*, where the hydrogen ion reduction reaction has the equilibrium potential close to -0.828 V SHE, only zinc, aluminium and magnesium can corrode among industrially used metals.

---

[4]With similar considerations discussed in note 3, it can easily be proved that the two oxygen reduction reactions ($O_2 + 4H^+ + 4e^- = 2H_2O$ and $O_2 + 2H_2O + 4e^- = 4OH^-$) are equivalent

Aerated solutions. In *acidic aerated solutions*, the equilibrium potential of the oxygen reduction process is +1.23 V SHE; therefore, only metals with a greater (more noble) equilibrium potential, such as gold or platinum, do not corrode, while all others do. In neutral aerated solutions (E_{eq} = + 0.815 V SHE), all metals less noble than silver can suffer from corrosion, while in alkaline aerated solutions (E_{eq} = + 0.401 V SHE) this occurs for all metals less noble than copper.

Oxidizing species. In solutions containing oxidizing species, things change. For example, in chlorine-containing solutions (E_{eq} = + 1.35 V SHE) silver can also corrode, and in those containing fluorine (E_{eq} = + 2.65 V SHE) even gold does.

Potentialof an electrochemical reaction

Example 1

Let's consider the electrochemical reaction of zinc dissolution to give zincate:

$$Zn + 2H_2O \rightarrow HZnO_2^- + 3H^+ + 2e^-$$

The equilibrium potential is given by the Nernst equation (taking into account water activity, $a_{H_2O} = 1$, as unitary):

$$E_{eq} = E^0 + \frac{2.3RT}{2F} \log \frac{a_{HZnO_2} \cdot a_H^3}{a_{Zn} \cdot a_{H_2O}} = E^0 + \frac{0.059}{2} \log a_{HZnO_2} - \frac{3}{2}0.059 \text{ pH}$$

Example 2

Let's consider the following redox reaction:

$$Mn^{2+} + 4H_2O \rightarrow MnO_4^- + 8H^+ + 5e^-$$

The equilibrium potential is given by the Nernst equation (taking into account water activity, a_{H_2O}, as unitary):

$$E_{eq} = +1.507 - 0.094 \text{ pH} + 0.012 \log([MnO_4^-]/[Mn^{2+}])$$

Example 3

Finally, let's consider the following redox reaction:

$$Fe^{2+} \rightarrow Fe^{3+} + e^-$$

The equilibrium potential is:

$$E_{eq} = 0.77 + 0.059 \log\left([Fe^{3+}]/[Fe^{2+}]\right)$$

3.7   Insoluble Products and Complexing Species

So far, we have considered metals which dissolve as ions while corroding. However, sometimes chemical species that form insoluble corrosion products—for example oxides, hydroxides or sulphides—are present. In these conditions, the extremely low concentration of metal ion in solution lowers the equilibrium potential from the standard potential drastically, changing the thermodynamic condition. This occurs, for instance, with gold, silver, copper and many other metals in cyanide or ammonia containing solutions and many other complexing species.

Let's take into consideration the case of silver in cyanide containing solutions, where the following complexing reaction takes place: $Ag^+ + 2CN^- \rightarrow Ag(CN)_2^-$. The silver ion, Ag^+, concentration is obtained from the complex stability constant:

$$K_s = [AgCN_2^-]/\left([Ag^+][CN^-]^2\right) = 10^{21.2} \qquad (3.29)$$

In a solution containing 0.1 mol/L of $AgNO_3$ and 1 mol/L of KCN, the complex concentration is 0.1 mol/L (i.e., the same as $AgNO_3$), therefore the cyanide concentration is $[CN^-] = 1 - 2 \cdot (0.1) = 0.8$ mol/L. The silver ion concentration in solution is given by the complex stability constant: $[Ag^+] = (0.1)/[(0.8) \times 10^{21.2}] = 10^{-22}$ mol/L.

The equilibrium potential is given by the Nernst equation applied to the electrochemical reaction $(Ag \rightarrow Ag^+ + e^-)$:

$$E_{eq} = 0.8 + 0.059 \log[Ag^+] = 0.8 + 0.059(-22) = -0.5 \text{ V SHE} \qquad (3.30)$$

In this condition, silver is less noble than copper and steel (iron).

3.8   Reference Electrodes

The potential is measured by connecting a voltmeter to the metal (or metal structure), also called working electrode and to a reference electrode which has the property of maintaining its potential constant (Fig. 3.6). The positive terminal has to be connected to the working electrode in order to obtain the correct sign of the reading value.

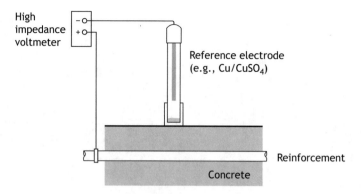

Fig. 3.6 Example of apparatus for potential measurement in concrete

Table 3.2 Reference electrodes used in laboratory and in industrial plants

Electrode	Description	Equilibrium reaction	E (V SHE)
Standard hydrogen electrode (SHE)	H_2(1 atm)|H^+(a = 1 M)	$2H^+ + 2e^- \rightarrow H_2$	0
Saturated calomel electrode (SCE)	Hg|Hg_2Cl_2, KCl (sat)	$Hg_2Cl_2 + 2e^- \rightarrow 2Hg + 2Cl^-$	+0.244
Silver/silver chloride (saturated)	Ag|AgCl, KCl (sat)	$AgCl + e^- \rightarrow Ag + Cl^-$	+0.200
Silver/silver chloride (seawater)	Ag|AgCl, seawater	$AgCl + e^- \rightarrow Ag + Cl^-$	+0.250
Copper/copper sulphate electrode (CSE)	Cu|$CuSO_4$ (sat)	$Cu^{2+} + 2e^- \rightarrow Cu$	+0.318
Zinc/seawater (ZN)	Zn|seawater	$Zn^{2+} + 2e^- \rightarrow Zn$	−0.800

By convention, the potentials are referred to the standard hydrogen electrode (SHE) taken as the zero reference. However, it is very unfeasible and therefore not used, either in laboratory testing or for industrial monitoring. The most used reference electrodes are reported in Table 3.2. Figure 3.6 shows the most used reference electrodes for potential measurements of structures buried in concrete, which is copper–copper sulphate.

Silver in presence of sulphides

In a sulphide containing solution, silver ions react with sulphides to give highly insoluble Ag_2S (solubility product $K_s = [Ag^+]^2 [S^{2-}] = 1.6 \times 10^{-49}$).

The Nernst equation for the electrochemical reaction $(Ag \rightarrow Ag^+ + e^-)$ when sulphide ions are present is:

$$E_{eq} = 0.8 + 0.059 \, \log[Ag^+]$$
$$= 0.8 + 0.059/2 \, \log(K_s/[S^{2-}])$$

If, for example, sulphide $[S^{2-}]$ concentration in solution is 1 ppm, corresponding to 3×10^{-5} M, it results: $\log (K_s/[S^{2-}]) = \log (1.6 \times 10^{-49}/3 \times 10^{-5}) = -44.3$. The equilibrium potential is then: $= 0.8 + (-44.3)$ $0.059/2 = -0.51$ V SHE, which is a value even more negative than the potential of active steel (iron).

This case study explains the tarnishing of silver in sulphide containing atmospheres.

Copper in concentrated hydrochloric acid

Copper immersed in concentrated pure hydrochloric acid corrodes rapidly with the evolution of gas bubbles. This is not contrary to thermodynamics as proved in the following explanation, making the necessary assumptions.

According to a chemical handbook's data, the concentration of "pure" HCl is 12.0 M. Assuming unitary activity coefficients, pH $= -\log[12] = -1.08$. The equilibrium potential of the hydrogen evolution reaction at pH -1.08 is -0.059 pH$= +0.064$ V SHE

The equilibrium potential of $+0.064$ V SHE for Cu^{+2}/Cu is reached if copper concentration drops to $[Cu^{2+}] = 7.56 \times 10^{-10}$ mol/L, as from Nernst equation:

$$E_{eq(Cu^{2+}/Cu)} = 0.064 = 0.334 + [0.059/2] \cdot \log[Cu^{2+}].$$

This occurs because of the formation of copper-chloride complexants.

3.9 Electrochemical Cells

When coupling two metal-solution systems—for example, zinc and copper—each one immersed in a solution of their sulphate salt with unitary concentration, separated by a porous membrane selectively permeable to the sulphate ions (this is the Daniell cell), the equilibrium potential (i.e., the tension measured at open circuit) is given by the difference of equilibrium potentials of each electrode.

By short-circuiting the two electrodes (Zn and Cu), the electrode reactions are, respectively:

$$Zn \rightarrow Zn^{2+} + 2e^- \tag{3.31}$$

$$Cu \rightarrow Cu^{2+} + 2e^- \tag{3.32}$$

the global reaction is:

$$Zn + Cu^{2+} \rightarrow Zn^{2+} + Cu \tag{3.33}$$

Standard potentials of the two electrodes, zinc and copper, respectively, are: $E° = -0.76$ V SHE (anode), and $E° = +0.34$ V SHE (cathode), therefore the open circuit potential (i.e., in equilibrium condition) is $(-0.76 - (+0.34)) = -1.1$ V. The driving voltage is, therefore, 1.1 V, which is available for the occurrence of the above overall reaction.

3.9.1   Concentration Cells

In some corrosion processes, anodic and cathodic reactions are the same. In these cases, the driving voltage derives from a chemical-physical unevenness that may occur: in solution, in gas phase in contact with the electrodes and on the electrodes.

Uneven solution. Let's consider a cell formed by two identical electrodes made of lead in contact with two aqueous solutions in which lead is soluble (for example lead perchlorate or sulphamate) having different lead concentrations:

I, Pb/lead salt diluited, $Pb_{(dil)}$//lead salt concentrated, $Pb_{(conc)}$/Pb, **II**

The equilibrium potential of the cell is [5]:

$$E_{eq\,I-II} = 0.059/2 \, \log[Pb_I]/[Pb_{II}] \tag{3.34}$$

Therefore, lead in contact with a diluted solution behaves as an anode, corroding; instead, lead in contact with a more concentrated solution acts as a cathode and then lead deposits on its surface. The system proceeds toward the reduction of the concentration difference between the two solutions.

---

[5]Equilibrium conditions are achieved only if a membrane selectively permeable to anions separates the two solutions. In the most common case in which the two solutions are in contact with each other, the potential of the cell is $E_{eq,I-II} = t_a \cdot 0.059/2 \, \log ([Pb_I]/[Pb_{II}])$, where t_a is the transport number of anions.

Uneven oxygen concentration in solution. Let's consider a cell formed by two identical electrodes made of a platinum wire immersed in two solutions with different oxygen content: C_1 and C_2 being $C_1 > C_2$. The electrode reaction is oxygen reduction, Eq. 3.27. The equilibrium potential of the cell:

$$\text{I}, \text{Pt}/\text{oxygen rich solution } C_1 // \text{oxygen poor solution } C_2/\text{Pt}, \text{II}$$

is as follows:

$$E_{eq,I-II} = 0.059/4 \, \log C_1/C_2 \tag{3.35}$$

The potential of electrode **I** is more noble than the potential of electrode **II**; therefore, electrode **I** is cathode and electrode **II** is anode.

Uneven electrode composition. Let's consider a cell formed by two electrodes made of cadmium amalgam; the first, **I**, with a higher Cd concentration and the second, **II**, with a lower Cd concentration in contact with a cadmium solution:

$$\text{I}, [\text{Cd}, \text{Hg}]_{\text{high Cd}}/\text{CdSO}_4 \text{ solution}/[\text{Cd}, \text{Hg}]_{\text{low Cd}}, \text{II}$$

The cell reaction is: $\text{Cd} \rightarrow \text{Cd}^{2+} + 2e^-$, and the cell potential is:

$$E_{eq,I-II} = 0.059 \log [\text{Cd}]_{\text{low}}/[\text{Cd}]_{\text{high}} \tag{3.36}$$

Since $[\text{Cd}]_{\text{low}} < [\text{Cd}]_{\text{high}}$, the equilibrium potential, $E_{eq,I,II}$, is negative. Therefore, there is a driving force because Cd of electrode **I** dissolves, and on electrode **II** it deposits from Cd^{2+} ions. All these examples show that the spontaneous tendency is the reduction of unevenness.

3.10   Questions and Exercises

3.1 Assuming standard conditions for reactants and products, determine the spontaneous direction of the following reactions by calculating the driving voltage, ΔE:

$$\text{Cu} + 2\text{HCl} \rightarrow \text{CuCl}_2 + \text{H}_2 \left[\text{Cu} + 2\text{H}^+ \rightarrow \text{Cu}^{2+} + \text{H}_2\right]$$
$$\text{Fe} + 2\text{HCl} \rightarrow \text{FeCl}_2 + \text{H}_2 \left[\text{Fe} + 2\text{H}^+ \rightarrow \text{Fe}^{2+} + \text{H}_2\right]$$
$$2\text{AgNO}_3 + \text{Fe} \rightarrow \text{Fe}(\text{NO}_3)_2 + 2\text{Ag}$$
$$\text{Ag} + \text{FeCl}_3 \rightarrow \text{FeCl}_2 + \text{AgCl}$$
$$2\text{Al} + 3\text{ZnSO}_4 \rightarrow \text{Al}_2(\text{SO}_4)_3 + 3\text{Zn}$$

3.2 Write the cathodic and anodic reactions occurring for the uniform corrosion of the following systems at standard conditions, where applicable:

    (a) Aluminium in oxygen-free sulphuric acid

    (b) Iron in oxygen-free ferric sulphate solution

    (c) Carbon steel in aerated seawater

    (d) Zinc-tin alloy in an oxygen saturated solution of $CuCl_2$, $SnCl_4$ and HCl

    (e) Copper in deaerated seawater.

3.3 ΔE as well as cell voltage is positive for spontaneous corrosion. Can you give an example of a cell with negative voltage? Describe the consequences of such condition.

3.4 Calculate the standard potentials E^0 and equilibrium potentials E_{eq} as function of pH for the following electrode reactions from standard ΔG^0:

    (a) $Zn^{2+} + 2e^- \rightarrow Zn$ ($\Delta G^0 = -147,000\,J$)

    (b) $HZnO^{2-} + 3H^+ + 2e^- \rightarrow Zn + 2H_2O$ ($\Delta G^0 = -10,400\,J$)

  Report results on a E-pH diagram, the so-called Pourbaix diagram.

3.5 Let's consider the following air-saturated solutions:

    (a) Diluted strong acid (pH = 3)

    (b) Alkaline solution (pH = 13)

  Indicate possible acids and alkalis making the above conditions, write cathodic reactions, calculate the equilibrium potential and plot it as a function of pH.

3.6 Calculate the difference between the equilibrium potential of reference electrode $Cu/CuSO_4$ (concentration 1 mol/L) and a carbon steel structure buried in soil (iron concentration 10^{-6} mol/L). Suppose to measure the free corrosion potential of the same structure in soil. Is this value different from the equilibrium potential? Why? [Hint: refer to Chap. 6 for the definition of free corrosion potential].

3.7 Gold does not corrode in aerated solutions, since its equilibrium potential is more noble than the equilibrium potential of oxygen reduction. How gold can be extracted from the sands that contain it?

3.8 Considering the following Table, check whether corrosion is possible or not [Hint: metal ions content is always 10^{-6} mol/L]. If the answer is positive, then estimate the driving force.

Metal	Environment	pH	Corrosion (Y/N)	Driving force
Fe	Aerated neutral soil	7		
Fe	De-aerated acidic water	4		
Fe	Aerated concrete	13		
Cu	De-aerated neutral water	7		
Al	Aerated neutral soil	7		

3.9 Calculate or give the general equation of the equilibrium potential of the following reactions:

(a) $Fe^{2+} \rightarrow Fe^{3+} + e^-$ ($[a_{Fe^{3+}}] = 0.5$; $[a_{Fe^{2+}}] = 10^{-6}$; $E^0 = +0.77$ V SHE)

(b) $O_2 + 4H^+ + 4e^- \rightarrow 2H_2O$ (pH = 3; pH = 7; pH = 12)

(c) $3Fe^{2+} + 4H_2O \rightarrow Fe_3O_4 + 8H^+ + 2e^-$ ($[a_{H_2O}] = 1$; $E^0 = +0.98$ V SHE)

(d) $Ni^{2+} + H_2O \rightarrow NiO + 2H^+$ ($[a_{H_2O}] = [a_{NiO}] = 1$)

(e) $Ti + H_2O \rightarrow TiO + 2H^+ + 2e^-$ ($[a_{Ti}] = [a_{H_2O}] = [a_{TiO}] = 1$; $E^0 = -1.306$ V SHE)

(f) $CuO + H_2O \rightarrow CuO_2^{2-} + 2H^+$ ($[a_{CuO}] = [a_{H_2O}] = 1$)

Consider $(2.3RT)/F = 0.059$ V.

3.10 A room temperature (25 °C) solution contains 10^{-3} mol/L of ferrous ions. If iron has a potential of −500 and −600 mV SHE, does it tend to corrode or deposit?

3.11 Calculate the equilibrium potential of the reaction $Fe^{2+} = Fe^{3+} + e^-$ knowing that: $[Fe^{3+}] = 0.5$ mol/L and $[Fe^{2+}] = 10^{-6}$ mol/L.

3.12 A carbon steel tank is filled with potable water. Based on the measured iron potential in solution, is it possible to state whether corrosion takes place?

Bibliography

Bardal E (2004) Corrosion and protection. Springer-Verlag London Limited, UK

Fontana M (1986) Corrosion engineering, 3rd edn. McGraw-Hill, New York

Piontelli R (1961) Elementi di teoria della corrosione a umido dei materiali metallici. Longanesi, Milano (in Italian)

Pourbaix M (1973) Lectures on electrochemical corrosion. Plenum Press, London

Roberge PR (1999) Handbook of corrosion engineering. McGraw-Hill, London

Shreir LL, Jarman RA, Burstein GT (1994) Corrosion. Butterworth-Heinemann, London

Winston Revie R (2000) Uhlig's corrosion handbook, 2nd edn. Wiley, London

Chapter 4
Pourbaix Diagrams

These diagrams embody a vast amount of pertinent information in a small place.

U. R. Evans

Abstract In 1945, Marcel Pourbaix (1904–1998) proposed in the *Atlas of Electrochemical Equilibria in Aqueous Solutions* the potential-pH diagram of elements in the presence of water, which is now called "*Pourbaix diagram*". It uses thermodynamic considerations to define potentials corresponding to the equilibrium states of all possible reactions between a given element, its ions, and its solid and gaseous compounds in aqueous solutions as a function of pH. This Chapter illustrates the basis of Pourbaix diagrams, how they are obtained, and shows examples for the most important metals. Three areas can be identified in the diagrams: immunity, corrosion and passivation, representing the fields of thermodynamic stability of the metal, of its ions and of its oxides and hydroxides, respectively.

Fig. 4.1 Case study at the PoliLaPP Corrosion Museum of Politecnico di Milano

© Springer Nature Switzerland AG 2018
P. Pedeferri, *Corrosion Science and Engineering*, Engineering Materials,
https://doi.org/10.1007/978-3-319-97625-9_4

Fig. 4.2 Pourbaix diagram,
E-pH, for water: *line a*,
hydrogen evolution; *line b*,
oxygen reduction

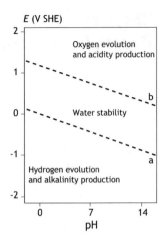

4.1   Oxygen Reduction and Hydrogen Evolution

Pourbaix diagrams graphically represent the thermodynamic conditions of immunity, corrosion (Fig. 4.1) and passivation. All Pourbaix diagrams report the equilibrium potential of the two most important cathodic reactions, i.e., oxygen reduction and hydrogen evolution, which are pH dependent, as the Nernst equation shows, and as already mentioned in Chap. 3:

$$E_{eq,H_2} = -0.059 \cdot pH \qquad (4.1)$$

$$E_{eq,O_2} = 1.229 - 0.059 \cdot pH \qquad (4.2)$$

Therefore, the two equations are represented by two parallel straight lines having a slope of -0.059 V/pH at 25 °C and spaced 1.23 V. Figure 4.2 shows these two lines denoted as *line a* and *line b* for hydrogen evolution and oxygen reduction, respectively. At potentials above *line b*, water dissociates producing oxygen and acidity (anodic reaction: $2H_2O = O_2 + 4H^+ + 4e^-$); whereas, below *line a*, water dissociates producing hydrogen and alkalinity (cathodic reaction: $2H_2O + 4e^- = H_2 + 4OH^-$). The area between the two straight lines is the one of electrochemical stability of water, in which the only possible reactions are: oxygen reduction and hydrogen oxidation (if these two gases are available at metal surface). The 1.23 V spacing between the two lines corresponds to the thermodynamic potential of water decomposition, according to the reaction $2H_2O = 2H_2 + O_2$.

4.2   Metal Immunity, Corrosion and Passivation

Referring to metals, the Pourbaix diagram (E-pH) displays the zones of stability of the chemical species involved as a function of potential and pH, namely: metal (*immunity* zone), metal ions (*corrosion* zone), oxides and hydroxides (*passivation*

Fig. 4.3 Pourbaix diagram,
E-pH, for metal dissolution
and passivation

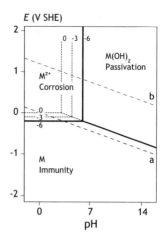

and *passivity* zones). To avoid misunderstandings, it is worth highlighting the difference between *passivation* (as Pourbaix calls it in his diagram) and *passivity*, as used later: *passivation* indicates, generically, the formation of oxides on the surface of a metal, while, if this oxide forms a continuous adherent and flawless layer, this condition is called *passivity*. Figure 4.3 shows the general potential-pH equilibrium diagram for the metal-water system, at 25 °C.

The equilibrium domains are defined by electrochemical or chemical reactions. A *chemical reaction* is a reaction in which only neutral molecules and positively or negatively charged ions take part, with the exclusion of electrons. Conversely, an *electrochemical reaction*, besides molecules and ions, involves electrons.

In practical non-equilibrium conditions, stability zones as well as kinetics can be different; for example, the Pourbaix diagram of iron as obtained experimentally in laboratory in agitated, oxygenated water shows a wider passivity zone than the one calculated from thermodynamics.

4.2.1 Equilibrium Between Immunity and Corrosion

On Pourbaix diagrams, the equilibrium potential of the dissolution process of a generic metal ($M \rightarrow M^{z+} + ze$) is given by the Nernst equation:

$$E_{eq,M} = E^0 + \frac{0.059}{z} \log a_{M^{z+}} \qquad (4.3)$$

This equation in the E-pH diagram gives a set of straight lines, parallel to the abscissa, where each line corresponds to a value of the parameter $\log a_{M^{z+}}$, hence is independent from pH as Fig. 4.4 shows. Each $\log a_M{}^{z+}$ value identifies a line of

Fig. 4.4 Pourbaix diagram, E-pH, for metal dissolution

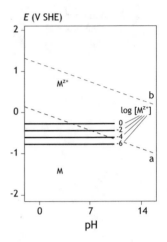

corresponding potential: for more noble potentials (i.e., above the line), metal oxidation is the spontaneous reaction, while below the line the opposite occurs, that is a cathodic reaction (reduction reaction) of an oxidized species of the metal, typically metal ions, M^{z+}, reduced to metal, when present in the electrolyte.

Among the straight lines set, Pourbaix proposed to represent the one corresponding to the concentration 10^{-6} mol/L of metal ions, M^{z+}, as separation boundary of metal corrosion from metal deposition, called *immunity*. Such concentration, which derived from the analytical limit threshold of metal ions in solution at the time these diagrams were first created, has the practical meaning of absence of metal ions in solution. For instance, a concentration of 10^{-6} mol/L for iron (steel) corresponds to 0.056 ppm, which, in the absence of a continuous renewal of the electrolyte, does not represent an appreciable loss of metal.

Metal ions with different oxidation numbers. There are electrochemical reactions that involve ions with a different oxidation number (or valence) as the following:

$$M^{x+} + ze^- \rightarrow M^{(x-z)+} \tag{4.4}$$

The corresponding equilibrium potential is given by the Nernst equation, as follows:

$$E_{eq,M} = E^0 + \frac{0.059}{z} \log \frac{a_{M^{z+}}}{a_{M^{(x-z)+}}} \tag{4.5}$$

Figure 4.5 shows the Pourbaix diagram of a metal with two oxidation states. The corresponding equilibrium potential is represented by a set of straight lines, again parallel to the abscissa, where each line corresponds to a value of the ratio between the activities of the two ions. The straight line corresponding to the unitary ratio identifies two zones: above this line, the species with a higher oxidation number is

Fig. 4.5 Pourbaix diagram, E-pH, for a metal with two oxidation states

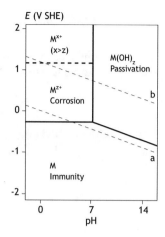

stable and the opposite applies below the line. For example, in the case of iron, where two oxidation states apply, $Fe^{2+} = Fe^{3+} + e^-$, the corresponding equilibrium potential is expressed by:

$$E_{eq,Fe^{3+}/Fe^{2+}} = 0.77 + 0.059 \log \frac{a_{Fe^{3+}}}{a_{Fe^{2+}}} \tag{4.6}$$

4.2.2 Equilibrium Between Immunity and Passivation

The dissolution reaction of metal, M, leads to the formation of hydroxides, especially in neutral or alkaline solutions, according to the electrochemical reaction:

$$M + zH_2O \rightarrow M(OH)_z + zH^+ + ze^- \tag{4.7}$$

The equilibrium potential is given by the Nernst equation:

$$E_{eq,M} = E^0 + \frac{0.059}{z} \log \frac{a_{M(OH)_z}}{a_M} - 0.059 \, pH \tag{4.8}$$

which shows that the equation of the equilibrium potential versus pH is a straight line having the same slope of the cathodic process (hydrogen evolution, *line a*; oxygen reduction, *line b*) as represented in Fig. 4.3. Below this straight line, there is the immunity zone, while above it there is the zone for the formation of the hydroxide, called *passivation*.

4.2.3   Equilibrium Between Corrosion and Passivation

The equilibrium condition between metal ions and hydroxides is defined by a chemical reaction where no electrons are involved, as follows:

$$M^{z+} + zH_2O \rightarrow M(OH)_z + zH^+ \tag{4.9}$$

The equilibrium condition of this reaction, once T and P are fixed and assuming $a_{M(OH)_z} = a_{H_2O} = 1$, is given by the following equation:

$$K(T, P) = \frac{(a_{H^+})^z}{a_{M^{z+}}} \tag{4.10}$$

which leads to:

$$\log a_{M^{z+}} = -\log K + z \cdot \log a_{H^+} = A - z \cdot pH \tag{4.11}$$

Taking $\log a_M{}^{z+}$ as the function parameter, the equilibrium condition (at 25 °C, 1 atm) is represented by a set of lines parallel to the ordinate axis. Let's consider the line corresponding to the value of the parameter $\log a_{M^{z+}}$ equal to −6 as shown in Fig. 4.3. For pH higher than the one given by the equation, the stable species is $M(OH)_z$, while for lower pH the stable species is the metal ion M^{z+}.

   The electrochemical passivation reaction (4.7) is the sum of two reactions: metal dissolution reaction ($M \rightarrow M^{z+} + ze^-$) and chemical passivation (4.9). In other words, the free energy of the metal-hydroxide equilibrium is the sum of the related energy of the metal dissolution reaction and equilibrium metal ions-hydroxide. The crossing point of the three domains on the Pourbaix diagram, as shown in Fig. 4.3, defines the equilibrium between metal, hydroxides and metal ions: any change of potential or pH causes the disappearance of one of them.

4.3   Amphoteric Metals

For amphoteric metals—for instance Al, Zn, Pb and others—metal dissolution occurs in both acidic and alkaline solutions, therefore the Pourbaix diagram of an amphoteric metal shows a further corrosion domain at alkaline pH, as Fig. 4.6 illustrates. Equilibrium lines ①, ② and ③ are defined according to the equations described in Sect. 4.2. Corrosion in strong alkaline solutions occurs by either an electrochemical or a chemical dissolution reaction, as discussed in the following.

Fig. 4.6 Pourbaix diagram, E-pH, for an amphoteric metal

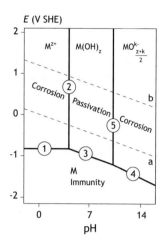

4.3.1 Electrochemical Dissolution in Alkaline Solution

The dissolution reaction of metal in alkaline solution (line ④ in Fig. 4.6) leads to the formation of metal ions, according to the following electrochemical reaction:

$$M + \frac{(k+z)}{2}H_2O \rightarrow MO_{\frac{k+z}{2}}^{k-} + (k+z) \cdot H^+ + ze^- \qquad (4.12)$$

The equilibrium condition is given by the following equation (at 25 °C, $a_{H_2O} = 1$ and $a_M = 1$):

$$E_{eq,M} = E^0 + \frac{0.059}{z}\log a_{MO_{\frac{k+z}{2}}^{k-}} - \frac{k+z}{z}0.059pH \qquad (4.13)$$

On the E-pH diagram, a straight line with slope $-0.059 \cdot (k+z)/z$ defines the equilibrium condition of reaction (4.12). For the system aluminium-water at 25 °C, the electrochemical dissolution in alkaline solution occurs according to the following reaction:

$$Al + 2H_2O \rightarrow AlO_2^- + 4H^+ + 3e^- \qquad (4.14)$$

with equilibrium potential $E_{eq} = -1.376 - (0.059 \cdot 4/3) \cdot pH = -1.376 - 0.079pH$ ($a_{H_2O} = 1$, $a_{Al} = 1$ and $a_{AlO_2^-} = 10^{-6}$).

The existence of an equilibrium potential would suggest the possibility to reach immunity by applying a cathodic polarization (i.e., by applying cathodic protection). Instead, practice has proved that by applying cathodic protection to aluminium, the amphoteric dissolution inevitably takes place by a chemical dissolution in accordance with the dissolution reaction (4.17).

4.3.2   Chemical Dissolution in Alkaline Solution

The equilibrium between the passivation and the dissolution domain at alkaline pH (line ⑤ in Fig. 4.6) is defined by the chemical reaction:

$$M(OH)_z \rightarrow MO^{k-}_{\frac{k+z}{2}} + kH^+ + \frac{(z-k)}{2} H_2O \tag{4.15}$$

The equilibrium condition is given by the following equation (at 25 °C, $a_{H_2O} = 1$ and $a_{M(OH)_z} = 1$):

$$pH = \frac{1}{k}\left(\log a_{MO^{k-}_{\frac{k+z}{2}}} - \log K \right) \tag{4.16}$$

Where K is the equilibrium constant of reaction (4.15). The equilibrium condition (at 25 °C, 1 atm and fixed metal ions concentration) is represented by a set of parallel lines to the ordinate axis. For the system aluminium-water at 25 °C, the chemical dissolution in alkaline solution occurs according to the following reaction:

$$Al(OH)_3 \rightarrow AlO_2^- + H^+ + H_2O \tag{4.17}$$

4.4   Pourbaix Diagrams of Some Metals at 25 °C

In this section the most relevant Pourbaix diagrams are reported and commented.

Iron. The Pourbaix diagram of iron is reported in Fig. 4.7. Corrosion is possible at low and high pH with formation of Fe^{2+} (or also Fe^{3+} at high potentials) and $HFeO_2^-$, respectively. Iron is stable in the immunity zone and can resist corrosion in passivation zone after the formation of Fe_3O_4 and Fe_2O_3 oxides, which form at low and high potentials, respectively. In the presence of some species, such as Ca^{2+}, Mg^{2+} or sulphate, SO_4^{2-}, the passivation zone broadens due to the formation of protective layers.

Gold. Its Pourbaix diagram (Fig. 4.8) clearly shows that even in the presence of oxygen, the stable species is gold as metal. This is because the equilibrium potential relevant to the dissolution reaction $Au^+ + e^- = Au$, once fixed the gold ion, Au^+, as 10^{-6} mol/L, is more noble than the equilibrium potential of the oxygen reduction reaction. The behaviour changes in the presence of complexing chemical species, as in the case of cyanides. Gold corrodes in cyanide solutions because of the formation of complex $Au(CN)_2^-$, which has a complex stability constant of 2×10^{38}. Gold ions in solution are so low to give an equilibrium potential below the one of oxygen reduction. This explains why sodium cyanide (NaCN) is used for exploiting gold mines.

Fig. 4.7 Pourbaix diagram,
E-pH, for iron

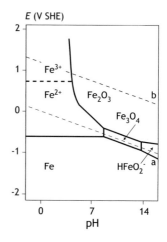

Fig. 4.8 Pourbaix diagram,
E-pH, for gold

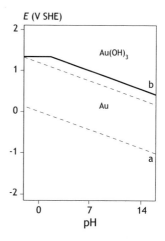

Chromium. Figure 4.9 shows the Pourbaix diagram of chromium. It appears that chromium shows a clear tendency to passivate (large passivation zone) and also a possible corrosion in acidic solutions, even close to neutrality, and at noble potentials. This behaviour extends to stainless steels also.

Copper and nickel. Pourbaix diagrams of copper and nickel are reported in Figs. 4.10 and 4.11, respectively. The behaviour of nickel is very similar to that of iron. As regards copper, it is important to highlight that it can corrode only in aerated solutions.

Aluminium and zinc. Figures 4.12 and 4.13 show the Pourbaix diagrams of aluminium and zinc, respectively. Both metals show an amphoteric behaviour at low and high pH, either in the presence of oxygen or in its absence, and have a passivation range around the neutrality.

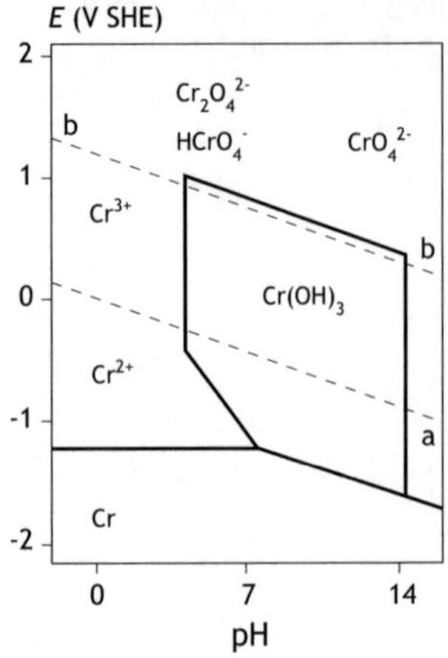

Fig. 4.9 Pourbaix diagram,
E-pH, for chromium

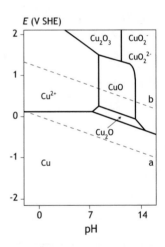

Fig. 4.10 Pourbaix diagram,
E-pH, for copper

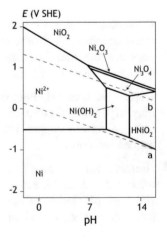

Fig. 4.11 Pourbaix diagram,
E-pH, for nickel

Fig. 4.12 Pourbaix diagram,
E-pH, for aluminium

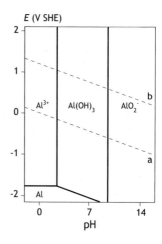

Fig. 4.13 Pourbaix diagram,
E-pH, for zinc

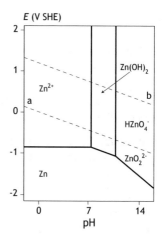

Immunity, Activity, Passivity

Immunity. Only gold can be collected in nature as nuggets. When gathered, its perfection and purity is recognized at first sight. No rust or other substance can affect its weight. (*Plinio*, Nat. Hist., 32, 62)

Activity. Iron is humankind's best and worst servant. It helps plough soil, plant trees, trim shrubs, rejuvenate grapes, build houses, cut stones and other; it is also a tool for weapons, outrages, and used in battles or as flying object thrown by hand or war machines. The most adbominious thing for the human spirit is when man provided it of wings for flying and hasten death. Its guilt is not on nature; instead, it is the goodness of nature, which limits its power, condemning it to rust and nothing gives more death than what is deadly for human beings. (*Plinio*, Nat. Hist., 34, 139)

> *Passivity*. Intelligent persons, unlike silly ones, rarely get ill and, if they
> do, they heal rapidly as the rusted Corinth bronzes. (*Cicero*, Tusc., 4, 32)

Lead. Pourbaix diagrams of lead depend on the solution composition. Figure 4.14
show the diagram in pure acidic or alkaline solutions. Figure 4.15 reports the
diagram in the presence of carbon dioxide and carbonic acid solution. The diagram
in the presence of sulphates is depicted in Fig. 4.16.

The comparison of those diagrams highlights the influence of carbonate ions on
passivation behaviour due to the formation of insoluble salts. In the absence of
carbonate ions, the corrosion zone extends from acidic to alkaline pH: this is typical
of lead exposed to pure or demineralized water. When carbonate ions are present,
there is a passivation zone around neutrality (close to pH 7) due to the formation of
lead carbonate, $PbCO_3$, which is highly insoluble. Because of this, lead ion, Pb^{2+},
concentration is below toxicity limits; for this reason, lead was used in the past for
household uses.

When sulphate ions are present, the passivation zone extends further although the
solubility product is sensibly lower than that of carbonate ($pK_{PbSO_4} = -8$ and
$pK_{PbCO_3} = -13$). At low pH, the corrosion zone disappears; at a very negative
potential when hydrogen evolution takes place, lead shows another form of
degradation due to the formation of hydrides; this also occurs on titanium.

Titanium. The Pourbaix diagram of titanium, shown in Fig. 4.17, indicates that Ti
would corrode in reducing acidic environments. In oxidizing solutions Ti passivates
through the full pH range.

Tungsten. The Pourbaix diagram for tungsten shows a different behaviour compared
to the others considered above, because in acidic media it forms a protective oxide,
while it dissolves in alkaline solution forming tungstate ions, as Fig. 4.18 shows.

Fig. 4.14 Pourbaix diagram,
E-pH, for lead

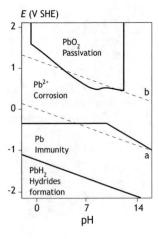

Fig. 4.15 Pourbaix diagram,
E-pH, for lead in CO_2
containing solutions

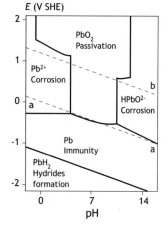

Fig. 4.16 Pourbaix diagram,
E-pH, for lead in
sulphate-containing solution

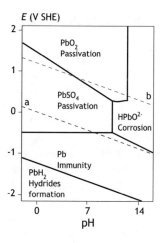

Fig. 4.17 Pourbaix diagram,
E-pH, for titanium

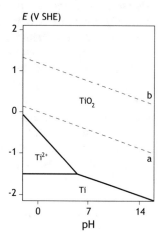

Fig. 4.18 Pourbaix diagram,
E-pH, for tungsten

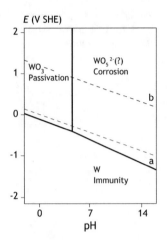

4.5 Final Remarks

Pourbaix diagrams provide a thermodynamic framework, which is of particular importance for the study of the aqueous corrosion of metals, because they provide fundamental information on the so-called immunity or, instead, the possible corrosion or activity condition or passivation state, i.e., the separation of oxides, hydroxides, alkaline salts with protection properties. In particular, the diagrams help understand how corrosion or protection conditions vary with potential and pH, and when a metal can be considered in immunity or in passivity conditions, and which conditions fit with an active-passive behaviour, or when an active-passive transition can take place, for example by varying the potential or instead the pH.

For the correct use of Pourbaix diagrams, the following limitations must be taken into account. First, they cannot provide any information on the kinetics of corrosion processes (i.e., the corrosion rate), since they represent chemical or electrochemical equilibrium conditions and the existence field of involved species, only. Second, the thermodynamic information can often be insufficient for practical applications. For example, activities should be taken into account and not the concentrations of species, therefore the diagrams are strictly applicable to dilute solutions and not to concentrate ones. Furthermore, particular attention must be paid when some complexing chemical species are present in the solution, which vary the concentration of metal ions in solution because of the formation of complex or insoluble compounds: in the presence of complex compounds the corrosion zone widens; instead, in the presence of insoluble compounds, the passivation zone widens. However, tailored diagrams are available for specific environments.

Finally, it has to be noted that the protection capacity of layers, even when their formation does occur, depends on uniformity, flawless structure, electrochemical properties and other, which does not allow to include in the Pourbaix diagrams, for instance, the specific action of some particular anions, such as chlorides.

Marcel Pourbaix on Pourbaix Diagrams

In the opening chapter of his book, *Lectures on Electrochemical Corrosion* (Plenum Press, New York, 1973), Pourbaix illustrates three examples on the matter. First example. The manager of a Danish laundry decided to soften the water with the aim to limit the consumption of soap. This novelty provoked the corrosion of the water supply piping, producing rust stains on clothes during washing. Second example. It went even worse in a hospital where it was decided to carry out the same treatment. The softened water provoked corrosion on the distribution piping, partially made of lead, which hitherto had resisted corrosion, aftermath leading to symptoms of lead poisoning in patients. Finally, the third example. Before transporting a series of lead-acid batteries, containing sulphuric acid, a moving company decided to replace the acid with distilled water in order to eliminate any hazard deriving from the presence of sulphuric acid. In a few hours the plates of the batteries were destroyed.

4.6 Questions and Exercises

4.1 In a corrosion process what is the range of potential by which a metal can corrode? Can this range be indicated on the Pourbaix diagram?

4.2 Explain briefly the meaning of lines on a Pourbaix diagram, as well as the regions between them and their practical utility.

4.3 Write Eqs. 4.15 and 4.16 for aluminium dissolution in alkaline solution. Calculate the pH threshold for amphoteric dissolution.

4.4 How can the passivation region be used for practical applications? Give some examples.

4.5 Calculate the concentration of sodium cyanide necessary for the gold dissolution (complex $Au(CN)_2^-$ has a dissociation constant of 2×10^{38}). How can this condition be shown on the Pourbaix diagram?

4.6 In natural environments (seawater, fresh water, soil, concrete, pure atmospheric dew) which cathodic reactions take place for the corrosion of steel? And for acid rains? Write the reactions. Find the corrosion status on Pourbaix diagram.

4.7 In natural environments (seawater, fresh water, soil, concrete, pure atmospheric dew) which cathodic reactions take place for the corrosion of copper? And for acid rains? Write the reactions. Find the corrosion status on Pourbaix diagram.

4.8 In natural environments (seawater, fresh water, soil, concrete, pure atmospheric dew) which cathodic reactions take place for the corrosion of lead? And for acid rains? Write the reactions. Find the corrosion status on Pourbaix diagram.

4.9   In natural environments (seawater, fresh water, soil, concrete, pure atmo-
      spheric dew) which cathodic reactions take place for the corrosion of alu-
      minium? And for acid rains? Write the reactions. Find the corrosion status on
      Pourbaix diagram.
4.10 In natural environments (seawater, fresh water, soil, concrete, pure atmo-
      spheric dew) which cathodic reactions take place for the corrosion of zinc?
      And for acid rains? Write the reactions. Find the corrosion status on Pourbaix
      diagram.

Bibliography

Pourbaix M (1973) Lectures on electrochemical corrosion. Plenum Press, New York
Pourbaix M (1974) Atlas of electrochemical equilibria in aqueous solutions, 2nd edn. NACE
    Cebelcor, Houston, TX

Chapter 5
Kinetics of Aqueous Corrosion

No more can we observe what's lost at any time,
When things wax old with eld and foul decay,
Or when salt seas eat under beetling crags.
Lucrezio, De Rerum Natura, Book I, 319–327

Abstract This chapter presents the forms of energy dissipation involved in a corrosion process when a positive driving voltage is available for corrosion to take place, as defined by thermodynamics, and how they concur to determine the overall corrosion rate. Dissipations are described in terms of overvoltage of the different processes involved, which are classified as activation overvoltage (metal corrosion, hydrogen evolution), represented by Tafel law, concentration overvoltage (oxygen diffusion) and ohmic drop (electrolyte resistivity). The trend of the overvoltage for an active-passive metal is also described in the different potential ranges, corresponding to immunity, activity, passivity and transpassivity (or localised corrosion).

Fig. 5.1 Pietro Pedeferri's drawing on titanium: Duomo of Milan (2006)

© Springer Nature Switzerland AG 2018
P. Pedeferri, *Corrosion Science and Engineering*, Engineering Materials,
https://doi.org/10.1007/978-3-319-97625-9_5

5.1 Driving Force and Corrosion Rate

The first question that thermodynamics asks for the prediction of corrosion occurrence on a metal exposed to an aggressive environment is: *"Is there an available driving force for the corrosion process?"* If the answer is no, thermodynamics excludes corrosion occurrence. This is the case with gold, which is more stable than its corrosion products (thankfully, otherwise we would not find nuggets!); it is also the case with metals such as silver, copper or their alloys, when oxygen is absent.

Instead, if the answer is positive, and this is the case for most widely-used metals in industry, corrosion may or may not occur significantly, depending on the intervention of some frictions, which can slow down corrosion processes, as for instance, when protective surface films can form (Fig. 5.1).

Therefore, when the answer to the first question is affirmative, there is the need to answer a second question, which we define as kinetic: *If there is a driving force, how fast is the corrosion process?* In the past, until the 1960s, the answer to this second question was based on knowledge from previous experiences or empirical evidence, while today it is based on knowledge of corrosion kinetics.

With reference to Fig. 2.2, the corrosion process is characterized by four partial processes:

- The *anodic process* which releases electrons in the metal, I_a
- The *cathodic process* which consumes those electrons, I_c
- *Electron transport* within the metal from the anodic region to the cathodic one, I_m
- *Current transport* within the electrolyte, I_{el}.

These processes takes place at same rate:

$$I_a = I_c = I_m = I_{el} = I_{corr} \qquad (5.1)$$

Hence, the corrosion rate is determined by the slowest of the three partial processes, out of the electron transportation in the metal. When the slowest process is the anodic one, for instance because of passivation effects promoted by the environment, corrosion is said to be under *anodic control*; conversely, when the cathodic process is the slowest one, corrosion is said to be under *cathodic control*; eventually, when the ohmic resistance in the electrolyte prevails, corrosion is said to be under *ohmic control*.

Fig. 5.2 Potential balance
for a corrosion process (from
Piontelli 1961)

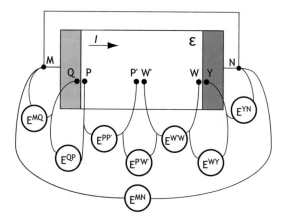

5.2   Dissipations in Corrosion Systems

Figure 5.2 helps locate driving force and dissipations which, as proposed by
Piontelli,[1] can be identified as follows:

$$E^{MN} = E^{MQ} + E^{QP} + E^{PP'} + E^{P'W'} - E^{W'W} - E^{WY} + E^{YN} = 0 \qquad (5.2)$$

E^{MN} is zero being M and N short-circuited and points are identified as:

M	anode connection
Q	on the side of the anodic metal phase, adjacent to the interface M/ε
P	on the side of the electrolyte, adjacent to the interface M/ε
P′ and W′	in an electrolyte region where chemical composition remains constant regardless the circulation of current
W	is as P on cathodic side
Y	is as Q on cathodic side
N	cathode connection.

By introducing the appropriate reference electrodes in those points, it is possible to
measure the relative tensions (i.e., the potential differences). Points P and P′ are
made of the same metal as electrode M and similarly for W′ and W which are equal
to N. These reference electrodes work in equilibrium conditions since they are not
affected by any exchanging current with the electrolyte (therefore there are no
dissipations). The significance of the measurements is the following:

- $E^{P'W'} = E^{MN}_{eq}$ equilibrium potential
- E^{MQ} and E^{YN} ohmic drop in the metallic conductor (often negligible)

---

[1]**Roberto Piontelli** (1909–1971) was an eminent Italian electrochemist.

- E^{QP} and E^{YW} overvoltages of processes occurring at the anode and cathode, respectively. Each overvoltage measures the dissipation inherent in anodic and cathodic processes, respectively
- $E^{PP'}$ (and similarly $E^{WW'}$) includes two terms: ohmic drop between PP′ (or WW ′) in the electrolyte and *concentration polarisation* due to chemical modifications caused by the circulation of current.

In summary:

- $(E^{MQ} + E^{YN})$ is the ohmic drop in metal (negligible)
- $(E^{QP} - E^{YW})$ is the algebraic sum of dissipations occurring at electrodes, called *activation overvoltage*, respectively anodic, $E^{QP} = \eta_a$, and cathodic, $-E^{WY} = E^{YW} = \eta_c$. These terms are associated with the *energy barrier* for the *charge transfer* of the electrochemical reaction
- $(E^{PP'} - E^{WW'})$ is the algebraic sum of dissipations occurring within the electrolyte (anodic and cathodic zone) which contains an ohmic drop contribution and a term due to the variation of chemical composition, called *concentration overvoltage*.

5.3　Activation Overvoltage

5.3.1　Exchange Current Density and Tafel Law

For an electrochemical reaction, for example $M = M^{z+} + ze^-$, equilibrium condition determines two key parameters: the *equilibrium potential*, E_{eq}, as thermodynamic parameter and the *exchange current density*, i_0, as the kinetic one. Both parameters can only be determined experimentally.

In an equilibrium condition, $E = E_{eq}$, the cathodic process occurs at the same rate of the anodic one, equals to the *exchange current density*:

$$i_a = i_c = i_0 \text{ at } E = E_{eq} \tag{5.3}$$

If the metal is brought to a potential, E, different from E_{eq}, the rates of anodic and cathodic processes, measured by current density, i_a and i_c, are different from the exchange current density.

With reference to the electrochemical cell shown in Fig. 5.3, with a current circulating between M and N, the dissipation which takes place at the anode is given by $E^{QP} = \eta_a$, where the point P within the electrolyte, ε, is very close at metal surface to zeroing the ohmic drop. P is a reference electrode made of the same metal of the electrode M (this configuration was named by Piontelli as *iso-electrodic electrode*). For the measurement, a high impedance voltmeter must be used to minimize the current flowing in the measurement circuit; the positive terminal must be connected to M.

Fig. 5.3 Measurement of
overvoltage in a corrosion cell
(from Piontelli 1961)

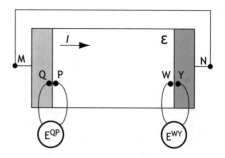

In practice, the measurement of overvoltage by means of an iso-electrode reference electrode is sometimes difficult because certain metals do not exhibit a stable potential. For this reason, *non-iso-electrodic* reference electrodes are used, such as standard reference electrodes as listed in Table 3.2. The overvoltage measured by means of a standard reference electrode (or non-iso-electrodic electrode) is given by:

$$\eta = E - E_{eq} \tag{5.4}$$

where E_{eq} is the metal equilibrium potential (given by the Nernst equation) versus the reference electrode used. According to Eq. 5.4, cathodic overvoltage is defined as negative, while anodic overvoltage is defined as positive:

$$\eta_c = E - E_{eq} < 0 \tag{5.5}$$

$$\eta_a = E - E_{eq} > 0 \tag{5.6}$$

The *activation* or *charge transfer overvoltage*, experimentally measured by the procedure discussed above, is associated with the transfer of a charge in the electrochemical reaction and requires the overcoming of an energy barrier as in kinetics of chemical reactions. Hence, the rate constant, $k(T)$, follows the Arrhenius equation:

$$k(T) = Z\, e^{\frac{-H}{RT}} \tag{5.7}$$

where H is activation energy (in J/mol), T is absolute temperature, R is gas constant and Z is a constant. Following the Arrhenius approach, Butler and Volmer[2] derived a general expression of current density as a function of overvoltage, called *Butler-Volmer equation*:

---

[2]**John Alfred Valentine Butler** (1899–1977) was an English physical chemist who developed kinetic theories of the origin of electrode potentials and developed the general theory of overvoltage with hydrogen and oxygen electrodes. **Max Volmer** (1885–1965) was a German physical chemist, who made important contributions to electrochemistry, in particular on electrode kinetics. He co-developed the Butler–Volmer equation.

$$i = i_0 \left(e^{\frac{(1-\beta)zF\eta}{RT}} - e^{\frac{-\beta zF\eta}{RT}} \right) \tag{5.8}$$

where η is overvoltage given by $\eta = E - E_{eq}$, i_0 is exchange current density as defined above, β (often taken as 0.5) is charge transfer coefficient, F is Faraday constant and z is reaction equivalence. The Butler-Volmer equation states that the exchange of current on the surface of an electrode takes place only if an activation energy is exceeded, hence, dissipating a portion of the driving voltage.

In the event that overvoltage is very low, taking $\beta = 0.5$ and remembering that $e^x \cong 1 + x$, the Butler-Volmer equation approximates to a linear relationship:

$$\eta = \frac{RT}{zFi_0} \cdot i = k \cdot i \tag{5.9}$$

In all other cases, however, the relationship between η and i is logarithmic and the equation shortens to the *Tafel law*[3]:

$$\eta = \pm a \pm b \log i \tag{5.10}$$

where the + sign applies to the anodic processes (η is positive), while the − sign is for cathodic processes (η is negative). Parameters a and b are defined as:

$$a = -\frac{2.3\,RT}{\beta\,zF} \cdot \log i_0 \tag{5.11}$$

$$b = +\frac{2.3RT}{\beta\,zF} \tag{5.12}$$

Parameter a is a constant which depends on the exchange current density i_0; parameter b is a positive constant that assumes the meaning of a straight-line slope of the function η−log i in a semi-logarithmic diagram, called *Tafel slope*. At room temperature (298 K) and considering $\beta = 0.5$ (symmetric behaviour), constant b has a value of 59 mV/decade for bivalent reaction ($z = 2$) and 118 mV/decade for monovalent reactions ($z = 1$), for example hydrogen evolution.

Overvoltage Correlations
Overvoltage depends on metal properties and temperature: for instance, it decreases as temperature increases. The corresponding correlations for normal metals involve, on the one hand, low melting temperature, low hardness and low mechanical properties, and on the other, a large crystal lattice size; conversely, inert metals are characterized by a high melting temperature, high

---

[3]**Julius Tafel** (1862–1918) was a German chemist who worked in electrochemistry with Wilhelm Ostwald.

hardness, high mechanical resistance and small crystal lattice size. This implies that there is a high affinity of atoms to the crystal lattice for inert metals, whilst normal metals show the opposite, i.e., low affinity. The same trend applies to the affinity of metal ions to the solution: weak for normal metal, and high for inert metals.

Electrochemical inertia derives from the bond existing between metal ions, produced at the electrode surface, either towards the crystal lattice or to the solution, hence determining the rate of the electrochemical process.

Indeed, the transition from an initial condition, for instance crystal lattice, to a final condition, that is, ionization, occurs through an intermediate configuration in which initial bonds are partially broken whilst final bonds are not yet completely defined. This corresponds to maximum energy, or the so-called *energy barrier*. If bonds of both initial and final conditions are weak, the intermediate configuration corresponds to a low energy barrier (level slightly higher than the initial and final ones) and conversely, when the bonds of both the initial and final conditions are very strong, the energy barrier is high.

5.3.2 Potential-Current Density Diagrams (or Characteristic Curves)

Figure 5.4 depicts the variation of overvoltage with current density according to the Butler-Volmer equation for a general electrochemical reaction. Such curves are called *characteristics*, *anodic* and *cathodic*, respectively. Figure 5.5 shows how characteristic curves can be represented: Fig. 5.5a in a linear scale where the anodic current is positive and the cathodic current is negative; Fig. 5.5b in a linear scale where the anodic and cathodic are overlapped (both positive); Fig. 5.5c semi-logarithmic scale, which is the most-used representation. By looking at the latter figure, it appears that curves become straight lines when far from the equilibrium potential (Tafel law).

When Tafel law applies, that is, for inert and intermediate metals, anodic and cathodic overvoltages can be written in terms of exchange current density as follows:

$$\eta_a = a + b \log i = b \log\left(\frac{i}{i_0}\right) \tag{5.13}$$

$$\eta_c = -a - b \log i = -b \log\left(\frac{i}{i_0}\right) \tag{5.14}$$

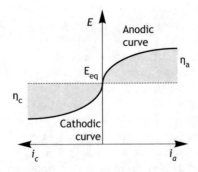

Fig. 5.4 Schematic trend of anodic and cathodic overvoltage of metals

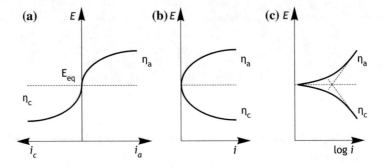

Fig. 5.5 Types of representation of characteristic curves

Figure 5.6 shows the plot of the potential as a function of experimentally measured current density, $i_{ext} = i_a - i_c$ or $i_{ext} = i_c - i_a$, and not as a function of i_a and i_c (which cannot be measured directly, see also Chap. 6). Far from equilibrium, i.e., for a potential shift higher than ±50 mV from the equilibrium potential, measured values match the extrapolated straight lines. Conversely, near equilibrium, i.e., for a potential shift lower than ±10 mV from the equilibrium potential, there is a deviation from the theoretical curve since the measured current, $i_{ext} = i_a - i_c$ or $i_{ext} = i_c - i_a$, strongly differs from either i_c or i_a. These experimental curves make it possible to obtain:

- Straight lines which give the potential, E, as a function of anodic or cathodic current densities, i_a and i_c, obtained by the extrapolation of the linear part of the curve
- Exchange current density, i_0, given by the crossing point of the two straight lines at the equilibrium potential
- The slope b which allows to calculate the overvoltage, for instance $\eta_a = b \log i_a / i_0$.

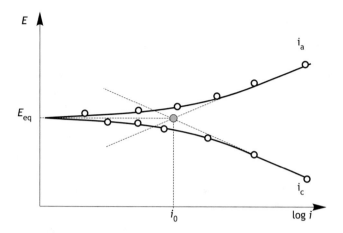

Fig. 5.6 Trend for the polarization curve to find the exchange current density, i_0, experimentally

Table 5.1 Tafel slope, b, and order of magnitude of exchange current density, i_0, for metals (from Piontelli 1961)

Metal classification		Exchange current density, i_0 (order of magnitude in mA/m^2)	Tafel slope, b (mV/decade)	
Inert	Pt	1	Monovalent	120
	Pd, Rh			
	W, Ta			
	Co, Ni			
	Fe		Bivalent	60
Intermediate	Cu, Ag	10	Monovalent	120
			Bivalent	60
Normal	Sn, Al, Be	10^4	$b \cong 0$ (linear trend)	
	Zn			
	Pb, Hg			

Table 5.1 reports exchange current density, i_0, and Tafel slope, b, for most common metals; it is worth noting that values of i_0 differ significantly from inert to normal metals.

5.3.3   Oxidation or Reduction of a Metal

The overvoltage associated to a metal dissolution process when immersed in a solution containing its ions (i.e., a soluble salt) depends on the nature of the metal. Its trend for metal working as anode (metal dissolution) or as cathode (metal

Fig. 5.7 Activation
overvoltage of metals as a
function of current density
(from Piontelli 1961)

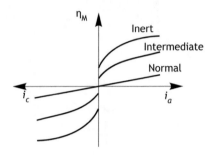

deposition) is symmetrical (Fig. 5.7). Piontelli proposed the following classification
on the basis of the value of overvoltage:

- *Normal* metals, for which even at high current density, from both the anode side
  and the cathode side, the overvoltage, η_M, is less than 10 mV (metals at low
  overvoltage). This class includes the <u>low melting temperature</u> metals: Cd, Hg,
  Sn, Pb, Mg, Al and Zn (the latter only at the anodic side)
- *Inert* metals, for which even at small current density and in a wide range of
  conditions, in the absence of polarizing current, overvoltage, η_M, is greater than
  100 mV (metals with high overvoltage). This class includes the <u>high melting
  temperature</u> metals: Fe, Co, Ni, Cr, Mo, Ti, metals of the platinum group and
  transition metals
- *Intermediate* metals, for which η_M is between the two above-mentioned limits.
  Examples of intermediate metals are: Cu, Au and Ag.

For normal metals, at low current densities (say, below 1 A/m^2), the overvoltage of
the metal dissolution process depends linearly on current density (Eq. 5.8); for
intermediate or inert metals in all other conditions, the overvoltage-to-current
density dependence is logarithmic and follows Tafel law (Eq. 5.9).

It must be emphasized that, with the exception of normal metals (for which η_M
zeros as i zeros), η_M is not nil for intermediates and especially inert metals, even for
values of current density close to zero (Fig. 5.7). Accordingly, electrodes of these
metals, even for infinitesimal current flow, require the overwhelming of a threshold,
like the static friction in mechanics.

5.3.4 Hydrogen Evolution (Activation Overvoltage)

The activation overvoltage of hydrogen evolution, η_{H_2}, can be measured as shown
in Fig. 5.3, as tension, E^{WY}, when on electrode N the cathodic reaction is hydrogen
evolution. This overvoltage is strongly dependent on the nature of the metal, as
experimentally observed: there is a clear inverse correlation between overvoltage,
η_M, related to the oxidation of metal, M, and overvoltage of hydrogen evolution,

Fig. 5.8 Reverse correlation between metal dissolution overvoltage and hydrogen evolution overvoltage (from Piontelli 1961)

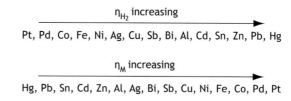

η_{H_2}, taking place on it. On normal metals, characterized by a low overvoltage, hydrogen evolution overvoltage is high, and the opposite occurs on inert and intermediate metals, as summarized in Fig. 5.8. Since normal metals have a low melting temperature and inert metals a high one, the following is a simple rule to establish overvoltage contributions:

- Metals with a low melting temperature have low dissolution-related overvoltage and high hydrogen evolution overvoltage
- Metals with a high melting temperature have high dissolution-related overvoltage and low hydrogen evolution overvoltage.

For this reason, Piontelli suggested calling this correlation *reverse correlation*. The strong influence of the nature of metal on hydrogen overvoltage explains the influence of impurities present on the metal surface on corrosion rate in acidic solutions, where the cathodic reaction is hydrogen evolution. For instance, pure low-melting temperature metals with high hydrogen overvoltage such as Zn, Pb and Al show an increase in the corrosion rate as the content of high-melting temperature impurities, like Fe, Ni and others, increases. Moreover, the presence of cold work deformation (by an increase in dislocation concentration on the surface) and surface finishing can influence hydrogen overvoltage. Hydrogen overvoltage follows Tafel law with a slope of 120 mV/decade; this leads to a variation of an order of magnitude of current density when potential changes by about 120 mV (Table 5.2).

Correlations for Hydrogen Overvoltage

Hydrogen reduction reaction involves three steps:

- $H_3O^+ + e^- + M \rightarrow M{-}H + H_2O$ with formation of atomic hydrogen adsorbed on metal M
- $H_3O^+ + e^- + M{-}H \rightarrow H_2 + H_2O + M$ (electrochemical *desorption*) or
- $2M{-}H \rightarrow H_2 + 2M$ (chemical *desorption*)

The slowest step determines the kinetic control of the overall hydrogen evolution process. If the slowest stage is the first, the Tafel slope is 120 mV/decade; if it is the second, i.e., an electrochemical process, the Tafel slope is also 120 mV/decade; if it is the third, i.e., chemical desorption, the Tafel slope is 30 mV/decade: the latter case occurs for metals of the platinum group.

Table 5.2 Order of magnitude of exchange current density, i_{0,H_2-M}, of hydrogen on metal, M, and order of magnitude of exchange current density, $i_{0,M}$, of metal M

Metal classification		Exchange current density, i_{0,H_2}, on metal, M (order of magnitude in mA/ m^2)	Exchange current density, $i_{0,M}$ (order of magnitude in mA/ m^2)
Inert	Pt	10^4	1
	Pd, Rh	10^3	
	W, Ta	10^2	
	Co, Ni	10	
	Fe	1	
Intermediate	Cu, Ag	10^{-1}	10
Normal	Sn, Al, Be	10^{-2}	$\approx 10^4$
	Zn	10^{-3}	
	Pb, Hg	10^{-4}	

The first stage applies when the M–H binding energy is high and, conversely, it is not favoured in the opposite case. It follows that hydrogen overvoltage is high (i.e., exchange current density, i_0, is low, as happens for Pb, Cd, Tl or In) because binding energy is low and the formation of M–H is not favoured.

However, when binding energy is very high, as in hydride-forming metals (Ti, Ta or Nb) overvoltage becomes high again because the desorption stage slows down. Figure 5.9 shows the so-called '*volcano plot*', well known in chemical catalysis.

Fig. 5.9 *Volcano plot* of exchange current density, i_0, of hydrogen evolution as a function of M–H binding energy (from Bianchi and Mussini 1993)

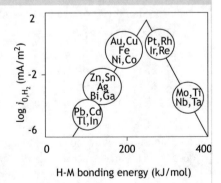

5.3.5 Oxygen Reduction (Activation Overvoltage)

The oxygen reduction process involves two dissipation contributions: one corresponding to the process of charge transfer to the metal surface and the other to the transport of oxygen in the solution. In this paragraph, the first contribution is discussed.

Overvoltage contribution associated with a charge transfer process depends, intrinsically, on metal and also pH, presence of surface layers, current density and follows Tafel law. Figure 5.10 shows the linear section of Tafel curves for various metals at different pH values. Unlike hydrogen overvoltage, the Tafel slope for oxygen reduction is generally higher. Therefore, it is in the order of hundreds of mV even for current density tending to zero and especially in acidic environments.

The presence of surface films increases oxygen overvoltage on chromium, and then on stainless steels, titanium, zirconium, as well as copper and nickel; instead, oxygen reduction takes place more easily on gold, palladium, platinum and graphite.

5.4 Concentration Overvoltage

Let's consider the galvanic cell $M/\varepsilon/N$ in Fig. 5.11 where the electrolyte, ε, is composed by a dissolved salt, MX. Any current circulation in the cell determines a variation of the chemical composition in the electrolyte because mant processes take place: electrochemical, electrophoretic, diffusive and convective. These changes in chemical composition generate an overvoltage contribution, called *concentration overvoltage*, and also *concentration polarisation*. Concentration overvoltage is given by the measurement of $E^{PP'}$, where P and P' are two

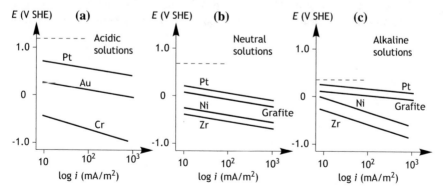

Fig. 5.10 Charge transfer overvoltage of oxygen reduction: **a** acidic, **b** neutral, **c** alkaline solutions

Fig. 5.11 Meaning and
measurements of oxygen
concentration overvoltage

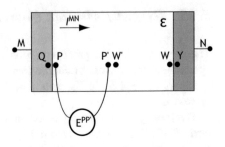

iso-electrodes made of metal M. Position P′ is not perturbed by current circulation. The measured potential, $E^{PP'}$, is the sum of two terms:

$$E^{PP'} = E^{PP'}_{I=0} + IR^{PP'} \tag{5.15}$$

where $E^{PP'}_{I=0}$ is concentration polarization which can be measured by zeroing the current in the cell; the measurement must be taken at the so-called instant-off condition, i.e., very soon after current switch off, and $IR^{PP'}$ is the ohmic drop in the electrolyte between P and P′; $R^{PP'}$ is the electrolyte's ohmic resistance and I is the current circulating in the cell.

Concentration polarization, $E^{PP'}_{I=0}$, depends on several factors: current density, elapsed time, initial composition of the electrolyte, ε, and any other factor influencing diffusive and convective processes (cell geometry, temperature, flow regime). Eventually, it depends also on the nature of reference electrode used for P and P′. As shown in Fig. 5.12, while applied current is constant, potential changes with time. When the current is switched-off, measured potential changes as follows:

- Ohmic drop $IR^{PP'}$ (between P and P′) zeroes in 10^{-6} s after switch-off. This property is the basis of the on-off technique often used in the laboratory and in the field for CP measurements
- Over a longer time, activation overvoltage, typically for hydrogen evolution, zeroes in 10^{-3} s after switch-off. Compared to concentration polarization, as in the point below, activation overvoltage fades practically instantaneously at ohmic drop

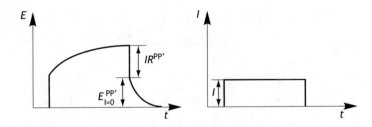

Fig. 5.12 Potential variation before and after current switch off

- Within this very short time, from 10^{-3} to 10^{-6} s, there is no modification of electrolyte composition as set up by the circulating current
- A slow potential modification takes place after current switch-off until the electrolyte has recovered the initial composition through a diffusive process in reverse direction. In other words, diffusive process is slow in nature.

In summary, according to the adopted definitions, overvoltage concentration in a galvanic cell as the sum of anodic $E_{I=0}^{PP'}$ and cathodic $E_{I=0}^{WW'}$ contributions, represents the instant variation of cell voltage before and after current circulation.

Polarity of Concentration Polarisation

If an external electromotive force, EMF, is applied to the galvanic cell, $M/\epsilon/N$, with a current flow from M to N (M works as the anode and N as the cathode), M^{z+} concentration grows at the anodic region and decreases at the cathodic one, while it remains unchanged at the intermediate region (P').

Now, by switching the external power supply off and short-circuiting the cell, an EMF is generated due to the concentration cell set-up because of different concentrations of M^{z+} in P and P' then producing a current circulation in the opposite direction. In fact, the electrode in contact with P (where the concentration of M^{z+} is increased) now tends to work as the cathode while N works as the anode. Therefore, an EMF is measured between P and P' which is in opposition to the previous external EMF; for this reason, it is called back-EMF (refer to Chap. 3.8.1 for the concentration cell).

It has to be emphasized that, while overvoltage ensures energy dissipation, which is necessary for the occurrence of electrode reactions, concentration polarization represents, instead, an accumulation of energy through the formation of a concentration cell. This stored energy, however, is not recovered as electrical work, because it is dissipated through diffusion phenomena that occur to make the solution more uniform.

Based on what is discussed above, both electrode overvoltage and concentration polarization contribute to increase cell voltage (i.e., absorbed energy) when working as the user, while they contribute towards decreasing cell voltage (i.e., obtained energy) when working as the generator.

5.4.1 Oxygen Reduction: Limiting Current

In the case of oxygen reduction as a cathodic process, concentration polarization (also called overvoltage concentration) caused by oxygen diffusion in the electrolyte can become particularly important, because of high concentration gradients set up in the diffusion layer close to the electrode surface.

Let's first consider the case of a stagnant solution without convective flows. In the electrolyte layer, a concentration gradient of oxygen forms next to the metal surface. In fact, oxygen is consumed on the metal surface due to the corrosion process. Hence, its concentration decreases on the metal surface, whereas it remains constant in the bulk. This concentration gradient is within the so-called diffusion layer or Nernst diffusion layer, δ, as shown in Fig. 5.13, between metal surface II and surface I in the electrolyte.

The concentration gradient in the diffusion layer gives rise to a concentration polarization with opposite polarity to EMF set up by the corrosion process; this overvoltage, η_{conc, O_2}, can be expressed as follows (see Sect. 3.10.1):

$$\eta_{conc, O_2} = E^{II,I} = -\frac{RT}{zF} \ln \frac{C_1}{C_2} = \frac{RT}{zF} \ln \frac{C_2}{C_1} \tag{5.16}$$

where C_2 is oxygen concentration on the metal surface and C_1 is the oxygen concentration in the bulk, i.e., outside the diffusion layer.

The oxygen diffusion rate, v_D (moles/(m^2 s)), in a stationary condition is governed by Fick law:

$$v_D = D\frac{(C_1 - C_2)}{\delta} \tag{5.17}$$

where D is the diffusion coefficient (m^2/s), δ is the diffusion layer thickness (m). The oxygen consumption rate, v_C (moles/(m^2 s)), is given by Faraday law:

$$v_c = \frac{i}{zF} \tag{5.18}$$

where i, z and F have the usual meanings.

In stationary conditions, oxygen consumption and diffusion rates are equal, therefore:

Fig. 5.13 Diffusion layer representation and concentration profile of reducing species

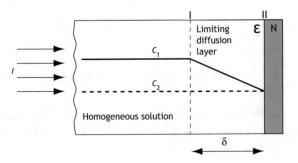

$$v_c = v_D = \frac{i}{zF} = D\,\frac{C_1 - C_2}{\delta} \tag{5.19}$$

$$i = \frac{D}{\delta}\,z\,F\,(C_1 - C_2) \tag{5.20}$$

As a result, the current density in stationary conditions is proportional to the diffusion coefficient, D, and the oxygen concentration gradient. Since, for a given system, the oxygen concentration in bulk, C_1, the diffusion coefficient, D, and the diffusion layer thickness, δ, are constant, current density, i, increases as oxygen concentration, C_2, decreases at the electrode surface. For metals with low potential, for instance Fe and metals below it, the oxygen consumption rate reaches a maximum when C_2 zeros. This maximum is the *oxygen limiting current density*, i_L, given by:

$$i_L = zF\,\frac{D}{\delta}\,C_1 \tag{5.21}$$

By introducing the relationship between concentrations C_1, C_2 in the Eq. 5.16, the current density, i, and the limiting current density, i_L $\left(C_2 = \frac{\delta}{zFD}\cdot(i_L - i)\right.$; $C_1 = \frac{\delta}{zFD}\cdot i_L\left.\right)$, overvoltage concentration, $\eta_{conc,\,O_2}$, is given by:

$$\eta_{conc,\,O_2} = \frac{RT}{4F}\ln\frac{C_2}{C_1} = \frac{RT}{4F}\ln\frac{i_L - i}{i_L} = 0.015\log\frac{i_L - i}{i_L} \tag{5.22}$$

where the constant 0.015 applies at 25 °C.

5.4.2 Total Oxygen Overvoltage

The cathodic process of oxygen reduction deals with the sum of activation overvoltage and concentration polarization; when the equilibrium potential of hydrogen evolution is reached ($E < E_{eq,H_2}$) the potential follows the Tafel law of hydrogen, therefore, the potential cannot tend to $-\infty$ as would be expected from Eq. 5.16 as shown in Fig. 5.14.

Analytically, the cathodic overvoltage of oxygen reduction process is as follows:

- In the range between E_{eq,O_2} and E_{eq,H_2}, (1.23 V), the cathodic current, i, varies from i_{0,O_2} to i_L and overvoltage is given by:

$$\eta = \eta_{act,O_2} + \eta_{conc,\,O_2} = -b\log\frac{i}{i_{0,O_2}} + 0.015\log\frac{i_L - i}{i_L} \tag{5.23}$$

Fig. 5.14 Cathodic
characteristic of oxygen
reduction process

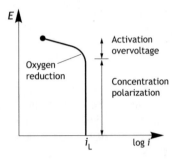

where i_{0,O_2} is the exchange current density for oxygen reduction which depends on metal, b is the Tafel slope for oxygen reduction, often taken as 100–120 mV/decade, and i_L is the oxygen limiting current density, which depends on the electrolyte and not on the metal

- For $E < E_{eq,H_2}$ when current density exceeds the oxygen limiting current density ($i > i_L$), the overvoltage is:

$$\eta = 1.23 + \eta_{act,H_2} = 1.23 - b\log\frac{i}{i_{0,H_2}} \tag{5.24}$$

where i_{0,H_2} is the exchange current density for hydrogen evolution which strongly depends on metal, b is the Tafel slope for hydrogen evolution and is 120 mV/decade, and i_L is the oxygen limiting current density which depends on the electrolyte and not on the metal.

It is worth noting that Eq. 5.24 applies to any pH. Figure 5.15 shows the resulting cathodic characteristic obtained by summing currents of occurring cathodic processes at each potential.

In summary, the cathodic curve is characterized by three intervals: $i < i_L$ where activation overvoltage prevails; i approaches i_L ($i \cong i_L$) under diffusion control and $i > i_L$, where hydrogen evolution becomes predominant.

Fig. 5.15 Resulting cathodic
characteristic of oxygen
reduction and hydrogen
evolution

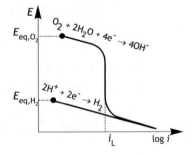

5.5 Other Cathodic Processes

Figure 5.16 reports cathodic curves for five cathodic processes, showing different oxidizing power:

- ① hydrogen evolution
- ②, ③ oxygen reduction
- ④, ⑤ more noble cathodic processes as reduction of ferric ion to ferrous ion, chlorine reduction or reduction of chromate.

The position and trend of curves depends on the metal and environment composition, the temperature and the flowing regime.

5.6 Passivation and Passivity

The corrosion resistance of metallic materials is strongly dependent on the surface condition through the presence of oxides and corrosion products which eventually define the corrosion behaviour in practice. There are many noteworthy examples of corrosion resistance by passivation or passivity:

Fig. 5.16 Qualitative trend of cathodic curves for: ① Hydrogen evolution; ②–③ Oxygen reduction; ④–⑤ Noble cathodic processes

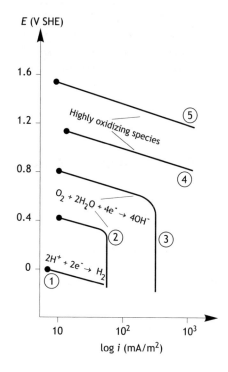

- Low nobility metals such as aluminium, chromium, titanium, zirconium, tantalum, stainless steels and other high alloy steels resistant to many aggressive environments
- Iron in concentrated but not in diluted sulphuric acid
- Titanium in oxygen-free sulphuric or hydrochloric acids but not in aerated ones
- Lead in diluted but not in concentrated sulphuric acid; in tap water but not in distilled water
- Aluminium and stainless steel in nitric acid, but not in hydrochloric acid; it is the opposite for silver
- Carbon steel in concrete but not in plaster (gypsum)
- Aluminium or lead in plaster, but not in concrete.

The formation of layers on the metal surface is called *passivation*; when the protection property of these layers leads to a practical halt of corrosion, this condition is called *passivity*. In practical terms, passivation involves the formation of thick layers, for instance in the order of tens of micrometres, while passivity occurs when the layer is much thinner, in the order of nanometres. In the case of passivity, the layer is called *passive film.*

Oxygen Limiting Current Density in Seawater

By applying Eq. 5.21 to seawater, where oxygen concentration does not exceed 11 ppm, the diffusion layer thickness, δ, varies in the range 0.1– 3 mm, depending on turbulence, and the diffusion coefficient, D, ranges between 1.3 and 2.5 $\times$ 10^{-9} m^2 s^{-1} for temperatures from 10 to 30 °C, *oxygen limiting current density*, $i_{L, seawater}$, ranges between 10 mA/m^2 (stagnant water and oxygen content about 1 ppm) and 2 A/m^2 (maximum turbulence and aeration, such as near ship impellers).

5.6.1   Film Formation Mechanisms

Protective films on metal surfaces can form following two mechanisms: by precipitation of insoluble corrosion products or directly as a result of the anodic reaction.

The first mechanism deals with metals of normal to intermediate electrochemical kinetic behaviour and gives rise to thick layers, characterized by some porosity, a defined crystallographic structure and poor conductivity. This behaviour is called *passivation* and is typical for:

- Lead in sulphuric acid which spontaneously forms a layer of lead sulphate or lead dioxide when an external voltage is applied

- Silver in chloride-containing solutions which forms a silver chloride layer
- Copper or bronze exposed to the atmosphere which form a basic copper carbonate layer, the so-called patina (from Latin, *patina nobile*); similarly, when exposed to sea water, the formation of copper oxi-chloride (*atacamite*) occurs. Patina can form spontaneously over a time span of possibly months to obtain protection films, or can be produced artificially through accelerated processes used in industry, such as phosphating of steel and zinc or patination of bronze.

The adhesion and protection properties of films depend on both metal and electrolyte composition as well as on the compatibility with other forming deposits.

The second mechanism is more important and leads to so-called *passivity*. It affects transition metals such as Fe, Cr, Mo, W, Ti, Zr, and alloys like stainless steels. Until the 1970s, two theories were proposed: the first suggested that passivity consists of a monoatomic layer of adsorbed oxygen; the second proposed the formation of a metal oxide film. Later, new surface analysis techniques such as AFM, Auger spectroscopy and electron microscopy clarified that, with few exceptions, passive films are an oxide-type layer, 3-5 nm thick, with semi-conductive properties.

For alloys, oxide film composition varies and rarely reproduces that of the alloy; most often one element prevails, determining passivity. For instance, on the surface of stainless steels, which have a minimum of 12% chromium content, there is an enrichment in chromium to form a passive film, composed mainly of chromium oxide (Cr_2O_3). Through this enrichment, stainless steels, although largely composed of iron, behave like chromium: for instance, the field for the formation of passive films widens as shown in Fig. 5.17, where Pourbaix diagrams of Fe and Cr are overlapped.

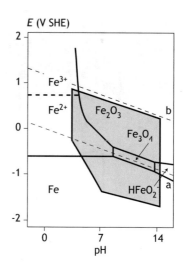

Fig. 5.17 Overlapping of Pourbaix diagrams of Fe and Cr (shadowed zone)

Anodizing can cause some metal oxides to grow thicker as in the case of Al, on which the oxide film thickness can be $20-30\,\mu m$, about three orders of magnitude thicker than spontaneous films. For Al, the thicker the passive film, the more resistant it is to corrosion. There are opposite cases, for instance for stainless steels where thicker film formed on the heat affected zone, HAZ, in welding (*tinted zone*) or during high temperature rolling, are less resistant than passive thin films.

Historical Notes

Towards the end of the nineghteenth century, it was noticed that iron was able to precipitate silver only from some solutions and not from others that are apparently similar. Faraday, among others, resumed this phenomenon around 1840 in relation to studies on the passivity of iron, then called the condition of non-reactivity of iron.

As represented in Fig. 5.18, the English scientist noticed how iron does not suffer corrosion in concentrated (fuming) nitric acid, while in diluted acid, obtained by adding the same quantity of water, it corrodes vigorously while also producing bubbles of gaseous nitrogen oxides. He also observed that in dilute nitric acid it did not corrode if previously immersed in concentrated nitric acid, provided that the iron surface was not mechanically scratched. He also noted that corrosion did not take place if connected to the positive pole of a battery in a galvanic cell. On the basis of these observations, Faraday attributed passivity to the presence of an oxide layer. In later decades it was discovered that this phenomenon applies also to other metals such as bismuth, tin and chromium.

It was also noted that chromium, unlike iron, only passivated by exposure to air. Finally, at the beginning of the twentieth century, in 1911, Monnartz highlighted that Fe–Cr alloys with a chromium content higher than 10.5%, i.e., stainless steels, behaved similarly. This discovery transformed the passivity from a scientific curiosity to a phenomenon with huge industrial implications.

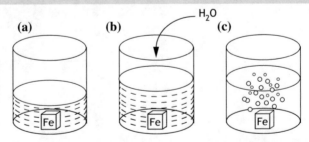

Fig. 5.18 Schematic illustration of Faraday experience on the passivity of iron: **a** nitric acid at 67%; **b** and **c** the solution obtained by diluting the same acid with an equal quantity of water

5.6.2   Oxide Properties

Passive films can be crystalline or amorphous, insulating or conductive (ionically or electronically). For instance, aluminium oxide is a good insulator and isolates the metal from the electrolyte; conversely, passive films formed on iron in alkaline solutions, as well as on chromium, nickel and stainless steels have a high electronic conductivity. Thus, they easily host the cathodic process, such as for instance, oxygen reduction.

Electrolyte composition influences film growth: for instance, halogens hinder formation of oxides, making them defective, other anions which produce insoluble products favour passivity (for example phosphates).

Potential and pH determine the field in which films form, as shown in the Pourbaix diagram. For instance, iron corrodes in gypsum plaster (pH < 7) and passivates in concrete (pH > 13); for lead it is the opposite, since it passivates at pH 7 and dissolves at pH 13, as occurs for amphoteric metals.

5.6.3   Active-Passive Metals

The typical anodic characteristic of an active-passive metal is shown in Fig. 5.19. Four zones are identified, indicating distinct corrosion behaviour:

- *Immunity*, $E < E_{eq}$, in which metal does not corrode; instead, it deposits from its ions if present
- *Activity*, $E_{eq} < E < E_p$, in which metal dissolves. In this range, E_{pp} is *primary passivation potential*, which copes with the maximum anodic current, i_{cp}, called *critical passivation current density* and E_p is *passivity potential*
- *Passivity*, $E_p < E < E_{tr}$, in which metal is passive. E_{tr} is *transpassivity potential* or *oxygen evolution potential* or *pitting potential*, E_{pit}. The dissolution rate corresponds to *passivity current density*, i_p, which is very low, about 10^{-6} lower than i_{cp}
- In oxidising-chloride containing environments, at potential higher than *pitting* potential ($E > E_{pit}$) passive film locally breakdowns and localised corrosion can initiate
- *Transpassivity* or *oxygen evolution* ($E > E_{tr}$). In this zone, noble anodic processes take place, such as oxygen evolution or the production of high valence ions such as chromate and bi-chromate ($Cr_2O_3 + 5H_2O \rightarrow 2CrO_4^{2-} + 10H^+ + 6e^-$).

If passive film is insulating, i.e., has ionic conductance only, it cannot transport electrons for electrodic processes like oxygen evolution. Therefore, passive film can grow, reaching very high (noble) potentials, for instance of 100 V and more. On

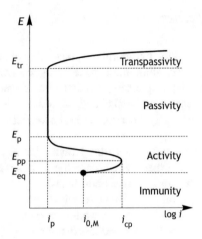

Fig. 5.19 Anodic characteristic of an active-passive metal

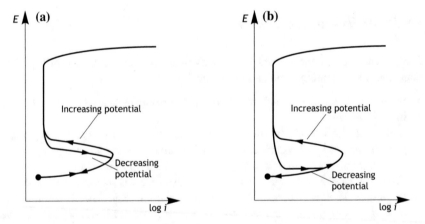

Fig. 5.20 Hysteresis effect on anodic characteristic for Fe–Cr alloys: **a** %Cr below 13%; **b** %Cr above 13%

titanium, very thick films can be obtained at 100–200 V in some electrolytes, giving rise to a new phenomenon produced by the perforation of the film (*anodic spark oxidation*).

The anodic curve varies in accordance with the potential sweeping direction, as shown in principle in Fig. 5.20a. Instead, Fig. 5.20b shows two different curves for Fe–Cr alloys with a Cr content higher or lower than 13%, respectively. Once formed, the passive film shows a persistency which helps reach protection conditions by applying cathodic protection by passivity (refer to Chap. 23).

5.6.4 Passivity-Related Parameters

The passivation and passivity behaviour of a metal depends on a number of parameters, which are derived from the anodic polarization curve obtained from testing. These parameters are:

- *Critical passivation current density*, i_{cp}; tendency to passivation increases as critical passivation current density decreases
- *Passivity current density*, i_p; passivity is more stable as the passivity current density lowers and the passivity interval widens
- *Primary passivation potential*, E_{pp}
- *Passivity potential*, E_p
- *Transpassivity potential*, E_{tr} (or *pitting potential*, E_{pit})
- *Passivity interval* ($E_{tr} - E_p$), which defines the extension of passivity.

Beyond the composition of the metal, environmental properties, such as temperature, acidity and chloride concentration, influence the curve shape as schematically depicted in Fig. 5.21. All these parameters must be measured at experimental conditions as close as possible to operating conditions, because they cannot be derived from metal composition (pure metal or alloy; annealed or cold worked; surface finishing), environment composition (in particular the presence of halogens) and temperature. In order to have a rough orientation of main parameters for stainless steels, a reference is Lazzari 2017.

Oxide Memory
Experience has shown that operating conditions at the moment of film formation from bare metal surface are of great importance to obtain a film with high and constant protection properties; in other words, a sound film formed at the beginning will also be the same later even if conditions change slightly;

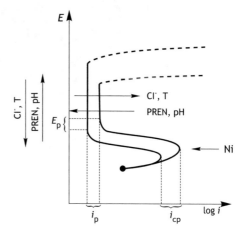

Fig. 5.21 Influence of the stainless steel composition of and environmental conditions (T, pH, Cl) on the shape of the active-passive anodic curve

conversely, a bad film will not improve successively even if new favourable operating conditions are set up. A remarkable example is given by the startup of a heat exchanger made of copper alloys where the cooling fluid is seawater: if the metal surface is bare, clean and cold with well aerated seawater, the passivation film forms with high protection properties; if startup is carried out on load (i.e., when operating at temperature) the film shows poor protection properties also in the next operating conditions.

Pietro Pedeferri exploited the memory of titanium oxides to obtain artistic and decorative results, as shown in Fig. 5.22: dark and silvery lines correspond to films with different properties, such as structure, corrosion resistance and colour, obtained by anodizing titanium surface at different *initial potentials* to imprint the early properties of the oxide.

As Pedeferri explained, this imprinting effect can be exploited for artistic purposes and for obtaining the required protection properties, since oxide film obtained at the beginning will also maintain its property thereafter.

Fig. 5.22 Artistic use of oxide memory of titanium by P. Pedeferri (1968)

5.7 Questions and Exercises

5.1 Draw the anodic curves for Fe and Zn dissolution in oxygen-free acidic solution and the cathodic curves for hydrogen evolution. Discuss whether they may change in time and how.

5.2 Suggest an interpretation why the passivity current density of stainless steels exposed to H_3PO_4 is lower than that in sulphuric acid.

5.3 Suggest an interpretation why the passivity current density of stainless steels exposed to strong alkali is higher than that in HCl.

5.4 A steel plate is exposed to seawater in equilibrium with the atmosphere at room temperature, pH 8.3, water velocity 1 m/s.

    (a) Determine the equilibrium potential of O_2 reduction reaction

    (b) Calculate the limiting current density of the oxygen reduction reaction when there is no deposit on the steel surface

    (c) Calculate the oxygen permeability (or also the deposit efficiency) after a couple of years, when a layer of rust and calcium carbonate, 1 mm thick, builds up if an average corrosion rate, determined by thickness measurement, is 0.02 mm/year.

5.5 Describe and explain how the corrosion rate of steel depends on temperature in the following systems:

    (a) Open system and equilibrium between water and atmosphere

    (b) Closed system (for instance a boiler circuit).

5.6 Consider the following statement: corrosion occurs if the sum of the moduli of anodic and cathodic overvoltages exceeds the driving voltage (i.e., the difference between equilibrium potentials of cathodic and anodic reactions). Do you agree with this statement?

5.7 Which parameter would you consider necessary to draw the anodic curve for an active, low melting temperature metal? And for a high melting temperature metal?

5.8 Which parameter would you consider necessary to draw the cathodic curve when only activation polarization applies? And when there is a concentration polarization such as in the case of oxygen reduction?

5.9 Which are the parameters you must know to draw the anodic curve of an active-passive metal?

5.10 Draw an anodic polarization curve for an active–passive metal. Define and indicate on the figure all characteristic potentials, potential regions and current densities.

Alessandro Volta and the Ohm's Law

In the letter sent to Van Marum (22nd June 1802 [4]), Volta showed that, not only he had clearly identified the physical factors that govern the circulation of current in the pile (i.e. the tension, the current and the conductivity), but also that he knew the quantitative relationship that linked them. In the same letter, Volta also described the principle of the first Faraday law (see box in Chap. 2).

In the best-known passage, that we quote here in the original text, is set out the Law that Ohm was to deduce more than twenty years later, exactly as we state it today.

«La rapidité de courant électrique est in raison composée de la tension électrique et de la liberté ou facilité du passage dans toutes le parties de la chaine ou cercle». («We can conclude that the speed of the electrical current and consequently the strength of the shock felt is in proportion to the electrical tension and to the freedom or ease of passage through all the parts of the chain or circuit»).

To demonstrate how clear in Volta's mind were the laws governing the functioning of galvanic chains, we quote another passage from the same letter.

«The electrical tension corresponds exactly, as our electrometric experiments demonstrate, to the number of metallic pairs of the pile, placed in a convenient order, at the rate of about 1/60 of a degree on my leaf-electrometer for each pair, if they are made of copper and zinc. The ease of passage of the electrical fluid depends on the permeability or the conductive capacity of the moistened disks made of pasteboard, cloth or similar material, interposed between the metallic pairs. Thus, assuming that the pile is formed of 120 pairs, it will always give a reading of 2 degrees on my electrometer, and will also charge to 2 degrees a Leyden jar and a battery of any size, whether the disks are rather dry or very wet, whether they are large or small, whether they are steeped in pure water or a saline solution, etc.

It will simply take the current a little longer to traverse these disks when pure water is used, and the smaller and drier these disks are. So, this delay, this decrease in speed of the current means that the shock given by the pile will likewise be less strong and either imperceptible or nonexistent.

Let's provide another example with this pile of 120 pairs: are the interposed pasteboard disks about one inch in diameter and not very moist? No noticeable shock will be obtained; nevertheless, it will charge the electrometer to 2 degrees, and in a few seconds it will charge to 2 degrees a battery which, thus charged, will provide a good shock. Now, let the interposed

---

[4]The letter is part of Volta's correspondence with the Dutch physics Van Marum, who has a powerful scrubbing electric machine in Rotterdam. Unfortunately, Van Marum, unlike what he did with previous letters, does not make it public. The letter was only disclosed in 1905 when J. Bosscha published the correspondence Volta-Van Marum.

pasteboard disks be sufficiently moistened with pure water: the electrical tension will still be 2 degrees but a weak shock will be felt beginning from the 20th pair.

Substituting the small pasteboard disks of about 1 inch in diameter with others of 8 or 10 inches which are well moistened with pure water (and so large metal plates are also used), the shock will be considerably stronger and already perceptible at the sixth or seventh pair: even so, the electrical tension has not increased; the larger size of the moistened disks has merely facilitated the passage of electrical fluid. Next, let small pasteboard disks be bathed in a saline solution, preferably ammonium muriate, a much less imperfect conductor than pure water (and so small metal plates are also used): an incomparably greater and almost unbearable shock will be obtained, even though the electrical tension is still 2 degrees, and a barely perceptible shock will be felt at the third or even second pair. Finally let the large disks be soaked in this same saline solution, and interposed between the large metallic pairs: nothing will be gained in electrical tension, which will still be 2 degrees for 120 of these pairs, but much will gained in the speed of the current, that finds the greatest ease of passage across these large and very good conductors. And from here we go on to the marvellous effects obtained from the scintillation and fusion of wires and metallic sheets subjected to this current, created from just a small number of these pairs, and to the most amazing results that you have achieved with a device of 200 pairs. But why does the shock, which heretofore had gained in strength thanks to the more complete soaking with water, to a greater extension of the disk surface soaked in this water, and above all to the substitution of water with a good saline solution, gain nothing or almost nothing more from the great size given to the disks soaked in this same liquid, while the speed of the electric current increases because of this, to the point at which it produces the fusions referred to? I explained this problem to you in my letter from Geneva, and I discussed it more thoroughly in a continuation of the memoir that I read to the Institute of Paris, only the first part of which was published in the Annales de Chimie, and the remainder will not be late in coming.

For the time being, I need only remind you that when one wants to prove a shock, the body of the man is inserted in the circuit and the longer is his body and the thinner his arms, the worse a conductor he will be, and he will be far less permeable to the electrical fluid than the disks of the pile soaked in salt water; the man's body, I say, greatly slows the electric current, which in this case is no longer capable of melting the metallic wires which are closed in a circuit by the man, who touches one end of the pile with one hand, and the opposite end with the other hand. I believe that the shock gains in strength as the electrical fluid passing through the moistened disks of the pile encounters less resistance, just as long, however, as the obstacle is greater than that of the human body that must be crossed. But when the greatest obstacle is the man's body, and it is the body which limits the speed of the current, this speed cannot be increased by further facilitating the current path in other parts of the circuit, namely in the moistened disks.

Pietro Pedeferri, *Pi*aneta Inossidabili, 4, 2002.

Bibliography

Bianchi G, Mussini T (1976) Elettrochimica. Masson Italia, Milano

Bianchi G, Mussini T (1993) Fondamenti di elettrochimica: teoria ed applicazioni. Masson Italia, Milano

Evans UR (1948) An introduction to metallic corrosion. Edward Arnold, London, UK

Fontana M (1986) Corrosion engineering, 3rd edn. McGraw-Hill, New York

Lazzari L (2017) Engineering tools for corrosion. Design and diagnosis. In: European federation of corrosion (EFC) series, vol 68. Woodhead Publishing, London, UK

Piontelli R (1961) Elementi di teoria della corrosione a umido dei materiali metallici. Longanesi, Milano (in Italian)

Pourbaix M (1973) Lectures on electrochemical corrosion. Plenum Press

Shreir LL, Jarman RA, Burstein GT (1994) Corrosion. Butterworth-Heinemann, London, UK

Chapter 6
Evans Diagrams

*Jupiter, father and king, I hope that my weapon is put to rest
and falls apart with rust.
And that nobody tries to hurt a peace-lover like me!*

Orazio, Serm., 2, 1, 43

Abstract This chapter deals with the potential-current density diagrams, also called Evans diagrams, which relate the variation of potential of a reaction—either anodic or cathodic—with the current density exchanged in the process, starting from its equilibrium potential in the corresponding environmental conditions. These diagrams allow to identify the corrosion working conditions, E_{corr} and i_{corr}, where the anodic and cathodic processes proceed with the same rate. The cases of active and active–passive metals are described. Ohmic drop can also be represented on the diagram, modifying corrosion rate. Both corrosion conditions and the imposition of a polarization are discussed with reference to the modification of the electrode working condition that they cause. Finally, experimental polarisation curves are introduced.

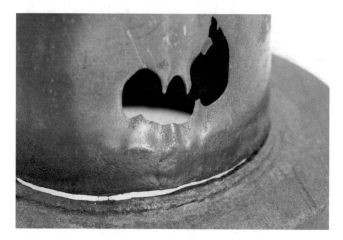

Fig. 6.1 Case study at the PoliLaPP Corrosion Museum of Politecnico di Milano

© Springer Nature Switzerland AG 2018
P. Pedeferri, *Corrosion Science and Engineering*, Engineering Materials,
https://doi.org/10.1007/978-3-319-97625-9_6

6.1 Introduction

When corrosion occurs at the surface of a metal, two electrochemical processes take place, an anodic and a cathodic process, each one characterized by a different equilibrium potential. The potential of the metal, called *corrosion potential*, E_{corr}, which can be easily measured, is necessarily intermediate between the two equilibrium potentials. At this potential, anodic and cathodic processes proceed with the same rate called *corrosion current* or *corrosion current density*, i_{corr}, when referred to a unitary surface area (Fig. 6.1).

The potential-current density ($E - \log i$) diagram that shows how potential changes as the current density of each process (anodic and cathodic) varies, hence identifying corrosion working conditions, is called the *Evans diagram*. It is named in honour of the English scientist Ulick Richardson Evans (1889–1980) who had first proposed it.

6.2 Evans Diagrams of Active Metals

Let's consider the Evans diagram shown in Fig. 6.2 for zinc corrosion in an oxygen-free acidic solution. The processes that take place at the surface of zinc are:

$$Zn \rightarrow Zn^{2+} + 2e^- \tag{6.1}$$

$$2H^+ + 2e^- \rightarrow H_2 \tag{6.2}$$

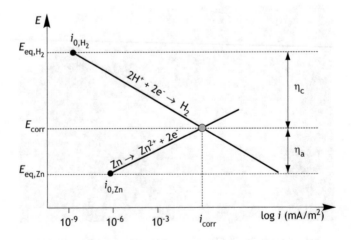

Fig. 6.2 Evans diagram of zinc in acid solution

This diagram summarizes that:

- The corrosion condition, corresponding to E_{corr} and i_{corr}, is identified by the crossing point of the characteristic overvoltage curves of the two processes
- The corrosion current density, i_{corr}, is determined by the condition at which the available driving voltage, ΔE, equals the sum of dissipations—or absolute values of overvoltage—taking place at the anode and the cathode, η_a and η_c:

$$\Delta E = E_{eq,c} - E_{eq,a} = E_{eq,H_2} - E_{eq,Zn} = \eta_a + |\eta_c| \qquad (6.3)$$

where $E_{eq,c}$ is the equilibrium potential, of the cathodic process (hydrogen evolution, E_{eq,H_2}); $E_{eq,a}$ is the equilibrium potential of zinc dissolution, $E_{eq,Zn}$. The ohmic drop in the solution, as well as in the metal, is negligible

- The free corrosion potential, E_{corr}, is closer to the cathodic equilibrium potential when dissipations take place mainly at the anode. The opposite occurs when dissipations prevail at the cathode.

Evans diagrams are very useful to illustrate the influence on the corrosion current density, i_{corr}, and the corrosion potential, E_{corr}, of the following factors:

- Figure 6.3 shows the influence of the oxidizing power. As it increases, both the potential and the corrosion rate increase. For example, with a 60 mV/decade slope of anodic overvoltage, an increase in corrosion potential by 120 mV, due to a more oxidizing cathodic process, cause the corrosion rate to increase by two orders of magnitude

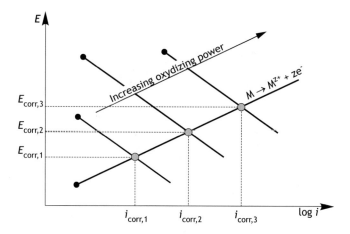

Fig. 6.3 Corrosion conditions of an active metal: influence of the oxidizing power

- Figure 6.4 shows the influence of the exchange current density, i_0, of the cathodic process. As it decreases, the corrosion rate is also lowered because the cathodic overvoltage increases
- Figure 6.5 shows the influence of the oxygen limiting current density, i_L. As the agitation or the fluid velocity of the electrolyte increases, both the corrosion potential and the corrosion rate increase.

Figure 5.16 of previous chapter, illustrates the cathodic curves of five cathodic processes with different oxidizing power, namely: ① hydrogen evolution; ②, ③ oxygen reduction for two different concentrations of oxygen; ④, ⑤ high oxidizing power processes, for example, reduction of chromates or reduction of ferric ions to ferrous ones. When overlapping these five cathodic curves on the anodic curve of an active–passive metal, as shown in Fig. 6.6, the resulting corrosion rate varies. For instance:

- In the absence of oxygen or other oxidizing species, hydrogen evolution is the only possible cathodic process; accordingly, the corrosion condition is represented by point A as the intersection of the anodic curve, active region, and cathodic curve ①
- In the presence of oxygen the cathodic process is given by curve ②. Two corrosion conditions can arise, as established by intersection points B and C: if metal is active when immersed in the solution, corrosion condition is B; conversely, if metal is previously passivated and then immersed in the solution, it works stably in point C. The intersection at Flade potential is disregarded since it does not represent any real corrosion condition
- By increasing oxygen content or turbulence of the solution, the oxygen limiting current density, i_L, can increase so significantly that the cathodic curve (curve ③) can intersect the anodic curve in the passive interval only (point D). It is

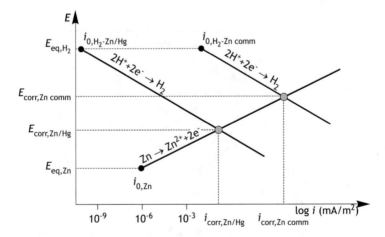

Fig. 6.4 Corrosion conditions of an active metal: influence of exchange current density, i_0, (amalgamated and commercial zinc in acidic solution)

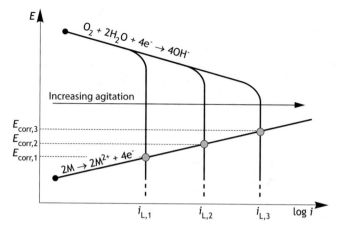

Fig. 6.5 Corrosion conditions of an active metal in an aerated solution: influence of agitation on i_L

Fig. 6.6 Possible corrosion conditions of an active–passive metal

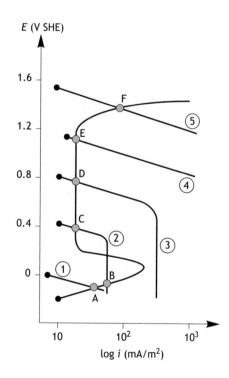

important to compare the two passive conditions indicated by C and D. Both are stable from the electrochemical point of view, but they are not equivalent from an engineering perspective. In fact, in the first case, if the protective film is locally damaged, for example, mechanically, the metal inside becomes active

(point B); instead, in the second case, the passive film heals spontaneously after local breakdown (point D)
- In the presence of strong oxidizing species (curve ④) the corrosion potential shifts to higher values, while remaining within the passivity range (point E)
- Corrosion occurs (point F) in the presence of stronger oxidizing species that are able to bring the material in the transpassive zone (curve ⑤).

6.3   Corrosion Conditions in the Presence of an Ohmic Drop

In a low conductivity environment, when anodic and cathodic surfaces are separated, the ohmic drop sensibly reduces the driving voltage and the corrosion current, as stated by the following balance:

$$\Delta E = E_{eq,c} - E_{eq,a} = \eta_a + \eta_c + \psi_{oh} = \eta_a + \eta_c + IR = f(I) \qquad (6.4)$$

as summarized in Fig. 6.7. The galvanic chain of Fig. 6.8 facilitates the understanding of how to measure the ohmic drop, ψ_{oh}, in the electrolyte, as well as the absolute values of electrode overvoltage, η_a and η_c. The ohmic drop, ψ_{oh}, can be calculated by the first Ohm's law, once the circulating current, I, and the electrolyte electrical resistance, R, are known.

In the presence of an ohmic drop in the electrolyte, corrosion potential is not uniquely defined, since it varies in the interval $(E_{eq,c} + \eta_c) - (E_{eq,a} + \eta_a)$ so that potential measurements depend on the reference electrode location: close to the anode, to measure anode potential, E_a, and conversely close to the cathode, to measure cathode potential, E_c. When a reference electrode is placed in between the

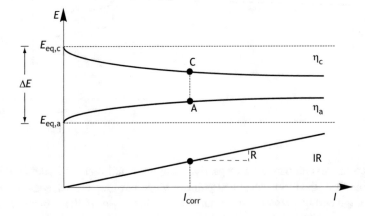

Fig. 6.7 Anodic and cathodic curves and ohmic drop in the electrolyte with constant resistance, R

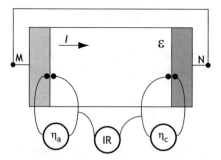

Fig. 6.8 Principle of short circuited galvanic chain

anode and the cathode, an intermediate value is measured (between A and C of Fig. 6.7).

The Evans diagram helps determine how driving voltage, $\Delta E = E_{eq,c} - E_{eq,a}$, is dissipated by examining the shape of overvoltage curves. Schematic cases depicted in Fig. 6.9 show which dissipation contribution determines the corrosion rate and the type of control: cathodic, anodic and ohmic, respectively.

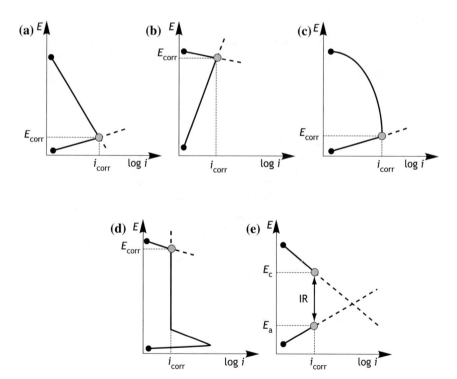

Fig. 6.9 Different types of kinetic control of a corrosion process: **a** cathodic overvoltage; **b** anodic overvoltage; **c** cathodic diffusion control; **d** anodic passivation; **e** ohmic control

6.4   Multiple Cathodic Processes

So far, corrosion processes have dealt with a single anodic and cathodic process; when multiple processes take place, the *mixed-potential theory* applies, which states that:

$$\sum I_a + \sum I_c = 0 \tag{6.5}$$

since there cannot be an accumulation of electric charges. Assuming the ohmic drop to be negligible, the corrosion potential is again the one at which Eq. 6.5 fits, i.e., where resulting anodic and cathodic curves cross. Figure 6.10 illustrates the corrosion condition when two cathodic processes take place: reduction of ferric ions and hydrogen evolution. Resulting curves are obtained by the following procedures based on the Eq. 6.5:

- The anodic curve starts from the lowest potential, i.e., the equilibrium potential of metal, M, and only fits metal oxidation. In principle, there could be the anodic process of hydrogen oxidation if hydrogen gas was present at the metal surface. The dashed line is the resulting anodic curve (sum of two processes)
- The cathodic curve starts from the highest potential, i.e., the reduction of ferric ions, then followed by the hydrogen evolution when hydrogen equilibrium potential is reached; metal deposition could also follow if metal ions were present in the solution. The dashed line is the resulting cathodic curve (sum of three processes).

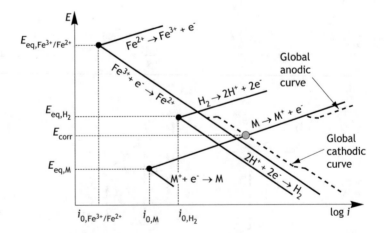

Fig. 6.10 Corrosion condition of a metal when cathodic processes are hydrogen evolution and reduction of ferric ions

Corrosion potential and corrosion rate are given by the crossing point of resulting anodic and cathodic curves.

6.5   Imposed Polarization

Let's consider a freely corroding metal, M, for example iron in an acidic solution. On its surface, both anodic and cathodic processes take place at the same rate, i_{corr}; the anodic process is iron dissolution (Fe $\rightarrow$ Fe^{2+} + 2e$^-$) and the complementary cathodic process is hydrogen evolution (2H$^+$+ 2e$^-$ $\rightarrow$ H$_2$). The corrosion potential, E_{corr}, is then given by the intersection point of the two characteristics, which follow Tafel law (Fig. 6.11).

Let's now consider what happens if an external current, i_e, is applied in the anodic or cathodic direction. The metal potential shifts to a potential higher, E_1, or lower, E_2, than E_{corr}, respectively. The anodic process rate, i.e., the dissolution of iron, $i_{a,Fe}$, and the cathodic process rate, i.e., hydrogen evolution, i_{c,H_2}, change from corrosion current density, i_{corr}, according to the electro-neutrality condition, as stated by the following relationships:

$$i_{c,H_2} = i_{a,Fe} + i_e \quad \text{for a cathodic polarization} \tag{6.6}$$

and

$$i_{a,Fe} = i_{c,H_2} + i_e \quad \text{for a anodic polarization} \tag{6.7}$$

If a cathodic external current, i_e, is applied, so that iron is cathodically polarized, the potential becomes more negative or less noble, and shifts to the potential value E_2

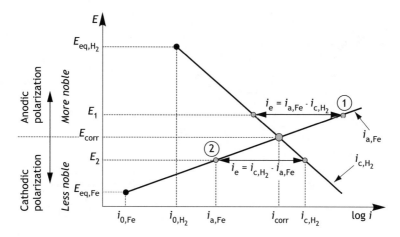

Fig. 6.11 Potential-current density plot, E-log$i$, in the presence of cathodic or anodic polarization as a result of the application of an external current

where the difference between the cathodic process (hydrogen evolution) rate and the anodic oxidation rate of the metal is given by:

$$i_e = i_{c,H_2} - i_{a,Fe} \tag{6.8}$$

Therefore, the corrosion rate of iron is now identified by point ②.

Similarly, if an anodic external current, i_e, is applied, so that iron is anodically polarized, its potential becomes less negative or more noble, then reaching potential E_1 to satisfy the relationship:

$$i_e = i_{a,Fe} - i_{c,H_2} \tag{6.9}$$

Therefore, the corrosion rate of iron is now identified by point ①.

In conclusion, by applying an external current, polarization is obtained in accordance with the current direction: anodic (i.e., $E > E_{corr}$) or cathodic (i.e., $E < E_{corr}$) causing an increase or a decrease of the corrosion rate, respectively.

6.6   Experimental Polarization Curves

Anodic and cathodic curves that have been described so far are not the ones that are obtained experimentally, although derived from them. In laboratory testing, it is not possible to separately measure the different anodic or cathodic currents involved, but only the current a metal coupon can exchange with the electrolyte; indeed, what can be measured is only the algebraic difference of the two currents, i.e., $I = \pm(i_a - i_c) \cdot S$, where S is the coupon surface area.

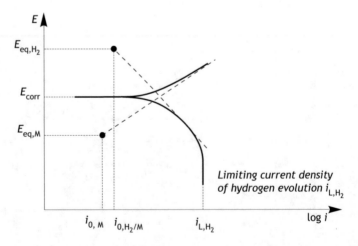

Fig. 6.12 Theoretical (dotted lines) and experimental polarization curves of an active metal in oxygen-free acidic solution

Figure 6.12 shows experimental curves and theoretical ones (dotted lines) for an active metal exposed to an oxygen-free acidic solution. Curves overlap when one of the two practically zeros, as easily shown by the relationship: $I = \pm(i_a - i_c) \cdot S$ when i_a or i_c fades. The corrosion rate is given by the crossing point of extrapolated lines from the Tafel region of measured curves. For comparison, experimental and theoretical curves (dotted lines) for an active metal exposed to an aerated acidic solution are reported in Fig. 6.13, where curves obtained in aerated neutral solution are plotted in Fig. 6.14.

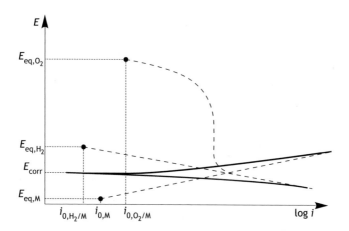

Fig. 6.13 Theoretical (dotted lines) and experimental polarization curves of an active metal as iron in aerated acidic solution

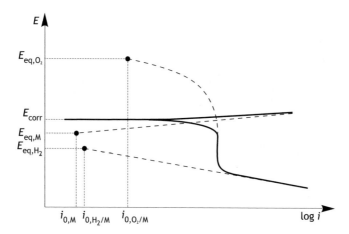

Fig. 6.14 Theoretical (dotted lines) and experimental polarization curves of an active noble metal as copper in aerated neutral solution

Figures 6.15, 6.16 and 6.17 show experimental curves and theoretical ones (dotted lines) for an active–passive metal when the cathodic process is more or less noble, then ranging from an active to a transpassive region. To obtain theoretical polarization curves from experimental plots, automatically gained by modern potentiostats, it must be predicted how a metal will behave when exposed to a particular environment.

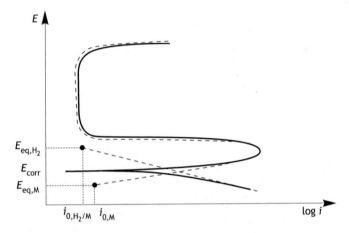

Fig. 6.15 Theoretical (dotted lines) and experimental polarization curves of an active–passive metal in oxygen-free neutral solution

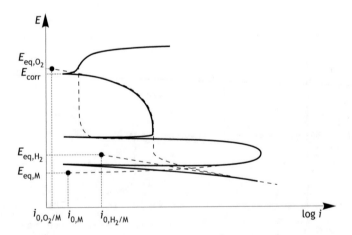

Fig. 6.16 Theoretical (dotted lines) and experimental polarization curves of an active–passive metal in aerated neutral solution

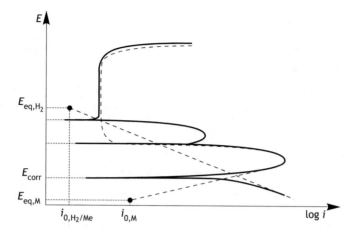

Fig. 6.17 Theoretical (dotted lines) and experimental polarization curves of an active–passive metal in oxygen-free acidic solution

6.7 Questions and Exercises

6.1 Draw the cathodic polarization curve in neutral aerated stagnant solution (T 25 °C; pH 7; 6 mg/L oxygen) from the equilibrium potential of oxygen reduction to −1.2 V SHE. Consider proper values of the exchange current density and Tafel slope.

6.2 Draw Evans diagrams and evaluate the corrosion potential and the corrosion rate for the following corrosion systems (consider the ohmic drop as negligible):

- Copper in seawater assuming that copper is active
- As above, instead assuming that copper passivates
- Iron (mild steel) in fresh water assuming iron is active
- As above, instead assuming that iron (as in stainless steel) passivates
- Copper in deaerated neutral soil
- Stainless steel in neutral deaerated freshwater
- As above, in deaerated seawater.

6.3 Draw the Evans diagram of iron (steel) in neutral aerated solution (T 40 °C; pH 7; v 2 m/s; 5 mg/L oxygen). Determine the corrosion potential and corrosion rate (in μm/y). Calculate anodic and cathodic overvoltages. Consider proper values of the exchange current density and Tafel slope.

6.4 Draw Evans diagrams of iron in a NaCl solution at pH 3 in two conditions:

- Aerated solution (air bubbling)
- De-aerated (oxygen free)

- Determine the corrosion current density and specify the cathodic processes in the two conditions. [Consider exchange current density, i_0, for Fe: 1 mA/m^2; H$^+$: 1 mA/m^2; O$_2$: 0.5 mA/m^2; oxygen limiting current density, i_L, 250 mA/m^2.]

6.5 Provide an example of each type of kinetic control of a corrosion process: (a) cathodic overvoltage; (b) anodic overvoltage; (c) cathodic diffusion control; (d) anodic passivation; (e) ohmic control

6.6 What are the relationships between I_a, I_c, E_c, E_a, i_a, i_c in a free corrosion condition in the presence of ohmic drop?

6.7 Draw Evans diagrams in the following case studies.

   a. Iron in 70% (15 mol/L) nitric acid, spontaneous passivation occurs (case I)

   b. Iron in 20% (4 mol/L) nitric acid, immersed as active (case II)

   c. Iron in 20% (4 mol/L) nitric acid, immersed as passive (previously immersed in 70% nitric acid (case III)

   d. Iron in 0.1 N = 0.1 mol/L nitric acid (case IV).
     Cathodic reaction is: HNO$_3$ + 3H$^+$ + 3e$^-$ = NO + 2H$_2$O, with standard reversible potential E^0 = 0.96 V (SHE). (For simplicity, assume that NO partial pressure is unitary). Assume same exchange current density and same Tafel slope.
     Which state do you think aluminium and chromium would adopt if they were exposed to dilute acid as in case IV? Which state would copper and nickel adopt?

6.8 Consider the experimental polarization curve of an active metal in oxygen-free acidic solution reported in Fig. 6.12. Why the curve deviates from the theoretical one?

6.9 Consider the experimental polarization curve of an active metal in oxygen-free acidic solution reported in Fig. 6.12. At which potential (with respect to free corrosion condition) Tafel slope of hydrogen evolution should be determined in your opinion?

6.10. With reference to Figs. 6.15, 6.16 and 6.17, please describe metal-to-environment characteristics deducible from the shape of the experimental/theoretical curve.

Ulick Richardson Evans
Evans was born in Wimbledon in 1889 and was educated at Marlborough College, 1902–1907, and King's College, Cambridge, 1907–1911, where he read for the Natural Sciences Tripos, specializing in chemistry. He then began research on electrochemistry at Wiesbaden and London, and after the First World War, he returned to Cambridge where he spent the rest of his life researching and writing prolifically on corrosion and the oxidation of metals.

U. R. Evans was described in the Biographical Memoirs of Fellows of the Royal Society as the "Father of the modern science of corrosion and protection of metals". His major contribution to the subject involved placing the electrochemical nature of corrosion on a firm foundation. His first paper in this area was published in 1923, which was followed in 1924 by his book "Corrosion of Metals", the first text book devoted to the subject. He continued to publish research papers for the next 50 years, as well as updating his classic text.

Excerpt from Ulick Richardson Evans, An Introduction to Metallic Corrosion, Edward Arnold, London, UK, 1948.

The chronological sequence of scientific discovery is rarely the logical one. To arrange the facts of metallic corrosion historically would conceal the true interconnection existing between them, and thus deprive them of significance. Nevertheless, in view of the prevailing interest in the History of Science, many readers may welcome a short narrative showing how knowledge of the subject discussed in this book has grown. The note which follows should serve to indicate some names and dates associated with the advance of understanding, but it must be remembered that the credit for any particular discovery cannot be assigned to a single year or to a particular person. If a recent investigator is cited as the discoverer, objection may fairly be raised by the quotation from older papers of passages which seem to contain the germ of the idea; yet to allot the entire credit to early investigators may be unjust to later ones, who have established as facts what had previously been mere suggestions. At the Dawn of History, the first metals to be used were those which were either found native, or could easily be reduced to the elementary state; such metals do not readily pass into the combined state, and their corrosion can have raised no serious problems.

But with the introduction of iron, the problem of its corrosion must have presented itself, although it is an undoubted fact that some of the iron produced in Antiquity is to-day more free from corrosion than much of that manufactured in later years. This may have been due partly to the fact that iron reduced with charcoal contained less sulphur than modern steel, but it may also be connected with the absence of sulphur compounds from the air in the days before coal was adopted as a fuel; for it is often the conditions of early exposure which determine the life of metal-work. Whatever the cause, ancient iron has in some cases remained in surprisingly good condition for many centuries; the Delhi Pillar is the example which has excited most interest, but others could be quoted.

Bibliography

Evans UR (1948) An introduction to metallic corrosion. Edward Arnold, London, UK
Hoar TP (1961) Electrochemical principles of the corrosion and protection of metals. J Appl Chem
    11:121–130
Hoar TP, Mears DC, Rothwell GP (1965) The relationships between anodic passivity, brightening
    and pitting. Corros Sci 5:279–289
Wagner C, Traud W (1938) On the interpretation of corrosion processes through the superposition
    of electrochemical partial processes and on the potential of mixed electrodes. Z Electrochem
    44:391

Chapter 7
Corrosion Factors

*In corrosion science the number of affecting factors is
often high and rarely easy to rank.*
Roberto Piontelli (1909–1971) Italian eminent electrochemist

Abstract Corrosion processes involve metal and environment properties through a
variety of factors. From the metal side, chemical composition and microstructure
play a major role. Chemical composition is also fundamental to define the electrolyte aggressiveness, together with its pH, temperature and the possible presence

Fig. 7.1 Case study at the PoliLaPP Corrosion Museum of Politecnico di Milano

© Springer Nature Switzerland AG 2018
P. Pedeferri, *Corrosion Science and Engineering*, Engineering Materials,
https://doi.org/10.1007/978-3-319-97625-9_7

of bacteria. Other relevant operating conditions are pressure, fluid velocity, presence of mechanical stresses and exposure time. Even though is not easy to depict a general trend on how these factors influence corrosion, a rational approach is proposed for the interpretation, based on thermodynamics (equilibrium potential) and kinetics (Evans diagrams).

7.1 Metal Affecting Factors

Various factors related to metals influence corrosion processes:

- Composition, presence of impurities, phases and constituents
- Crystalline structure, constituent phases, lattice defects, surface finishing grain boundary precipitates and mechanical stresses (Fig. 7.1).

In the following, also the influence of modification of chemical composition on surface is considered.

7.1.1 Modification of Metal Surface Composition

Let's consider an alloy consisting of two metals M and N exposed to an electrolyte in which M^{z+} and $N^{z'+}$ ions are present. Anodic processes are the following:

$$M \rightarrow M^{z+} + ze^- \tag{7.1}$$

$$N \rightarrow N^{z'+} + z'e^- \tag{7.2}$$

Equilibrium potential of each metal is respectively:

$$E_{eq}^M = E_M^0 + \frac{RT}{zF} \ln[a_{M^{z+}}] \tag{7.3}$$

$$E_{eq}^N = E_N^0 + \frac{RT}{z'F} \ln\left[a_{N^{z'+}}\right] \tag{7.4}$$

The two anodic processes take place at an equilibrium condition, which implies the same potential:

$$E_{eq}^M = E_{eq}^N \tag{7.5}$$

Assuming same valence, i.e., $z = z'$, it results that:

$$E_M^0 - E_N^0 = \frac{RT}{zF} \ln \frac{[a_{N^{z+}}]}{[a_{M^{z+}}]} \tag{7.6}$$

Fig. 7.2 Anodic polarization curve of Fe and Cr in sulphuric acid 1 N

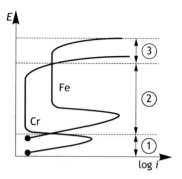

It results that differential corrosion of the two metals occurs until equilibrium condition is reached, given by the above relationship, through the consumption of less noble metal. For example, for copper–zinc alloys the difference between standard potentials is 1.1 V. Therefore, Zn to Cu ion concentration ratio at equilibrium is of the order of $\approx 1/10^{-37}$, hence, only Zn corrodes.

Kinetics may change the practical behaviour. Let's consider an iron/chromium alloy, which forms a solid solution. To predict corrosion in sulphuric acid, anodic overvoltage can be considered, as shown in Fig. 7.2. In the potential interval ①, chromium is active and less noble, then corroding; in interval ③, chromium is transpassive and less noble, then again corroding. Conversely, in interval ②, the passivity current of chromium is much lower than that of iron, therefore, the iron dissolution rate is higher; thus, accordingly, the alloy surface enriches in chromium, thereby strengthening passivity.

7.1.2 Nobility by Alloying

Corrosion resistance can be improved by changing the metal composition in order to increase nobility, or increase the overvoltage of cathodic and anodic processes, as discussed below. This paragraph does not address other interventions to increase resistance to localized corrosion, such as pitting, intergranular, SCC, hydrogen damage and flow-induced corrosion that are considered elsewhere.

The addition of a more noble element increases the nobility of an alloy. An example is Monel, a nickel-based alloy containing about 30% copper, which increases the equilibrium potential from −0.25 V SHE of pure nickel to about −0.07 V SHE, since copper is more noble. Prediction of this alloying effect is given by the calculation of equilibrium potential as the weighted average. For example, the standard potential of α-brass, a solid solution copper–zinc alloy with composition about 70% Cu and 30% Zn, standard potential is: $E^0_{\alpha\text{-brass}} \cong +0.34 \cdot 0.7 - 0.76 \cdot 0.3 \cong +0.01$ V SHE.

Table 7.1 Corrosion rate of commercial aluminium as a function of iron content in 20% HCl solution at 26 °C

Fe content (%)	Corrosion rate	
	$(g/m^2 \ d)$	(mm/y)
0.002	6	0.8 ($\approx$1 μm/d)
0.01	112	15 ($\approx$1 μm/h)
0.03	6500	880 ($\approx$1 μm/min)
0.12	36,000	4,860 ($\approx$10 μm/min)
0.8	190,000	25,690 ($\approx$1 μm/s)

7.1.3 Overvoltage of Cathodic Processes

Metal composition strongly influences the overvoltage of hydrogen evolution in acidic solutions for normal or intermediate metals. The addition of impurities or noble precipitates having low hydrogen overvoltage, as in the case of Al–Cu alloys (3–5% Cu) used in the aerospace industry, causes a localized attack when intermetallic $CuAl_2$ precipitates at grain boundaries. Zn, Al and Mg alloys exhibit same behaviour with high melting temperature impurities with a low hydrogen overvoltage (Table 7.1); conversely, to reduce corrosion, high hydrogen overvoltage elements like Cd, Sn or Hg, even if only on the surface, as obtained by metal displacement, should be added.

7.1.4 Cathodic Alloying

In the 1950s, Nikon D. Tomashov (1905–1990) proposed an elegant and intriguing method for setting up an anodic protection effect (ref. Chap. 19) on titanium by adding a noble metal to form a solid solution. This is known in literature as "*noble-metal alloying*", and in the following, it is called *cathodic alloying*. Figure 7.3 shows the passivating effect of the addition of platinum on titanium in an acidic, oxygen-free solution. The galvanic effect is achieved by adding a small amount of Pd, 0.5% max, in a solid solution of titanium, which brings free-corrosion potential in the passive interval even in an oxygen-free solution. Figure 7.3 shows how cathodic alloying works: by reducing the overvoltage of hydrogen evolution through the higher exchange current density of Pt or Pd, the critical passivation current density is also exceeded. This is called *anodic protection effect by cathodic alloying* and is applicable if:

- The base metal is an active–passive one
- Equilibrium potential of hydrogen evolution, E_{eq,H_2}, is more noble than the passivation potential, E_p, of the base metal
- Cathodic alloying elements are more noble than the base metal and they form a solid solution.

Based on the above requirements, titanium and chromium alloys, for instance stainless steels, are possible candidates with Pt and Pd.

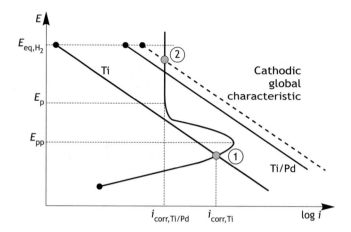

Fig. 7.3 Anodic protection effect by *cathodic alloying* with Pd (or Pt)

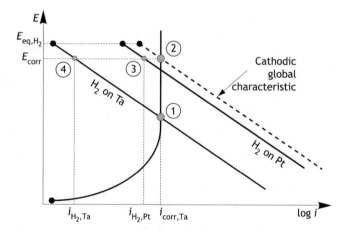

Fig. 7.4 Comparison between pure tantalum and Pt-alloyed tantalum

Similar cathodic alloying is applied to tantalum to avoid hydrogen-related damage instead of reducing corrosion. Ta suffers from hydrogen embrittlement (see Chap. 14) when exposed to highly corrosive environments in which the corrosion rate is considered acceptable, for instance less than 10 μm/y. By alloying with Pt, the corrosion condition moves from ① to ② as shown in Fig. 7.4, so that molecular hydrogen evolution takes place on Pt, mainly (point ③), while the corrosion rate of Ta remains unchanged (i.e., passivity current density) and the hydrogen evolution rate on Ta fades (from ① to ④), then avoiding hydrogen embrittlement. A small Pt content is sufficient because of the huge difference in the exchange current density, $i_{0,H}$, for hydrogen evolution on Ta versus Pt.

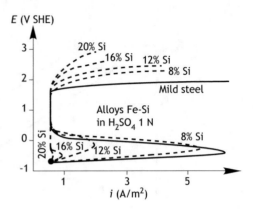

Fig. 7.5 Silicon content influence on the anodic behaviour of iron in sulfuric acid 1 N at 25 °C

7.1.5 Reduction of Anodic Areas

In a non-homogeneous metal, the anodic component should be dispersed in a cathodic matrix so that after the initial rapid corrosion attack, the surface enriches with the cathodic component or passivating element. For example, this effect occurs in grey cast iron where after an initial corrosion of iron, which is anodic, the surface enriches with silicon, enhancing passivation. Also selective corrosion attack may lead to a similar effect as in the case of brass on which zinc corrodes initially hence the surface enriches with copper (more noble and with passivating propriety). The opposite does not work, because a cathodic component dispersed in an anodic matrix cannot spread on the surface as the anodic component continues corroding.

7.1.6 Passivation Induced by Alloying

When adding a component easy to passivate to an alloy, the alloy does the same. Two typical examples are the addition of chromium or silicon to iron. Figure 7.5 illustrates the effect of the addition of Si on steel in sulphuric acid. Figure 7.6 shows the influence of the addition of some elements to iron on susceptible parameters, as i_p, i_{cp}, E_{pp}, E_p e E_{tr}; the influence depends on content.

7.2 Environment Affecting Factors

From a corrosion viewpoint, most relevant environment-related properties, either in bulk or on a metal surface, are:

- Conductivity (determined by total dissolved solids)
- pH

Fig. 7.6 Influence of
alloying elements on the
anodic characteristic of iron in
sulfuric acid solution 1 N at
25 °C

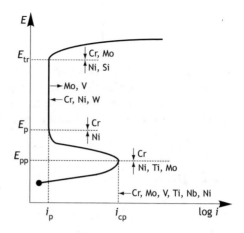

- Oxygen content
- Other oxidizing species (for instance chlorine)
- Bacteria.

Typically, acidity and oxygen content in bulk are important for sustaining the
cathodic process, while pH at metal surface determines the passivating tendency.

It is worth mentioning that some cathodic processes would be possible in
principle because of their noble potential, but exhibit very slow kinetics. These
processes are:

- $SO_4^{2-} + 4H^+ + 2e^- \rightarrow H_2SO_3 + H_2O$ with $E° = +0.17$ V SHE (in acidic
  solutions)
- $ClO_4^- + 2H^+ + 2e^- \rightarrow ClO_3^- + H_2O$ with $E° = +1.19$ V SHE (in acidic
  solutions)
- $2NO_2^- + 3H_2O + 4e^- \rightarrow N_2O + 6OH^-$ with $E° = +0.15$ V SHE
- $2NO_3^- + 4H^+ + 2e^- \rightarrow N_2O_4 + 2H_2O$ with $E° = +0.81$ V SHE.

Although the reduction of NO_2^- is a less noble reaction than the reduction of NO_3^-,
it is kinetically more active.

7.2.1   *Conductivity*

An electrolyte has an electrical conductivity due to the presence of ions (anions and
cations) as a result of the dissociation of dissolved salts. For fresh water, in practice,
conductivity is a function of TDS (total dissolved solids) and temperature through
the following practical relationship (Lazzari 2017):

$$\sigma\left(\frac{S}{m}\right) = \frac{1}{\rho(\Omega m)} \cong \frac{(1+0.02\Delta T)\cdot TDS(g/L)}{9} \qquad (7.7)$$

where TDS is total dissolved solid (g/L) and ΔT is temperature variation from 25 °C.

Conductivity has a direct influence on the corrosion rate, especially on localized corrosion where a macrocell mechanism sets up: the higher the conductivity, the higher the corrosion rate, provided a cathodic process is present.

7.2.2 pH

For various metals, the corrosion rate is strongly pH dependant as Fig. 7.7 shows. Type a trend characterizes amphoteric metals such as Al, Zn, Pb and Sn, which suffer corrosion either in acidic solutions (for instance, aluminium and zinc form Al^{3+} and Zn^{2+} cations, respectively) or in alkaline solutions (aluminium and zinc dissolve as AlO_2^-, and ZnO_2^{2-} anions, respectively). The type b trend applies to metals such as Fe, Ni, Co, Cr, Mn, which passivate in neutral or higher pH solutions. Finally, the type c trend applies to noble metals, such as Au or Pt, which resist corrosion in both acidic and alkaline solutions.

7.2.3 Differential Aeration

Evans' experience helps illustrate the influence of non-homogeneous oxygenation. A cell composed of two compartments, each hosting an iron or mild steel strip,

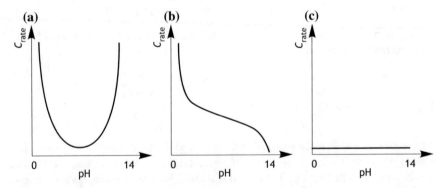

Fig. 7.7 Trend of corrosion rate of metals with pH; **a** amphoteric metals (Al, Zn, Pb, Sn); **b** metals passivating in alkali (Fe, Ni, Co, Cr, Mn); **c** noble metals (Au, Pt). Adapted from Speller (1926) and Piontelli (1961)

communicating through a porous plug, is filled with a neutral solution (Fig. 7.8).
Oxygen and nitrogen bubble separately in each compartment. By connecting the
two strips by an ammeter, the current flows from the strip of the oxygenated
compartment (positive pole, cathode) to the other (negative pole, anode). The
cathodic strip, because of the alkalinity produced by the cathodic reaction (i.e.,
oxygen reduction) passivates, as shown in Fig. 7.9. Initial conditions, before the
short-circuiting of the two strips, are ① for a low oxygen zone and ② for a high
oxygen zone. After coupling: zone ① starts working as the anode (i.e., less noble
electrode) and zone ② as the cathode. Point ① moves to point ③ as the final stable
anodic corroding zone while point ② shifts to ④ passivating and becoming a stable
final cathodic zone for oxygen reduction.

By interchanging the two gas flows, corrosion reverses: the passivated one starts
corroding and the one that was corroding passivates. This type of corrosion, called
differential aeration, occurs where oxygen concentration zeros, such as under
deposits or in cavities, because the solution cannot be continuously replaced. It
often takes place on buried structures (pipings, tanks) where soil is characterized by

Fig. 7.8 Evans experiment
of differential aeration

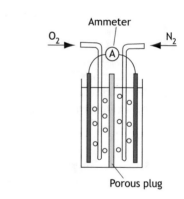

Fig. 7.9 Evans diagram for
differential aeration ① low
oxygen zone (initial); ② high
oxygen zone (initial); ③
anodic corroding zone (final)
④ cathodic zone (final)

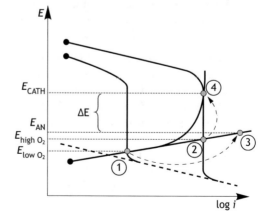

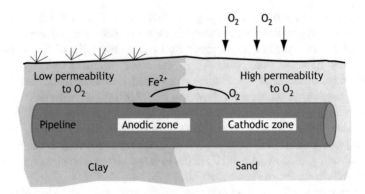

Fig. 7.10 Differential aeration corrosion on a buried structure in non-homogeneous soil

a different permeability to oxygen (Fig. 7.10): in zones where soil is less permeable to oxygen (clayey soil) corrosion occurs, while aerated zones (sandy soil) behave as a cathode.

Although not strictly a differential aeration, it is worth mentioning the case of so called *corrosion at liquid line*, which occurs on a partially immersed metal or inside partially empty pipes: corrosion attack localizes at the three-phase, metal-electrolyte-air contact, i.e., at the liquid line, where oxygen is easily available and corrosion products are less protective due to the non-homogeneous exposure condition. Corrosion attack is more intensive when liquid height changes, for example in frequently operated tanks because oxygen access is facilitated.

7.2.4   Salt Formation/Precipitation

The corrosion process leads to a production of metal ions at the anode and hydroxyls at the cathode; accordingly, low solubility salts may precipitate on the anode and the cathode if specific ions are present in the aggressive environment. The following compounds may form:

- Soluble corrosion products, such as *chlorides* or *alkali sulphates,* which contribute to increase corrosion rate through the lowering of electrolyte resistivity and because there is no protection effect by the corrosion products being soluble
- Insoluble salts on the **cathode**, such as *zinc, calcium and magnesium* hydroxides or basic salts, which contribute towards decreasing the corrosion rate by a barrier effect
- Insoluble salts on the **anode**, such as *phosphates and carbonates*, which contribute towards passivating the metal, and then contributing towards decreasing the corrosion rate by passivation.

7.2.5 Cation Displacement

Cations present in the electrolyte can give rise to a corrosion attack because of:

- Self-displacement if metal cation concentration is not uniform
- Displacement, if noble metal cations are present.

Self-displacement occurs when metal ion concentration is not uniform (for example, due to the presence of cracks, cavities or recesses, in which the solution has concentrated). Metal potential is more positive where the solution is more concentrated, therefore an electromotive force, *EMF*, sets up, given by:

$$EMF = \Delta E = E_{C_1} - E_{C_2} = \frac{RT}{zF} \ln \frac{C_1}{C_2} \qquad (7.8)$$

where E_{C_1} and E_{C_2} are metal potential in the concentrated solution, C_1, and in the diluted one, C_2, respectively, and z is cation valence. Accordingly, a current flows from the diluted less noble (i.e., anodic) zone to the concentrated more noble (i.e., cathodic) zone. As illustrated in Fig. 7.11a, self-displacement shows the tendency to fill a cavity rather than to deepen it. For comparison, Fig. 7.11b shows the current path of a corrosion attack due to differential aeration: in this case, the cavity is anodic.

A curious case of self-displacement occurs in concentrated solutions of heavy metals, lead or tin, as they tend to stratify by gravity. For example, by dipping a wire of lead in a stagnant concentrated solution of lead sulphamate or perchlorate, where previously lead ions stratified (i.e., higher concentration at the bottom) an attack occurs on the upper side where the solution is diluted, and metal deposition takes place at the bottom where the solution is more concentrated. This phenomenon, known for centuries, gives rise to attractive dendritic-type deposits called by alchemists *Saturn's tree* for lead and *Diana's tree* in the case of tin. Given the small available *EMF*, this phenomenon only occurs with normal metals, such as lead and tin.

Displacement occurs when a metal, M, is immersed in a solution containing cations of a more noble metal, N: according to thermodynamics, noble metal deposits (cathodic reaction $N^{z+} + ze^- = N$) and those less noble dissolve (anodic reaction $M = M^{z+} + ze^-$). This displacement reaction occurs easily for normal metals; for example, cadmium displaces tin; cadmium and tin displace lead; cadmium, tin and lead displace with mercury; aluminium can displace copper.

Instead, at least at room temperature, the displacement process of inert or intermediate metals does not occur, even when a significant driving voltage is available. For example, let's consider the series zinc ($E_{Zn}^0 = -0.76$ V SHE), nickel ($E_{Ni}^0 = -0.25$ V SHE) and copper ($E_{Cu}^0 = +0.34$ V SHE). At standard conditions, zinc should displace nickel and nickel should displace copper; instead, because of overvoltage on nickel, displacement does not occur. In practice, nickel's behaviour is more noble with copper, thereby not displacing copper, and less noble than zinc,

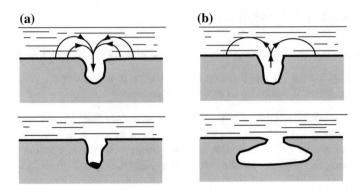

Fig. 7.11 Morphology of the attack caused by: **a** cation concentration cells; **b** differential aeration

thereby not being displaced by zinc. In short, in cation displacement, what matters is practical nobility given by the expression:

$$E = E^0 + \frac{RT}{z\mathrm{F}} \ln \frac{a_{\mathrm{M}^{z+}}}{a_{\mathrm{M}}} \pm \eta_{\mathrm{I}=0} \qquad (7.9)$$

$\eta_{\mathrm{I}=0}$ is anodic (+) or cathodic (−) overvoltage at zero current, having the meaning of a starting friction loss, which is negligible for normal metals while it can exceed 100 mV for inert ones; on the other hand, at elevated temperatures, displacement reaction becomes possible since friction loss decreases.

In chemical plants, concerns about displacement reactions arise when ions of mercury, silver or, more often, copper are present. Mercury forms from salts used as catalysts as a result of cathodic reduction supporting the anodic process of iron dissolution. Copper ions can form as copper or its alloys corrode in some parts of the plant. Copper deposits on a less noble material, in particular, of aluminium components where it triggers localized attacks by galvanic corrosion. In some circumstances, copper can deposit by the effect of CP.

7.2.6 Microorganisms

Corrosion caused by microorganisms, called MIC, *Microbiologically Induced Corrosion*, is often encountered in several industrial plants: production, transport and storage of hydrocarbons; fire-fighting systems; water cooling circuits; sewage treatment plants; marine and buried structures, beneath fouling or in clayey and swampy soils.

The first step of this corrosion process is the formation of so-called *biofilm*, which consists of microorganism colonies stuck on metal surfaces by self-produced gel, which locally modifies pH and oxygen availability to enhance conditions for bacteria to thrive. Biofilms are a few tens of microns thick, generally characterized by two layers, where the inner layer, adherent to metal, is almost oxygen-free. Bacteria can be divided into two families:

- Aerobic such as *Cladosporium resinae, Thiobacillus thiooxydans, Thiobacillus ferroxidans, Gallionella, Sidercapsa, Spheaerotilus,* which lower pH thereby promoting acid-related corrosion attacks
- Anaerobic as *Desulfovibrio desulfuricans* also called sulphate-reducing bacteria, SRB.

Among aerobic bacteria, it is worth mentioning *Cladosporium resinae* that causes corrosion of aluminium fuel tanks in the presence of condensate water by lowering pH then causing aluminium passivity breakdown. *Thiobacillus thiooxydans* and *Thiobacillus ferroxidans* are oxidizing bacteria, which oxidize sulphur, sulphides and other sulphur containing compounds to give sulphuric acid at concentrations as high as 3%.

The most common anaerobic MIC takes place on carbon steels and low alloy steels due to sulphate-reducing bacteria, SRB (*Desulfovibrio desulfuricans,* Fig. 7.12), which catalyse the reduction of sulphates to sulphides in a local oxygen-free condition and a sulphate content exceeding 100 ppm. This model was firstly proposed by Von Wolzogen and Van der Vlugt (1934).

> **Frequent Case Study of MIC**
> It is often reported that MIC is the root cause of localized attacks on stainless steels, typically in hydro-testing of piping and equipment. Caution should be taken before attributing the cause of corrosion to MIC because the morphology is very similar to the one in the absence of bacteria. Generally, there are two possible scenarios: piping remained full after hydro-testing or was drained.
>
> In the former, MIC would appear to be the only possible cause, since corrosion was unexpected based on operating conditions (namely, low chloride content in the water used for testing). In the latter, instead, if

Fig. 7.12 MIC mechanism on carbon steel in the presence of SRB (in anaerobic waters and soils)

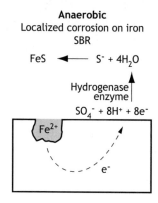

drainage was not accurate, most likely the cause was pitting due to a local increase of chloride concentration in stagnant water in the plant and high oxygen availability from the entrapped air.

A Unified Model for MIC

MIC has been recognized as responsible for localized corrosion on either active or active–passive metals. In both cases, a necessary condition is required: the occurrence of a cathodic process.

For localized corrosion of active metals, typically carbon and low alloy steels in anaerobic environments, the cathodic process is the reduction of sulphate ion to sulphide. This cathodic reaction, although as noble as oxygen reduction in neutral solutions, cannot take place spontaneously if not catalysed because of its slow kinetic. The most known catalyser is given by SRB metabolism, as proposed by Von Wolzogen and Van der Vlugt in 1934. In short, an enzyme, idrogenase, allows the reduction of sulphate to sulphide: $SO_4^{2-} + 8H^+ + 8e^- \rightarrow S^{2-} + 4H_2O$. Indeed, this reaction also occurs in nature in sulphate enriched anaerobic environments, where the oxidation reaction is carbon to carbon dioxide; if SRB find metallic iron available, the thermodynamically preferred reaction becomes iron oxidation.

For localized corrosion of stainless steels, i.e., the active–passive metals suffering a MIC attack, two more conditions are necessary besides the requirement of a cathodic process, i.e., a chloride content sufficient to locally breakdown the passive film and a cathodic potential exceeding pitting potential. As far as the cathodic process is concerned, oxygen reduction is the one recognized in aerated waters where, in seawater, biofilm causes cathodic potential ennoblement. In aerated freshwater, where biofilm does not form, the potential ennoblement, mandatory for pitting initiation, is exerted by manganese oxidizing bacteria, MOB, which colonize Mn^{3+}/Mn^{2+} by their metabolism. The standard redox potential of Mn^{3+}/Mn^{2+} is +1.51 V SHE, i.e., more noble than oxygen standard potential, is a value easily exceeding the pitting potential of most used stainless steels in the presence of a very low level of chlorides (for instance a few tens of ppm).

From Lazzari (2017)

Stainless steels (Figs. 7.13 and 7.14) may suffer pitting corrosion in chloride containing waters when contaminated by bacteria in an aerobic condition, where oxygen is necessary for the pitting propagation. In seawater, aerobic bacteria that colonize the surface of stainless steels form a *biofilm* that raises the potential by about 200–300 mV compared to that in sterilized seawater, then allowing pitting initiation ($E > E_{pitting}$). In freshwater, what increases the potential is the presence of aerobic *Leptospirillum oxidans* bacteria, called MOB (manganese oxidizing

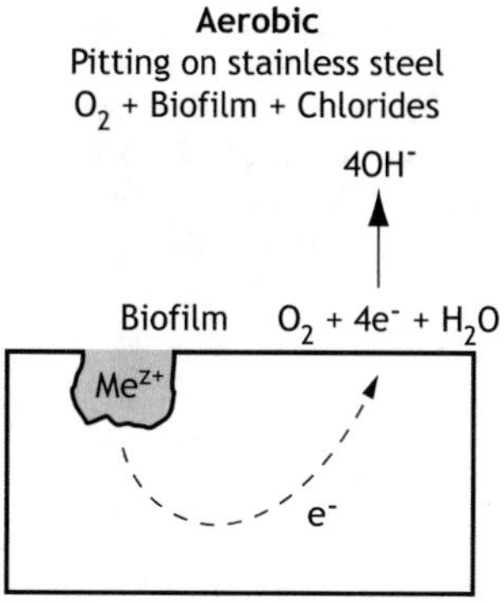

Fig. 7.13 MIC mechanism on stainless steels in the presence of biofilm (typically in seawater)

Aerobic
Pitting on stainless steel
O_2 + Biofilm + Chlorides

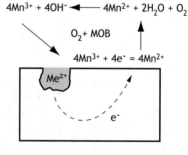

Fig. 7.14 MIC mechanism on stainless steels in the presence of MOB (typically in freshwater)

bacteria) which produce the manganese couple, Mn^{3+}/Mn^{2+}, having a noble standard potential +1.51 V SHE sufficient to trigger pitting initiation. A manganese concentration as low as 2 µg/L and a chloride content of only 40 mg/L is sufficient for that bacterium to initiate pitting of AISI 304 stainless steel, as has occurred in many European rivers (Rhine, Seine and others).

7.3 Metal/Environment Affecting Factors

7.3.1 Temperature

Although a corrosion process is the result of a series of elementary processes, namely, electrochemical (electrode reactions), chemical (homogeneous reactions) and physical (solubility and diffusion), the corrosion rate depends on temperature through a complex law, neither exponential, as typical for chemical reactions (reaction rate doubles every 10 °C increase in temperature), nor linear as in some physical processes. Indeed, when the prevailing elementary process is gas solubility in the solution, the corrosion rate may even decrease as the temperature increases.

Fig. 7.15 Corrosion rate of
mild steel with temperature
in: ① water in equilibrium
with atmosphere; ② in closed
circuit with constant oxygen
content (Speller 1926)

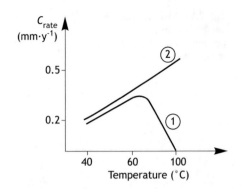

We have already seen that as the temperature increases the diffusion coefficient increases while the oxygen solubility in water decreases, then zeroing at boiling temperature. Accordingly, in the case of oxygen-related corrosion, the corrosion rate follows the trends shown in Fig. 7.15: (curve ①) for open circuits (water exposed to the atmosphere) and (curve ②) for closed circuits where oxygen remains entrapped in the plant and therefore, available even at high temperature. In closed circuits, the corrosion rate doubles about every 25 °C increase.

In acidic solutions, where the cathodic process is hydrogen evolution, the corrosion rate increases as the temperature increases by following an exponential trend. For active–passive metals, dissolved oxygen helps passivity therefore, an increase in temperature reducing the dissolved oxygen contributes to weaken the passivity film. For example, this is the case of stainless steel, titanium or other active–passive metals in non-oxidizing acids, such as sulphuric or hydrochloric acid.

Non-uniform temperature conditions favour the localization of corrosion on higher temperature zones which become anodic, as typically observed in boilers and heat exchangers.

7.3.2   Condensation

Local variations of physical-chemical conditions are important from a corrosion viewpoint. Most important is water condensation caused by temperature decrease or pressure increase which leads to severe corrosion, as in chimneys where carbonic acid and also sulphuric acid form, when sulphur is present in fuel. Another common example is water condensation in gas wells containing carbon dioxide where the corrosion rate can be very high. Critical parameter is dew point temperature/pressure, which determines the water condensation condition. In the case of sulphuric acid formation, dew point temperature at atmospheric pressure can be much higher than boiling water temperature, for instance around 140 °C because of the formation of concentrated sulphuric acid.

7.3.3 Corrosion Products and Deposits

The presence of deposits (scales, corrosion products, debris, dirties) may be either beneficial or harmful. The protective action is linked to the ability to form a barrier which separates the metal surface from the environment. The protection properties of such a barrier depend on a variety of characteristics: solubility, state (uniform, non-uniform, crystalline or colloidal), porosity, hygroscopic nature (in the case of atmospheric corrosion) and electrical conductivity. The latter is the most important since:

- Electronic conductivity, as shown by magnetite and sulphides, can cause galvanic-like corrosion
- Ionic conductivity, as in the case of cuprous oxide of copper, almost neutralises the barrier effect
- Insulating properties are always beneficial: in this case the metal surface can scarcely support anodic or cathodic processes. For example, this is the case of Al alloys whose corrosion resistance enhances when thick oxide films form.

On the other hand, a scale can enhance crevice corrosion or differential aeration corrosion, or corrosion-erosion when a local turbulence is set up downstream or upstream of the deposit, or local overheating as is typical in heat exchangers.

7.3.4 Flow Regime

Flow regime affects corrosion in different ways. In oxygen diffusion control, as the flow rate increases, for instance from v_1 to v_4 as shown in Fig. 7.16, the corrosion rate of an active metal increases up to a limit represented by point ④. Conversely, as shown in Fig. 7.17, for an active–passive material, an increase in the flow rate can facilitate passive film formation. Indeed, the corrosion rate increases from point ① to ② and up to ③, where the corrosion rate reaches a maximum, then drops to point ④ when the flow rate exceeds v_4 and metal passivates. An increase of flow rate is also beneficial when it contributes to avoid stagnant conditions that are hazardous for pitting. More generally, the high flow rate regime is dangerous when leading to corrosion-erosion conditions which are enhanced in the presence of suspended hard solids (mechanical wear effect) (see Chapter 16).

7.3.5 Active–Passive Related Parameters

Whether an active–passive metal, for example stainless steel, operates in a passive or active condition depends on both the metal and the environment and it can be

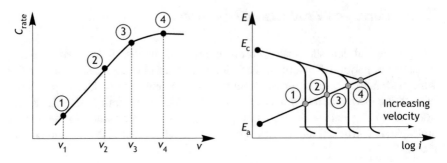

Fig. 7.16 Effect of electrolyte velocity on the corrosion rate of an active metal

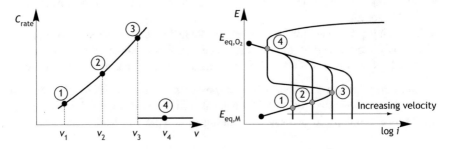

Fig. 7.17 Effect of electrolyte velocity on the corrosion rate of a passivating metal

determined by examining anodic curves on Evans diagrams. As reported previously in par. 5.6.3 and 5.6.4, parameters defining active–passive corrosion behaviour are:

- *Passive interval*, between passivity and transpassivity potentials, E_p and E_{tr}, respectively
- *Passivity current density*, i_p, which measures the corrosion rate within passive interval
- *Critical passivation current density*, i_{cp}, which defines the ease to passivate.

Preferred corrosion behaviour is when the passivity current density, i_p, is low, passive interval, E_{tr}–E_p, wide and critical passivation current density, i_{cp}, is low. The importance of these parameters is illustrated by the two examples depicted in Figs. 7.18 and 7.19.

Figure 7.18 schematically shows the anodic curves of some iron-chromium alloys in a dilute acidic solution exposed to air; it is also reported the cathodic curve of the oxygen reduction process in a stagnant condition. By increasing the chromium content from 10 to 12%, the corrosion condition shifts from point ① to point ② with a corrosion rate reduction of about three orders of magnitude. It is clear that passivity is achieved if the chromium content exceeds the threshold of about 12%, which is historically the condition to assess stainless steel.

Fig. 7.18 Evans diagram for
Fe–Cr alloys in dilute aerated
sulphuric acid solution

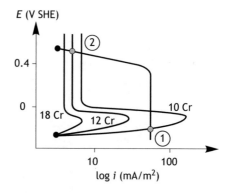

Fig. 7.19 Evans diagram of a
ferritic and an austenitic
stainless steel in aerated 5%
sulphuric acid solution

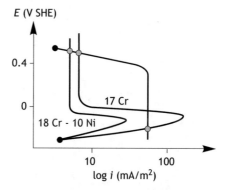

Let's evaluate the corrosion resistance of two stainless steels, one ferritic (17% chromium) and the other austenitic (18% chromium, 10% nickel), in an aerated 5% sulfuric acid solution with or without the addition of oxidants. As depicted in Fig. 7.19, both stainless steels show a comparable passivity current density; however, ferritic stainless steel shows a much higher critical passivation current density (a big nose). Therefore, to set up passivity, a robust cathodic process is required that enables the surpassing of the nose at the primary passivation potential. In practice, in dilute sulphuric acid, oxygen alone makes austenitic stainless steel passive, while this is not the case for a ferritic one; in addition to oxygen, a stronger oxidizer is necessary, for example, Fe^{3+} or NO_3^-, even in low concentration.

Wollaston's and Pallade's Jokes
One of the first tasks of electrochemists was the ranking of metals by nobility, far before it was linked to the corrosion behaviour. Volta in 1793 proposed a first rank and a few years later Ritter discovered that alloying even with small content changed the nobility of a metal. He wrote: *"By dissolving a thin leaf of tin in mercury, its nobility moves from high position between gold and silver to low position where reactive metals fall"*. He then concluded: *"It*

seems difficult to establish the content limit below which tin addition doesn't change the position of mercury in the potential series".

Ostwald reported that Ritter was pushed to study the influence on nobility by an anonymous advertisement in an 1803 London newspaper, which announced the discovery of a new metal, not yet named, sold by a well-known metal shop at the prize of one shilling per grain.

Chevenix, a famous chemist of the period most likely for his emphatic writing rather than his research, confirmed the properties of the new metal. However, he added, that it was not new, instead a Pt–Hg alloy, where mercury was so strongly linked that it could not be freed even by extreme heating. Moreover, he said that it was easily produced from a platinum solution, neutralized by mercury oxide, reduced by ferrous sulphate to a black powder and eventually melted, and voila, the new metal. As the anonymous discoverer was told that, he published a second advertisement promising 20 lb to whoever was able to obtain the new metal from platinum and mercury. Obviously, nobody claimed the prize—not even Chevenix.

Now, Wollaston, a Davy's assistant, revealed that he was the discoverer and the person responsible for the advertisement, and how during the experiments to produce malleable platinum, he obtained the new metal along with another metal, which he had already named rhodium. Chevenix had to admit his mistake, but not before trying to give vent to righteous indignation on Wollaston's behaviour not befitting a scientist. Ritter broke into this controversy. He found that the metal provided by Chevenix stood over platinum in the nobility scale and not between platinum and mercury as one might have expected from their alloy. He also found that it behaved exactly like the new metal on sale in the mineral shop. In short, he found that the material supplied by the arrogant chemist for examination did not come from its mergers but from the shop where he had bought it.

Like the asteroid discovered just a year before, the new metal was named Palladium, in honor of the Greek goddess Pallas who was produced from Zeus' brain. Besides rhodium, Wollaston also discovered the metal which is able to give titanium the precious feature to withstand both oxidizing and reducing environments, and a few years ago to have been raised to the honour (or perhaps to the dishonour?) of the news because of nuclear cold fusion. And if I were not convinced that some hitherto unknown phenomenon to some extent really happens in this ghost process, and had I not had respect for Fleischmann and Pons, particularly the former who I got to know personally, it would be easier to joke and say that this metal, as well as Chevenix, mocked even the two "inventors" who believed that humanity's energy problem was solved. Unless it was Pallas, the goddess of science, to have done so.

Special Corrosion Products

In 1967, under the direction of Professor Roberto Piontelli I was investigating the corrosion behaviour of copper in copper sulfamate solutions because measured corrosion rates were much higher than expected. While we were looking for possible root causes of this anomaly, we obtained white cuprous corrosion products, as needle-like crystals, as soon as sodium or potassium ions were added. Surprisingly, they did not alter in air: a real rarity, because cuprous salts, in general, are not stable.

X-ray diffraction analysis and chemical composition revealed two new compounds: copper–sodium and copper–potassium sulfamate, respectively. We found how to produce them in large quantities at low cost by stimulating copper corrosion by injecting a direct or alternating current, then we decided to file a series of patent applications, including their use, in place of copper sulphate in fighting the blight of vineyards. "The fungicide activity of cuprous ions is certainly greater than cupric ones," we said, and "by using potassium salt, the new product, once the primary fungicide action is finished, could turn into a fertilizer." In short, a brilliant idea, way more, a sure hit, and definitely also a case of beneficial corrosion!

Because in sulfamic acid, unlike sulphuric acid, it was possible to soak hands safely, no doubts arose about any possible dangerous action to vegetables. In any event, we had sent samples for testing to the Institute of Pathophysiology of the University of Pavia: in any case, an official certification would have been required to start commercialization and collect royalties. Meanwhile, to shorten experimentation time, we decided to hold our own testing. I prepared some salts and we planned homework: Professor would take them and convince his gardener to sprinkle them on hydrangeas in his villa in Santa Margherita Ligure, Italy; I, more modestly, had to give them to my uncle, Antonio, and have them tried on one row of vineyards and on a potato field in Mese, Valchiavenna, Italy.

We succeeded, but a week later a disaster happened.

The new product destroyed Professor's hydrangeas and my uncle's vineyards and potato field in a few days. Indeed it also burned the zucchini in the garden because, either to please his nephew or for the pride of collaborating in a prestigious research study of the glorious Politecnico, my uncle, careless, decided to widen the testing—that can really be called "in field". After one month, the now useless response of the University of Pavia was received on a letterhead and with stamp duty: the new salts were officially declared lethal even towards more resistant shrubs. In short, new stuff that was able to compete with defoliants which, in those days, were used in Vietnam.

The gardener and my uncle were not surprised by the later verdict from the University of Pavia because they immediately understood that the fungicide was actually a herbicide. Unfortunately, the idea that the advancement of science may also require personal sacrifice did not even touch them. On the contrary, from the day of the foul incident, they were bitter and did not hide it. The first repeatedly informed the professor of what he thought. Every time he met him he added a sarcastic: "Here is the Professor" to a sort of greeting. Dearest uncle never missed the chance to ask his nephew sarcastically if the research at the Politecnico was always so interesting and useful. And when one day the naive nephew told him that potato cultivation in Valtellina and Valchiavenna was probably introduced by Alessandro Volta, as had occurred near Como, he replied: "It could be; it means that at that time electrochemists helped to raise potatoes. Today, grass does not grow where they walk."

So, the Professor and I for some months, every Monday, on arrival at the Institute after the weekend, confessed to each other with complicity what he in Santa Margherita, and I in Valtellina, had to suffer over the weekend. And there were jokes about it. In the end, it could have been worse. If instead of a fungicide we had discovered, say, an anti-flu, we could have been tempted to test it on them. Indeed, seeing as they were feeling so bad, perhaps we had found it! And we laughed. A little tight-lipped because of the truth.

Pietro Pedeferri's memory

7.4 Questions and Exercises

7.1 Discuss the effect of the addition of platinum on titanium in acidic, oxygen free solution by means of Evans diagram.

7.2 Consider the values of corrosion rate of commercial aluminium in hydrochloric acid reported in Table 7.1. Propose an interpretation.

7.3 A controlling corrosion factor is the presence of corrosion products. What properties are typical for a passivating oxide film? Mention examples of various types of surface films/surface layers.

7.4 Demonstrate and comment this sentence: "Given the small available driving force, this phenomenon (*Saturn's tree* or *Diana's tree*) occurs with normal metals, such as lead and tin, only".

7.5 Discuss the mechanism of a cation displacement reaction.

7.6 Discuss the effect of electrolyte velocity on an active metal and on a passivating metal.

7.7 Inside a water pump, maximum corrosion rate was found when the flow rate was increased to about 0.1 m^3/s. At higher flow rates, the corrosion rate decreased. What is your interpretation?

7.8 Based on experience, the corrosion rate on carbon steel caused by sulphate-reducing bacteria, SRB, (for instance in some soils or in pipes containing stagnant water) can reach 1 mm/y. Try to justify this value based on an electrochemical mechanism.

7.9   What is the effect of manganese oxidizing bacteria on stainless steels corrosion in freshwater?

7.10 Corrosion by differential aeration happens on iron (i.e., mild steel) and not on copper and copper alloys. Give a comprehensive interpretation/justification.

Bruno Mazza

Bruno Mazza (1936–2004) was undoubtedly the person who most influenced the corrosion group that has formed in the mid-60s at Politecnico di Milano, with Sinigaglia and Pedeferri. It was not because of his works in corrosion, still important but not as much as those in electrochemistry, or because he taught for a couple of years a corrosion course, so his lesson imprinting is echoed in the first part of this handbook: Bruno was for thirty years the moral guide of the group. He graduated in 1961 (gold medal as the best Italian graduate of the year). In 1965 he was appointed as Lecturer of Electrochemistry and in 1968 the Board of Faculty of Engineering asked the Ministry of Education to assign a full professor position for him, who was nearly 30. But 1968 student movement came so his idyll with Faculty turned into conflict. What happened with academic authorities can be understood by taking into account that Faculty rejected any comparison with students and, in opposite direction, Bruno was fascinated of their demand for change and hope for a world of justice and solidarity. They decided to cut off his career by withdrawing the position. He could maintain the teaching because Parliament approved the conservation of the status quo. After nearly two decades he had the satisfaction of having the chair previously denied, he became director of the department and was called to fill some of the highest offices of the University. All the group remembers of Mazza the scientist, the serious and charm teacher, the commitment, the honesty, the courage, the consistency, the availability, the gentleness, the respect of persons. And also the masterful lesson, handouts on which many learned electrochemistry, the patience for explaining and making things clear, so hundreds of engineers remember his teaching.

Roberto Piontelli

Professor Piontelli (1909–71) was an eminent electrochemist of international prominence who contributed the foundations of modern electrochemistry. In 1949, together with major European and American scientists of this sector, he founded the CITCE, later named ISE, the International Electrochemical Society. M. Pourbaix, who was the first chairman, recognized in his paper "The birth of CITCE, in Electrochimica Acta, XVI (1971), pp. 173–175) the key role of Piontelli. Pourbaix asked Piontelli to write the preface of his famous book Atlas of electrochemical equilibria in aqueous solutions, published in 1966. Since he graduated in 1930s, he greatly contributed to simplify the electrochemistry that was, as often he said, entangled by conventions of sign instead interested of the interpretation and understanding of complex phenomena. In 1948, as director at Politecnico di Milano (Milan, Italy), he set the discipline on a chemical-physical basis and improved his experimental methods. During 1950s, with reference to Linus Pauling approach on electronegativity of metals, he proposed a theory on the correlation between structure and electrochemical behaviour of metals, now known as anti-correlation. On the experimental side, he determined the anodic and cathodic behaviour of polycrystalline and monocrystalline metals, investigating the associated corrosion and protection problems. On the theoretical side, he gave an important contribution to thermodynamics and kinetics of the pile and wet corrosion processes. It is interesting to read what Piontelli wrote on the research methodology in the field of electrochemistry and corrosion, mirroring the Baconian philosophy. *"Only the activity of the bee is suitable for electrochemistry,"* paraphrasing the English philosopher, hence there is no need for *"ant-type researchers"* who *"collect"* and *"use"* the measurement data without bothering to understand what happens in their systems, neither "spider-types", who build their theories as their own nets regardless the nearby reality. The *"bees suck nectar from flowers of gardens and fields"* and then rework it to transform it into honey. *"In a transversal discipline such as electrochemistry and corrosion, only on a solid phenomenological platform both the tower of the most daring theoretical speculation and the more modest yet robust and efficient building of rational technology can be erected"*. He proceeded by building the mosaic of the electrochemical behaviour of metals composed of generalization of observation results, calculations, studies, meditations, recognition of essential factors and their interrelationships in a final rational frame. His motto was *"it is preferable to make a small contribution to a great problem that the most complete success in responding to an occasional question"*. Then he entered into specific problems of the industry and electrochemical applications. Convinced that—in sectors such as these *"born on empiricism and often still his devoted subjects"*—the solution of fundamental problems must rest *"on chemical-physical and theoretical premises"*, he worked to transform dominant technology based on experience and common sense in *"rational technology"*. Finally, closing the cycle, suggestions or cues are taken from the industrial reality to reset the teaching and research work. The volume *"Elements of the theory of wet corrosion of metallic materials"* (Edition Longanesi, Milan, Italy, 1961) was the first major Italian volume on corrosion, often considered as beautiful as difficult to understand. In addition to his book on corrosion, he published Lezioni di termodinamica chimica, Milano, 1961 as well as more than three hundred papers, collected in R. Piontelli, Scientific papers: 1935–1971, Milano 1974. In Piontelli, the scientist and the researcher of electrochemical science coexist, enriched by the teaching in an engineering faculty. This gave him the best position to verify the gap between the ants operating in the industrial world without knowing what happens in their cells and spiders nested in the academies often ignoring what was going on in the real world. This condition allowed him to transfer knowledge or methodologies from the laboratory to the industry and from field experiences to the research. In 1958, some colleagues of the CITCE, through their respective national academies, proposed him for the nomination of Nobel Prize for Chemistry.

Bibliography

Lazzari L (2017) Engineering tools for corrosion. Design and diagnosis. European Federation of Corrosion (EFC) Series, vol 68. Woodhead Publishing, London, UK

Piontelli R (1961) Elementi di teoria della corrosione a umido dei materiali metallici. Longanesi, Milan, Italy (in Italian)

Speller FN (1926) Corrosion. Causes and prevention. McGraw-Hill, London, UK

Von Wolzogen Kuhr CAV, Van der Vlugt SS (1934) Graphitization of cast iron as an electrochemical process in anaerobic soil. Water (Den Haag) 18:147–165

Winston Revie R (2000) Uhlig's corrosion handbook, 2nd edn. Wiley, London, UL

Chapter 8
Uniform Corrosion in Acidic and Aerated Solutions

Memories bring diamonds and rust.

Joan Baez

Abstract In this Chapter, the causes and consequences of uniform (or generalized) corrosion are described. This is the simplest form of corrosion, which affects the whole metal surface, and is characterized by the spatial coincidence of anodic and cathodic areas. Corrosion rates range in a very large interval, depending on the environmental conditions, and the phenomenon is easily observed and easily predictable, especially if compared with localized corrosion forms. Here the main conditions leading to

Fig. 8.1 Case study at the PoliLaPP Corrosion Museum of Politecnico di Milano

© Springer Nature Switzerland AG 2018
P. Pedeferri, *Corrosion Science and Engineering*, Engineering Materials,
https://doi.org/10.1007/978-3-319-97625-9_8

uniform corrosion are detailed, from acidic environments to aerated neutral solutions, with reference to the different classes of active and active–passive metals. Moreover, examples of algorithms used to express corrosion rate are provided.

8.1 Introduction

Uniform corrosion, also called generalized corrosion, affects the entire exposed surface of active metals in contact with an aggressive environment because anodic and cathodic zones coincide. The main cases of metals suffering generalized corrosion are:

- Carbon steel when exposed to the atmosphere, or immersed in neutral or acidic solutions, in carbonated concrete, in soil and sea water
- Aluminium in low and high pH solutions
- Stainless steel in acidic solutions (Fig. 8.1)
- Carbon steel in CO_2-containing environments (sweet corrosion)
- Zinc in acidic solutions
- Lead throughout the entire pH range when insoluble corrosion products, such as carbonates or sulphates, do not form.

Despite the name, this corrosion form is often not uniform either at the microscopic or at the macroscopic scale. For example, internal corrosion in carbon steel pipelines carrying hydrocarbon containing CO_2, called *mesa corrosion* because it is reminiscent of the mesa landscape in the region across Texas and Mexico; or the case of antiquities on which generalized corrosion highlights the inhomogeneous structure. Corrosion rate varies widely from very low values, some μm/y, to tens of mm/y depending on the metal and the aggressive environment. Figures 8.2, 8.3 and 8.4 show some examples of uniform corrosion.

Fig. 8.2 Generalized corrosion occurred in 24 h on a carbon steel tube erroneously etched with a non-inhibited 10% HCl solution

Fig. 8.3 Uniform corrosion of a carbon steel nail, which remained for 150 years in a wooden beam. The curved zone outside the beam suffered a more pronounced thinning

Fig. 8.4 The effect of expansion of corrosion products by generalized corrosion

In some cases uniform corrosion is beneficial, for example in:

- Surface roughening of orthopaedic implants to facilitate bone integration (Fig. 8.5)
- Metallographic attack by etching to highlight a microstructure (Figs. 8.6, 8.7 and 8.8)

Fig. 8.5 Roughness on
titanium orthopaedic implant
obtained by generalized
corrosion

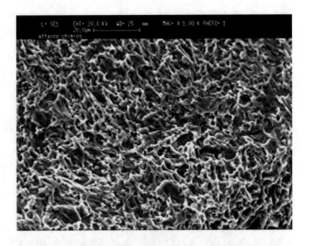

Fig. 8.6 Picture of a curious
dendritic structure of a silver
sample obtained by etching

- Surface polishing
- Etching for metal carving to produce moulds, prints and matrices for artistic use
- Pickling for metal surface cleaning (for removal of oxides and some of first
  metal layer).

Some Consequences of Uniform Corrosion

Although generalized corrosion causes the greatest amount of corrosion
products, it is in general less insidious than localized corrosion because
corrosion rate (i.e., thickness loss rate) is often low and predictable, with good
accuracy, and is easily monitored during operating; nevertheless, attention
should be paid to its consequences.

For example, corrosion products exert an expansive action because their
volume is much greater than that of the corroded metal. This is shown in
Fig. 8.4, where corrosion products between two steel profiles led to the

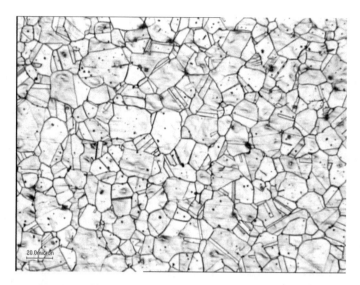

Fig. 8.7 Micrograph of an AISI 304 stainless steel after metallographic etching

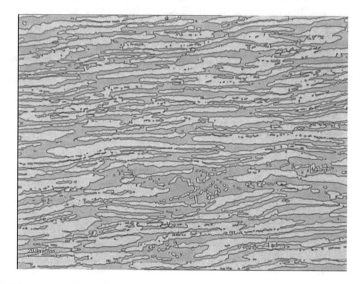

Fig. 8.8 Micrograph of a duplex stainless steel after metallographic etching

failure of welds, and the same happens to carbon steel inserts in ceramic or stone, thus causing cracking. Some historians attributed the fall of the Roman Empire to the consequences of lead corrosion.

8.2   Acidic Solutions

Tafel law allows the calculation of a metal corrosion rate with good approximation, for instance in the case of ferrous alloys (carbon and low alloy steels) and zinc in acidic solutions, where the predominant cathodic reaction is hydrogen evolution. Typical acidic solutions present in industry are: strong acids, carbonic acid, hydrogen sulphide and organic acids.

The model proposed here, already adopted in Lazzari (2017), can be named *"Tafel-Piontelli model"* (see De Giovanni 2017) for active metals such as iron and zinc in acidic solutions and where the cathodic process is hydrogen evolution; it simply consists of considering the Evans diagram on the basis of two conditions:

- Anodic overvoltage of active metal dissolution is negligible (according to the Piontelli's classification)
- Cathodic process of hydrogen evolution follows Tafel law.

Figure 8.9 shows an example of the Evans diagram for two metals in an acidic solution with the same pH (hence, the same equilibrium potential of hydrogen evolution reaction) and assuming the Tafel slope to be near zero ($b \cong 0$) for metal dissolution reaction. This means that the free corrosion potential of the metal is taken as its equilibrium potential. This approximation is acceptable when the cathodic reaction is hydrogen reduction only (i.e., oxygen-free/chlorine-free acidic solutions).

The equilibrium potential at room temperature of the metal is given by Nernst equation:

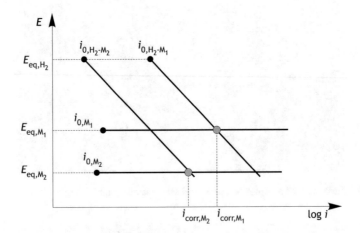

Fig. 8.9 Evans diagram to determine corrosion rate in acidic solutions based on Tafel-Piontelli model

$$E_{\text{eq,M}} = E^0_{\text{M}^{z+}} + \frac{0.059}{z} \log a_{\text{M}^{z+}} \tag{8.1}$$

where M could be Fe or Zn or other active metals, for instance Cu in Cu-complexant solutions. The cathodic curve, i.e., the Tafel straight line for hydrogen evolution is given by Tafel law:

$$\eta = b_{\text{H}_2} \log \frac{i}{i_{0,\text{H}_2}} \tag{8.2}$$

which can be rewritten as follows:

$$i = i_{0,\text{H}_2} \cdot 10^{\frac{\eta}{b_{\text{H}_2}}} \tag{8.3}$$

where b_{H_2} is the Tafel slope for hydrogen evolution equal to 0.12 V/decade, i_{0,H_2} is the exchange current density for hydrogen evolution on the metal (see Table 5.2), η is given by $E_{\text{eq,H}_2} - E_{\text{corr}}$, where E_{corr} is taken as $E_{\text{eq,M}}$, hence:

$$\eta = \Delta E = -0.059 \, \text{pH} - \left(E^0 + \frac{0.059}{z} \log[\text{M}^{z+}]\right) \tag{8.4}$$

An empirical equation for iron can be derived taking into account the influence of temperature and velocity of the fluid as follows (Lazzari 2017; Kreysa and Schütze 2006a, b):

$$C_{\text{rate,acid}} = 1.2 \cdot i_{0,\text{H}_2} \cdot (1+v) \cdot 2^{\frac{T-25}{20}} \cdot 10^{\frac{-0.059 \, \text{pH} - \left(E^0 + \frac{0.059}{z} \log[\text{M}^{z+}]\right)}{0.12}} \tag{8.5}$$

where the required parameters are:

- i_{0,H_2} (exchange current density of hydrogen evolution on the considered metal, in mA/m^2)
- v (fluid velocity in m/s)
- pH of the acid solution
- a_{M}^{z+} (metal ion concentration in the diffusion layer, i.e., at the interface)
- T (temperature in °C).

To use Eq. (8.5) in practical applications, there is the need to input metal ion concentration, a_{M}^{z+}, which is in practice the only unknown or uncertain variable. For different types of acidic solutions, the suggested a_{M}^{z+} values are reported in Table 8.1. The model is the same for organic and complexant acids; metal ion concentration is derived from the complex constant. Since metal ion concentration in equilibrium with the complexant is quite low, the corrosion rate is high although the pH is close to neutrality.

Table 8.1 Parameters for the calculation of corrosion rate in different acids (from Lazzari 2017)

Active metal or alloy	Exchange current density of hydrogen evolution on metal M, $i_{0,H_2-M}(mA/m^2)$	Metal ion M^{2+} concentration $[a_M^{2+}]$ mol/L			
		Strong acids	Organic acids	Hydrogen sulphide (H$_2$S)	Carbonic acid (H$_2$CO$_3$)
Zn	10^{-3}	$0.1-1^a$	$\approx 10^{-10}$	$\approx 10^{-12}$	10^{-6}
Cu	10				
Fe (ferrous alloys)	1				$10^{-6}-10^{-9}$

[a]In the presence of HCl, copper-chloride complexes form, after which the concentration of copper ions drops to 10^{-15} mol/L

8.2.1 Strong Acids

In the presence of strong concentrated acid solutions, such as nitric acid, sulphuric acid, or hydrochloric acid (pH lower than 1), very high corrosion rates are expected on Zn and Fe.

Table 8.2 shows calculated values taking the metal ion concentration to be in the range 0.1–1 mol/L in stagnant conditions for Zn, Fe and ferrous alloys. Copper is corroded only in pure hydrochloric acid (HCl 12 mol/L, pH < 0), since copper-chloride complexes form, reducing the concentration of copper ions to 10^{-15} mol/L. Corrosion rates of some hundreds of micrometres per year up to 1 mm/y are expected.

8.2.2 Carbonic Acid

Carbon steel and low alloy steels corrode in carbonic acid containing media, as is well known in the oil and gas industry. The approach used is typically empirical and equations used to estimate the corrosion rate, starting from the very first by de

Table 8.2 Calculated corrosion rates for iron and zinc in strong acids and copper in HCl

	Temperature °C	Fe	Zn	Cu
E^0 (V SHE)		-0.44	-0.76	0.34
a_M^{2+} (mol/L)		$1-0.1$	$1-0.1$	10^{-15}
M, $i_{0,H_2-M}(mA/m^2)$		1	0.001	10
C_{rate} (mm/y) pH = 0	25	$5-8$	$2-4$	0.07
	50	$9-18$	$7-14$	0.08
	100	$39-80$	$85-174$	0.12

Waard and Milliams, are derived from laboratory testing. The de Waard and Milliams base equation (dWM) is the following:

$$\log C_{rate,dWM} = 5.8 - \frac{1710}{T+273} + 0.67 \cdot \log p_{CO_2} \qquad (8.6)$$

where $C_{rate,dWM}$ is corrosion rate (mm/y), T is temperature (°C) and p_{CO_2} is carbon dioxide partial pressure (bar). Some empirical coefficients are used (see Chap. 24) to either mitigate or increase the calculated corrosion rate when temperature exceeds a so-called scaling temperature and pH is lower than the equilibrium value.

The general Eq. (8.5) can be used by introducing the pH of carbonic acid solution given by the following relationship as a function of CO_2 partial pressure, p_{CO_2}:

$$pH = -\log\left[10^{-4}(p_{CO_2})^{0.5}\right] = 4 - 0.5\log p_{CO_2} \qquad (8.7)$$

where p_{CO_2} is expressed in bar. By introducing in Eq. (8.5) $i_{0,H2} = 10^{-3}$ A/m^2 and a_{Fe}^{2+} close to 10^{-7} mol/L and pH in accordance with Eq. 8.7 and reversing in a logarithm form, the following simplified equation for low velocity flows is obtained:

$$\log C_{rate,CO_2} \cong 0.015(T - 25) + 0.8\log p_{CO_2} \qquad (8.8)$$

where C_{rate} is corrosion rate (mm/y), T is temperature (°C), v is fluid velocity (m/s) and p_{CO_2} is carbon dioxide partial pressure (bar). This equation is valid for temperature up to about 80 °C because iron passivates at higher temperature.

Table 8.3 compares results obtained by the Tafel-Piontelli model and de Waard and Milliams base equation for temperature below 100 °C. The good match confirms the reliability of the model because results of de Waard and Milliams equation have been confirmed experimentally.

Corrosion Mechanism in Carbonic Acid
There is a variety of references on this matter. In short, the most cited and accepted mechanism for the corrosion of carbon steels in carbonic acid is the following:

- Corrosion rate in carbonic acid, although it is a weak acid, is one order of magnitude higher than the one in a strong acid at the same pH (3–6)
- This is attributed to the presence of more than one cathodic reaction besides hydrogen reduction
- Other cathodic reactions would be the direct reduction of either bicarbonate ion or undisassociated carbonic acid:

Table 8.3 Comparison of the calculated corrosion rates of steel in carbonic acid obtained from the model ($a_{Fe}^{2+} = 10^{-6}$ mol/L) and the de Waard and Milliams base equation in stagnant conditions

	$p_{CO_2} = 1$ bar		$p_{CO_2} = 2$ bar		$p_{CO_2} = 5$ bar		$p_{CO_2} = 10$ bar	
T (°C)	Model (mm/y)	deW & M (mm/y)	Model (mm/y)	deW & M (mm/y)	Model (mm/y)	deW & M (mm/y)	Model (mm/y)	deW & M (mm/y)
25	1.0	1.2	1.7	1.8	3.6	3.4	6.3	5.4
50	2.4	3.2	4.1	5.1	8.6	9.4	15.0	15
75	5.6	7.7	9.8	11.5	20.4	22.6	35.5	36

$$H_2CO_3 + 2e^- = H_2 + CO_3^{2-}$$

$$2HCO_3^- + 2e^- = H_2 + 2CO_3^{2-}$$

As shown in previous paragraphs, the same corrosion rates as those from testing (from which the de Waard and Milliams equation was derived) are obtained on the basis of the corrosion theory:

- The cathodic reaction is hydrogen reduction (to give hydrogen gas)
- The free corrosion potential approximates iron equilibrium potential, governed by the solubility product of FeCO$_3$ which is 2×10^{-11} at room temperature and decreases strongly as temperature increases.

In conclusion, it seems that there is no need to cite exotic mechanisms or empirical models derived from testing. Indeed, results obtained from experimental tests confirm those calculated by a theoretical model, valid for strong or weak or organic acid, hydrogen sulphide and carbonic acid.

from Lazzari (2017)

8.2.3   Hydrogen Sulphide

Hydrogen sulphide (H_2S) dissolves in water forming a weak acid which decreases the solution pH to about 6. With this pH, the corrosion rate of ferrous alloys would be low to negligible in an oxygen-free solution, while experience shows that the corrosion rate is considerable because insoluble FeS forms. By inputting $a_{Fe}^{2+} = 10^{-12}$ mol/L into Eq. 8.4, corrosion rate would be surprisingly high (Table 8.4) since the solubility constant is 10^{-24}. Observed corrosion rates are obtained by inputting $a_{Fe}^{2+} = 10^{-9}$ mol/L; this takes the protection effect (barrier type) of the FeS layer into account.

Table 8.4 Calculated corrosion rates for ferrous alloys obtained at pH = 6 in the presence of H$_2$S

		a_M^{2+}	T	C_{rate}
		(mol/L)	(°C)	(mm/y)
Hydrogen Sulphide (H$_2$S)		10^{-12}	25	5.4[a]
			50	9.6[a]
			100	30[a]

[a]True corrosion rates are much lower because of the formation of FeS which is a protective product

8.2.4 Organic Acids

Organic acids (for example, formic acid, acetic acid, citric acid) severely corrode ferrous alloys, with very high corrosion rates—although the pH is close to neutrality. Even stainless steels in reducing conditions (i.e. in the absence of oxygen) may suffer a strong corrosion rate. Organic acids form complexes with the metal resulting in metal ion concentration, derived from the complex constant, in equilibrium with the complexant being quite low, in the range of 10^{-10} mol/L. The proposed model predicts well the expected corrosion rate values (Table 8.5 for ferrous alloys).

8.2.5 Corrosion of Passive Metals

In acidic solutions, passive metals can show:

- Depassivation (i.e., oxide dissolution) then behaving as an active metal following the model summarized in Fig. 8.9
- Passivity without risk of localized corrosion (i.e., chloride content is below the critical threshold content and free corrosion potential is below pitting potential)
- Passivity with the potential risk of localized corrosion (i.e., chloride content is above the critical threshold content but free corrosion potential is below pitting potential).

Uniform corrosion may occur if depassivation takes place. This occurrence is possible if pH is below depassivation pH which, for low grade stainless steels, is

Table 8.5 Calculated corrosion rates for ferrous alloys in organic acids (obtained at pH = 6)

		a_M^{2+}	T	C_{rate}
		mol/L	°C	mm/y
Organic acids		10^{-10}	25	1.8
			50	2.8
			100	7.3

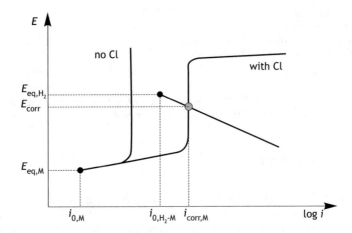

Fig. 8.10 Evans diagram of passive metals in acidic solutions

about 2, and much lower for higher nickel-based alloys. The uniform corrosion rate can be predicted by the model discussed above.

As illustrated in Fig. 8.10, two conditions exist if there is no de-passivation:

- *Absence* of chlorides. The passivity current density of most common passive metals (i.e., stainless steels, nickel-based alloys, titanium alloys and others) is lower than the exchange current density of hydrogen evolution. Therefore, the free corrosion potential is the equilibrium potential of hydrogen evolution (i.e., -0.059 pH) and the driving voltage for hydrogen evolution is zero (this means that the hydrogen evolution rate equals the passivity current density)
- *Presence* of chlorides. Due to the presence of chlorides, the passivity current density may exceed the exchange current density of hydrogen evolution, leading to a free corrosion potential that is lower than the equilibrium potential of hydrogen evolution. As a rule of thumb, since the passivity current density in the presence of chlorides is in the same range as the exchange current density of hydrogen evolution, the free corrosion potential is as a maximum about 100 mV more negative than the hydrogen equilibrium potential (Tafel slope is 120 mV/decade).

8.3   Aerated Solutions

The cathodic processes in aerated near-neutral solutions are oxygen reduction, followed by hydrogen evolution when potential drops below its equilibrium potential; this occurrence depends on the metal involved. For instance, if the metal is noble, like copper or silver, the only possible cathodic reaction is oxygen reduction (if chlorine is also present, this would be first) even in acidic solutions; if the metal is iron and pH exceeds neutrality, again the only practical cathodic process is oxygen reduction (and also chlorine reduction, when present); if the metal is more electronegative, such as in the case of zinc, both reactions, i.e., oxygen reduction and hydrogen evolution, take place.

8.3.1 Oxygen Limiting Diffusion Current

When oxygen reduction takes place below the field of activation overvoltage, diffusion is the control factor of oxygen availability through the oxygen limiting diffusion current density, i_L, which is governed by Fick law, as seen in Chap. 5:

$$i_\mathrm{L} = 4\,\mathrm{F}\frac{D}{\delta}C_1 \tag{8.9}$$

where D is diffusion coefficient, F is Faraday constant, C_1 is oxygen concentration in the bulk and δ is diffusion layer thickness. The latter parameter does not depend on the metal; rather, it depends on the turbulence of the solution. In aerated near-neutral solutions, the corrosion rate for mild steel coincides with i_L, hence, its knowledge is of primary importance in applications.

Oxygen dissolves in aqueous solutions when in contact with the atmosphere; in natural waters it is also present due to photosynthesis. Oxygen solubility in water varies and depends on temperature and salinity. Oxygen content in seawater (salinity about 35 g/L) varies from 9 mg/L at 0 °C, to 6 mg/L at 30 °C, 3 mg/L at 60 °C, and zero at 100 °C (at the pressure of 1 bar). As salinity increases, oxygen solubility decreases until zero above 150 g/L; in the Dead Sea, which is saturated with salt (more than 200 g/L), there is no dissolved oxygen and therefore there is no life and no iron corrosion. In natural water, photosynthesis and fouling may determine anaerobic, over- or under-saturation local conditions; the absence of oxygen, which would be ideal to impede corrosion, instead favours microbiological-related corrosion.

The values of parameters of Eq. 8.9 can be approximated as follows:

- The diffusion layer thickness, δ, in a *stagnant* condition varies between 0.5 and 3 mm for minimum and maximum oxygen content (from 1 to 11 ppm)
- The diffusion layer thickness, δ, in a *flowing* condition varies by a parabolic law with water velocity, v
- The diffusion coefficient D (m^2/s) varies with temperature by the following relationship: $\log D$ [cm^2/s] $= -4.410 + 773.8/T - (506.4/T)^2$. As rule of thumb, it doubling every about 25 °C of temperature increase starting from 2.25×10^{-9} (m^2/s) at 25 °C (1.97 at 20 °C; 4.82 at 60 °C)
- Faraday constant, F = 96485 C.

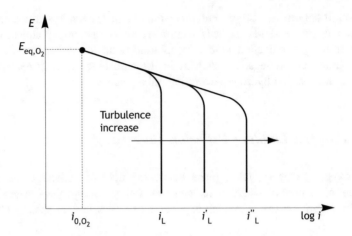

Fig. 8.11 Qualitative representation of the influence of turbulence on the cathodic curve

By introducing the relevant parameters in Eq. 8.9, i_L (mA/m^2) is expressed by the following empirical equation:

$$i_L \cong 10 \cdot 2^{\frac{T-25}{25}} \cdot [O_2] \cdot \left(1 + \sqrt{v}\right) \qquad (8.10)$$

where [O$_2$] is the oxygen content in water in mg/L ($\approx$ ppm), v is water velocity (m/s) and T is temperature (°C). Changes of oxygen content in water determine a double variation:

- Oxygen limiting current density increases as turbulence (i.e., water velocity) increases (Fig. 8.11)
- Equilibrium potential changes by about 50 mV every 1 ppm variation of oxygen content, through Henry law for oxygen partial pressure (Fig. 8.12).

8.3.2   Presence of Chlorine

The presence of chlorine gives rise to a more noble cathodic process, which takes place first:

$$Cl_2 + 2e^- \rightarrow 2Cl^- \qquad (8.11)$$

chlorine, like oxygen, is a gas that dissolves in water, but unlike oxygen it partially dismutes. Therefore, the fraction available for diffusion is about 30% (chlorine diffusion coefficient is 1.38×10^{-9} m^2/s at 25 °C and, like oxygen, it doubles every 25 °C). According to this, Eq. 8.10 can be revised by also introducing the chlorine

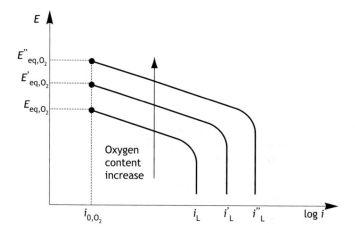

Fig. 8.12 Qualitative representation of the influence of oxygen content on the cathodic curve

content, taking into account the different diffusion coefficient and valence (8 g of oxygen are equivalent to 35 g of chlorine); eventually, it becomes:

$$i_L \cong 10 \cdot 2^{\frac{T-25}{25}} \cdot \{[O_2] + 0.04 \cdot [Cl_2]\} \cdot \left(1 + \sqrt{v}\right) \tag{8.12}$$

where oxygen limiting current density, i_L, is in mA/m^2, $[O_2]$ and $[Cl_2]$ are oxygen and chlorine concentrations in water in mg/L ($\approx$ppm).

8.3.3 Dimensionless Number Approach

In a flowing condition, the diffusion layer thickness, δ, can be calculated by using the classic hydrodynamic approach based on the Sherwood (or Nusselt) dimensionless number, which gives:

$$Sh = \frac{\phi}{\delta} = i_L \frac{\phi}{4FD[O_2]} \tag{8.13}$$

where ϕ is called the characteristic dimension, for example, the pipe diameter; the meaning of other parameters is known. The Sherwood number is a function of Reynolds (Re) and Schmidt (Sc) dimensionless numbers, as follows:

$$Sh = 0.023 \cdot Re^{0.87} \cdot Sc^{0.33} \quad Re = \frac{\phi v}{v} \quad Sc = \frac{v}{D} \tag{8.14}$$

where ϕ is characteristic dimension (m); v is water velocity (m/s); υ is kinematic viscosity (m²/s) and D is diffusion coefficient (m²/s). Oxygen limiting current density, i_L (A/m²) is given by the Sherwood number as follows:

$$i_L = 4FD[O_2]\frac{Sh}{\varphi} \tag{8.15}$$

It is of practical interest the comparison of results obtained by applying the empirical Fick equation (Eq. 8.10) and the dimensionless number approach. For instance, on the basis of the following input data:

- Oxygen content 10 ppm (mg/L) = 0.3 mol/m³
- Temperature: ambient
- Water velocity: 1; 2; 3 m/s
- Viscosity: 0.001 m²/s
- Diffusion coefficient: 2×10^{-9} m²/s
- Size of characteristic dimension: 0.5 and 1 m.

The comparison of the results is very good (scattering below ±20%). For example, 241 mA/m² against 212 and 232 for a velocity of 2 m/s and size diameter 1 or 0.5 m respectively.

8.3.4   Corrosion of Noble Metals

Figure 8.13 shows some examples of the corrosion behaviour of noble metals (i.e., with an equilibrium potential that is more noble than hydrogen equilibrium potential) in aerated solutions. Metal ③, which is the least noble, practically works under diffusion control; hence, the corrosion rate coincides with oxygen limiting current density.

Metals ① and ② work in the region of activation overvoltage: in these cases, the corrosion rate is determined by the overvoltage once the equilibrium potential of oxygen is fixed (which is given by pH and oxygen concentration), and hence by the nature of metal involved. In Fig. 8.13 the working condition is represented by points ① or ①′ depending on overvoltage: for instance, for metal 1 the free corrosion potential is high when oxygen overvoltage is low (i.e., a high exchange current density) or, instead, is low when oxygen overvoltage is high (i.e., a low exchange current density).

8.3.5   Corrosion of Non-noble Metals

Metals are defined as "non-noble" when the free corrosionpotential of the active behaviour is below the equilibrium potential of hydrogen evolution. In practice, the

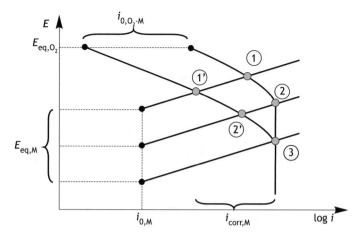

Fig. 8.13 Corrosion rate representation of noble metals in aerated solutions

non-noble metals start from lead, and are then followed by tin, nickel, iron, zinc, aluminium and magnesium. Chromium and titanium should also be considered if they were active; instead, since they passivate, then they work like noble metals. In aerated solutions, non-noble metals work below the activation overvoltage region as shown in Figs. 8.14 and 8.15.

The working conditions represented in Fig. 8.14 by points ①, ② and ③ are determined by the nobility of the metal: metals ① and ② are sufficiently noble not to work in hydrogen evolution conditions. Instead, metal ③ can support both cathodic reactions, that is, oxygen reduction first followed by hydrogen evolution. Figure 8.15 shows how the corrosion rate for metal ① increases as the oxygen limiting current density increases.

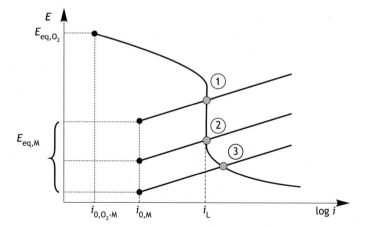

Fig. 8.14 Corrosion rate representation of non-noble metals in aerated solutions

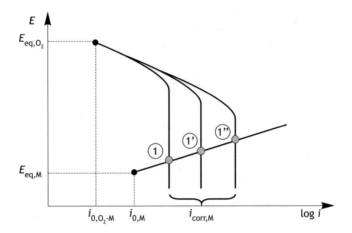

Fig. 8.15 Corrosion rate of non-noble metal in aerated solutions at increasing oxygen limiting current density

8.3.6 Corrosion of Passive Metals

Unless localized corrosion occurs, such as in pitting, crevice, interstitial and stress corrosion cracking, passive metals in aerated, near-neutral or alkaline solutions behave like noble metals, thus exhibiting a noble free corrosion potential and a very low corrosion rate coinciding with the passivity current, i_p, as depicted in Fig. 8.16. When critical passivation current density, i_{cp}, exceeds oxygen limiting current density, i_L, there is the possibility of another "stable" working condition, represented by point ②, as depicted in Fig. 8.17. Working conditions ① and ② represent two final stable and mutually exclusive conditions determined by the initial passive or active condition: in fact, the initial condition they are in at the start will be maintained. This behaviour can be interpreted in the light of catastrophe or chaos theory, by which the system evolves towards two opposite stable conditions, passive or active, as a function of the initial state only.

The measurement of potential gives a clear indication of the metal state: passive if potential is noble, or active if near the equilibrium potential. In aerated solutions, passive metals exhibit a noble potential and for this reason, this state is said to be of practical nobility.

When Incorrect Repairs Cause Damage
Considering the damage caused by the corrosion of iron inserts inside stones, one of the most commonly reported examples is the damage that occurred on the monuments of the Acropolis in Athens after some irresponsible restoration works that were carried out at the turn of the XX century up to the Second World War. In 1943 John Meliades, Superintendent of the Acropolis monuments, wrote: *"The damage to monuments at the Acropolis, and in*

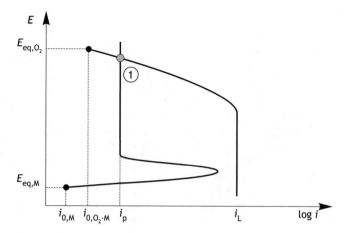

Fig. 8.16 Corrosion conditions of passive metals in aerated solutions

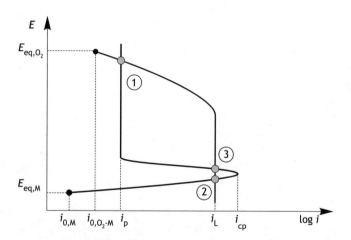

Fig. 8.17 Corrosion conditions of passive metals with a critical passivation current density higher than the oxygen limiting current density

particular the Parthenon, is due to the criminal way in which marble elements were assembled with unprotected iron inserts."

This type of attack, unfortunately, also affected Italian artefacts. In an article entitled "Restoration of the Cathedral" (in La lettura, No. 1, 1936, pp. 61–65, Milan, Italy), Carlo Emilio Gadda, an Italian writer, wrote: "*At the top of the Duomo di Milano, a harmonious command gathered and lined up on the aligned cusps: a Donatellian cohort of saints, white martyrs [...] and Filipino-designed ogives for all the dreams and prayers over the centuries. At the moment, the ogives are not a cause of concern, but the saints are*

anchored to the spire's capital through an iron pin (which sustains them like
a plinth). The weather, filtering through the commissure support, is causing
the iron to oxidize and swell up. By doing so, the plug has functioned as a
wedge and has cracked and sometimes split the pedestal: theoretically, the
saints' statues can fall off with a stronger than usual gust of wind and the
disintegration of the specific section. [...] The same occurrence is prevalent
in other decorative elements. The specious fastigium with ogival arches,
ending in an acute tasselled triangle, interspersed with cusps (the pediment
which in jargon is called "falconatura") is maintained by a continuous key
iron ligament: this key or pass-through, like a long stick on which stuffed
birds of gentle beak to lardons and sage leaves were alternately placed, has
swelled over the years like rusty iron forgotten in roofs and is splitting the
load-bearing parts of the ornamentation's statically vital points. This makes
it necessary to change the tunnel elements: iron is the Cathedral's disease."

Today titanium is chosen for metal inserts to be used on important works, not only for its excellent resistance to corrosion but also for its coefficient of expansion approximately equal to that of stone and about half that of austenitic stainless steels. These are, however, employed for the consolidation of historic buildings or for the scheduling of hangings to a building outer surface.

Game of Triangles

The three vertices (A, B and C) of an equilateral triangle are marked on a paper together with a fourth point denoted by P1 in the middle of the triangle. Now, pull a random vertex among A, B and C. Let's suppose that we select A. We then connect the point P1 with A, and identify the midpoint of the segment P1-A which we call P2. Let's make another extraction. Suppose that this time we select B. Draw then the point P2 with B and identify the midpoint of the segment P2-B, which we call P3. Continuing with the extractions, we identify the points P4, P5, P6, P7, P8, P9, P10, P11, P12...... Pn. Once scored within the ABC triangle, there should be a sufficiently large number of points and once the first 8 points P1 to P8 are deleted, the result is not a chaotic set of points, as would be expected. Instead, we get the Sierpinsky triangle, which is a fractal figure spotted by Sierpinsky in 1915 (Fig. 8.18).

And what does corrosion matter?

Curiously, a metallographic etching of the plane (1 1 1) of a monocrystal of silver (Fig. 8.19) looks like the Sierpinsky fractal.

Fig. 8.18 Fractal Sierpinsky triangle

Fig. 8.19 Metallographic etching of the plane (1 1 1) of a monocrystal of silver

8.4   Questions and Exercises

8.1   Since about 30 years, the exhaust of cars is made of stainless steel. Previously, it was made of carbon steel and frequently suffered premature perforation: for this reason, the material was changed. Make a corrosion assessment, indicate the condition required for corrosion occurrence and try to estimate the time-to-perforation.

8.2   The material selected for heat exchanger tubes in a food industry plant was Ti. The choice was suggested on the idea to prevent food contamination by metals. After six months of operating, tubes were perforated. Indicate possible causes for failure occurrence, if fluid process was slightly acidic.

8.3   To store concentrated sulphuric acid, carbon steel is used. Conversely, transfer piping and pumps are made of stainless steel. Why? Estimate corrosion rates.

8.4   Estimate the corrosion rate during an acidizing job carried out in an oil well with a tubing made of low alloy steel if acidic solution is 20% HCl or 20% formic acid at 50 °C. Compare the corrosion rates and suggest possible strategies for such operating.

8.5   Suggest which metals can give working conditions represented in Fig. 8.14 by point ①, ② and ③ respectively.

8.6   Suggest which metals can give working conditions represented in Fig. 8.15 by point ①, ② and ③ respectively. Give a relationship with water velocity for working conditions ①, ①′ and ①″.

8.7   Corrosion rate of steels in aerated waters depends on total dissolved solids, TDS, although oxygen limiting current density does not. Give an explanation and discuss how TDS affects corrosion rate.

8.8   Corrosion rate of cast iron in aerated waters is much lower than that for mild steel. Give an explanation and explain by using Evans Diagrams.

8.9   A water-carrying pipe (100 mm in diameter) showed an average corrosion rate of 0.2 mm/year. What is the expected corrosion rate in an extension smaller pipe (50 mm in diameter) if exposed surfaces are clean (i.e., without surface deposits) and cathodic reaction is oxygen reduction?

8.10 On the stay-vane in a water turbine drum the maximum corrosion rate was found to occur at approximately 30 m/s. At even higher velocities the corrosion rate was low. What is your interpretation?

8.11 A reinforced concrete structure is totally immersed in seawater. If reinforcements are active, calculate the corrosion rate due to oxygen reduction.

8.12 Comment and suggest the philosophy behind the techniques adopted for preventing uniform corrosion listed below:

- Corrosion allowance, i.e., an extra thickness to be consumed within design life, when expected/calculated corrosion rate is moderate
- Selection of appropriate metals (for example, stainless steel instead of carbon steel for oxygenated low chloride containing waters)
- Environment conditioning by removing corrosive agents (for example by de-oxygenation) or by injecting corrosion inhibitors
- Use of organic or metallic coatings or painting and/or cathodic protection on structures exposed to soil and waters.

Bibliography

De Giovanni C (2017) Validation of a model Tafel-Piontelli for the calculation of corrosion rate of metals in acidic solutions. Application to sweet corrosion of carbon steel. MS Thesis 2016–2017, Politecnico di Milano

De Waard C, Lotz U, Milliams DE (1991) Predictive model for CO_2 corrosion engineering in wet natural gas pipelines. Corrosion 47(12):976

De Waard C, Lotz U, Dugstad A (1995) Influence of liquid flow velocity on CO_2 corrosion: a semi-empirical model, Corrosion, 95, paper n. 128, NACE, Houston, TX

De Waard C, Milliams DE (1975) Carbonic acid corrosion of steel. Corrosion 31(5):131

Lazzari L (2017) Engineering tools for corrosion. Design and diagnosis. European federation of corrosion (EFC) Series, vol 68. Woodhead Publishing, London, UK

Kreysa G, Schütze M (eds) (2006) Carbonic acid, chlorine dioxide, seawater. In: DECHEMA corrosion handbook, corrosive agents and their interaction with materials, vol 5. Wiley-VCH, Weinheim

Kreysa G, Schütze M (eds) (2006) Chlorinated hydrocarbons. DECHEMA corrosion handbook, corrosive agents and their interaction with materials, vol 8. Wiley-VCH, Weinheim

Kermani MB, Smith LM (eds) (1997) A working party report on CO_2 corrosion control in oil and gas production. Institute of Materials, London

Piontelli R (1961) Elementi di teoria della corrosione a umido dei materiali metallici. Longanesi, Milano (in Italian)

Chapter 9
Macrocell Corrosion Mechanism

Rust in Peace… Polaris.

Megadeth

Abstract When a macrocell is formed in a corrosion process, an electrical field is established in the environment because a net current flows from the anode to the cathode, which are physically separated. This situation occurs in galvanic corrosion, differential aeration, localized attacks such as pitting and crevice, and cathodic protection. Potential and current distributions are extremely important because they determine the corrosion rate. Analytical solutions of electric fields exist only for very simple geometry and simplified conditions. In the last two decades, the use of numerical calculations based on Finite Element Methods (FEM) has overcome these difficulties.

Fig. 9.1 Case study at the PoliLaPP Corrosion Museum of Politecnico di Milano

© Springer Nature Switzerland AG 2018
P. Pedeferri, *Corrosion Science and Engineering*, Engineering Materials,
https://doi.org/10.1007/978-3-319-97625-9_9

This Chapter gives an overview of macrocell electrical field and current distribution, giving analytical solutions for both quantities in simple geometries, such as inside and outside a pipe. In some of these geometries the throwing power is also evaluated.

9.1 Electrical Field in Uniform Corrosion

In uniform corrosion (Fig. 9.1), each point of the corroding metal surface is simultaneously anode and cathode and, therefore, there is no net current flowing in the environment. In other words, there are many minute microcells whose anodes and cathodes are continuously and dynamically exchanging each other. The working condition of the corroding system is represented by the Evans diagram as the crossing point of anodic and cathodic curves. When measuring the potential, there is one fixed value only, regardless the position of the reference electrode in the electrolyte. In particular, as soon as the reference electrode is progressively moved away from the surface, the potential does not change, as depicted schematically in Fig. 9.2. According to this, an experimental confirmation for a uniform corrosion condition is a constant and homogeneous potential mapping.

9.2 Electrical Field in a Macrocell

When a macrocell forms on a metal or bimetal surface, a potential distribution on the exposed metal surface exists (the so-called potential mapping) and its influence is extended within the electrolyte. Qualitatively measured, the potential changes as a function of the increasing distance from the metal surface, as shown in Fig. 9.3. Beyond a certain distance from the metal surface, the measured potential is constant and it is called *remote potential*.

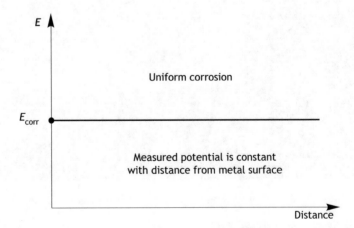

Fig. 9.2 Potential trend with distance from a metal surface corroding uniformly

More generally, the electric field and all field-related problems are governed by the Poisson-Laplace *quasi-harmonic equation*, which, for stationary phenomena independent of time, assumes the general form[1]:

$$\nabla(\kappa\nabla E) + Q = 0 \tag{9.1}$$

where E is potential function, Q is electric charge flux and κ is conductivity. If κ is constant and independent of direction, Eq. 9.1 changes to:

$$\kappa\nabla^2 E + Q = 0 \tag{9.2}$$

The derivative equation for an electrodic system becomes the current density:

$$i = -\kappa\nabla E \tag{9.3}$$

Based on Eq. 9.3, equipotential lines and current density lines are orthogonal to each other.

9.2.1   Pure Ohmic Systems

In a purely ohmic conductor with resistivity ρ ($\rho = 1/\kappa$), the term Q is nil and the field equation is simplified to Laplace equation:

$$\nabla(\kappa\nabla E) = 0; \quad \text{or} \quad \nabla^2 E = 0 \tag{9.4}$$

which gives Ohm's law:

$$i = -\kappa\nabla E = -\frac{1}{\rho}\nabla E \cong \frac{\Delta E}{\rho} \tag{9.5}$$

Boundary conditions are reduced to $\partial E/\partial n = 0$, which means that insulated surfaces do not exchange current.

---

[1]Examples of field problems are: heat transmission (E = temperature, Q = heat, κ = transport coefficient), mass transport (E = pressure, Q = mass, κ = transport coefficient) and the transportation of current as in the case of cathodic protection, where E is electric potential, Q is electrical charge and κ is conductivity.

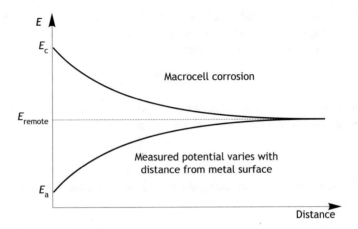

Fig. 9.3 Potential trend with distance from a metal surface where a macrocell works

9.2.2   Two-Electrode Macrocell

In a macrocell, there is an electrolyte and two or more electrode surfaces among which the current is exchanged through electrode reactions, an oxidation reaction at the anode and a reduction one at the cathode.

The system can schematically be divided into two distinct domains: the *electrolytic solution* (the bulk) assumed to be a homogeneous ohmic conductor and the *double layer* at the electrode surface. By assuming the double layer negligible, the bulk is then governed by the Laplace equation, as an ohmic system:

$$\nabla^2 E_\varepsilon = 0 \tag{9.6}$$

where E_ε is the electrolyte potential at the electrode–electrolyte interface given by:

$$E_\varepsilon = E_m + \eta \tag{9.7}$$

where E_m is the potential of a metal surface, which is uniform and constant (equipotential electrodes), and η is the overvoltage, which is positive for the anode and negative for the cathode. Overvoltage η is function of current density, i, which is exchanged with the electrolyte.

9.3   Current Distribution

Two types of current distribution are observed in electrochemistry: *primary distribution*, when the overvoltage is negligible and *secondary distribution*, when the overvoltage applies.

9.3.1 Primary Current Distribution

When overvoltage is negligible, Laplace equation solutions show that:

- Current distribution depends on geometry only, and not on electrolyte resistivity
- The ohmic drop is concentrated close to electrodes. As a rule of thumb, the smaller the electrode, the higher the electrode ohmic drop contribution.

Current flows in the electrolyte from anode to cathode. The ohmic drop of a general flux tube, nth, is given by:

$$\Psi_{\mathrm{ohm,n}} = I_n R_n = I_n \rho \int_n \frac{\mathrm{d}L}{S} = i_n S \rho \int_n \frac{\mathrm{d}L}{S} \tag{9.8}$$

where R_n is electrical resistance, I_n is circulating current in flux tube nth of length L and section S. Both i_n and S vary with L.

Table 9.1 reports Laplace equation solutions to calculate the primary current distribution for simple geometry; for more details refer to Kasper (1940), Wagner (1952) and Newman (1974).

By comparing two flux tubes, which have the same ohmic drop because overvoltage is negligible, the current density ratio is a function of geometry, only:

$$\Psi_{\mathrm{ohm,1}} = \Psi_{\mathrm{ohm,2}} \tag{9.9a}$$

$$\frac{i_1}{i_2} \simeq \frac{\int_2 \frac{\mathrm{d}L}{S}}{\int_1 \frac{\mathrm{d}L}{S}} \tag{9.9b}$$

Generally, with an increase in flux tube length, there is a decrease in current density and therefore, the system is characterised by poor *throwing power*. Conditions of uniform primary distribution are created only by particular geometries where the term $\int \mathrm{d}L/S$ is constant for all current paths, that is, in practice, when electrode systems are made of large parallel plates, concentric spheres or coaxial cylinders.

9.3.2 Secondary Current Distribution

When electrode polarisation is established, solution of Laplace equation has to take into account boundary conditions between the overvoltage and the current density. In general, the secondary current distribution can be expressed as a function of the following type:

Table 9.1 Primary current distribution for simple macrocell geometry

	Potential	Current	Resistance
Parallel plates (large surface area) d = distance from plates S = surface area of plates	$E_z = \rho I \frac{z}{S}$ $E_{x,y} = $ const	Uniform $i_a = i_c = \frac{I}{S}$	$R = \rho \frac{l}{S}$
Spheric (concentric spheres) r_a; r_c = radius of two spheres r = distance from centre	$E_r = \frac{\rho I}{4\pi}\left(\frac{1}{r_a} - \frac{1}{r}\right)$	Uniform $i_a = \frac{I}{4\pi r_a^2}$ $i_c = \frac{I}{4\pi r_c^2}$	$R = \frac{\rho}{4\pi}\left(\frac{1}{r_a} - \frac{1}{r_c}\right)$ $R \approx \frac{\rho}{4\pi r_a}$ if $r_a \ll r_c$
Cylindrical (coaxial cylinders) $L \gg r_c$	$E_r = \frac{\rho I}{2\pi L}\ln\frac{r}{r_a}$	Uniform $i_a = \frac{I}{2\pi L r_a}$ $i_c = \frac{I}{2\pi L r_c}$	$R = \frac{\rho}{2\pi L}\ln\frac{r_c}{r_a}$
Infinite plate (cathode) (small anode at distance d on Z axis) r = distance of point P (x, y) from anode projection	$E_z = -\frac{\rho I}{2\pi}\left[\frac{d}{d^2 - z^2}\right]$	NOT uniform $i_{max} = \frac{I}{2\pi d^2}$ $i = i_{max}\left(\frac{d}{\sqrt{d^2+r^2}}\right)^3$	$R \approx \frac{2\rho}{\pi d}$
Infinite plate (cathode) (linear anode at distance d on Z axis and infinite length) r = distance of point P (x, y) from anode projection	$E_z = \frac{\rho}{\pi}\left(\frac{I}{L}\right)\ln\frac{d+z}{d-z}$ $E_{y,x} = \frac{\rho}{2\pi}\left(\frac{I}{L}\right)\ln\frac{(d+z)^2+x^2}{(d-z)^2+x^2}$	NOT uniform $i_{max} = \frac{1}{\pi d}\left(\frac{I}{L}\right)$ $i_x \approx i_{max}\frac{d^2}{d^2+x^2}$	$R_L \approx \frac{\rho}{\pi L}$ if $d \gg$ anode diameter
Flat disk (cathode radius r_o) Anode at infinite distance d = distance from disk	$E_d \approx \frac{\rho I}{4\pi d}$ with $r_o \ll d$	Approximately uniform $i \approx \frac{I}{2\pi r_o^2}\sqrt{1 - \frac{r^2}{r_o^2}}$	$R \approx \frac{\rho}{4\pi r_o}$

$$\frac{i}{i_{av}} = f\left(\Gamma, \rho, \frac{d\eta}{di}\right) \tag{9.10}$$

where i_{av} is the average current density, Γ a geometric factor, $d\eta/di$ the overvoltage function and ρ the electrolyte resistivity.

Therefore, unlike primary distribution, the secondary distribution does not depend on geometry only and is, in particular, electrolyte resistivity dependent. As a general conclusion, secondary distribution is more uniform than primary, because overvoltage effects that spread current distribution overlap geometric factors determining the primary distribution.

When the cathodic reaction is hydrogen evolution, overvoltage is low (Tafel slope is about 120 mV/decade) so that current distribution is closer to the primary than the secondary distribution. Conversely, when the cathodic reaction is oxygen reduction, overvoltage is much higher in the order of 1 V, so that secondary distribution prevails. This typically occurs in galvanic corrosion and cathodic protection in natural environments, especially in low resistivity solutions such as seawater, where overvoltage largely exceeds ohmic drop.

9.4 Throwing Power

The term *throwing power* refers to the ability of the current to reach areas distant from anodes. It is thus related to both primary and secondary current distributions. A macrocell generally has poor throwing power, although for different reasons: in high resistivity electrolytes because primary distribution applies and in low resistivity electrolytes because of geometry (reduction of the term S in the integral $\int dL/S$).

To calculate, or rather, to estimate the throwing power, ohmic drop, ΔV, settled in the electrolyte must be considered assuming constant and uniform overvoltage contributions: in other words, the driving voltage available for overwhelming the ohmic drop in the electrolyte is considered known. Based on this assumption, the ohmic drop can be written as follows:

$$\Delta V = I \cdot R = \rho \int_n i_n \, dL \tag{9.11}$$

where symbols are known. Throwing power, L_{max}, as the maximum distance between the two electrodes assuming a constant current density, i, within the electrolyte, can be obtained from Eq. 9.11 and approximated to the following two forms:

$$L_{max} \cong k \frac{\Delta V}{\rho \cdot i} \qquad (9.12a)$$

$$L_{max} = k \sqrt{\frac{\Delta V \, \phi_k}{\rho \, i}} \qquad (9.12b)$$

where ϕ_k is the characteristic dimension and k is an appropriate constant for each geometry. Table 9.2 summarizes the empirical equations possibly used for the calculation of throwing power for typical geometries.

9.5   Typical Geometries

Equations reported in Table 9.2 for simple typical geometries, such as inside a pipe, on a plate or on external surface of a pipe, which deals with pitting, crevice, differential aeration and galvanic corrosion, as well as galvanic anode cathodic protection systems, are obtained in the following paragraphs, assuming a uniform cathodic current density, i_c.

9.5.1   Inside a Pipe

With reference to Fig. 9.4, the current supplied by the anode that crosses the pipe section is a function of the distance from the anode, according to the following expression:

$$I_x = i\pi\emptyset(L_{max} - L_x) \qquad (9.13)$$

Table 9.2 Throwing power for typical macrocell geometry (from Lazzari 2017)

Geometry	Throwing power
Pitting on a plate like geometry	$L_{max} \cong \frac{\Delta V}{2\rho i}$
	$L_{max} \cong \sqrt{\frac{2\Delta V \emptyset_{pit}}{\rho i}}$
Inside a pipe	$L_{max} \cong \sqrt{\frac{\Delta V \emptyset_{pipe}}{2\rho i}}$
Outside a coated pipeline	$L_{max} \cong \frac{\Delta V}{2\rho i}$
	$L_{max} \cong 10\sqrt{\frac{\Delta V \emptyset_{pipe}}{\rho i}}$

L_{max}: Throwing power (m); ϕ_{pit}: diameter of the localised corrosion attack (m); ϕ_{pipe}: diameter of the pipe (m); i: cathodic current density (A/m^2); ρ: resistivity (Ω m); ΔV: ohmic drop (V) Applicable conditions: $1 < \rho \, (\Omega \, m) < 10^3$; $0.01 < i \, (A/m^2) < 1$

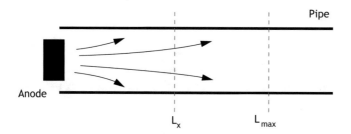

Fig. 9.4 Throwing power inside a pipe

where I_x is current that crosses the tube section at distance L_x, i is uniform current density over the pipe's internal surface, ϕ is internal diameter and L_{max} is maximum distance or throwing power, which can be derived taking into account that at a distance L_x from the anode, the ohmic drop and resistance are given by:

$$\partial V = I_x \partial R \tag{9.14a}$$

$$\partial R = \frac{\rho}{S} \partial L \tag{9.14b}$$

where ρ is electrolyte resistivity and S ($=\frac{1}{4}\,\pi\,\phi^2$) is the section area. By substituting I and ∂R, it changes to:

$$\partial V = \frac{4\rho i \pi \varnothing (L_{max} - L_x)}{\pi \varnothing^2} \partial L = \frac{4\rho i (L_{max} - L_x)}{\varnothing} \partial L \tag{9.15}$$

Integrating the expression from 0 to ΔV (ohmic drop) with L between 0 and L_{max}, Eq. 9.15 becomes:

$$\Delta V = \frac{2\rho i L_{max}^2}{\varnothing} \tag{9.16}$$

From Eq. 9.16 L_{max} can be calculate:

$$L_{max} \cong \sqrt{\frac{\Delta V \varnothing_{pipe}}{2\rho i}} \tag{9.17}$$

where ϕ_{pipe} is the internal tube diameter or equivalent tube diameter in the case of different cross section fluxes.

9.5.2 *Outside a Pipeline*

The throwing power can be estimated by assuming a spherical electrical field as depicted in Fig. 9.5. Like in the Wenner method measurement, as illustrated in Chap. 19, where the electrical field in considered hemispherical, the following linear relationship applies:

$$\Delta V \cong k\rho i L_{max} \tag{9.18}$$

where ΔV (mV) is the driving voltage totally consumed as ohmic drop in the electrolyte, i (mA/m^2) is cathodic current density (assumed constant), ρ (Ω m) is electrolyte resistivity, L_{max} (m) is throwing power and k (adimensional) is a constant.

The throwing power, L_{max}, is given by:

$$L_{max} \cong \frac{\Delta V}{2 \cdot \rho \cdot i} \tag{9.19}$$

The constant k is often taken as being equal to 2, as FEM simulations confirmed for a pipeline in electrolytes with resistivity greater than 1 Ω m and cathodic current density greater than 0.1 mA/m^2.

A more general approach can start from Eq. 9.17, then introducing the pipeline diameter, $\varnothing_{pipe}$, as characteristic dimension. From FEM simulation, the equation can be approximated to the following:

$$L_{max} \cong 10\sqrt{\frac{\Delta V \varnothing_{pipe}}{\rho i}} \tag{9.20}$$

This equation is applicable under the following conditions: pipeline diameter: $0.1 < \varnothing_{pipe}$ (m) < 1; electrolyte resistivity: $1 < \rho$ (Ω m) < 1000; cathodic current density: $0.01 < i_C$ (A/m^2) < 1.

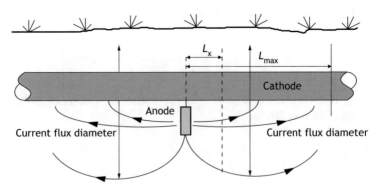

Fig. 9.5 Throwing power outside a tube

9.5.3 On a Plate

Two approaches can be adopted: (1) one based on the assumption that the electrical field is hemispherical around the anode as in the case of pitting corrosion (as shown in Fig. 9.6); (2) one based on the evaluation of the ohmic resistances on both sides, anode and cathode.

Method # 1

Similar to a pipeline, throwing power can be estimated by assuming a spherical electrical field around the anode (for instance a pit) as shown in Fig. 9.6; taking into account FEM simulations, the following can be obtained:

$$L_{max} \cong \frac{\Delta V}{k \cdot \rho \cdot i} \tag{9.21}$$

where ΔV (mV) is the driving voltage totally consumed as ohmic drop in the electrolyte, i (mA/m^2) is cathodic current density (assumed constant) and ρ (Ω m) is electrolyte resistivity. The constant k is often taken as being equal to 2, as FEM simulations confirmed for electrolytes with resistivity in the range $1 < \rho\,(\Omega\,m) < 1000$ and current density i_C in the range $0.01 < i_C\,(A/m^2) < 1$.

Method # 2

By this second approach, let's assume that the anode (e.g. a pit) is a flat disk with diameter ϕ fixed on a plate working as the cathode and the electrolyte has a large domain. The macrocell current flows from the anode, a circle-shaped area, to the cathode with the radius equal to the throwing power, L_{max}, as shown in Fig. 9.6. The ohmic drop, ΔV, already given in Eq. 9.16, is given by two terms physically located at the electrodes (the smaller the electrode, the closer the ohmic drop to the electrode).

The resistance of a disk-shaped anode can be approximated to the following:

$$R_a \cong \frac{\rho}{2\pi\varnothing} \tag{9.22}$$

where ρ is electrolyte resistivity and ϕ is anode diameter. The resistance of the cathode can be expressed by a similar equation also because of a disk-shaped type with a diameter of twice the throwing power. As ϕ is generally much smaller than

Fig. 9.6 Throwing power on a plate

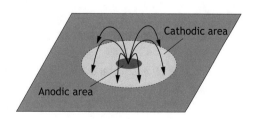

Cathodic area

Anodic area

the expected throwing power, L_{max}, only the ohmic drop at the anode applies, therefore:

$$\Delta V = IR \cong I(R_a + R_c) \cong I \frac{\rho}{2\pi \varnothing_{pit}} \tag{9.23}$$

The current, I, can be easily calculated by assuming the cathodic current density as constant:

$$I \cong i\pi L_{max}^2 \tag{9.24}$$

and

$$L_{max} \cong \sqrt{\frac{2 \cdot \Delta V \varnothing_{pit}}{\rho i}} \tag{9.25}$$

Accordingly, the throwing power depends on pit size, electrolytes resistivity and cathodic current density $(1 < \rho \ (\Omega \ m) < 1000; \ 0.01 < i_C \ (A/m^2) < 1)$.

9.6 Maximum Surface Area Ratio

In a macrocell, throwing power inherently defines the surface area ratio. As experience has proved, parameters determining the throwing power affect the surface area ratio setting up the macrocell. Accordingly, a similar relationship used for throwing power applies for maximum surface area ratio as follows:

$$\left(\frac{S_c}{S_a}\right)_{max} \cong \left(1 + \frac{S_c}{S_a}\right)_{max} \cong k \sqrt{\frac{\Delta V}{\rho \cdot i_C}} \tag{9.26}$$

where symbols are known. In near neutral-to-alkaline electrolytes the cathodic process is oxygen reduction, accordingly the cathodic current density is the oxygen limiting current density, i_L $(0.01 < i_L \ (A/m^2) < 1)$, and the constant k $(m^{-1/2})$ is about 20 when ohmic drop ΔV is expressed in V and resistivity ρ is Ω m $(1 < \rho < 1000)$.

9.7 Questions and Exercises

9.1 Consider a macrocell mechanism on a metal surface with formation of anodic and cathodic zones. How does the potential change by increasing the distance from the metal surface where the macrocell works?

9.2 What is the difference between primary and secondary current distribution?

9.3 Consider the statement: "when the cathodic reaction is hydrogen evolution, current distribution is closer to the primary than secondary distribution. Conversely, when the cathodic reaction is oxygen reduction, secondary current distribution prevails". Explain.

9.4 What is the throwing power? What is the effect of electrolyte resistivity and current density on throwing power?

9.5 Evaluate the throwing power of a macrocell, which can setup in a heat exchanger between the tube-plate made of cupronickel and the tubes made of titanium. The fluid is aerated seawater. By painting internally the titanium tubes, how does the throwing power change? Derive a relationship as function of the coating efficiency.

9.6 Estimate the driving voltage of a galvanic corrosion macrocell for the following case studies in seawater:

- Carbon steel and copper
- Zinc and carbon steel
- Carbon steel and stainless steel.

For each case study, indicate the anode, the cathode and the pertinent reactions.

9.7 Estimate the driving voltage of the corrosion macrocell established in the crevice and pitting corrosion of stainless steel in seawater. Compare the value with the driving voltage of a galvanic corrosion macrocell between carbon steel and stainless steel in seawater (previous exercise).

9.8 A pipe system carrying aerated water consists of a section made of stainless steel separated from a carbon steel section by a polymeric composite spool, 0.1 m long. Try to determine the throwing power on the macrocell on either anodic side or cathodic side. (Note. Both section are connected to the grounding system). What is the influence of the nature of the metals involved? What would be the difference if metals are copper and aluminium?

9.9 Calculate the throwing power inside tubes of a heat exchanger, of ½ inch in diameter, when the fluid is seawater (resistivity $0.2\ \Omega$ m) and fresh water (resistivity $20\ \Omega$ m), considering two separated case studies: tubes made of noble metal and made of a low nobility metal, then corroding.

9.10 Estimate the maximum surface area ratio, $(S_c/S_a)_{max}$, for the macrocell established by a galvanic coupling in soil between carbon steel and a copper made grounding system. Assume a well-aerated soil.

Bibliography

Lazzari L, Pedeferri P (2006) Cathodic protection. Polipress, Milan, Italy
Lazzari L, Pedeferri P (1981) Protezione catodica. CLUP, Milano, Italy

Lazzari L (2017) Engineering tools for corrosion. Design and diagnosis. European federation of corrosion (EFC) Series, vol 68. Woodhead Publishing, London

Kasper C (1940) The theory of the potential and the technical practice of electrodeposition. Trans Electr Soc 77:365–384

Newman J (1974) Mass transport and potential distribution in the geometry of localised corrosion, NACE-3, 45–61, Houston TX

Wagner C (1952) Contribution to the theory of cathodic protection. J Electr Soc 99:1

Chapter 10
Galvanic Corrosion

The world is holding back
The time has come to galvanize.

The Chemical Brothers

Abstract Galvanic corrosion occurs when two or more metals with different practical nobility are electrically connected and immersed in the same environment: the less noble metal experiences an increase in corrosion rate due to the presence of the more noble one. The effects of this coupling on the less noble and the more noble metal are discussed in this chapter and represented by Evans diagram. The main factors influencing the extent of corrosion rate increase are analysed: availability of a

Fig. 10.1 Case study at the PoliLaPP Corrosion Museum of Politecnico di Milano

© Springer Nature Switzerland AG 2018
P. Pedeferri, *Corrosion Science and Engineering*, Engineering Materials,
https://doi.org/10.1007/978-3-319-97625-9_10

driving force for galvanic corrosion, its possible dissipation in cathodic overvoltage and ohmic drop in the electrolyte, ratio between cathodic and anodic areas. Finally, prevention of this localized corrosion phenomenon is briefly described.

10.1   Effects on Metal Corrosion

Galvanic corrosion, also called *bimetallic corrosion*, occurs when two metals immersed in an electrolyte are in electrical contact and are characterized by a different practical nobility, i.e., a different free corrosion potential. The less noble metal, M, with more negative potential, works as anode and its corrosion rate is accelerated by the coupling (Fig. 10.1). The noble metal, N, with more positive potential, behaves as cathode, hence its corrosion rate decreases up to a halt. In addition, materials with electronic conductivity can work as cathode, such as magnetite that forms near welds, calamine in hot rolling, magnetite in boilers, and also graphite and sulphides in industrial processes. A current, I, so-called galvanic current or macrocouple current, circulates inside the electrolyte from less noble metal to more noble and within metallic circuit in opposite direction (Fig. 10.2).

The first mention of the galvanic effect can be dated on 1799 by Giovanni Fabbroni, who wrote on Journal de Physique: "[…] *I have observed that the alloys used for soldering roof copper plates of the Observatory of Florence have rapidly transformed, altered to a white oxide, only at the junctions* […]". A few years later on 1805, Davy suggested to prevent corrosion attack through a galvanic coupling, by which cathodic protection was born.

Besides an acceleration of the corrosion rate of less noble metal, there may be side effects such as:

- Possible passivation of less noble metal (titanium in reducing environments)
- Hydrogen evolution on more noble metal with possible hydrogen embrittlement of susceptible metals (high strength steels, titanium)
- Reduction of oxides, sulphides or other chemical species present on cathode surface.

By introducing the surface area of the anodic and cathodic metals, S_M and S_N respectively, referring to Fig. 10.3, the macrocell current, I, is equal to the anodic current exiting the less noble metal, I_M, and to the cathodic current, I_c:

$$I = I_M = I_a = I_c \tag{10.1}$$

Fig. 10.2 Galvanic corrosion of metal M (less noble) by coupling with more noble metal N

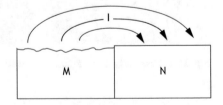

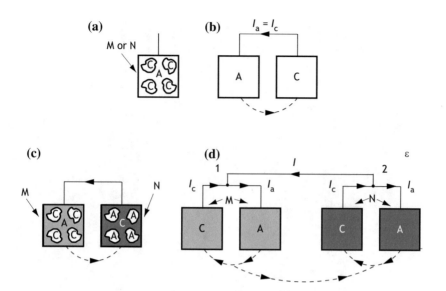

Fig. 10.3 Scheme of galvanic current between metal M, less noble, and metal N, more noble

The anodic current exiting the metal M is:

$$I_a = i_{corr} \cdot S_M \qquad (10.2)$$

where i_{corr} is the corrosion rate on the less noble metal M of exposed surface S_M. The cathodic current is the product of the cathodic current density, i_c, by the total surface where the cathodic process occurs:

$$I_c = i_c \cdot (S_M + S_N) \qquad (10.3)$$

By inserting Eqs. 10.2 and 10.3 in Eq. 10.1, the corrosion rate on the less noble metal M in a galvanic coupling is obtained:

$$i_{corr} = i_c \cdot (1 + S_N/S_M) \qquad (10.4)$$

Galvanic corrosion rate is proportional to the cathodic current density, i_c, and to the surface area ratio $(S_M + S_N)/S_M$. When $S_N \gg S_M$ the ratio approximates to S_N/S_M. To show the practical effect of surface area ratio, Evans proposed his students to carry out the experiment illustrated in the following box.

Figures 10.4 and 10.5 show two examples of galvanic attack in industrial applications caused by the coupling with more noble materials, as graphite used as a gasket and magnetite that forms near welds.

Fig. 10.4 Galvanic corrosion of stainless steel, AISI 304 grade, coupled with graphite in aerated flowing acid solution

Fig. 10.5 Galvanic corrosion of weld in the presence of a magnetite layer

Evans' Experience on Galvanic Corrosion

To give students a practical demonstration of the effect of surface area ratio in galvanic corrosion, Evans invited them to carry out the following experiment.

Prepare a carbon steel plate of surface area 1 m^2, one face coated, and dip it in seawater in a harbour zone, practically in stagnant condition. The mass loss after 1 year exposure is 780 g. Corrosion rate is then 100 μm/year (the reader is asked to verify this calculation).

Now, prepare another twin carbon steel plate and apply a gold plating on half of the bare surface, then dip it in same harbour zone. The mass loss after 1 year exposure is as before, 780 g. Root cause is because of same oxygen availability on 1 m^2 plate (no matter if the surface of the plate is half steel and half gold). But, corrosion rate of steel is 200 μm/year.

Now, prepare a third carbon steel plate with ¾ of steel bare surface gold plated and again dip it in the same harbour zone. The mass loss after 1 year exposure is again 780 g. So the corrosion rate of exposed steel (1/4 of the 1 m^2 surface area) is 400 μm/year.

And so on, by preparing other carbon steel plates, gold plated for 7/8 or 9/10 of the bare surface, then obtaining a corrosion rate 8 times and 10 times higher, respectively. The influence of the surface area ratio was then evident.

Silverware Cleaning

A curious household beneficial effect of galvanic coupling is the method used to clean silverware which blackens by the presence of traces of H_2S in the atmosphere to form a thin film of brownish silver sulphide. The procedure consists of putting silverware in a sodium bicarbonate solution in an aluminium pot; to speed up the process, near boiling water is suggested. After a while, black spots disappear magically, even in recesses.

What happened?

Simply, the effect of the galvanic coupling between aluminium, the pot, and silver, the more noble metal, on which the cathodic reaction (hydrogen evolution) occurs then reducing silver sulphide to silver; on the anodic side a slight surface corrosion of aluminium takes place. Instead of an aluminium pot, which is now difficult to find in a kitchen, then an aluminium foil on the bottom of any pot type can be used. To increase the cleaning effect, magnesium can be used, if available, instead of aluminium: still it is a matter of galvanic coupling.

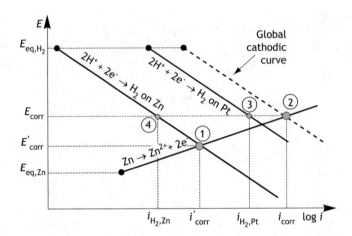

Fig. 10.6 Simplified Evans diagram for galvanic coupling of Zn and Pt in an acid solution

10.2   Galvanic Effects on Less Noble Metal

Let's consider a galvanic coupling in acidic solution between an active metal, for example zinc, and a more noble metal, for example platinum. On zinc, both anodic and cathodic processes occur simultaneously, whilst on platinum only a cathodic process takes place. A simplified E-log$i$ plot, shown in Fig. 10.6, can conveniently represent corrosion conditions, set before and after coupling, where current coincides with current density and ohmic drop is disregarded.

Point ① represents free corrosion condition (i'_{corr} and E'_{corr}) of zinc before coupling, point ② corrosion condition (i_{corr} and E_{corr}) of zinc after coupling. The final condition is obtained by summing cathodic current density ③ on platinum and the one of point ④ on zinc, which equals anodic current density on zinc (condition of electro-neutrality). The dotted line represents the global equivalent cathodic curve. It appears that, after coupling, corrosion rate of zinc increases from i'_{corr} to i_{corr} and hydrogen evolution splits on platinum and zinc, where on the latter it decreases (compare ① and ④).

Let's consider now an active–passive metal as titanium coupled with platinum in a de-aerated acidic solution. As for previous case study, Fig. 10.7 shows a simplified E-log$i$ plot of corrosion conditions before, point ①, and after coupling, point ②. Titanium in point ① is on active zone, showing a corrosion rate, i'_{corr}; in point ② titanium is in the passivity zone and the corrosion rate, i_{corr}, drops to passivity current (at least, one order of magnitude lower). According to this result, in this case the galvanic coupling produces a protection effect, quantified by the difference $i'_{corr} - i_{corr}$. Point ③ and point ④ indicate hydrogen evolution on platinum and titanium, respectively. Potential moves from E'_{corr} to E_{corr}.

In summary, by a galvanic coupling with a noble metal, corrosion rate of an active less noble metal increases, conversely the one of an active–passive metal may decrease.

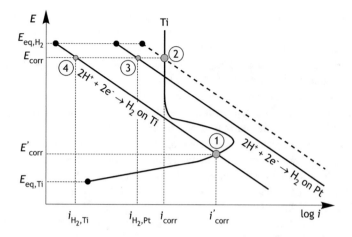

Fig. 10.7 Simplified Evans diagram for galvanic coupling of Ti–Pt in deaerated acid solution

10.3 Galvanic Effects on More Noble Metal

Let's consider an opposite case study as seen above, where a metal, for example iron (i.e., mild steel) is coupled with a less noble metal, for example zinc. On iron, both anodic and cathodic processes occur simultaneously, while on zinc the anodic process takes place, only.

Again, assuming that current coincides with current density, the simplified E-$\log i$ plot of Fig. 10.8 refers to the galvanic coupling in an acid solution. Point ① represents free corrosion condition ($i'_{corr,Fe}$ and E'_{corr}) of iron before coupling, point ② corrosion condition (i_{corr} and E_{corr}) of iron after coupling. The final condition is

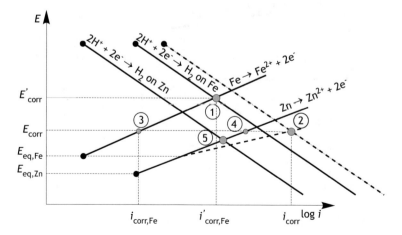

Fig. 10.8 Simplified Evans diagram for galvanic coupling of Fe–Zn in an acid solution

obtained by summing anodic current density ③ on iron and the one of zinc, point
④, which equals cathodic current on iron (condition of electro-neutrality). The
dotted line represents the global equivalent anodic curve (of iron plus the one of
zinc). It results that, after coupling, corrosion rate of zinc increases and corrosion
rate of iron decreases (compare ① and ③).

In practice, by changing the anodic to cathodic area ratio, dotted line can cross
cathodic line in point ② at a potential below the equilibrium potential of iron: when
this condition sets up, the corrosion rate of iron zeros.

Figure 10.9 shows the iron-zinc galvanic coupling in oxygen-containing neutral
solution: the corrosion condition changes from ① (free corrosion of iron) to point
② that is close to free corrosion condition of zinc, so that corrosion of iron stops.
This is the so-called *immunity condition* that is achieved by applying cathodic
protection (see Chap. 19).

When the more noble metal is active–passive, again coupled with zinc and
operating in high oxidizing condition, as shown in Fig. 10.10, potential moves from
corroding point ① to point ③ within the passive range, established by the corrosion potential of zinc (point ②). At point ③ corrosion rate, $i_{corr,M}$, is the passivity
current density, hence negligible. This effect is named *cathodic protection by
passivity* (see Chap. 19); typical examples are the galvanic coupling of stainless
steel with zinc or also iron in seawater and galvanic coupling between passive steel
and zinc in concrete.

Figure 10.11 shows the hypothetical simplified E-log$i$ plot when the active–
passive metal works in passive state (point ①) and is coupled with a less noble
metal as Zn. Operating condition of this galvanic coupling would be point ②, at
which cathodic process occurs on active–passive metal, and global anodic process
would be the sum of corrosion rate of zinc (point ③) and the one of noble metal in

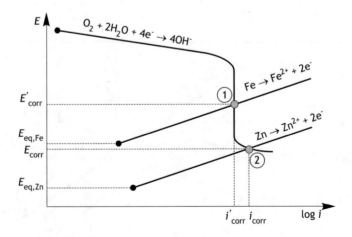

Fig. 10.9 Simplified Evans diagram for galvanic coupling of Fe–Zn in a neutral aerated solution

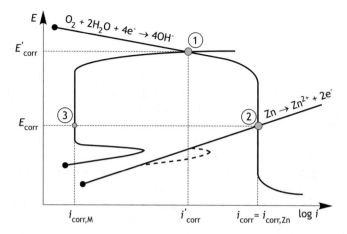

Fig. 10.10 Principle of cathodic protection by passivity through a galvanic coupling with a less noble material

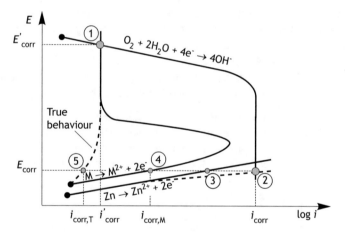

Fig. 10.11 Simplified Evans diagram for a galvanic coupling between passive metal and a less noble metal

active condition (point ④). This representation is misleading since it implies that the passive metal would become active and corrosion rate would increase from i'_{corr} to $i_{corr,M}$, while the noble passive metal remains passive because the alkalinity produced by the cathodic process (i.e., oxygen reduction) impedes the activation, hence the corrosion condition moves to point ⑤, $i_{corr,T}$. Accordingly, the dotted line represents the true behaviour.

10.4   Galvanic Coupling Representation by Evans Diagrams

Simplified Evans diagrams considered so far in this Chapter (Figs. 10.6, 10.7, 10.8, 10.9, 10.10 and 10.11) are used because they help understand the effects of a galvanic coupling. However they are misleading. In fact, when measuring the potentials of the metals involved in the coupling, either in laboratory or on field, the operating conditions predicted by these representations, i.e. measured working potential, do not fit. In fact in all environments, even in high conductive ones as seawater, the potential of the two coupled metals is different and never given by the crossing point of anodic and cathodic curves.

There is a twofold reason: first, anodic and cathodic current densities cannot be considered equal because their ratio, i_c/i_a, is as a minimum 2. Second, and most important, the macrocell current established by the galvanic coupling causes an ohmic drop, IR, in the electrolyte; in other words, to have the macrocell current circulation in the electrolyte, the potential of the cathode (positive pole of the established cell) must be higher than the anode (negative pole). The set up of the ohmic drop in the electrolyte determines the designation of the two following parameters: the potential of the anodic metal, E_{AN}, and the potential of the cathodic metal, E_{CATH}, after the galvanic coupling.

In the following graphs (Figs. 10.12, 10.13, 10.14, 10.15, 10.16 and 10.17) the real working conditions are reported. In each graph, only involved anodic and cathodic curves are reported. Highlighted points have the following meaning:

- Point 1: free corrosion potential of more noble metal
- Point 2: free corrosion potential of less noble metal
- Point 3: corrosion potential of more noble metal after coupling, E_{CATH}
- Point 4: corrosion potential of less noble metal after coupling, E_{AN}.

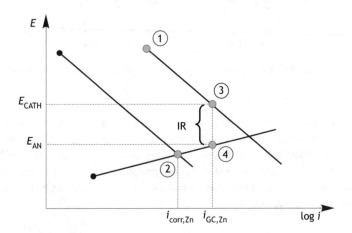

Fig. 10.12 Simplified Evans diagram for galvanic coupling of Zn and Pt in an acidic solution taking into account the ohmic drop

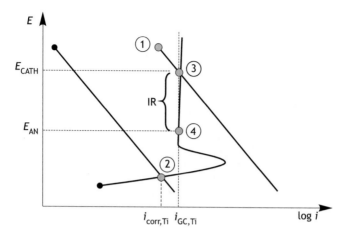

Fig. 10.13 Simplified Evans diagram for galvanic coupling of Ti and Pt in de-aerated acidic solution taking into account the ohmic drop

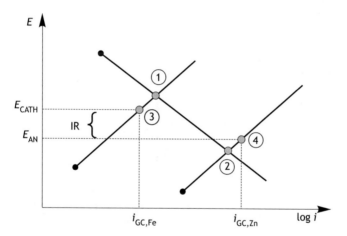

Fig. 10.14 Simplified Evans diagram for galvanic coupling of Fe and Zn in an acidic solution taking into account the ohmic drop

To calculate the potential difference between the two metals, which is also the driving voltage, ΔV, of the cell, reference has to be made to Chap. 9. In short, the following relationship applies:

$$\Delta V = E_{\text{CATH}} - E_{\text{AN}} = IR \cong i_{\text{c}} \cdot \rho \cdot \left(\frac{S_{\text{c}}}{k \cdot S_{\text{a}}}\right)^2 \tag{10.5}$$

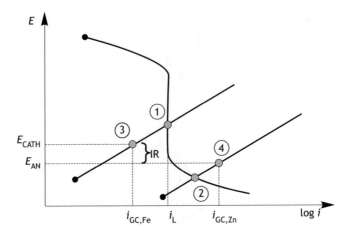

Fig. 10.15 Simplified Evans diagram for galvanic coupling of Fe and Zn in a neutral aerated solution taking into account the ohmic drop

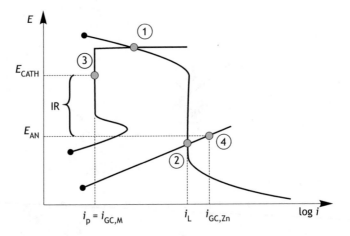

Fig. 10.16 Principle of cathodic protection by passivity of an active–passive metal through a galvanic coupling with a less noble metal

where symbols are known and k is a constant, about 20 m$^{-0.5}$ for plate-like geometry. It is important to note that when using this relationship, surface areas S_a and S_c cannot be considered independent, as discussed in Paragraph 10.5.3.

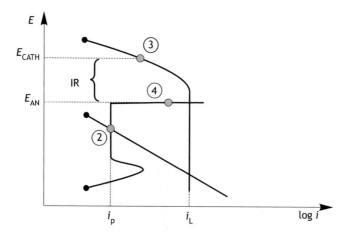

Fig. 10.17 Simplified Evans diagram for galvanic coupling of stainless steel and graphite (example reported in Fig. 10.4) taking into account the ohmic drop

10.5 Four Main Factors

In summary, galvanic coupling depends on the following main four factors:

- Practical nobility which determines potential difference between coupled metals
- Overvoltage of cathodic reaction on more noble metal
- Surface area ratio
- Electrolyte conductivity.

10.5.1 Practical Nobility

The driving voltage set by the galvanic coupling is the difference between the two free corrosion potentials in the environment, which depends on nature, composition and structure of the metal, presence of oxide films or other compounds on metal surface, composition, temperature and oxidizing power of the electrolyte. The rank of practical nobility of metals depends on the environment to which the coupling is exposed: aerated, stagnant or turbulence conditions. Table 10.1 shows the practical nobility rank of metals in seawater. To illustrate how practical nobility varies in a wide range as electrolyte condition changes, let's consider a stainless steel immersed in seawater:

- If seawater is aerated, stainless steel remains passive and free corrosion potential is close to equilibrium potential of oxygen reduction process, then approaching the equilibrium potential of copper; instead, as soon as corrosion starts, potential drops to a more negative system, approaching the equilibrium potential of iron

Table 10.1 Ranking of metals in seawater based on practical nobility (from LaQue 1975)

More noble	
Platinum	Brass
Gold	Hastelloy B (60 Ni, 32 Mo, 6 Fe, 1 Mn)
Graphite	Nickel (active)
Titanium	Tin
Silver	Lead
Hastelloy C (62 Ni, 18 Cr, 18 Mo)	Stainless steel (active)
Stainless steel (passive)	Cast iron
Nickel (passive)	Mild steel
Soldering alloy	Cadmium
Monel (60 Cu, 40 Ni)	Aluminium (commercially pure)
Copper nickel	Zinc
Bronze	Magnesium and alloys
Copper	Less noble

- In seawater with low oxygen content or even deaerated, potential drops from a noble value due to passive condition, then close to copper, to a value typically close to active conditions, then close to iron, even if it is in passive state.

Mix-Ups

Metals may change position in the rank of practical nobility by changing environmental conditions. In some cases, there may be a real reversal. There are considerable potential variations from standard condition when anions forming complexant or insoluble salts are present (thermodynamic issue, see Chap. 3); also kinetics effects give similar changes. For example, in general, zinc is less noble than iron and then protecting iron in a galvanic coupling (i.e., by cathodic protection). However, at temperatures above 40 °C the formation of an oxide with properties of a semiconductor makes zinc in absence of chlorides more noble than iron, then causing corrosion rather than protection of iron. Similarly, tin is generally cathodic against iron, instead in presence of food substances which passivate iron and form complex with tin, there is a reversal so that tin becomes anodic and iron cathodic. Similarly, iron is less noble than copper, nevertheless in phosphate containing environments iron passivates, then becoming more noble than copper.

Influence of Surface Condition

Sometimes, to predict galvanic corrosion, the *practical nobility* to be considered is not the one of initial surface condition, instead it is the stationary one established by reactions occurring on early exposure stage. This is the case of brass when dezincification takes place, so that surface composition changes strongly, giving a thin copper-rich layer, with practical nobility as copper. A similar behaviour occurs on grey cast iron in some electrolytes where iron is selectively etched so that a graphite film forms on its surface. The opposite occurs in case of local destruction of a surface film due to high turbulence or abrasion effects, which brings active the initially passive metal, hence changing practical nobility.

10.5.2 Cathodic Overvoltage on More Noble Metal

Catalytic properties that noble metals exert on the cathodic reaction influence the corrosion rate in galvanic coupling. This is the reason why often passive metals (which are covered with oxide films) are not very effective in accelerating the galvanic attack of less noble metals because the most frequent cathodic process, i.e., oxygen reduction, takes place slowly. For example, let's consider galvanic coupling of commercial aluminium with copper or with stainless steel in seawater: although copper and stainless steel show roughly same nobility (i.e., same free corrosion potential), hence same driving voltage, the corrosion rate of aluminium is much lower in stainless steel-aluminium coupling than in copper-aluminium one, for about an order of magnitude. This different behaviour depends on different over-voltage of oxygen reduction, which is much higher on stainless steel than copper; instead, overvoltage on titanium is even higher than stainless steels and is much lower on magnetite (iron oxide that forms a passive film on carbon steel).

10.5.3 Surface Area Ratio and Maximum Corrosion Rate

An important factor in defining corrosion rate in galvanic coupling is *surface area ratio*, as Evans' experience, reported in the box, clearly demonstrated. Accordingly, the corrosion rate in the presence of a galvanic coupling, $C_{rate,GC}$, is given by:

$$C_{rate,GC} = \beta \cdot i_c \cdot \frac{S_c + S_a}{S_a} \cong \beta \cdot i_c \cdot \frac{S_c}{S_a} \tag{10.6}$$

where the conversion factor, β, in the case of iron is 1.2 (1 A/m^2 = 1.17 mm/year $\cong$ 1.2 mm/year); i_c (A/m^2) is the current density of the cathodic reaction, and S_c and S_a are the cathodic and anodic surface areas, respectively.

However, unlike Evans' experiment, surface area ratio is not always the one we think or simply we measure, because it is determined by the electrical field setup by the galvanic macrocell; in other words, it is determined by the throwing power, L_{max}, of the macrocell, as discussed in Chap. 9. It can be stated that the maximum surface area ratio is given by Eq. 9.26 as follows (from Lazzari 2017):

$$\left(\frac{S_c + S_a}{S_a}\right)_{max} \cong \left(\frac{S_c}{S_a}\right)_{max} \cong k\sqrt{\frac{\Delta V}{i_c \cdot \rho}} \tag{10.7}$$

where ΔV (V) is ohmic drop or driving voltage (i.e., the practical nobility difference of the two metals), ρ (Ω m) is electrolyte resistivity, i_c (A/m^2) is cathodic current density and k is an experimental constant. By comparing Eqs. 10.6 and 10.7, it results:

$$C_{rate,GC} \cong k\sqrt{\frac{i_c \cdot \Delta V}{\rho}} \tag{10.8}$$

Practical experiences and FEM simulations have shown that Eq. 10.8 is applicable for the following intervals: $1 < \rho$ (Ω m) $< 10^3$ and $0.01 < i_c$ (A/m^2) < 1.

In summary, surface area ratio depends on:

- Driving voltage
- Cathodic current density which generally varies for oxygen limiting current density in the range $0.01 < i_L$ (A/m^2) < 1
- Electrolyte resistivity ($1 < \rho$ (Ω m) $< 10^3$)
- Geometry of the domain, through constant k, generally taken as 20 m$^{-0.5}$.

10.5.4   Electrolyte Resistivity

As stated by Eq. 10.8 corrosion rate of a galvanic coupling depends strongly on electrolyte resistivity; assuming the same driving voltage, corrosion rate decreases as resistivity increases; for instance, in high resistivity electrolytes like fresh water galvanic effects are often negligible, whilst in high conductivity ones like seawater corrosion rate is at least two orders of magnitude greater.

Electrolyte resistivity determines also the throwing power, therefore the extension of the attack on the anodic (i.e., less noble) metal. As resistivity increases, anodic and cathodic processes tend to localize close to the coupling boundary (low throwing power) and the opposite occurs in high conductivity electrolytes (high throwing power).

As a rule of thumb, resistivity influences galvanic coupling in an open domain as follows:

- Distilled water ($\rho \approx 2000$ Ω m or $\sigma \approx 5$ μS/cm): affected zones do not extend beyond some tenths of a millimetre
- Fresh waters ($\rho \approx 20$–50 Ω m or $\sigma \approx 200$–500 μS/cm): affected zones do not extend beyond a few centimetres
- Seawater ($\rho \approx 0.2$ Ω m or $\sigma \approx 5$ S/m): galvanic effect extends to distances of the order of meters.

10.5.5 Geometry of the Domain

Besides electrolyte resistivity, domain geometry is another important factor influencing galvanic corrosion: for example in small diameter tubes or thin film like electrolyte, as it happens on atmospherically exposed surfaces, the extension of affected (working) areas is greatly reduced; however, strong condensation in marine atmospheres leads to a galvanic corrosion not negligible as it would be in less aggressive atmospheres.

10.6 Prevention

With reference to Figs. 10.18, 10.19 and 10.20, prevention of galvanic corrosion is achieved by:

- Avoiding dangerous couplings with a choice of metals close in scale of practical nobility
- Separating coupled metals, for example, by insulating flanges
- Taking care that the anodic-to-cathodic area ratio is not unfavourable ($S_a \gg S_c$)
- Applying paints on both surfaces or only on the cathodic one. Avoid painting of anodic metal, only
- Applying cathodic protection.

Fig. 10.18 Example of galvanic corrosion control by insertion of a sacrificial replaceable unit

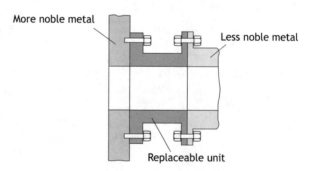

Fig. 10.19 Insulating flange

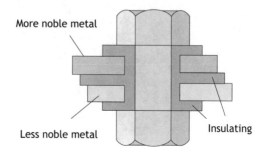

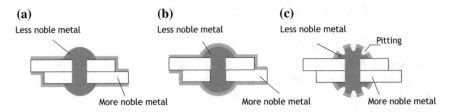

Fig. 10.20 Example of corrosion prevention by protective coatings: **a** correct; **b** correct; **c** incorrect

Electrical insulation as prevention method for galvanic coupling is not always easy to obtain in practice. For example, consider a pump and relevant piping. The electrical contact takes place not only through the connecting flanges, but also through the structural supports, other pipes and most likely through the grounding system.

Painting may be used as prevention method, provided its application on cathodic areas, i.e., more noble metal, or on whole coupling, i.e., both anodic and cathodic areas. Painting of less noble metal only (as sometimes occurs in trying to protect the zone that corrodes) is very dangerous because, in the presence of defects, surface area ratio is largely unfavourable (Fig. 10.20c).

The Professor and Seadog

The writer heard this story from Professor Hoar during a lecture on galvanic corrosion in Cambridge, UK, mid 1970s. As professor Hoar was a brilliant actor, Pedeferri thought that the story could not be true, nonetheless very instructive for students. In 1940 in Virginia, USA, in spite of echoes of war coming from Europe, the owner of a wealthy tobacco farm decided to purchase a 24-m long yacht, named *Seadog*, made of an exotic material for those times: the monel, a nickel-copper alloy (about 65% Ni and 30% Cu). Indeed, he could afford it.

Not to lose time while designing the boat, he started purchasing hull sheets, meanwhile Pearl Harbor attack forced the U.S. to enter the war. Monel became strategic material, so it was no longer possible to purchase rivets made of monel and steel was then proposed and agreed. Since galvanic corrosion was a concern, a professor of a nearby University was asked to investigate the matter. The professor, scrupulously, connected a monel sample with a rivet and measured the galvanic current once immersed in seawater. Such current was small and then the professor concluded that galvanic corrosion was acceptable and steel rivets would have solved the problem for at least two decades. "It is enough" said the "tobacconist" and the yacht construction began. The *Seadog* was built in record time, then launched and after six weeks sank.

Hoar concluded: "In front of the coast of Virginia, plates of monel of the poor *Seadog* still lie". What happened? The Professor perfectly measured the current flowing between monel and steel, but the geometry or surface area ratio was much different, so corrosion rate in seawater was some orders of magnitude higher than what measured in laboratory. Truly, the story was referred by Speller (1926) some years before with some small different details.

Draft Beer

This case study is taken from Fontana's book (1986). In the 1950s in United States, in beer brewery industry, although beer is not a particularly aggressive solution, tanks were made of carbon steel, internally coated with phenolic resin, to avoid contamination by corrosion products, which alter beer taste. The protective coating served the purpose provided no mechanical damage was present, especially on the bottom. To overcome this inconvenience, a company decided to change the bottom only and not the wall of tanks, by replacing phenolic resin coating with a stainless steel clad, AISI 304 grade. After a few months of service, pinhole corrosion, never seen before, appeared in a narrow band above the weld, near stainless steel, because of galvanic corrosion through coating porosity and the bare stainless steel surface.

Galvanic Corrosion of Artefacts Exposed to the Atmosphere

In dry environment and in the absence of condensed water, as often happens inside buildings, galvanic corrosion is not a concern for atmospherically exposed artefacts. The iron crown kept in a reliquary in the Cathedral of Monza, Italy, is an ideal example to prove it. The crown is a ninth century artefact and, according to tradition, crowned the kings of Italy in the middle age and Napoleon two hundred years ago. It consists of a circle of gold studded with gems and diamonds, carrying inside a thin strip of iron, which, according to tradition, was fashioned with one of the Holy Cross nails. (There are many doubts about the nature of the metal because in 1985 it was found that the nail is not magnetic, so it may not be iron but zinc or tin or silver; nevertheless, we continue to believe in it). A perfect galvanic coupling, which has never worked for a thousand and more years because exposed to dry atmosphere: the iron strip is perfectly preserved so far.

Going from inside to outside the dome, things change. Figure 10.21 shows a detail of a damage occurred on a monument dedicated to heroic soldiers due to galvanic corrosion. In fact, to fix labels of soldier names, made of bronze, carbon steel rivets were erroneously used. The rainwater film, made acidic by pollution during '60–'70 of XX century, produced a galvanic attack of the rivet. The picture shows also the area that has worked as cathode.

Fig. 10.21 Example of galvanic coupling: a bronze plate fixed with carbon steel rivets

Particular and unique examples of galvanic corrosion of gilded bronze artefacts exposed to the atmosphere are Marco Aurelio in Campidoglio, Rome; the Grifo in Perugia; the doors of the Baptistery of Florence (two by Ghiberti, one is named Paradise, and one by Pisano); the horses of Venice. Indeed, these artefacts are batteries ready to work, although the driving force is modest. Corrosion of bronze occurs at gold foil pinholes, but for almost five centuries, no damage took place because corrosion products were insoluble and protective by plugging the pinholes. Unfortunately, since '50s of 20th century, things have changed because due to acid rains corrosion products became soluble. The writer in the mid-70s could see under the belly of one of the horses of Venice, at that time still exposed in S. Marco Square, blue corrosion products of copper sulphate, a clear sign of occurring corrosion. Although in the other mentioned cities the environment is less critical than that of Venice, wisdom administrators decided to preserve such monuments into a museum and copies were exposed.

Also the Statue of Liberty in New York suffered galvanic corrosion and a repair was necessary in early 1990s. The copper sheets of the statue are internally fixed to a carbon steel frame, which corroded due to the formation of a layer of condensed water. Carbon steel was replaced with stainless steel, then limiting the galvanic effect of the coupling.

10.7   Questions and Exercises

10.1   What are the four main factors affecting galvanic corrosion?

10.2   Represent by means of Evans diagram the electrochemical free corrosion condition of the following materials exposed to seawater, making a ranking based on their practical nobility: copper, super-austenitic stainless steel, AISI 304 stainless steel, titanium, zinc, and mild steel.

10.3   Discuss the effect of electrolyte resistivity on the maximum corrosion rate in galvanic coupling.

10.4   Corrosion rate of commercial aluminium is higher when coupled with copper rather than with stainless steel, although copper and stainless steel show roughly same free corrosion potential. Why?

10.5   Discuss by means of Evans diagram the electrochemical behaviour of stainless steel, Pt enriched on surface, to have stable passive behaviour.

10.6   Give an exhaustive comment of Fig. 10.16. Compare galvanic coupling effects described in Figs. 10.15 and 10.16.

10.7   Write a testing procedure for measuring anode and cathode potential as well as potential distribution on a galvanic coupling.

10.8   A carbon steel plate (1 m^2 exposed surface area) and a zinc plate (16 cm^2 exposed surface area) are coupled in stagnant seawater containing 10 mg/L of oxygen.

- Which is the anodic material?
- How is current direction in seawater?
- Determine corrosion rate of the anodic material
- Determine the time for total consumption of the anodic material if initial thickness is 10 mm.

10.9   A carbon steel plate, 5 mm thick, is coated with a copper layer, 0.1 mm thick and is completely immersed in fresh water ($\rho = 20$ Ω m). A localised defect of copper layer, 10 cm^2 large, is present.

- Which corrosion form is possible?
- Which is the anodic area? And the cathodic area?
- Draw schematically corrosion current path
- If oxygen limiting current density is 20 mA/m^2, when will the plate be perforated?

10.10 A localised corrosion attack occurred on a stainless steel plate (square shape, 1 m wide; 5 mm thick) immersed in seawater. The anodic area was 1 cm^2. Draw corrosion current path and estimate corrosion rate. What would have been the corrosion rate in fresh water assuming same oxygen limiting current density of 50 mA/m^2?

Giovanni Fabbroni

He was a prominent Florentine intellectual across XVIII–XIX centuries and for many years deputy director of the Imperial Royal Museum of Physics and Natural History in Florence. Vivid animator of Florentine cultural life, he showed interest in many different fields, from economics to chemistry and also agrarian and justice. In 1792 at the Academy of Georgofili, in Florence, he presented a paper on the action of metals when coupled, which was published only in 1799 in *Journal de Physique* (49, 348, 1799) with title: *"Sur l'action chimique des différent métaux entr'eux, à la température commune de l'atmosphère; et sur l'explication de quelques phénomènes galvaniques,* (About the chemical action of different metals coupled at atmospheric temperature; and on the explanation of some galvanic phenomena). Piontelli wrote that with this work Fabbroni *"founded the chemical theory of galvanism and laid the foundations of galvanic corrosion theory ten years before Volta's invention."*

His paper is primarily the result of acute observations. For example, he wrote: *"I noticed that alloys used to solder copper plates on the mobile roof of the Observatory of Florence had rapidly transformed, altered into white oxide, right at contact with copper plates. I also knew that iron nails that fasten copper sheets of ships hulls rusted so much that their stem expanded even to exceed their head size."*

It is also the result of a series of ingenious experiences such as the following: *"In a pot filled with water I put some golden foils, in another pot, silver and in a third copper and then in others, tin, lead and so on. In other pots I put the same metals two by two, separated by a small glass plate, one more and the other less oxidizing.*

Finally, in a third series of glasses the same couples of metals in contact each other. The first two series showed no change, whereas in the third, the more oxidizing metal became visibly oxidized immediately after being put in contact with another metal and the oxide grew gradually. This phenomenon began, albeit imperceptibly, as contact was made [...] but after a month I observed that different metals not only became oxidized but on their surface were formed even small salt crystals of different shapes. It seemed, therefore, that a chemical action [between metals] took place in a clear manner." He further stated: *"I believe that from these and other observations we have to recognize that metals in these cases exert a reciprocal action that is the cause of phenomena that occur following their joining or when they come in contact."*

In the past, scientists' opinion on these statements has been contradictory. Someone have acknowledged Fabbroni's observations on corrosion of coupled metals the embryo of the chemical theory of batteries; others have even ignored that he certainly marked the beginning of the correlations between corrosion and galvanic coupling. Today, nobody put in doubt that Fabbroni's paper has been one major scientific event.

P.S. The beautiful Museum of History of Science in Florence, heir to the Museum where Fabbroni worked, which devotes an entire room to his former deputy director, does not mention this important contribution. We hope there will be a remedy.

Alessandro Volta, the Practical Potential Ranking and the Driving Force

In 1792 the future inventor of the pile decided to classify the metals in relation to the greater or lesser capacity of the bimetal arc formed by them *to excite stronger convulsions in the frog* and to give acidic or rather basic flavors on the tongue tip. With these two techniques, Volta was able to evaluate the extent and direction of the *electric current around the circuit* consisting of the bimetallic arc and the body of the frog, and then he decided which of the two metals is more noble. Then, comparing all the metals two by two, he built up *the scale of the conductors of the first class, which possess a different power to push the electric fluid and drive it forward in the wet conductors, or second class*: in practice a sort of scale or series of potentials, indeed the prototype of the series of potentials that will come later. And so he wrote: *one can comfortably split the metals in three categories, placing tin and lead in the lower one, iron, copper and brass in the middle, and gold, silver and platinum in the upper one. So then it is more useful to oppose to one of the lower rank, that is to lead or tin, one of the higher rank, gold or silver and the latter maximum.*

This is not yet the scale that Volta will establish the following year, which concerns a greater number of metals and alloys and different *"mines"*, including manganese dioxide and carbon. Volta rightly held a lot on this scale and he claimed its priority. He wrote: "*Practically I had already sketched it at the beginning of 1793 [...]. It differs little from the other scale or series that Dr. Pfaff gave us in 1793.*

One is admired by the work that, seven years before the invention of the pile, Volta was able to accomplish, but also by the sharpness of his observations. For example, the scientist from Como notes the importance of the nature, the composition, the structure of metals in defining the electrochemical behavior. He wrote: *The small movements of the frog are obtained even with an electrical arc apparently constituted by a single metal because [...] even accidental small differences between the two ends of the metallic arc (differences in alloy, in hardening, in tempering, in heat and perhaps other modifications that we do not know) are enough to give movement to the electric field. [...] We can not even trust two similar coins, as there may be some difference between them.*

Volta introduced another important concept. I merely recall an investigation by Varney and Fisher on the authorship of the concept of driving force or electromotive force, or "driving virtue" to use Volta's words: obviously it is a concept of paramount importance in electrochemistry and even in corrosion (above all in galvanic coupling). The two scientists wrote in an article with a significant title: Electromotive Force: Volta's Forgotten Concept: "*We have examined over a hundred references in which electromotive force is mentioned. They include beginning as well as advanced texts, scientific dictionary, encyclopedias and papers. None credits Volta with originating the term*".

These (and other) oversights show that evidently the inventor of the pile, "*the apparatus that gave the physicists so much amazement*", has clouded the scientist who, in the nine years preceding the invention and in the two that follow it, has operated with success in experimental and theoretical electrochemistry, with very successful raids also in corrosion.

Pietro Pedeferri

Bibliography

Fabbroni G (1799) Sur l'action chimique des diff érents métaux entr'eux, à la temperature commune de l'atmosphère; et sur l'explication de quelques phenomènes galvanique. Journal de Physique 49:348

Fontana M (1986) Corrosion engineering, 3rd edn. McGraw-Hill, New York, NY

LaQue FL (1975) Marine corrosion, The Electrochemical Society monograph series. Wiley, New York, NY

Lazzari L (2017) Engineering tools for corrosion. Design and diagnosis. European Federation of Corrosion (EFC) Series, vol 68. Woodhead Publishing, London, UK

Speller FN (1926) Corrosion. Causes and prevention. McGraw-Hill, London, UK

Chapter 11
Pitting Corrosion

A thousand times better petting than pitting.

Peter Ironfoot

Abstract This chapter describes a localized corrosion attack called pitting, which is typical of active-passive metals in oxidizing chloride-containing environments: the passive film breaks locally, then corrosion proceeds at the damaged spot, few millimetres wide or even less, creating a macrocell with the surrounding intact passive metal. The influence of metal composition and environmental parameters on corrosion, pitting and repassivation potential for stainless steel in chloride containing environments is shown. Empirical parameters such as Pitting Resistance Equivalent Number (PREN) are discussed and correlated to the likelihood of pitting occurrence. The use of Pedeferri's diagram, a potential vs chloride content diagram, is also introduced, as a tool to assess corrosion conditions of an active passive metal in chloride-containing environments.

Fig. 11.1 Case study at the PoliLaPP Corrosion Museum of Politecnico di Milano

© Springer Nature Switzerland AG 2018
P. Pedeferri, *Corrosion Science and Engineering*, Engineering Materials,
https://doi.org/10.1007/978-3-319-97625-9_11

11.1 Pitting Morphology

Stainless steels, nickel-based alloys, copper and copper alloys, aluminium alloys, titanium and titanium alloys, carbon steel in concrete and other metals can be employed because a nanometre-scale oxide layer forms spontaneously on their surface, the passive film, which greatly reduces corrosion rate. When passive film breaks, localised corrosion of the underlying metal occurs, often accelerated (Fig. 11.1).

Pitting is a severe localized corrosion which produces a deep penetrating attack, the *pit*, with diameter less than a few millimetres, occurring most often isolated in a number varying from a few to several hundred per squared meter. The term pitting is often used to simply indicate a localized corrosion attack, however, it should be used more properly for the typical localized attack occurred on active-passive metals in oxidizing chloride containing environments. According to the ASTM G46 standard, a pit is defined extensive shallow or narrow deep or even elliptical, transverse, sub-skin, vertical or horizontal, as depicted in Fig. 11.2. Some examples of pitting attack are shown in Figs. 11.3, 11.4, 11.5 and 11.6.

The severity of pitting is twofold: once started, penetration rate is so high that it affects the whole metal thickness in short time; on the other hand, the attack is intrinsically of stochastic nature on either initiation time or localization; hence, prediction is a matter of probabilistic approach.

Pitting propagation is the result of a macrocell mechanism: the anodic area is inside the pit while the cathodic zone is the external surrounding passive area, where oxygen reduction is the most common cathodic process. Penetration rate is high because cathodic to anodic area ratio is as high as 100 in high conductivity electrolytes and noble cathodic process.

Fig. 11.2 Typical forms of pitting attack (from ASTM G46)

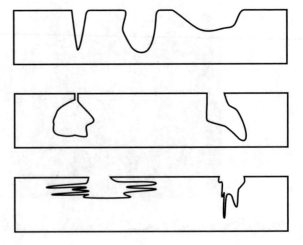

Fig. 11.3 Pitting corrosion
on a AISI 304 stainless steel
plate in the presence of
chlorides

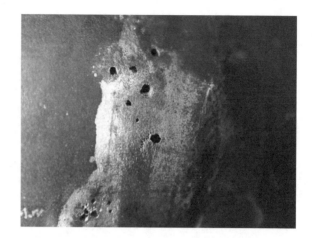

Fig. 11.4 Pitting corrosion
on a AISI 304 stainless steel
plate of a heat exchanger due
to the presence of chlorides

The circulation of macrocell current gives rise to a series of reactions and chemical modifications: inside the pit pH decreases and chloride content increases to further stimulate the anodic attack, and on the external passive surface pH increases, then helping strengthening passivity. Inside the pit, conditions necessary to trigger stress corrosion cracking may establish (see Chap. 13).

11.2 Pitting Mechanism

Pitting corrosion follows two distinct stages: pit initiation and pit propagation (Fig. 11.7).

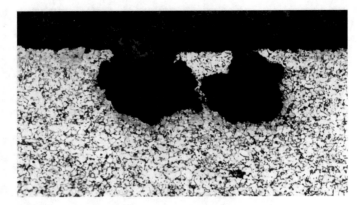

Fig. 11.5 Cavernous pits on AISI 304 stainless steel in the presence of chlorides

Fig. 11.6 Pitting-like corrosion on a carbon steel pipe transporting formation water

11.2.1 Pit Initiation

The initiation stage is the time required for the local breakdown of passive film, which is produced by the action of specific chemical species present in the environment, such as chloride ions (Cl^-) and to a lesser extent halides F^-, Br^- and I^-. It is agreed that the necessary electrochemical condition required to locally breakdown the passive film is that the cathodic process has to be more noble than a specific operational parameter, named *pitting potential*, which depends both on metal and environment properties.

This step might last a few weeks up to several months, depending on metal and operating conditions:

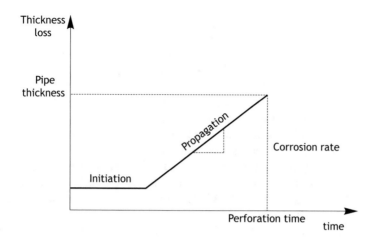

Fig. 11.7 Stages of pitting corrosion: initiation and propagation

- Strength of passive film, related to the metal chemical composition
- Inclusions
- Surface finishing
- Stagnant or turbulent fluid conditions
- Presence of biofilm and MIC
- Chloride (or halide) content
- Oxidizing species
- Continuous exposure time (wetting permanency)
- Horizontal versus vertical surfaces.

Typically, pits start where passive film is weaker or flawed (for example, near welding because of depletion of some elements, or because oxide film is too thick on work-hardened zones) or where the local environment is more aggressive due to an increase in temperature or concentration of aggressive species.

Impurities and inclusions present on metal surface can perturb passive film formation, which results weaker and thinner and also mechanically stressed, hence favouring pitting initiation.

Surface finishing strongly influences pitting initiation: smooth surfaces are more resistant or result into few, large pits, while rough surfaces experience easier initiation of numerous small pits.

In order to strengthen the passive layer, in industrial plants passivation treatments are performed. A typical treatment consists of acid pickling followed by immersion in a passivating solution, and a final rinsing in NaOH and water.

Stagnant condition favours pitting initiation, while agitation or turbulence help inhibit it. For example, for AISI 304 stainless steel, chlorides content to trigger pitting at 20 °C in agitated solutions is about 200 ppm and sometimes even more, and drops to 100 ppm or even less in stagnant solutions.

Another important aspect related to pit initiation regards bacterial activity. In 1976 the first paper on the subject appeared by the Italian researchers Mollica and Trevis (1976), and it is now well known that bacterial activity, caused by either iron bacteria or sulphur-oxidizing ones, raises the potential of the cathodic process occurring on stainless steels. This potential increase takes place after the formation of oxidizing species, such as hydrogen peroxide, or in fresh water, when manganese ions are present, due to the formation of a redox couple.

In addition, the presence of oxidizing species, which leads to a cathodic process with a noble potential, such as the presence of free chlorine, promotes pitting initiation. This experimental evidence is used in laboratory to enhance pitting initiation on stainless steels by using a solution of ferric chloride, which exhibits a noble potential as high as about +0.4 V SCE due to the redox couple Fe^{3+}/Fe^{2+} (see also the Pourbaix diagram of iron).

Another practical important factor is the continuous exposure time (or wetting permanency) as illustrated by a household example. Common cookware is made of austenitic stainless steel (typically 18-8, or AISI 304 grade); although often in contact with a chloride-containing environment, the cooking food, which is theoretically harmful for pitting initiation (even if the environment is bacteria free) pitting does not start. This behaviour depends on the limited exposure time, i.e., few minutes to few hours, which is not enough to locally destroy the passive film. Furthermore, operations as washing and drying reset the initiation time countdown. Same considerations apply to the behaviour of batch reactors made of stainless steels in food and chemical industry, which work substantially in the same way as cookware.

Finally, since inside pits the electrolyte concentrates, then increasing its density, pits can grow only on horizontal surfaces; curiously, on vertical surfaces it can initiate, then growing inside wall thickness proceeding downward vertically.

Pitting Initiation Theory

For materials that initially are typically passive, e.g. aluminium and stainless steels, it is assumed that pitting is initiated by adsorption of halide ions that penetrate the passive film at certain positions. This happens at weak points of the oxide film, e.g. at irregularities in the oxide structure due to grain boundaries or inclusions in the metal. Absorption of halide ions causes strong increase in ionic conductivity in the oxide film so that metal ions can migrate through the film. In this way, localized dissolution occurs, and intrusions are subsequently formed in the metal surface.

Another theory is that the initial adsorption of aggressive anions at the oxide surface enhances catalytically the transfer of metal cations from the oxide to the electrolyte and thus causes successive local thinning of the oxide film.

> A third possibility is that the attacks start at fissures in the passive layer. Which of the mechanisms is the most effective depends on both material and environment.
>
> from Bardal (2004)

Pitting Potential

A key parameter for pitting occurrence is *pitting potential*, E_{pit}. It is an empirical, operational parameter, i.e., not defined as the equilibrium potential or the corrosion potential, instead is obtained through a laboratory measurement procedure, now standardized (see ASTM G61), because its value is influenced by how the measurement is carried out, for instance by potential scan rate. Pitting potential, E_{pit}, which is the upper limit of passivity range, lowers as chloride concentration in the electrolyte increases as shown in Fig. 11.8.

Above E_{pit} the protective film is perforated, making the corrosion process possible. The pitting attack can therefore occur only if the free corrosion potential, E_{corr}, namely the potential of metal before pit initiation, is greater than E_{pit}; this happens when a passive metal is in contact with solutions of sufficiently high oxidizing power. Conversely, if free corrosion potential, E_{corr}, is below pitting potential, E_{pit}, $(E_{corr} < E_{pit})$ pitting does not start.

It is therefore evident that as pitting potential decreases pitting initiation becomes possible also in gradually less oxidizing electrolytes. On the other hand, as cathodic potential decrease because of reduced oxidizing power, the number of metals that can withstand pitting widens.

Fig. 11.8 Potentiodynamic curve for determination of pitting potential and repassivation potential

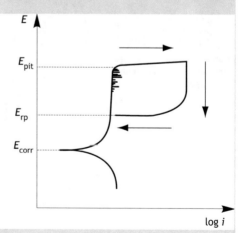

Pitting as Stochastic Event

Filed experience and laboratory testing confirmed that pitting initiation is random in nature; moreover, pit in a very initial stage should be considered unstable or metastable and can re-passivate or propagate on a stochastic or probabilistic base. Figure 11.9 shows an example of metastable pits observed during laboratory testing for pitting potential determination.

To determine pitting potential, which seems to follow a normal distribution, it can be assumed as the value at which the probability of pitting potential is 0.5, after a total number, N, of measurements obtained from the cumulative function:

$$P\left(E_{\text{pit}}\right) = \frac{n}{N+1} \tag{11.a}$$

where n is the number of samples that had pitted at a potential value, E_{pit}, under potentiostatic test condition.

The *induction time* at a given potential, E, is the measured time after the application of potential, E, at which n samples have initiated pits. Survival probability, $P(t)$, is the cumulative function:

$$P_{(E)}(t) = \frac{n}{N+1} \tag{11.b}$$

where n is number of samples that initiate pits by time t after application of potential, E. The pit generation rate, λ, given by:

$$\lambda(t) = -\frac{\mathrm{d}}{\mathrm{d}t} \ln P_{(E)}(t) \tag{11.c}$$

Fig. 11.9 Potentiodynamic curve showing metastable pit formation

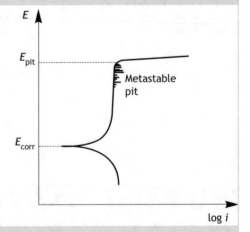

11.2.2 Propagation of Stable Pits

Once a stable pit initiates, a macrocell current starts flowing, as shown in Fig. 11.10, between the anode (i.e., where passive film has broken down and metal dissolves) and the surrounding passive zones acting as cathode. Inside pit, the solution becomes gradually more aggressive as hydrolysis reaction of metal ions proceeds, hence acidification increases and pH drops to values close to 3–4:

$$M^{z+} + zH_2O \rightarrow M(OH)_z + zH^+ \tag{11.4}$$

Conversely, on cathodic zones outside the pit, pH increases, then passive film strengthens and other pits cannot initiate within.

As corrosion proceeds, metal cations migrate and diffuse towards the pit mouth, then reacting with hydroxyl ions and precipitating as hydroxide: this configuration is called *occluded cell*. Figure 11.11 depicts an example of occluded cell for copper in hard water containing traces of chlorides, showing how hydroxide precipitates as series of layers of different oxides and salts.

Another important consequence of macrocell current flow is that chloride concentration increases inside pit; in the case of stainless steels, chloride concentration can increase more than one order of magnitude compared to bulk concentration. According to this mechanism, called *autocatalytic*, pitting corrosion proceeds inside the metal and does not spread on the surface.

Fig. 11.10 Schematic representation of an occluded cell

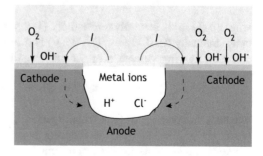

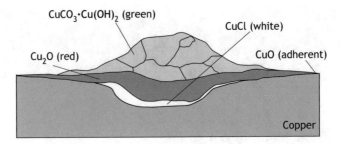

Fig. 11.11 Schematization of operculum that may form on the pit of copper tubing in contact with water (from Pourbaix 1973)

11.2.3 Corrosion Rate of Stable Pits

The corrosion rate of a pitting attack is given by the macrocell current, I_{MC}, which can be written as follows:

$$I_{MC} = I_a = I_c = i_{corr} S_a = i_c(S_c + S_a) \tag{11.5}$$

where meaning of symbols is known. Accordingly, corrosion rate, which is *pitting penetration rate*, $C_{rate\text{-}pit}$, is given by:

$$C_{rate-pit} = i_{corr} = i_c \, (S_c + S_a)/S_a \tag{11.6}$$

where surface ratio is determined by the electric field established by the macrocell. Rearranging above equation, as shown in Chap. 9, taking into account the throwing power and oxygen reduction as cathodic process, pitting penetration rate, $C_{rate\text{-}pit}$ (mm/y), has the following expression (Lazzari 2017):

$$C_{rate,pit} \cong k \sqrt{\frac{i_L \Delta V}{\rho}} \tag{11.7}$$

where ρ (Ω m) is electrolyte resistivity ($1 < \rho < 10^3$), i_L (A/m^2) is oxygen limiting current density ($0.01 < i_L < 1$), ΔV (V) is driving voltage and k is an experimental constant. Driving voltage is in general given by:

$$\Delta V = (E_c - E_{pit}) - IR \tag{11.8}$$

where E_c is potential of the cathodic process, E_{pit} is pitting potential and IR is the ohmic drop. The latter can be reduced to two main contributions: ohmic drop at anode (because of its tiny size) and ohmic drop at cathode due to the oxide film resistance. If the oxide film has good insulating properties, the macrocell is drastically reduced: this is the case of aluminium in seawater, where pit growth is quite slow because aluminium oxide is an insulator. In summary, driving voltage, ΔV, is up to 1 V for stainless steels and 0.1 V in the case of aluminium.

11.3   Pitting on Stainless Steels in Chloride-Containing Solutions

For stainless steels, pitting in chloride containing solutions is one of most threatening localized corrosion attacks. To forecast, hence to prevent, pitting corrosion of stainless steels, four parameters have to be considered:

- PREN index
- Free corrosion potential
- Pitting potential
- Repassivation potential.

11.3.1   PREN Index

In case of stainless steels and nickel alloys, experience and laboratory testing have shown the influence of metal composition on pitting susceptibility. An index called PREN (*Pitting Resistance Equivalent Number*) has been proposed and is currently used. The main agreed definition of PREN is the following:

$$PREN = [\%Cr] + 3.3 \cdot [\%Mo] + 16 \cdot [\%N] \tag{11.9}$$

As rule of thumb, stainless steels with PREN lower than 18 (as 13 or 17% Cr, or 18-8, i.e. AISI 304 type) are recommended in the presence of low chloride content or under special conditions as discontinuous operation, absence of oxygen and other oxidants, cathodic protection or favourable galvanic coupling or at high pH, such as in concrete.

Molybdenum containing stainless steels as AISI 316 type, with PREN 26, can be used for non-acidic brackish waters with chloride content up to 1 g/L, at temperature not exceeding 30–40 °C; conversely, in seawater they can suffer pitting.

Higher PREN stainless steels, such as 35–40 or higher, resist pitting attack in seawater, provided there are no galvanic couplings with carbonaceous materials and they are not anodically polarized and without chlorination treatment. For best performance, even in presence of chlorine, the use of stainless steel with a PREN greater than 45 is mandatory (such as superaustenitic steels or superduplex: for example, alloys with 6% molybdenum, such as the alloy 254 SMO).

11.3.2 Free Corrosion Potential

The free corrosion potential, E_{corr}, of a passive stainless steel, that is, before corrosion initiation, mainly depends on the oxidizing power of the solution, and then increases with the content of oxygen or other oxidizing species that may be present, such as chlorine, ferric and cupric ions.

In the case of stainless steels in seawater at temperatures below 30–40 °C, the presence of bacterial activity leads to the formation on the surface of a film consisting of biological substances, the so-called *biofilm*, which catalyses the reduction of oxygen and increases E_{corr} by more than 300 mV.

E_{corr} increases spontaneously in the presence of a galvanic coupling with more noble metals or graphite, or by an anodic polarization due to a stray current interference, and decreases under cathodic protection or cathodic polarization conditions (for instance in contact with less noble metals such as zinc, aluminium or carbon steel).

11.3.3 Pitting Potential

Pitting potential, E_{pit}, depends on both stainless steel composition and environmental conditions, namely, chloride content, pH and temperature. Figure 11.12 shows the qualitative influence of chloride content on pitting potential. Figure 11.13 summarizes the anodic behavior of two austenitic steels in a solution at a fixed Cl^- content: the first (AISI 304, PREN 18) does not contain molybdenum, the second (AISI 316, PREN 24–28) does, and shows the best behavior. Pitting potential depends also on surface finishing and, in particular, on the conditions of the passivating layer; for example, a significant reduction of pitting potential is found in so-called colored zones (*tinted zones*) composed of mixed oxides that are formed on *heat-affected-zones* of welds performed in non-controlled atmosphere or during hot forming. The original passivity is regained by removing oxides by pickling and by repassivation.

For an estimation of the pitting potential as function of stainless steel composition, chloride content, temperature, flowing conditions and electrolyte composition, reference can be made to Lazzari (2017).

11.3.4 Repassivation Potential

Once pitting has initiated, it proceeds even at lower potentials than E_{pit}; however, if potential is decreased below a value called *repassivation potential*, E_{rp}, where $E_{rp} < E_{pit}$, pits stop growing, as shown in Fig. 11.14 (Pourbaix 1973).

Fig. 11.12 Influence of chloride content on pitting potential

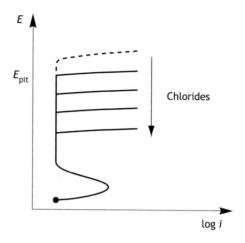

Fig. 11.13 Example of influence of stainless steel composition on pitting potential

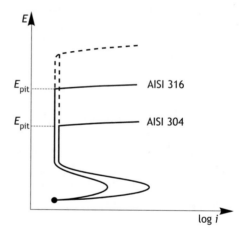

Fig. 11.14 Anodic curve of an active-passive material identifying pitting and repassivation potentials

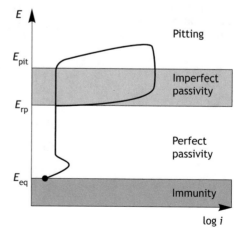

Pitting potential, E_{pit}, and repassivation potential, E_{rp}, indicatively about 300 mV lower, identify three potential ranges:

- $E > E_{pit}$: the attack starts and proceeds
- $E < E_{rp}$: *perfect passivity*, the attack cannot start and, if already started, stops
- $E_{rp} < E < E_{pit}$: *imperfect passivity*, the attack does not start and, if already started, it proceeds.

11.3.5   *Pedeferri's Diagram*

For each stainless steel, that is, for each PREN, pitting and repassivation potentials depend on chloride content, as proved by laboratory testing and experience. Pedeferri proposed a potential-chloride, E-[Cl⁻], diagram (Fig. 11.15) that has now his name. Pedeferri proved the diagram for passive carbon steel in concrete and forecasted its extension to stainless steels in chloride containing solutions. Pedeferri's diagram helps understand the *cathodic prevention* technique he invented (Pedeferri 1995).

11.3.6   *Pitting Induction Time*

From experience, the time required for pitting initiation, which is the time required to locally breakdown the passive film, once established the electrochemical condition $E_c > E_{pit}$, where E_c is potential of the cathodic process and E_{pit} is pitting potential, is considered by many authors as a stochastic variable.

Fig. 11.15 Pedeferri's diagram for carbon steel in concrete

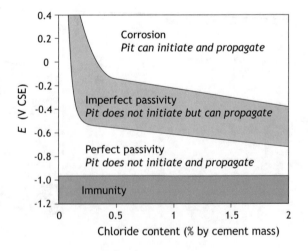

Indeed, induction time for a specific stainless steel (in other words, for a fixed PREN) depends on many factors, as:

- Chloride content
- pH
- Temperature
- Fluid velocity
- Potential of the cathodic process.

The key parameter is the *driving voltage*, ΔE_{pit}, as difference between the potential of cathodic reaction occurring on passive film, E_c, and pitting potential, E_{pit}. The driving voltage summarises all the influencing factors. The higher the driving voltage available, the lower the pitting induction time.

An innovative approach is proposed in Lazzari (2017) where an estimation of the pitting-induction-time, PIT (in h), is based on experimental data, which can be summarized as follows:

$$PIT = k \cdot 10^{\left(\frac{PREN}{2 \cdot \log[Cl^-]} \left(1 - \Delta E_{pit} \right) \right)} \tag{11.10}$$

where k is an experimental constant close to 1 (in h) and other symbols are known. As rule of thumb, if PIT exceeds about 10^4 h, pitting attack does not initiate if operating conditions do not change. This behaviour is typical of phenomena which are characterized by so-called infant mortality, i.e., should the passive film failure occur, it occurs early after exposure or never.

It appears that the cathodic potential is of primary importance for the estimation of the induction time. In general, the cathodic potential derives from the following three conditions occurring in most industry related environments:

- Oxygen reduction in sterile electrolyte
- Oxygen reduction in the presence of biofilm
- Chlorine reduction.

In addition, the potential of ferric chloride solutions used in testing should be considered.

The potential of oxygen reduction in a sterile electrolyte is simply the potential obtained by Nernst equation, therefore function of pH and oxygen concentration. An empirical equation is reported in Lazzari (2017) as follows:

$$E_{O_2} = 1.23 - 0.33 \cdot \log \left(\frac{50}{[O_2]} \right) - 0.059 \cdot pH \tag{11.11}$$

where $[O_2]$ is oxygen concentration in ppm.

Potential of Stainless Steels in Seawater

Typical values of free corrosion potential, E_{corr}, of stainless steel before pitting initiation in seawater are as follows:

- Deaerated: about -0.5 V SSC (Silver/Silver Chloride reference electrode)
- Aerated, sterile seawater (no bacterial activity): around 0 V SSC
- Aerated, with biofilm (bacterial activity as natural seawater): up to $+0.3$ V SSC
- Aerated with chlorine injection (about 0.5–1 ppm to reduce bacterial activity): $+0.6$ V SSC
- Galvanic coupling with iron (or carbon steel): -0.4 V SSC
- Galvanic coupling with zinc or aluminium: from -0.8 to -1.0 V SSC

SCC has a potential $+0.25$ V SHE. If SCE (saturated calomel electrode) is used, its potential is $+0.24$ V SHE.

In the presence of biofilm, such as in seawater, the potential of oxygen reduction can be expressed simply by adding 0.3 V to the potential in absence of biofilm. This 300 mV increase was measured by Mollica and Trevis (1976) for the first time, and then confirmed by other researchers, therefore:

$$E_{O_2/bio} = E_{O_2} + 0.3\,\text{V} \tag{11.12}$$

In the presence of chlorine, such as in treated or sanitized waters, the potential of cathodic reaction of chlorine reduction is more noble than oxygen reduction, therefore pitting initiation can occur also in absence of oxygen. The potential of chlorine reduction is obtained from Nernst equation as follows:

$$E_{Cl_2} = 1.36 + 0.6 \cdot \log\frac{[Cl_2]}{36} \tag{11.13}$$

where $[Cl_2]$ is chlorine concentration in ppm (>0.1).

Stainless steels in seawater can experience a pitting induction time varying from a few days for AISI 304 grade to about a month for AISI 316 grade when biofilm forms. As rule of thumb, because pitting is an infant-related phenomenon, in practice induction time lasts less than a year. In other words, if pitting has not started within a year from exposure time it will not occur anymore if operating conditions remain unchanged.

11.4 Pitting Susceptibility

To assess pitting susceptibility on stainless steels, parameters as pitting potential, pitting critical temperature, critical chloride concentration and PREN are used.

11.4.1 Critical Pitting Temperature and Critical Pitting Chloride Concentration

Critical Pitting Temperature, CPT, is the minimum temperature at which stainless steel resists pitting attack, once fixed potential and environmental conditions. Similarly, *Critical Pitting Chloride Concentration*, CPCC, is a threshold below which pitting does not initiate.

Laboratory testing are performed based on international standard, such as ASTM G48, ASTM G150 and ASTM F 746-04.

From testing results, CPT in a 6% ferric chloride solution, as often used for comparison to rank stainless steels, is a function of PREN through the following empirical relationship:

$$CPT(^{\circ}C) \cong 3.3 \cdot PREN - 58 \qquad (11.14)$$

For instance, CPT is 0 °CC for AISI 304, 20 °C for AISI 316, 75 °C for 254 SMO, and 100 °C for 564 SMO.

Similarly, from laboratory testing results, CPCC is a function of stainless steel composition, i.e., PREN, and operating conditions, namely pH and temperature, as follows (Lazzari 2017):

$$\log[CPCC]_{critical} \cong \left[\frac{PREN}{9} - \frac{7-pH}{5} - \frac{T-25}{50} \right] \qquad (11.15)$$

where parameters are known.

For reinforcing carbon steel in concrete structures exposed to the atmosphere, CPCC ranges between 0.4 and 1% by cement weight; for galvanized steel it is about 1%, whereas for stainless steel AISI 304 or 316 is in the range 5–8%, which decreases to only 3% on welded zones, covered with coloured oxide.

11.5   Pitting on Carbon Steel in Chloride-Contaminated Concrete

Carbon steel reinforcements in sound concrete (pH > 13 and no chlorides) are passive. Passivity breakdowns when chloride content at steel surface exceeds a critical content. Pedeferri's diagram helps understand the influence of potential and chloride content. Three regions can be identified (Fig. 11.15):

- Corrosion condition (pit initiates and propagates)
- Imperfect passivity (pit does not initiate, instead it can propagate if started)
- Perfect passivity (pit does neither initiate nor propagate).

Pitting and repassivation potential curves depend on chloride content. At any potential the critical chloride content is determined. It can be noted that as chloride content increases, potential decreases. Reinforcement of concrete structures, exposed to the atmosphere, shows a pitting potential usually around +0 V SCE then critical chloride content is in the range 0.4–1% by cement weight. In the case of structures immersed in water, where oxygen diffusion is impeded, and therefore characterized by a corrosion potential lower than a few hundred mV, critical chloride content is much higher. Repassivation potential, E_{rp}, is approximately 300 mV more negative than pitting potential. For more details, refer to Bertolini et al. (2013).

11.6   Pitting on Aluminium Alloys

Although aluminium passivates as stainless steel by forming an oxide layer on the metal surface as soon as it is exposed to an electrolyte, pitting-like corrosion differs strongly, mainly for propagation rate. As already mentioned, aluminium oxide is a good insulator, therefore electrons are strongly impeded to flow from the metal to the oxide-electrolyte interface.

Pitting initiation requires the presence of an anion able to locally breakdown the passive film: typically, it is again chloride or more generally halogens. The most important condition for pitting initiation is chloride concentration, which follows, approximately, the same trend as stainless steels. One factor that strongly enhances pitting initiation is the presence on the aluminium surface of metallic copper, as small particles deposited from copper ions accidentally present in the solution: the galvanic effect of copper on aluminium determines the passive film breakdown.

Because of the insulating properties of the oxide film, the macrocell set up by the pit on aluminium alloys has a low throwing power, even in highly conductive electrolytes as seawater: hence, many pits form, which is the opposite of what happens on stainless steel, where only a few isolated pits forms.

The propagation rate of the numerous pits is much lower than that for stainless steel because of the insulating properties of the oxide film. Instead of using the approach adopted for stainless steel, which could be used again, an empirical equation is proposed as follows:

$$y_{\text{pit}-\text{depth}-\text{Al}} \cong k \cdot t^{0.33} \tag{11.16}$$

where the constant, k, is averagely 0.75 for pit depth in mm, considering time in year. Accordingly, corrosion rate is given by:

$$C_{\text{rate}-\text{pit}-\text{Al}} \cong \frac{0.25}{t^{0.66}} \tag{11.17}$$

C_{rate} is in mm/y and time, t, in year. The above equations are derived from laboratory testing results, confirmed by field experiences (Godard 1967).

11.7 Pitting as Markovian Process or Prevention of Pitting

Copper is used in water circuits, with either freshwater or seawater. In rare cases, in freshwater, copper suffers localized corrosion with a morphology and mechanism of pitting. Copper and copper alloys resist corrosion in aerated waters because they passivate, although the nature of passive film is coarse if compared with the passive film of stainless steels. Therefore, in this case, it is more appropriate to call this condition passivation instead of passivity. The passivation of copper is caused by the formation of corrosion products, such as copper oxy-carbonate, $Cu_2(OH)_2CO_3$. Pitting initiation can follow two distinct mechanisms: by the first, initiation is triggered by the presence of carbonaceous particles produced during drawing manufacturing from the decomposition of lubricants and not removed by successive proper chemical etching. The second one, discussed in literature, is somewhat evanescent because there is no specific recognized condition for prediction of pitting occurrence, unless again the presence of some noble particles.

Pitting propagation follows the macrocell mechanism and, therefore, general equations apply, taking into account the following:

- The anodic process is copper dissolution, which occurs at quite noble potential. Since oxygen reduction (i.e., cathodic process) occurs at potentials relatively more noble than equilibrium copper potential, the driving voltage is much lower than the one in case of pitting on stainless steels. For calculations, driving voltage in practice is not exceeding 0.2 V
- The cathodic current density in the activation overvoltage interval is about one order of magnitude lower than oxygen limiting current density. For calculations, current density should not exceed 50 mA/m².

In practice, the maximum corrosion rate is about 1 mm/year (oxygen current density 50 mA/m², driving voltage 0.2 V and water resistivity 20 Ω m).

Pitting Initiation for Stainless Steel as a Markovian Process

Pitting occurrence is stochastic in nature. Almost a century ago, Mears and Evans introduced the concept of "probability of corrosion" and emphasized the practical importance of a statistical assessment of localized corrosion (Mears 1935), and starting from late 1970s probabilistic models to predict pitting initiation were proposed. An interesting probabilistic approach is based on Markov chain theory (Provan and Rodriguez 1989), valid for memoryless processes. Indeed, pitting on stainless steel can be considered a memoryless process, since if it happens, this is when critical conditions are present, regardless any previous ones.

A Markov chain is a memoryless stochastic process that undergoes transitions from one state to another through a finite number of possible states, until it stops at the so-called *absorbing state*. Each transition is characterized by a transition probability given by the Markov transition matrix. The model, described in Brenna et al. (2018), starts from a metastable state where metapits form, then evolving toward stable pit (pitting occurrence) or passive condition through a repassivation process. Figure 11.16 shows how the model works based on five states. From a metastable condition, two competitive processes can be recognized: (a) the breakdown of the passive film with the formation of a stable occluded cell and (b) the formation of a passive layer on the metal surface and death of metastable pits. Transition probabilities are indicated as m, p and r, respectively from metastable to metapassive, metapitting to pitting and metapassive to passive. There are two absorbing states: stable pitting, characterized by the formation of a macrocell, and stable passive state.

From the initial metastable condition, the system transforms necessarily to one absorbing state after a finite (countable) number of transitions. To determine the final probability toward which absorbing state the system evolves (i.e., pitting or passive state) probability p and r are the output of the matrix calculation, once known the initial probability. The higher the final probability r, the higher the resistance to pitting corrosion of the metal.

Initial probabilities are determined by the actual conditions, as depicted in Fig. 11.17. In practice, to input the initial transition probabilities, m, p and r, are calculated from parameters indicated in figure, namely: PREN, pH,

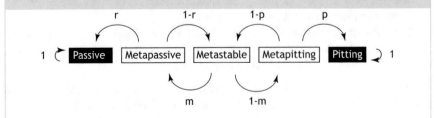

Fig. 11.16 Five model states for pitting corrosion

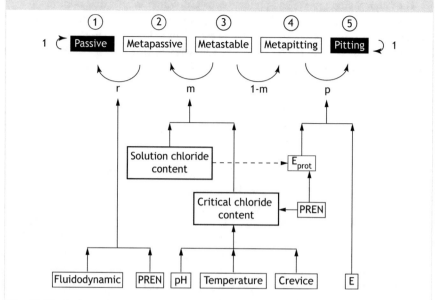

Fig. 11.17 Factors influencing the transition probabilities

temperature, chloride content, cathodic process or redox potential and fluid velocity, then final probabilities, r and p, are obtained. When the final probability, p, exceeds 50%, it could be concluded that susceptibility of metal to pitting corrosion is too high for applications.

11.8   Prevention of Pitting Corrosion

To prevent pitting on susceptible metals, two strategies are followed based on the evidence that it is difficult to stop a pit once started, if deeply penetrated: in the latter case, a drastic grinding action is necessary, often almost impossible to put into practice. Only shallow pits, less than 0.3 mm deep, can be recovered by washing with alkaline, chloride-free solutions (for example sodium carbonate).

These two strategies are:

- Selection of resistant metals (for stainless steels, PREN is used as guide for proper and safe choice)
- Application of cathodic protection.

Material selection in design phase has to take into account expected operating condition for the entire design life (for instance, chloride content, oxidizing power, acidity, bacterial activity, surface condition).

The second strategy is the application of CP which is effective both to prevent pitting initiation (in this case, Pedeferri named it *cathodic prevention*, CPrev), and to stop pitting propagation. In CPrev, it is sufficient to lower the potential below the pitting potential, E_{pit}, in the second, it is necessary to drop the potential below the repassivation potential, E_{rp}.

11.9 Applicable Standards

- ASTM G 46—Standard Guide for Examination and Evaluation of Pitting Corrosion, West Conshohocken, Pa.: American Society for testing of Materials
- ASTM G 48—Standard Test Methods for Pitting and Crevice Corrosion Resistance of Stainless Steels and Related Alloys by Use of Ferric Chloride Solution, West Conshohocken, Pa.: American Society for testing of Materials
- ASTM G 61—Standard Test Method for Conducting Cyclic Potentiodynamic Polarization Measurements for Localized Corrosion Susceptibility of Iron-, Nickel-, or Cobalt-Based Alloys, West Conshohocken, Pa.: American Society for testing of Materials
- ASTM G 150—Standard Test Method for Electrochemical Critical Pitting Temperature Testing of Stainless Steels, West Conshohocken, Pa.: American Society for testing of Materials
- ASTM F 746—Standard Test Method for Pitting or Crevice Corrosion of Metallic Surgical Implant Materials, West Conshohocken, Pa.: American Society for testing of Materials
- ISO 8993—Anodizing of aluminium and its alloys—Rating system for the evaluation of pitting corrosion. Chart method, International Standard Organization, Geneva, Switzerland
- ISO 8994—Anodizing of aluminium and its alloys - Rating system for the evaluation of pitting corrosion. Grid method, International Standard Organization, Geneva, Switzerland
- ISO 11463—Corrosion of Metals and Alloys—Evaluation of Pitting Corrosion, International Standard Organization, Geneva, Switzerland
- ISO 15158—Corrosion of Metals and Alloys—Method of measuring the pitting potential for stainless steels by potentiodynamic control in sodium chloride solution, International Standard Organization, Geneva, Switzerland
- ISO 17864—Corrosion of Metals and Alloys—Determination of the critical pitting temperature under potentiostatic control, International Standard Organization, Geneva, Switzerland.

11.10 Questions and Exercises

11.1 Pitting potential of stainless steel AISI 304 and AISI 316 in seawater at 20 °C is −0.1 V SCE and +0.2 V SCE, respectively. Establish which material suffers pitting in deaerated, natural aerated, chlorine containing and sterile seawater. Can pitting corrosion initiate when stainless steel is coupled with iron, zinc, or aluminium?

11.2 A tank designed to store natural, i.e. not treated, seawater was made of stainless steel, 18-8 grade (AISI 304) with 4 mm thick bottom plate. Can you predict perforation time if plant is in Norway, Italy and Persian Gulf? What would you expect if waters were sterilized? Suggest remedial actions.

11.3 In a heat exchanger tube, seawater flows at a velocity of 1 m/s. Predict pitting occurrence if tube is made of: (a) AISI 304 stainless steel; (b) AISI 316 stainless steel; (c) high-alloy austenitic stainless steel with 6% Mo. As second choice, consider water velocity of 2 m/s and shutdown time (for maintenance) of 2 weeks.

11.4 A plate made of stainless steel, grade AISI 304 (18-8 Cr–Ni), was immersed in the water of a swimming pool. Pitting corrosion occurs on the plate corresponding to some welds. Find most likely root cause for pitting corrosion. Estimate pitting initiation time and pitting propagation rate. Suggest practical solution, either in new design or for intervention.

11.5 In a case of pitting corrosion on an aluminium sheet in seawater, the largest pit depth is 200 μm after 2 months. What will be the maximum depth expected after 1 year? After 10 years?

11.6 In a piping system, cold seawater flows slowly, i.e. water velocity is lower than 1 m/s. Which stainless steel would you recommend? Conventional AISI 304 stainless steel or AISI 316 or high-alloy austenitic stainless steel with 6% Mo?

11.7 Laboratory testing demonstrated that addition of sulphate ions (for instance as Na_2SO_4) to a NaCl solution increases pitting resistance as follows: 18-8 steel (AISI 304, PREN 18) behaves like AISI 316 (PREN 25) in absence of sulphate. Suggest an interpretation.

11.8 A localized corrosion attack occurred in the centre of a stainless steel plate grade AISI 304 L (PREN 18). The anodic area (the pit) can be estimated in 1 cm². The oxygen limiting current density is 50 mA/m². How the current will flow in the electrolyte? Evaluate the corrosion penetration rate in the following conditions: fresh water (resistivity 20 Ω m), brackish water (resistivity 5 Ω m). Refer to Chap.9 for the equation that defines the throwing power on a plate geometry.

11.9 Cold seawater is flowing slowly through a pipe with a joint between stainless steel and carbon steel pipes. Discuss if corrosion of the stainless steel pipe can occur in the three following condition: (a) use of stainless steel AISI 304; (b) use of stainless steel AISI 316; (c) use of high-alloy austenitic stainless steel with 6% Mo.

11.10   For the following common stainless steels (AISI 304, AISI 316, AISI 430, AISI 904, duplex 2205, duplex 2507) calculate the induction time in the following working condition: seawater, lake water, drinking water (assume proper values for the affecting parameters). Comment on the results.

Bibliography

Bardal E (2004) Corrosion and protection. Springer, London, UK

Bertolini L, Elsener B, Pedeferri P, Redaelli E, Polder P (2013) Corrosion of steel in concrete: prevention, diagnosis, repair. 2nd edn. Wiley-VCH, Weinheim

Brenna A, Bolzoni F, Lazzari L, Ormellese M (2018) Predicting the risk of pitting corrosion initiation of stainless steels using a Markov chain model. Mater Corros 69:348–357

Frankel GS (1998) Pitting corrosion of metals. A review of the critical factors. JES 145:2186–2198

Godard HP (1967) The corrosion of light metals. Wiley, New York, NY

Lazzari L (2017) Engineering tools for corrosion. Design and diagnosis. European Federation of Corrosion (EFC) Series, vol 68. Woodhead Publishing, London, UK

Mears RB, Evans UR (1935) The "probability" of corrosion. Trans Faraday Soc 31:527–542

Mollica A, Trevis A (1976) The influence of the microbiological film on stainless steels in natural seawater. In: Proceedings of the 4th International Congress on Marine Corrosion and Fouling, paper n. 351, Juan-les Pins, France

Pedeferri P (1995) Cathodic protection and cathodic prevention. Constr Build Mater 20:12–20

Pourbaix M (1973) Lectures on electrochemical corrosion. Plenum Press, New York-London, UK

Provan JW, Rodriguez ES III (1989) Part I: development of a Markov description of pitting corrosion. Corrosion 45:178–192

Chapter 12
Crevice Corrosion

Rust never sleeps.

Neel Young

Abstract Crevice corrosion is a form of localized corrosion related to the presence of sub-millimetric interstices (gaps, screens, deposits) on the surface of a metal. The mechanism involves the consumption of oxygen in the gap and the impossibility to restore it, with consequent setup of a differential aeration macrocell. This chapter presents the main aspects of metal and environment composition affecting the onset of crevice corrosion and—like in the previous chapter—proposes empirical parameters, as Critical Crevice Gap Size (CCGS), which help predict it.

Fig. 12.1 Case study at the PoliLaPP Corrosion Museum of Politecnico di Milano

© Springer Nature Switzerland AG 2018
P. Pedeferri, *Corrosion Science and Engineering*, Engineering Materials,
https://doi.org/10.1007/978-3-319-97625-9_12

12.1   Definition

The presence of cracks, gaps, screens or deposits on a metal surface can give rise to a localized corrosion form, called *crevice corrosion* or *interstitial corrosion* and *corrosion under deposit* (Fig. 12.1). Crevice corrosion is a concern in many environments for active-passive alloys as stainless steels, nickel alloys and titanium. Typically, stainless steels suffer from crevice corrosion in seawater or in chloride-containing solutions, present in most of industrial plants as in chemical, petrochemical, pharmaceutical, food processing, as well as in biomedical, nuclear and civil engineering.

Crevice corrosion produces local thinning or even perforation, with risk of out of service of equipment and pollution of fluids; in some situations, the corroded area can create conditions for stress corrosion cracking occurrence (see Chap. 13).

12.2   Crevice Critical Gap Size (CCGS)

The key parameter of crevice corrosion is the *critical crevice gap size* (or critical interstice size), defined as the minimum that allows the aggressive environment to enter the interstice but impedes the diffusion of oxygen. Critical gap size is between 0.1 μm and 0.1 mm, depending on metal composition. From literature data, in particular from Oldfield and Sutton (1978), the critical crevice gap size for stainless steels, assuming a crevice depth of 5 mm, can be estimated by the following equation:

$$\text{CCGS} \cong \left(\frac{17}{PREN} \right)^2 \tag{12.1}$$

where CCGS is in μm and *PREN* is pitting resistance equivalent number, which is calculated from stainless steel composition (see Chap. 11). Crevice corrosion occurs when gap size is smaller than the critical one. In general, the narrower and deeper the gap the higher the risk of crevice occurrence.

Figure 12.2 illustrates typical conditions that give rise to crevice corrosion as follows:

- Cracks in the metal, typically due to lack of penetration in welds
- Surface overlapping as in joints and threaded connections. If metals are different, galvanic effects have to be evaluated
- Interstices between metal and non-metallic materials (plastic, rubber, glass or wood) as typically in flanges and sealing gaskets
- Presence of deposits or scales or corrosion products or *fouling*.

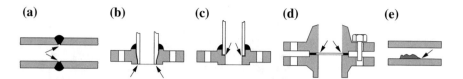

(a)   **(b)**   **(c)**   **(d)**   **(e)**

Fig. 12.2 Types of crevices due to: **a** Incomplete weld penetration; **b** and **c** joints; **d** seals; **e** presence of a probe (from Shreir et al. 1994)

Fig. 12.3 Crevice corrosion on AISI 321 stainless steel after a few months in seawater

For example, in a heat exchanger, critical situations are: interstices between plates, tubesheet and tube, tube and diaphragm, welding defects, supports, spacers, joints (bolted or riveted or forced), under gas bubbles (gas-liquid-metal three-phase contact), under deposits and porous coatings.

Severe crevice occurs in conditions of high heat flux as, for example, in gaps that form between tubes and tubesheet in boilers; this situation worsens by the formation of deposits and the increase in concentration of aggressive species.

Figures 12.3 and 12.4 show examples of crevice corrosion of stainless steel in seawater.

12.3 Mechanism of Crevice Corrosion

The mechanism of crevice corrosion of stainless steels in chloride-containing solutions follows three stages.

Fig. 12.4 Crevice corrosion
on AISI 316 stainless steel in
seawater

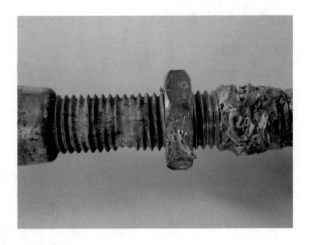

12.3.1   First Stage

It is also called incubation or oxygen depletion stage, during which oxygen inside
the gap is consumed through the corrosion reactions occurring on passive stainless
steel, namely, oxygen reduction and passive film growth, as follows:

$$O_2 + 2H_2O + 4e^- \rightarrow 4OH^- \tag{12.2}$$

$$xM + yH_2O \rightarrow M_xO_y + 2yH^+ + 2ye^- \tag{12.3}$$

The rate of reactions equals the slowest one, which is the *passivity current density*
of the metal, i_p (mA/m^2).

How long this stage takes depends on two parameters: passivity current density,
i_p, and crevice volume, i.e., maximum available oxygen mass inside crevice. It
appears evident that the most influencing and decisive parameter is the passivity
current density, which in turn is function of composition of stainless steel—in short,
of PREN, pH and temperature (see Chap. 5, Fig. 5.21).

It has to be considered that when passivity current density is very low, in spite of
the tiny gap, some oxygen diffusion, i.e., oxygen renewal, becomes possible,
therefore the incubation has never an end: accordingly, crevice can never start.

For an estimation of the duration of this stage reference can be made to Oldfield
and Sutton (1978a, b) and to Oldfield (1987).

12.3.2   Second Stage

The second stage starts once oxygen is completely depleted in the crevice. The
elimination of oxygen inside the crevice brings stainless steel in *active* condition, as
schematically shown in Fig. 12.5.

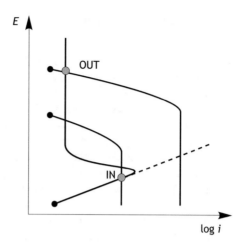

Fig. 12.5 Potential change inside crevice during first stage: conditions inside (IN) and outside (OUT) the interstice

During this stage, two important processes take place: inside the crevice, metal ions concentration can exceed 1 M, so hydrolysis reactions take place:

$$M \rightarrow M^{x+} + xe^-\tag{12.4}$$

$$yM^{x+} + xH_2O \rightarrow M_y(OH)_x + yxH^+\tag{12.5}$$

and pH drops to very low values, such as below 2 (mainly for the contribution of chromium), hence impeding repassivation; outside crevice, oxygen reduction increases alkalinity, therefore strengthening passivity.

12.3.3 Third Stage

Third stage is when macrocell current becomes stationary. Macrocell current can be calculated as reported in Lazzari (2017) as seen for pitting in Chap. 11. It is important to emphasize that for crevice corrosion the driving voltage is expected to be lower than that of pitting due to the higher ohmic drop contribution through the narrow gap. As rule of thumb, it could be considered about a half of that adopted for pitting corrosion; hence, a value as low as 0.2 V. In aerated seawater, crevice corrosion rate varies between 1 and 5 mm/year regardless the stainless steel grade.

For several aspects, crevice corrosion can be compared to pitting and some authors consider crevice a particularly severe form of pitting or a pitting just initiated; others consider pitting a particular form of crevice corrosion, arguing that the main difference is a matter of size (crevice corrosion in something macroscopic whilst pitting is microscopic). In any case, experience tells that environmental conditions, which cause pitting, cause also crevice corrosion, instead the opposite is not always true.

12.4   Metal Composition

Crevice occurrence depends on many factors, primarily on those related to nature, composition and structure of metal. In the case of stainless steels, the increase in chromium content and even more the presence of molybdenum and nitrogen is beneficial in promoting a stable passive film.

The resistance to crevice corrosion of stainless steels and other passive metals is determined by the passivity current density, therefore it depends on PREN (see Chap. 11). For example, in seawater stainless steels with PREN in the range 35–40 are not subject to pitting (at least for temperatures below 30 °C and no biofilm formation), instead they can suffer crevice.

Even the type of crevice has an important influence. For *piping* systems used in oil & gas industry operating at temperatures above 60 °C, superaustenitic steels with 6% Mo (type 254 SMO—PREN 43) are suitable for flanges with O-ring seal type, but not for threaded joints. For the latter, a superaustenitic steel with more than 7% Mo (type 654 SMO—PREN > 50) has to be used.

Other parameters, in analogy to those used for pitting to determine whether metals resist, are the critical crevice temperature (i.e., maximum temperature without crevice attack for each gap size) and the critical potential. Both are smaller than those for pitting and depend on crevice size; for example, crevice critical temperature is about 15 °C lower than pitting critical temperature.

12.5   Environmental Factors

Factors favouring crevice are chloride content, acidity, temperature, potential and bacterial activity: as each of these factors increase, crevice likelihood increases. For example, incubation time for stainless steel AISI 316 grade is a few weeks in seawater (pH around 8, chloride content 20 g/L) and increases to several months in industrial waters (same pH and chloride content below 1 g/L). However, with same chloride content, if pH drops from 8 to 5 or if the solution is contaminated by bacteria, incubation time decreases by one order of magnitude.

The influence of potential is summarized by comparing Figs. 12.6 and 12.7, which show the different crevice attack on two stainless steel plates in a test performed with the multiple crevice assembly, which consists of a Teflon segmented washer having a number of grooves and plateaus pressed on the metal surface. An AISI 316 stainless steel plate freely exposed in seawater, showing a potential between +0.2 and +0.4 V SCE, was compared with a plate of the same material welded to carbon steel, hence operating at a more negative potential by some hundreds mV. After an exposure of six months, the first specimen showed crevice attacks as deep as 0.5 mm, while the second one showed no corrosion and the formation of some calcareous deposit as consequence of the galvanic coupling with carbon steel.

Fig. 12.6 Results of crevice attack obtained with multiple crevice assembly on an AISI 316 plate

Fig. 12.7 Results of crevice attack obtained with multiple crevice assembly on an AISI 316 plate welded to carbon steel

12.6 Prevention of Crevice Corrosion

The prevention of crevice corrosion has to be carried out, primarily, in design and construction phases in order to avoid crevices as cracks, gaps and deposits. Sometimes, it is sufficient to change design through just a simple trick, as in the case of heat exchangers, by choosing the aggressive fluid, for example sea water, to pass inside tubes where crevice conditions are absent, unlike the shell side where crevices are inevitably present (between tubes and tubesheet and tube and diaphragms).

When crevice conditions are inevitable, its prevention follows two ways: selection of resisting material or by cathodic protection (in the case of stainless steels, iron anodes are often used). The conditioning of the electrolyte, as the use of corrosion inhibitors or by removal of oxygen, is not recommended because risky: if a crevice attack starts during the accidental suspension of the treatment, it cannot be halted by the treatment restart.

12.7   Crevice-Like Corrosion of Active Metals

The presence of interstices is always an aggravating corrosion factor also for active metals. The incubation mechanism is different, instead propagation follows again a macrocell mechanism. Typical cases are accumulation or entrapment of aggressive liquids in gaps, temperature increase in screened zones, non-homogeneous zones.

Differential aeration is a typical case of localized corrosion for carbon and low alloy steels as active metals: the anode is inside a recess or interstice or beneath a deposit where oxygen supply is reduced or even hampered and the cathodic zone is where oxygen is available. Corrosion rate is ohmic drop controlled and determined by the oxygen limiting current density; the anodic surface area is the gap or the screened surface area and the effective cathodic surface area is determined by throwing power (see Chap. 9). Some crevice-like attacks on active metals occur on atmospherically exposed structures in absence of electrolyte outside the crevice, as discussed in the following.

12.7.1   Corrosion Under Insulation

Under insulated surfaces of equipment operating in marine atmosphere there is often a high chloride concentration, carried by percolating rainwater which has washed polluted surfaces; beneath the insulation, high temperatures facilitate the evaporation of water, then leading to high chloride concentrations. Similarly, in chemical plants, the leakage of saline solutions favours contamination under insulation.

12.7.2   Automotive Related Corrosion

Some hidden parts of cars chassis, made of steel sheets, present interstices, especially where sheets overlap, as well as in joints made by spot welding or by stapling or sheet folding. Corrosion of these hidden parts is called *in-out corrosion*. Junctions are located mostly inside boxed parts, as for example doors or bonnet, which are made with two stapled sheets, one internal and the other external: inside these boxes the environment is different from the external one. In these areas, almost inaccessible for painting protection, corrosion proceeds invisibly until interesting full plate thickness, then degrading aesthetics and jeopardizing structural strength. Since about two decades, this corrosion has been neutralized by improving design, simplifying and reducing the number of traps and overall by adopting galvanized steel sheets.

12.7.3 Riveted Structures

Riveting or bolting of plates forms interstices, which retain moisture or electrolytes of different origin, with risk of corrosion beneath because any prevention measure is difficult. Corrosion products, which occupy a much larger volume than the corroded metal (iron), provoke a significant distortion on coupled plates.

12.7.4 Stored Plates

A crevice-like corrosion takes place on galvanized or aluminium sheets used in civil and furniture structures when stacked in high humidity storage environment. Moisture condenses between overlapped surfaces and generates aesthetically unacceptable surface alterations. For galvanized sheets, this phenomenon is known as *white corrosion* for the whitish colour of the corrosion products. Best practice for storage of these artefacts is dry condition or packaging in sealed polyethylene sheaths.

12.8 Filiform Corrosion

A particular type of crevice corrosion, called *filiform corrosion*, occurs beneath paints or lacquers on a coated metal surface. Corrosion attack starts from coating defects and grows as thin-looking grooves or wires, a few tenth of a millimetre wide, which affect only superficially the metal surface, for example a few tenth of a micron deep. Corrosion products, having a volume larger than that of the corroded metal, cause local coating disbonding and formation of clearly visible linear bulges, with the curious trend described in Fig. 12.8 and shown in Fig. 12.9 after a laboratory testing. Each "wire" propagates in a straight line until it crosses another one already developed; at this point, the wire proceeds in a mirrored or a parallel direction, depending on the incidence angle, greater or lower than a critical value (if greater there is reflection, if lower there is parallelism or merging), producing a labyrinth-type array. Metals typically showing this attack are steel, aluminium and magnesium when covered by coatings that are permeable to moisture, placed in an atmosphere with relative humidity above 70%.

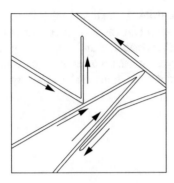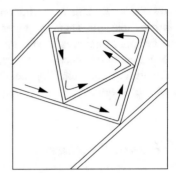

Fig. 12.8 Scheme of filiform corrosion

Fig. 12.9 Filiform corrosion
obtained during testing

The mechanism of this corrosion attack was studied through a transparent lacquer: in case of steel, the wire head became green because of ferrous ions while the body became red because ferric ions formed by oxidation of previous ferrous ones. Therefore, the anodic process is located in the head, where there is a lack of oxygen and acidity by hydrolysis is maintained, and the cathodic areas, where oxygen is more easily available, are localized behind (see sketch in Fig. 12.10). The alkalinisation of the cathodic area explains the formation of the array: the acidic head of a new wire is neutralized when crossing an old one, therefore it deviates.

Filiform corrosion stops as humidity decreases or coatings imperviousness increases.

12.9 Applicable Standards

- ASTM G 48—Standard Test Methods for Pitting and Crevice Corrosion Resistance of Stainless Steels and Related Alloys by Use of Ferric Chloride Solution, West Conshohocken, Pa.: American Society for testing of Materials.

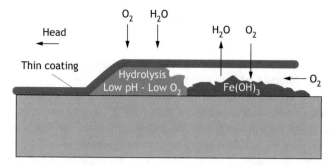

Fig. 12.10 Filiform corrosion mechanism

- ASTM G 78—Crevice Corrosion Testing of Iron-Base and Nickel-Base Stainless Alloys in Seawater and Other Chloride-Containing Aqueous Environments, West Conshohocken, Pa.: American Society for testing of Materials.
- ASTM G 192—Standard Test Method for Determining the Crevice Repassivation Potential of Corrosion-Resistant Alloys Using a Potentiodynamic-Galvanostatic-Potentiostatic Technique, West Conshohocken, Pa.: American Society for testing of Materials.
- ASTM F 746—Standard Test Method for Pitting or Crevice Corrosion of Metallic Surgical Implant Materials, West Conshohocken, Pa.: American Society for testing of Materials.
- ISO 18070—Corrosion of metals and alloys—Crevice corrosion formers with disc springs for flat specimens or tubes made from stainless steel, International Standard Organization, Geneva, Switzerland.
- ISO 18089—Corrosion of metals and alloys—Determination of the critical crevice temperature (CCT) for stainless steels under potentiostatic control, International Standard Organization, Geneva, Switzerland.

12.10   Questions and Exercises

12.1   Can a galvanic coupling influence the crevice initiation? Explain what would be the expectation in case of coupling with a less noble metal or instead a more noble one. Make an example.

12.2   A stainless steel is exposed to seawater with the presence of interstices (for example, a plate-tube assembly without welding sealing). In design phase, the coupling with titanium is analysed. What is your opinion about this choice? Would you approve such a choice? Explain.

12.3 To prevent crevice of stainless steels, cathodic protection can be adopted. Can you suggest protection conditions, as protection potential and protection current?

12.4 Crevice corrosion does not occur on aluminium alloys although their active-passive behaviour. Suggest an interpretation.

12.5 Crevice corrosion on titanium alloys takes place in acidic chloride containing solution at temperature above 70 °C. Try to investigate the mechanism involved through two phases: initiation and propagation.

12.6 In the case of stainless steels in seawater, crevice corrosion rate does not depend on stainless steel grade or composition. Give an extensive explanation.

12.7 Consider the previous exercise. What is the effect of a galvanic coupling with titanium on crevice corrosion rate?

12.8 List the affecting parameters of crevice corrosion. Rank by your opinion such a list from the most important to the lesser one and give a justification.

12.9 Crevice corrosion of stainless steels in seawater is influenced by the presence of biofilm. Because biofilm cannot grow inside the interstice, how do you explain such influence? Can a treatment with chlorine, which avoids the formation of biofilm, reduce the risk of crevice?

12.10 Once crevice corrosion of stainless steels in seawater has started, someone suggests cleaning and neutralizing treatments to stop the corrosion propagation. Discuss critically such recommendation, indicating the associated risks, if any.

Bibliography

Lazzari L (2017) Engineering tools for corrosion. Design and diagnosis. European Federation of Corrosion (EFC) Series, vol 68. Woodhead Publishing, London, UK

Oldfield JW (1987) Test techniques for pitting and crevice corrosion resistance of stainless steels and nickel alloys in chloride containing environments. Int Mater Rev 32:153–170

Oldfield JW, Sutton WH (1978a) Crevice corrosion of stainless steels: I. A mathematical model. II. Brit Corros J 13:13–22

Oldfield JW, Sutton WH (1978b) Crevice corrosion of stainless steels: II. Experimental studies. Brit Corros J 13:104–111

Shreir LL, Jarman RA, Burstein GT (1994) Corrosion. Butterworth-Heinemann, London, UK

Chapter 13
Stress Corrosion Cracking and Corrosion-Fatigue

> *Stress corrosion and hydrogen-induced cracking resemble the fable of the blind men and the elephant. Investigators have tended to perceive only single aspects of the problem and to design experiments in which important variables were either not appreciated, not controlled, or not measured.*
>
> R. A. Oriani

Abstract When a susceptible metal is in contact with a specific environment, in presence of a tensile stress exceeding a threshold, the corrosion-enhanced formation of cracks and catastrophic failure is called stress corrosion cracking (SCC). Although infrequent, consequences are so dangerous that it deserves to be described in details: in this chapter, after the introduction of the SCC mechanism, mechanical, metallurgical and environmental factors at its basis are explained, and specific preventative measurements are suggested. Another form of degradation that links mechanical stress and corrosion, i.e., corrosion-fatigue, is also described.

Fig. 13.1 Case study at the PoliLaPP corrosion Museum of Politecnico di Milano

© Springer Nature Switzerland AG 2018
P. Pedeferri, *Corrosion Science and Engineering*, Engineering Materials,
https://doi.org/10.1007/978-3-319-97625-9_13

13.1   Definitions

Stress Corrosion Cracking (SCC) together with *Corrosion-Fatigue* and *Hydrogen Embrittlement* belongs to so-called *Environmentally Induced Cracking* or *Environment Sensitive Cracking*. This chapter refers exclusively to SCC of metals taking place through the anodic dissolution process (Fig. 13.1), whilst hydrogen embrittlement is discussed in Chap. 14. In the past, SCC was known with different names depending on metals and environments involved, for instance: caustic embrittlement of carbon steel, seasonal cracking of brass, nitrate cracking, liquid metal embrittlement and many others. To have SCC, a secific combinations *metal-environment-tensile stress* as depicted by a Venn diagram (Fig. 13.2), that result into the formation of cracks, is necessary. This condition is intrinsically rare as the three factors are in an AND logic relationship (i.e., there is the need for the simultaneous presence of all three conditions: a susceptible metal, a specific environment and a tensile stress regime exceeding a threshold). Nevertheless, when it happens, the resulting cracking has catastrophic consequences. For this reason it has to be avoided.

Typical case studies of SCC reported in literature deals with:

- Carbon steel and low alloy steel in boilers, chemical equipment, piping, in the presence of nitrate or alkaline solutions or hydrogen sulphide as in oil & gas facilities
- High strength ferritic steels in aqueous environments
- Stainless steels in chloride containing solutions (Figs. 13.3 and 13.4) and high pH solutions in chemical and petrochemical plants or in nuclear reactors exposed to pure water at elevated temperatures (above 290 °C)
- Copper alloys in presence of ammonia
- Aluminium in chloride containing solutions.

Fig. 13.2 Representation of required conditions for SCC occurrence by means of a Venn diagram

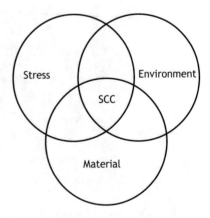

Fig. 13.3 Example of SCC-related failure of a carbon steel pipeline

Fig. 13.4 SCC of high strength low-alloy steel tool joint in an oil and gas well

As said, SCC is produced only for specific metal-environment combinations. For example:

- Austenitic stainless steels suffer SCC in chloride containing solution or in hot alkaline solutions and not in ammonia as copper alloys or nitrate solutions as carbon steel
- Carbon steel in nitric acid or hot alkaline solutions but not in solutions with chlorides or ammonia
- Copper alloys in ammonia containing solutions and not in solutions with chlorides

- High strength steels can crack simply in water. In this case, the specificity of the environment is less significant.

SCC occurs in electrolytes, organic liquids, molten salts, liquid metals, gaseous atmospheres, under scales. For example, titanium can crack in methanol, under sodium chloride scale in hot or gaseous nitrogen tetroxide, N_2O_4. Even non-metallic materials suffer cracking in specific environments; for instance, polyethylene or natural and synthetic rubbers crack when mechanically stressed in environments in which they are not soluble.

13.2   SCC Mechanisms

Several hypotheses of SCC mechanism have been proposed. Following a simplified approach, there are two basic mechanisms, as described in Fontana (1986): one as anodic and another one as cathodic. Accordingly, the growth of cracks takes place according to one of the following mechanisms:

- Crack tip dissolution (called *slip-dissolution*) is based on the anodic dissolution of the metal
- Crack tip rupture caused by atomic hydrogen, produced by the cathodic process, which enters the metal and accumulates at crack tip (called *Hydrogen Embrittlement, HE*).

The first mechanism claims a continuous crack growth by anodic dissolution of the crack tip, which is active, while crack wall surface and surface outside the crack remain passive, then a macrocell sets up (Fig. 13.5a). The reason why the crack tip is active is that the stress field around it, representing a sharp defect, produces a slip exposing *new* bare metal surface, which does not passivate. To match this critical

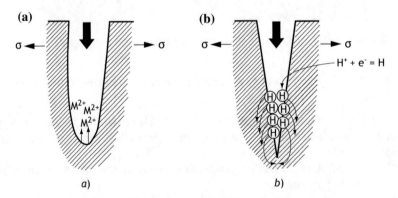

Fig. 13.5 Simplified representation of SCC mechanisms: **a** slip dissolution; **b** hydrogen embrittlement (HE). Adapted from Fontana (1986)

condition, it is required that the rate of formation of new bare metal surface by slip deformation at crack tip (which equals strain rate caused by the tensile load) is of the same order of magnitude of the passivation rate, which depends on electro-chemical conditions.

When the mechanism is HE, as discussed in Chap. 14, crack grows by successive mechanical ruptures at crack tip due to a decrease of metallic bond strength that is caused by the accumulation of the atomic hydrogen originated by the cathodic process (Fig. 13.5b). Steps of the mechanism are: the cathodic process produces atomic hydrogen, which enters the metal and diffuses to the dislocations at the crack tip, then provoking interatomic de-cohesion and crack growth; at the new crack tip, new dislocations form and new atomic hydrogen arrives and so on cycling in a discontinuous manner.

Several experiments support the two mentioned mechanisms. The most important one is the influence of potential: a cathodic polarization decreases corrosion rate and increases hydrogen evolution, therefore crack growth rate reduces when the first mechanism applies, while it increases if it fits the second one; the opposite happens when potential varies toward the anodic direction.

However, the two mechanisms do not explain the morphology of cracks either in the case of slip-dissolution or HE when, in the latter, cracks are surprisingly intergranular; furthermore, they do not predict when it may or may not be produced, also taking into account that similar failure occurs in non-metallic materials. For these reasons, other theories have been proposed to try and to put in one frame all environment assisted cracking phenomena, corrosion-fatigue included.

13.3 Morphology and Conditions of Occurrence

SCC related failures often appear without plastic deformation, then at first sight it could be mistaken for a fracture of brittle material; instead, metals suffering SCC are normally ductile. Cracks form and grow in perpendicular direction to maximum tensile stress and do not show visible corrosion products. Figure 13.6 depicts the morphology of cracks which depends on metal, environment and entity and

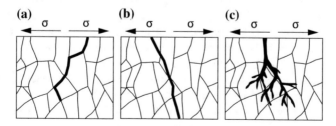

Fig. 13.6 Depict of the appearance of cracking propagation in SCC: **a** intergranular; **b** trans-granular; **c** delta river transgranular

(a)   **(b)**

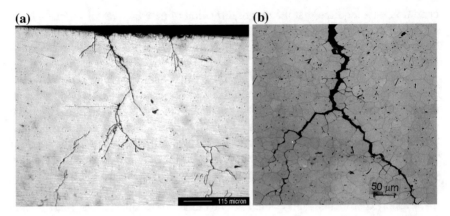

Fig. 13.7 **a** Transgranular cracks due to SCC observed on an AISI 316 stainless steel exposed to seawater at 70 °C; **b** intergranular cracks from SCC on AISI 304 stainless steel in caustic soda, 200 °C

distribution of stresses, cracks are mainly inter-crystalline (intergranular) or trans-crystalline (transgranular) and are more or less branched. Figure 13.7 shows some examples of cracking morphology for different metals.

SCC follows three steps:

- Initiation, or incubation
- Propagation
- Final mechanical failure.

13.3.1   Crack Initiation

Starting from a smooth surface, a certain time is needed before the first micro-crack can be detected, this time period, called incubation time for crack initiation varies from a few minutes to several years, depending on mechanism of crack nucleation, metal structure and environment properties such as salinity, pH and oxygen content or other oxidizing species, open circuit potential, temperature, static or variable applied stress. For example, austenitic stainless steels in boiling magnesium chloride solution show cracks after only a few hours of exposure; instead, nickel super-alloys, used in nuclear reactors in contact with pure water at high temperature (290–320 °C), show cracking after several years. In any case, crack nucleation is a stochastic phenomenon and consequently incubation time is characterized by a high scattering also influenced by the fluctuation of affecting factors.

The presence of notches as welding defects or mechanical grooves favours crack initiation. However, SCC also occurs on smooth surfaces, free from macroscopic defects. As often happens, the exposure environment is, per se, SCC safe, but as

Fig. 13.8 SCC starting at the bottom of a pit

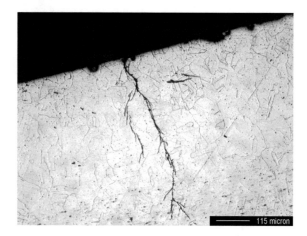

a consequence of local change, it may become harmful. This typically occurs inside pits and crevices or under scales, because of concentration processes, as, for example, in distillation columns of crude oil. Figure 13.8 shows cracks originated from a pit bottom. In other situations, aggressiveness increases because temperature increases, as in heat exchangers; to reduce this risk in household boilers made of austenitic stainless steel, where SCC may occur beneath the calcareous deposit that precipitates in high hardness waters, cathodic protection is adopted.

13.3.2 Crack Propagation

Once a crack has initiated, its propagation takes place by the combined action of a corrosion mechanism and the applied tensile stress, at a crack growth rate, which varies in a wide range between 10^{-6} m/s ($\cong$31 m/y) and 10^{-11} m/s ($\cong$0.3 mm/y), the latter being close to laboratory measurement limit. This crack growth stage is called *subcritical* or *stable propagation stage*. For example, SCC growth rate is close to the upper limit for:

- Austenitic stainless steels in chloride containing solutions
- Copper alloys in ammonia
- Carbon steels in nitrate solutions.

Conversely, crack growth rate is close to the lower limit for example for:

- High nickel alloys, as alloy 600 used in nuclear plants in contact with pure water at 290 °C
- welded carbon steel in liquid ammonia with traces of water.

Eventually, as crack size comes across a critical value in accordance with fracture mechanics, as discussed later, crack propagates at a very fast rate under the action

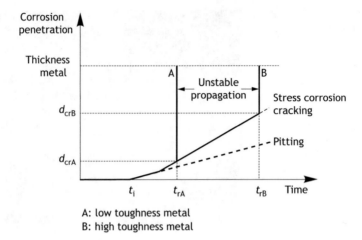

A: low toughness metal
B: high toughness metal

Fig. 13.9 SCC time to failure for two metals, A and B, with different fracture toughness with crack initiation from a pit (t_{rA} and t_{rB} service life of A and B, respectively; d_{cr} is critical defect size). Adapted from Brown (1968)

of purely mechanical stress to the final rupture, that can be brittle or ductile. This behaviour is called *instable crack propagation stage*.

Time-to-failure of SCC is the sum of crack initiation time and crack growth time. Figure 13.9 shows schematically the behaviour of two metals, A and B, with different fracture toughness, both suffering SCC from an initial pitting attack. It is possible to observe the pitting initiation time, t_i, and the SCC propagation time, whose sum gives the time needed to reach unstable propagation, i.e. the end of service life for material A, t_{rA}, and B, t_{rB}. Fracture toughness of metal A is lower than the one of metal B, so that critical crack size for A, d_A, is smaller than the one for B, d_B, and accordingly, time-to-failure for A is lower than the one for B.

13.4 Mechanical Aspects

The conventional approach for studying SCC is, once identified the metal-environment coupling susceptible to SCC, the experimental determination of the time-to-failure obtained on smooth specimens by varying the applied nominal stress, σ, as affecting mechanical parameter. Although the nominal stress does not reflect the real stress at crack tip (because of the intensity factor), by this approach a *threshold stress*, σ_{th}, is obtained, below which SCC does not take place. This parameter, which is regarded as a metal characteristic, is affected by a high scattering because measured on smooth specimens. For this reason, it cannot be used for design purposes; instead, it is useful to carry out a ranking of candidate metals through tailored laboratory tests.

Since mid-1970s, SCC studies adopted concepts, parameters and methods of fracture mechanics, assuming that sharp defects were, cautionary, always present in metals, then enabling to trigger crack initiation; this, indeed, is realistic because defects are always present in raw metals or generated in construction and also during operating.

13.4.1 Stress Intensity Factor, K_I, and Fracture Toughness, K_{IC}

The stress at the tip of a sharp notch increases as tip radius lowers. It is demonstrated that in elastic behavior and without plasticization at the notch tip, the stress field at the tip is represented by a parameter called *stress intensity factor*, K_I, expressed by the following relation:

$$K_I = \beta \sigma \sqrt{\pi a} \tag{13.1}$$

where β is dimensionless geometry correction factor, whose value depends on geometry of component and defect (typically in the range 1–2), σ is a characteristic stress and a is a characteristic crack size.

From a purely mechanical point of view, the stress intensity factor, K_I, allows to specify the condition for unstable fracture propagation in low ductility metals, which takes place when K_I reaches a critical value, K_{IC}, called *fracture toughness*. Fracture toughness, K_{IC}, is an inherent material property. Since SCC occurs for an applied tensile stress below the yield strength (i.e., in the elastic range), K_I, among various parameters of fracture mechanics, resulted as the most appropriate.

13.4.2 Crack Growth and K_{ISCC}

Experience has shown that there is a finite crack propagation rate of SCC, for which the term *subcritical growth* is used, when K_I has values between the limits given by two parameters:

- A threshold value, called K_{ISCC} that indicates the value of the stress intensity factor below which cracks cannot grow by an SCC mechanism
- K_{IC}, the fracture toughness.

As crack grows, nominal stress σ as well as K_I varies, either increasing or decreasing. When increasing, that is under constant load condition, subcritical crack growth proceeds until K_I equals K_{IC} with final sudden rupture; in the second case, which occurs under constant strain condition, crack propagation stops as K_I decreases to K_{ISCC}.

While K_{IC} is an intrinsic parameter of materials, similarly K_{ISCC} is an intrinsic parameter of metal-environment coupling and, in principle, can be used as design parameter. In practice, this approach is not followed because of a twofold reason: the scattering of results in laboratory testing and its variability even with little variations of either metal or environmental characteristics, for instance, impurities in metal composition and changing of environmental parameters.

13.4.3   Crack Growth Rate and K_I

Relationship between crack growth rate, da/dt, and stress intensity factor, K_I, is shown in a semi-logarithm plot in Fig. 13.10. In the range K_{ISCC} to K_{IC}, three intervals can be found:

- **Interval I**: crack growth rate depends strongly on K_I, whereby a small increase of K_I produces orders of magnitude increase of crack growth rate
- **Interval II**: two possible behaviours

    - Curve (A) for slip-dissolution mechanism: crack growth rate is roughly independent on K_I and generally coincides with the corrosion rate at crack tip. Accordingly, rather than mechanical, controlling factors are those associated with the corrosion process as driving voltage, current density of cathodic process and environment resistivity
    - Curve (B) for hydrogen embrittlement mechanism, where mechanical factors generally govern the crack growth.

- **Interval III**: crack growth rate increases rapidly as K_I approaches fracture toughness, K_{IC}.

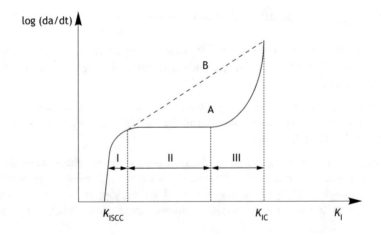

Fig. 13.10 Schematic trend of crack growth rate with stress intensity factor

13.4.4 Crack Growth and Strain Rate

At the beginning of 1970s, Parkins showed that SCC does not depend only on stress level, instead on strain rate and it seems erroneous to assume that there is a critical stress threshold, σ_{th}. Parkins showed precisely that SCC occurs only when strain rate falls within a critical interval that depends on metal/environment coupling and applied potential. As shown later, this behavior is typical for metals following the slip-dissolution mechanism, where crack grows by the dissolution of the active crack tip, while other surfaces are passive. In short, the mechanism claims that as first step crack tip becomes active due to the formation of new bare surface caused by the strain, followed by a second step by which the corrosion attack takes place during the repassivation time. This cycle film rupture-corrosion -repassivation starts again and so continuing. Figure 13.11 shows examples of SCC behaviour for different strain rates in various environment compositions and applied potentials; Fig. 13.12 illustrates also how strain rate influences stress-strain curves. Eventually, Fig. 13.13 shows that small, slow fluctuations of applied load, unable to sustain fatigue or corrosion-fatigue phenomenon may be sufficient to widen the range of conditions for crack to grow.

13.4.5 Test Methods—SSRT

When testing SCC susceptibility, an important parameter is the strain rate range that must be specified. Parkins developed and suggested a test method at constant strain rate, called as *Slow Strain Rate Test*, SSRT, consisting of the application on either

Fig. 13.11 Effect of strain rate on SCC occurrence on Mg–Al alloys in a 20 g/L chromate containing solution with different amounts of chlorides (from Parkins 1973)

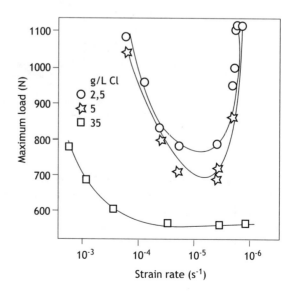

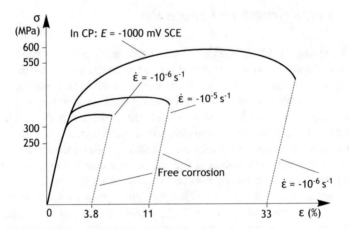

Fig. 13.12 Stress-strain curves of an austenitic stainless steel containing nitrogen for different strain rates in free corrosion and under cathodic protection in magnesium chloride boiling solution (from Magnin 1996)

Fig. 13.13 Effect of small fluctuations of tensile load on intergranular SCC occurrence on C-Mn steels in carbonate/ bicarbonate solutions at 82 °C and −0.65 V SCE (from Parkins 1972)

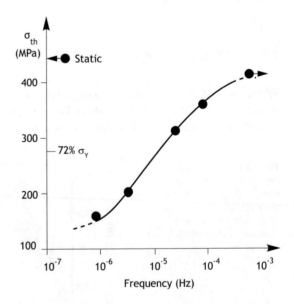

pre-cracked or smooth specimens of a constant strain rate, in the interval 10^{-6} to 10^{-4} s^{-1}, which determines a time to failure ranging from a few minutes to a few weeks. The selected strain rate depends on environment aggressiveness and passivation tendency of the metal. In the SSRT the stress-strain curve obtained in the environment under examination is compared with the one obtained on the same material in an inert environment, e.g., vacuum, dry air or oil, when the material is susceptible to SCC, the maximum applied load is lower than that required in the inert environment at same testing conditions. SCC occurrence is also evaluated by metallographic observations of fracture surface in combination with the ductility decreasing measured as reduction of fracture surface area and percentage elongation on failed specimen.

13.5 Environment-Related Parameters

Table 13.1 shows a partial list of combinations of metal-environment that give SCC; the list has grown with time making clear the concept of specific environment for a metal. Based on that, corrosion engineers have become aware that only precise environments can promote SCC on a metal.

As SCC starts, there is the need, at least for the anodic mechanism, that the reaction occurring at the crack tip is faster than repassivation, otherwise either a general attack or pitting would take place. Accordingly, an environment is *specific* when it enables to ensure the passivity of exposed surfaces and crack wall surfaces, and at the same time makes the crack tip active.

In most cases, the environment should exhibit the combined presence of factors that act in opposite directions: one favours passivation and the other does not. For

Table 13.1 Metal/environment combinations susceptible of SCC

Metal	Environment (solutions)	Temperature (°C)
Carbon steel	Caustic	>80
	Carbonate-bicarbonate	$\geq 50/60$
	Nitrate	$\geq 50/60$
	Phosphate	$\geq 50/60$
	Liquid ammonia with traces of water $CO/CO_2/H_2O$	T_{amb} All
Austenitic stainless steel	Neutral aerated containing chloride	>80/100
	Acid containing chloride	$\geq T_{amb}$
	H_2S containing chloride	$\geq T_{amb}$
	Caustic Oxygen containing pure water	$\geq 80/120$ ≥ 100
Nickel alloys	Caustic	>100/200
Nickel alloys (Cr < 30%)	Water with dissolved H_2	>250/280
Sensitized stainless steel and nickel alloy	Polythionates and thio-sulphates	$\geq T_{amb}$
Copper alloy	Ammonia containing	T_{amb}
Aluminium alloy	Chloride containing	T_{amb}
Titanium alloy	Alcoholic containing chloride	T_{amb}

example, magnesium alloys suffer SCC in chromates-chlorides mixtures, that is, in the presence of a passivating agent (i.e., chromates) and a depassivating one (i.e., chlorides), but not when only one of these species is present.

In general, it is observed that metals highly resistant to corrosion because protected by a passive film such as titanium, aluminium, nickel, chromium and stainless steels, suffer SCC when a chemical species, for example chlorides, that enables to breakdown the passive film is present.

Conversely, metals easily corroding, for example magnesium alloys and carbon steels, suffer SCC only if a passivating chemical species is present, such as hydroxides, carbonates and nitrates.

For the above reasons, SCC occurs only in narrow ranges of potential, across the active/passive boundary or the passive/transpassive one (Figs. 13.14 and 13.15): therefore, the cathodic process is important, as well as the presence of galvanic couplings that determine the working potential. Accordingly, oxygen content or the presence of oxidizing species can move the working potential inside or outside the critical range. For examples: austenitic stainless steels suffer SCC in pure water containing traces of chlorides at temperature above 200 °C only if oxygen is present; conversely, in carbonates and bicarbonates solutions the presence of oxygen keeps carbon steel potential outside hazardous conditions.

An important variable is *temperature*: as it increases, SCC occurrence increases (except for HE, see Chap. 14). There is often a temperature threshold (Fig. 13.16). As temperature changes, environment composition may change, making SCC possible; for example austenitic stainless steels suffer SCC at temperatures around 60 °C with a few tens of ppm of chlorides, while at temperatures close to 300 °C, typical of nuclear plants, critical chloride concentration drops below 1 ppm; eventually, below 50 °C, SCC does not occur at any chloride content, at least in neutral and alkaline solutions.

Fig. 13.14 Potential intervals for SCC occurrence. Adapted from Staehle (1977)

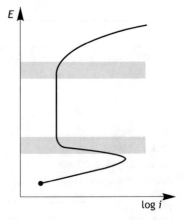

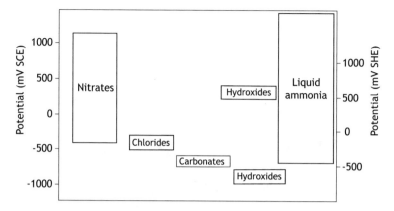

Fig. 13.15 Potential intervals for SCC occurrence with slip-dissolution mechanism of carbon steel in different electrolytes (after Parkins)

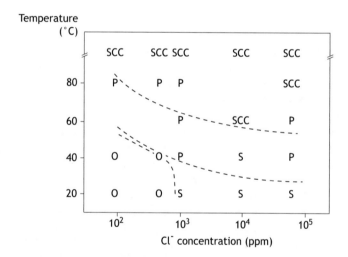

Fig. 13.16 Temperature and chloride content mapping for AISI 304 stainless steel: SCC, pitting (P), rouging (S), no corrosion (O) (Sedriks 1996)

13.6   Metallurgical Factors

13.6.1   Composition

Composition and structure of the metal, in combination with environmental conditions, influence the occurrence of SCC. For example, the addition of nickel to steel improves SCC resistance in alkaline environment, while it does not in nitrate

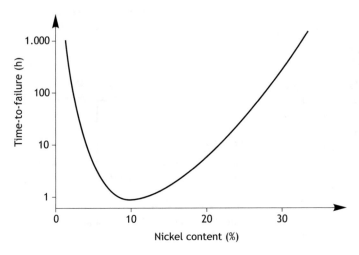

Fig. 13.17 SCC resistance of Ni containing ferrous alloys in MgCl$_2$ boiling solutions (Copson 1959)

containing solutions, and decreases in presence of chlorides. Similarly, the addition of molybdenum to ferritic steels is beneficial in carbonates/bicarbonates containing environments but has an opposite effect in alkaline solutions. Nevertheless, a general trend exists for predicting SCC.

Pure metals (99.99% grade or above) have a very high resistance to SCC; however, just a small level of impurities such as in commercially pure metals (for example 99.5%) is sufficient to decrease the resistance to SCC. An interpretation is that selective corrosion attack, caused by the impurities as inclusions or intermetallic phases, triggers SCC initiation.

The addition of an element to an alloy as solid solution changes SCC resistance: sometimes as content increases, resistance decreases to a minimum and then growing again. For instance, stainless steels exhibits the minimum resistance to chloride-induced SCC when nickel content is about 10% (Fig. 13.17) which matches the one of most used austenitic stainless steels, as AISI 304 and 316 stainless steel. These grades suffer SCC in chloride containing solutions, even at low stress level (100–150 MPa), when temperature exceeds 50–60 °C.

Stainless steels with low to zero nickel content, such as duplex type (with austenitic–ferritic structure) with about 4–7% Ni and ferritic type, which contain no nickel, resist much better to SCC. Furthermore, SCC resistance increases when nickel content exceeds 20% (so-called alloys 20 and superaustenitic stainless steels) or even better when it exceeds 40% (nickel based alloys) as depicted in Fig. 13.18.

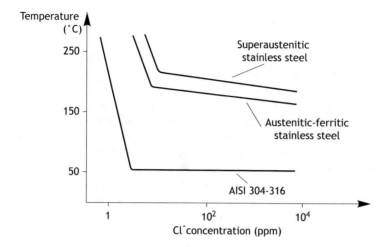

Fig. 13.18 Chloride induced SCC in neutral solution (above lines) for austenitic, superaustenitic and duplex steels. Adapted from Denhard (1960)

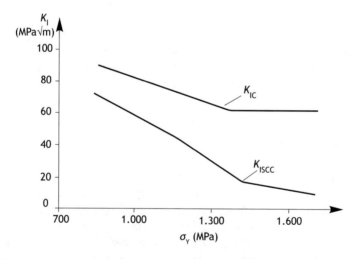

Fig. 13.19 K_{IC} and K_{ISCC} of AISI 4340 low alloy steel as function of yield strength (Brown 1968)

13.6.2 Mechanical Strength

The mechanical strength of a metal, regardless how it is obtained, influences SCC susceptibility through K_{ISCC}. Figure 13.19 shows for a low alloy steel how the yield strength, σ_Y (obtained at different tempering temperatures) influences K_{IC} and K_{ISCC} which both decrease as yield strength increases; K_{ISCC}, i.e., SCC susceptibility, is more strongly influenced.

13.6.3   Sensitization

Sensitization plays a very important role in SCC susceptibility. Let's consider two case studies regarding austenitic stainless steels, which sensitize in the temperature range 500–850 °C, see Chap. 15.

The first case study, which caused in the past major problems in petrochemical industry, deals with the exposure of sensitized austenitic stainless steels to sulphurous acid, thiosulphates or polythionic acids (these latter form by action of moisture and oxygen on sulphide scales) causing SCC also at room temperature. Stainless steel components may suffer SCC after a shutdown when production restarts because of two occurrences: first they sensitize during operating if exposed to the critical temperature interval and secondly polythionic acids form from sulphides during shutdown. This phenomenon was reproduced in laboratory.

The second case study deals with caustic soda solutions. The use of conventional austenitic stainless steels for treating high purity soda up to concentrations of 50% is an optimal choice from a technical-economical perspective, if temperature is never higher than 60–70 °C: this strong limitation depends on the fact that at higher temperatures a slight sensitization, possibly present near welds, is sufficient to cause SCC.

Other factors affecting SCC occurrence are the presence of precipitates, their distribution and orientation after a plastic deformation process as shown by the anisotropic resistance to SCC of aluminium alloys after lamination, which is greater when the applied stress is parallel to the rolling direction.

13.7   SCC Prevention

There are two general approaches for SCC prevention: the so-called *safe-life* and *fail-safe*. The first one, which is adopted in the vast majority of applications, checks that candidate material, environment composition, operating conditions (temperature, potential, tensile stress level) do not match the requirements for crack growth. Only in rare cases the *fail-safe* philosophy is adopted: that is, when crack growth rate is very low (for example cracks on welds in liquid ammonia containing tanks) and monitoring is possible and reliable.

In practice, the prevention of SCC is obtained through:

- Reduction of either mechanical tensile stresses, in particular residual stresses, or defect size so that the stress intensity factor, K_I, is always lower than the threshold value, K_{ISCC}
- Control of metallurgical, environmental and electrochemical (i.e., potential) influencing factors.

13.7.1 Reduction of Stress and Defect Size

Stress level that triggers SCC is the sum of operating stresses due to internal pressure and external loads, residual stresses as consequence of manufacturing process, cold working, heat treatment, welding, construction stresses due to the matching of the different parts to be joined during plant construction when components do not perfectly fit; thermal stress for thermal expansion during operating. Accordingly, the elimination or reduction of these stresses is different; in manufacturing it consists of stress relieving heat treatments, for example one hour at 300 °C for copper alloys, or one hour at 500 °C for stainless steels, in construction by improving design and welding procedure control, in operating by avoiding thermal inhomogeneity.

When high-strength alloys are employed because of the need to stand high loads, SCC prevention must be based on fracture mechanics, by calculating the maximum defect size tolerable at the nominal stress applied using the stress intensity factor, K_{ISCC}, so that SCC cannot propagate. If critical defect size is detectable by a non-destructive method, prevention is based on routine inspection; if not, design must be changed by selecting another material or different heat treatment or by changing environmental conditions or reducing stress level. Finally, as general warning, it is necessary to bear in mind that SCC susceptibility increases as mechanical strength increases, regardless how it is obtained.

13.7.2 Control of Environment, Metallurgy and Polarization

Environment control for SCC prevention consists of eliminating or reducing the content of chemical species that trigger SCC initiation, either in bulk or locally as in interstices or under deposits.

A polarization can bring the potential outside critical intervals, for instance by the application of cathodic protection, as in the case of AISI 321 stainless steel heating elements in household boiler by magnesium anodes.

Most often, to avoid risk of SCC, a resistant material is chosen. For example, in the presence of chlorides at temperatures above 60 °C, the use of austenitic stainless steels has to be discarded, and it is necessary to switch to highest grade such as nickel-based alloys, or, conversely, to less expensive metals, such as carbon steels, which suffer generalized corrosion but not SCC in these environments.

13.8   Corrosion-Fatigue

Metals subjected to a cycling variable tensile load can suffer a phenomenon of crack formation and propagation, called *fatigue*, which may lead to rupture although the applied load is lower than tensile strength. Figure 13.20 shows a typical mechanical fatigue fracture surface, where beach marks (striations), as typical fingerprint of fatigue, are easily recognized. The presence of an aggressive environment may accelerate fatigue crack propagation so the phenomenon is called *corrosion-fatigue* (C-F). The cracks are usually numerous, although branching typical of SCC do not appear, and are predominantly transcrystalline on surfaces which are perpendicular to the tensile stress direction.

Before reviewing the corrosion-fatigue, it is worth refreshing the pure mechanical fatigue.

13.8.1   Mechanical Fatigue

The classical approach for studying fatigue is through σ–log N diagrams (called S-N diagrams or Wöhler diagrams), illustrated schematically in Fig. 13.21. The number of cycles to failure increases as the oscillation amplitude of applied load, $\Delta\sigma$ decreases. For carbon and low alloy steels a stress threshold exists, called *fatigue limit*, σ_f, below which cracks do not grow.

S-N curves are affected by a high statistical dispersion as cycle number to failure for a given value of σ includes both crack initiation and propagation phases.

Fig. 13.20 Typical mechanical fatigue failure (case study at the PoliLaPP corrosion Museum of Politecnico di Milano)

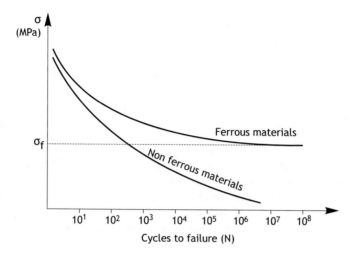

Fig. 13.21 σ-log N diagram for fatigue

Typically, a crack starts from notches, surface micro defects as non-metallic inclusions and dislocation slips. To study more accurately the crack propagation phase, a fracture mechanics approach has to be used. Then, crack growth rate is plotted against load amplitude, ΔK_I, defined as difference between maximum and minimum stress intensity factor during a load cycle:

$$\Delta K_I = K_{I,max} - K_{I,min} = \beta \left(\sigma_{max} - \sigma_{min} \right) \left(\pi a \right)^{0.5} = \beta \Delta \sigma \left(\pi a \right)^{0.5} \qquad (13.2)$$

where β is a geometry factor, $\Delta\sigma$ is the difference between maximum stress, σ_{max}, and minimum stress, σ_{min}, and a is crack size. Figure 13.22 shows the plot of

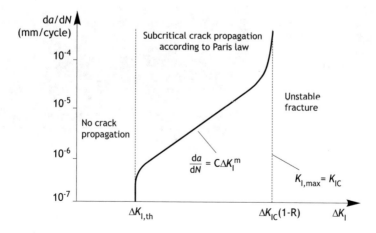

Fig. 13.22 Crack growth rate as function of ΔK_I

fatigue crack growth rate per cycle (da/dN) as a function of ΔK_I in inert environments. Crack growth proceeds according to the following equation:

$$\frac{da}{dN} = C \cdot \Delta K_I^m \tag{13.3}$$

called *Paris law*, where C and m are two constants of the metal and ΔK_I, (Eq. 13.2) can also be written as $K_{Imax} (1 - R)$ where R is K_{Imin}/K_{Imax}. Relevant parameters are:

- $\Delta K_{I,th}$, or threshold value, below which a crack, even pre-existing, for example in welds or wrought, does not propagate
- ΔK_{IC}, as maximum value reached: when $K_{I,max} = K_{IC}$, the crack becomes unstable and propagates at high velocity
- Module, m, of *Paris law* which gives the slope of the fatigue crack growth rate as a function of ΔK_I in the interval $\Delta K_{I,th} - \Delta K_{IC}$.

13.8.2 Influencing Factors

An aggressive environment can enhance fatigue damage rate by reducing the initiation time e.g. by pitting, and accelerating crack propagation rate, this phenomenon is called *corrosion-fatigue*. Figure 13.23 shows a case study of C-F in seawater.

Variables involved are many and interrelated, including metal and environment properties as well as the mechanical stress level. It is important to note that on one side fatigue is a damage phenomenon controlled by the number of cycles

Fig. 13.23 C-F failure of a carbon steel chain in seawater (case study at the PoliLaPP corrosion Museum of Politecnico di Milano)

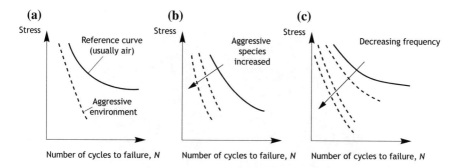

Fig. 13.24 S-N curves for steel: **a** influence of aggressive environment; **b** influence of aggressiveness; **c** influence of load frequency

independently of the frequency or of the cycle period, on the other side corrosion is a time dependent phenomenon. Figure 13.24 shows by means of σ-log N diagrams how the presence of an aggressive environment and frequency of load variation influence general behaviour of corrosion-fatigue. It appears that the fatigue limit vanishes in presence of an aggressive environment and the higher the frequency, the lower the environment influence; hence, accelerated testing of corrosion-fatigue is not possible.

13.8.3   Corrosion-Fatigue and Fracture Mechanics

$da/dN - \Delta K_I$ curves depend on whether the metal is susceptible or not to SCC in the considered environment. When the metal is not susceptible to SCC, fatigue behaviour is called *True Corrosion Fatigue* (TCF) and, conversely, when it is susceptible, *Stress Corrosion Fatigue* (SCF).

13.8.4   True Corrosion Fatigue

Figure 13.25 shows an example of TCF on a platform node near a weld. When TCF applies, da/dN − ΔK_I plot in semi-logarithmic scale is the one shown in Fig. 13.26. In the presence of an aggressive environment, ΔK_{Ith} reduces, ΔK_I range widens, crack growth rate, da/dN, increases in the entire ΔK_I range and da/dN − ΔK_I relationship does not change, therefore Paris law applies again:

$$\frac{da}{dN} = C^* \cdot \Delta K_I^{m^*} \qquad (13.4)$$

Fig. 13.25 Example of cracking near a weld on a platform node due to true corrosion-fatigue (case study at the PoliLaPP corrosion Museum of Politecnico di Milano)

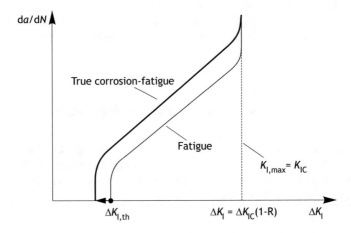

Fig. 13.26 Comparison of fatigue and TCF

where constants, C^* and m^*, no longer depend on metal only, instead also on environment and load variation frequency. In particular, C^* decreases as frequency increases until it reaches, at highest frequencies (above 10 Hz), the same value C governing fatigue in air. Figure 13.27 shows for a carbon manganese steel that as frequency increases, crack propagation rate in seawater and in air tends to coincide.

The waveform has no influence on crack growth rate for fatigue in air, instead it does influence for corrosion-fatigue, because the effect of environment is produced, at least for most metal–environment couplings, only during the tensile load phase, i.e., while plastic deformation occurs at crack tip.

The most important and studied case is corrosion-fatigue on submerged nodes of offshore platforms, not cathodically protected, in harsh environments such as North

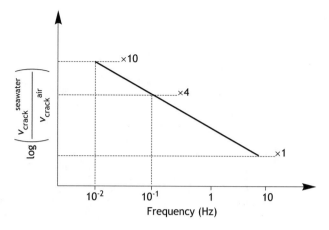

Fig. 13.27 Influence of frequency on TCF of carbon-manganese steel in seawater, expressed as ratio between crack propagation rate in seawater and in air

Sea and Alaska, where sea also provides high cyclic stress/load variations especially during the frequent and persistent winter storms. In these cases, the cracks propagate from weld beads of nodes or tubular joints.

13.8.5 Stress Corrosion Fatigue

When metal–environment coupling is susceptible to SCC, crack growth rate increases remarkably as $K_{I,max}$ exceeds K_{ISCC}. This type of attack is called *Stress Corrosion Fatigue* to indicate that cracking follows an SCC mechanism although stress is variable with time. Figure 13.28 shows schematically how crack growth rate increases as ΔK_I varies: generally, ΔK_{ISCC} is higher than $\Delta K_{I,th}$ so that TCF occurs first, then once $K_{I,max}$ equals K_{ISCC} there is a sharp increase in crack growth rate because SCC prevails, then the typical plateau appears when the slip-dissolution mechanism applies.

Which Potential?
Since the 1970s, when offshore structures operating in North Sea experienced several failures studies on fatigue behavior in seawater of carbon–manganese steels with ferritic–pearlitic structure proliferated. In particular, the behavior of welded joints of platforms, from which cracks originated, was studied in free corrosion and cathodic protection (CP) conditions, at different potentials. The set of data collected shows the complexity of this phenomenon. Let's consider, for example, only the influence of potential.

For low values of ΔK_I, the potential condition that makes the crack growth rate minimum, several times smaller than that measured in free corrosion and

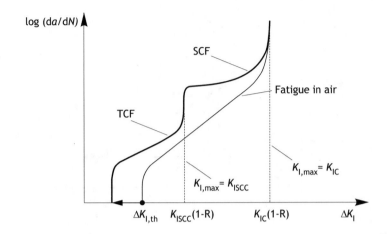

Fig. 13.28 Crack growth rate versus ΔK_I for SCF

nearly the one measured in air, is -0.8 V SCE. Instead, crack growth rate increases hugely at potential more negative than -1.3 V SCE (overprotection condition).

For high values of ΔK_I, crack growth rate is minimum at -0.7 V SCE, i.e., near free corrosion condition, probably because a minimum of corrosion favours the blunting of crack tip. At recommended CP conditions, i.e., potential of -0.8 V SCE, crack growth rate increases slightly and remains stable up to -1.1 V SCE, then increases again.

So, which protection potential has to be adopted? Working condition consists for a small percentage of the structure life of high ΔK_I, which produces very deep cracks, and for most of the time low ΔK_I. By applying the Miner's law (hypothesis of effect addition) corrosion-fatigue at low ΔK_I weighs much more than that at high ΔK_I. Therefore, for cathodically protected offshore structures the best value of protection potential to limit damages caused by corrosion-fatigue seems to be the same one recommended to prevent general corrosion (-0.8 V SCE), provided that overprotection condition is carefully avoided.

13.8.6 Prevention of Corrosion-Fatigue

The strategy for C-F prevention depends on mechanism: TCF, SCF by slip-dissolution (anodic) and SCF by HE (cathodic).

For TCF, the strategy is based on the increase in $\Delta K_{I,th}$ for instance by means of surface treatments, such as rolling, hammering or peening, which generate a compressive stress surface layer, 50–75 μm thick.

For SCF by slip-dissolution (anodic), two interventions help increase SCC resistance: an increase of $\Delta K_{I,th}$ as above and the reduction of crack tip corrosion rate (given by the plateau) by injecting inhibitors or by applying cathodic protection, for instance with the application of anodic metallic coatings (for example, zinc on steel). Conversely, cathodic coatings (for example, nickel on steel) are dangerous since the presence of porosity or defects enhances fatigue on the base metal.

For SCF by HE (cathodic), the strategy is based on the increase in $\Delta K_{I,th}$ only. Attention should be paid on cathodic protection because helpless and even dangerous as soon as potential lowers from the standard protection level (overprotection).

Sea Gem and Alexander Kielland Platforms

Before oil production booming in North Sea, most of offshore platforms of the world (about a thousand in 1960) were in the Gulf of Mexico. Many structural failures, which occurred in that area during the 1950s on platforms weakened by corrosion, were attributed to temporary overloads caused by tornados, so, accordingly, a new platform design was adopted to withstand maximum loads generated by so-called 100-year storm, associated with mandatory CP.

This design philosophy, validated in tropical seas where storms are an exception, was extended to North Sea in the early 1960s, which, instead, is a calm sea only exceptionally. Because of this, failure mechanism becomes fatigue or corrosion-fatigue rather than overloads. Furthermore, CP design philosophy didn't change because it was ignored that passing from a warm and calm tropical sea to a stormy, cold sea, as North Sea is, the protection current at least doubles (higher oxygen limiting current density) and seawater resistivity increased, too. So, severe corrosion-fatigue soon appeared although same structures had long operated without problems in the Gulf of Mexico. The Sea Gem platform sank on Dec. 27th 1965 with 13 workers dead, and many other failures occurred without so dramatic consequences. The last disaster was in 1980, when the Alexander Kielland platform wrecked causing 123 deaths.

During the 1970s, design philosophy changed again, taking into account corrosion-fatigue and introducing a redundant design to avoid that a component failure would jeopardize the security of the structure. In parallel, based on what learned from experience, CP design and CP criteria for offshore structures were upgraded; in other words, CP in seawater advanced from an empirical practice to a rational engineering.

13.9   Some Conclusions

As conclusion of this chapter on environmental assisted cracking of metals, it could be useful to summarize, in a unified framework, conditions for crack propagation due to purely mechanical or jointly mechanical and environmental actions.

Critical parameter used for the classical approach to failure analysis is the nominal stress measured on a smooth specimen for the following conditions:

- σ_R for mechanical failure under static load
- σ_{th} for crack initiation and propagation under SCC
- σ_{CF} for fatigue on corrosion-fatigue

When linear elastic fracture mechanics approach is adopted, the following three critical parameters are referred to:

- K_{IC} for mechanical failure
- K_{ISCC} for SCC
- K_{th} for fatigue or corrosion-fatigue.

These six parameters (σ_R, σ_{th}, σ_{CF}, K_{IC}, K_{ISCC} and K_{th}) can be represented on a two axes diagram in double logarithmic scale, nominal stress, σ, and crack length, a (Fig. 13.29).

The three conditions of classical approach are three straight lines parallel to the abscissa, while critical conditions expressed in terms of fracture mechanics through the parameter K_I, are straight lines with slope $-1/2$, according to equation: $K_I = \beta \, \sigma \, (a)^{1/2}$.

To represent on this diagram also conditions of initiation and propagation of fatigue cracks in terms of K, it has to consider that:

$$\Delta K = K_{Imax}(1 - R) \tag{13.5}$$

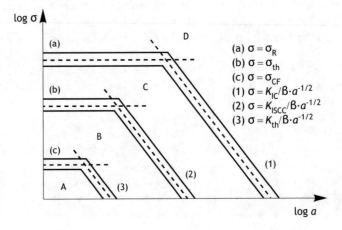

Fig. 13.29 Schematic representation of critical conditions for crack propagation. Adapted from Sinigaglia et al. (1979)

where R is K_{Imin}/K_{Imax} ratio. If $R = 0$, K_{Imin} is 0, then $\Delta K = K_{Imax}$.

In practice, it is not possible to determine rigorously critical parameters because data on geometry, stress level and environment are scattered; accordingly, on the plot a band rather than a line is reported. Each point of the diagram represents a structure or a part of it, subjected to a nominal stress, σ, and containing defects smaller than a maximum value a_{max}, which is defined as minimum defect size determined by applicable non-destructive testing. Four zones are identified:

- **Zone A**: it is a safe condition (life structure is unlimited, no subcritical propagation of defects is possible)
- **Zone B**: defects can propagate through fatigue or corrosion-fatigue to reach size to trigger SCC
- **Zone C**: defects propagate through SCC under static or variable load, then leading to final mechanical fracture
- **Zone D**: structure is mechanically unstable.

13.10 Applicable Standards

- ASTM G 30, Making and using U-bend stress-corrosion test specimens, American Society for testing of Materials, west Conshohocken, PA.
- ASTM G 36, Standard practice for evaluating stress-corrosion-cracking resistance of metals and alloys in a boiling magnesium chloride solution, American Society for testing of Materials, west Conshohocken, PA.
- ASTM G 37, Standard practice for use of Mattsson's solution of pH 7.2 to evaluate the stress-corrosion cracking susceptibility of copper-zinc alloys, American Society for testing of Materials, west Conshohocken, PA.
- ASTM G 38, Standard practice for making and using C-ring stress-corrosion test specimens, American Society for testing of Materials, west Conshohocken, PA.
- ASTM G 49, Standard practice for preparation and use of direct tension stress-corrosion test specimens, American Society for testing of Materials, west Conshohocken, PA.
- ASTM G 58, Standard practice for preparation of stress-corrosion test specimens for weldments, American Society for testing of Materials, west Conshohocken, PA.
- ASTM G 129, Standard practice for slow strain rate testing to evaluate the susceptibility of metallic materials to environmentally assisted cracking, American Society for testing of Materials, west Conshohocken, PA.
- ISO 6957, Copper alloys. Ammonia test for stress corrosion resistance, International Standard Organization, Geneva, Switzerland.
- ISO 7539, Corrosion of metals and alloys. Stress corrosion testing, International Standard Organization, Geneva, Switzerland.

- NACE TM 0177, Laboratory testing of metals for resistance to sulfide stress cracking and stress corrosion cracking in H_2S environments, NACE International, Houston, TX.

13.11   Questions and Exercises

13.1   Describe the crack growth mechanism by tip dissolution. Provide an example.

13.2   In your opinion, what is the effect of metal microstructure on hydrogen embrittlement susceptibility?

13.3   What is the effect of a cathodic polarization on stress corrosion cracking?

13.4   Consider this statement: "in general, it is observed that metals highly resistant to corrosion because protected by a passive film such as titanium, aluminium, nickel, chromium and stainless steels, suffer SCC when a chemical species, for example chlorides, that enables to breakdown the passive film is present". Justify this sentence according to SCC mechanism.

13.5   What is the effect of nickel content on SCC of stainless steel in hot chloride containing solution? Which stainless steels do you suggest for such applications?

13.6   What is the difference between K_{IC} and K_{ISCC}? Do these parameters depend on the environment?

13.7   An AISI 316 stainless steel pipe (18% Cr, 8% Ni, 2% Mo) transports a chloride-containing solution at 80 °C. The pipe (wall thickness 30 mm), due to the internal pressure, suffers a tensile stress. By non-destructive testing, a crack-like defect (length 1 mm), transversally oriented with respect to the applied stress, has been detected on the metal surface. Calculate: a) the maximum tensile stress without leading to stress corrosion sub-critical crack propagation; b) the maximum crack length before critical crack propagation if a tensile stress of 700 MPa is applied.

13.8   Discuss the effect of the frequency on the applied tensile load on SCC.

13.9   Discuss the effect of sensitization of austenitic stainless steels on SCC.

13.10   In a hypothetical metal-environment combination, $K_{ISCC} < K_{Ith}$. Suggest a possible da/dN $-$ ΔK_I plot and comment the expected behaviour. Suggest a possible practical example.

Dany Sinigaglia - Dany Sinigaglia (1936-83) graduated from Politecnico di Milano (Milan, Italy) in 1962 receiving the gold metal for the best thesis. He started research activities at the Politecnico, where he was nominated lecturer in Metallurgy in 1969 and associated professor in 1982. He was a very good and tireless researcher. Since the mid-sixties he started the study of localized corrosion with methodologies and techniques absolutely innovative for that time. He was one of the first to propose theoretical models of corrosion interpretation in occluded cells, to develop calculation methodologies to highlight the influence of electrochemical, geometrical and environmental factors in the onset and development of localized attacks. Later, he was involved in fracture mechanics approach of stress corrosion cracking. Who worked with Dany could appreciate his qualities as a researcher and a teacher (he held the metallurgy course) but unfortunately, a premature death prevented him from fully showing what temper he was done, because he certainly would have become an important reference in the field of corrosion and metallurgy. He published books on metallurgy, fracture mechanics and environmental assisted cracking, together with about 100 publications on journal and congress proceedings.

Bibliography

Brown BF (1968) The application of fracture mechanics to stress corrosion cracking. Met Rev 156:55

Brown BF (1971) The theory of SCC in alloys. Im: Scully JC (ed) NATO Scientific Affairs Division, Brussels

Brown BF (1977) SCC control measures. NBS Monograph 156, National Bureau of Standard, Washington DC

Copson HR (1959) Physical metallurgy of stress corrosion fracture. Interscience, New York, p. 247

Denhard EE (1960) Effect of composition and heat treatment on the stress corrosion cracking of austenitic stainless steels. Corrosion 16(7):359t–370t

Fontana M (1986) Corrosion engineering, 3rd edn. McGraw-Hill, New York. ISBN 0-07-100360-6

Henthorne M (2016) The slow strain rate stress corrosion cracking test—a 50 year retrospective. Corrosion 12(72):1458–1518

Magnin T, Chambreuil A, Bayle B (1996) The corrosion-enhanced plasticity model for stress corrosion cracking in ductile FCC alloys. Acta Mater 44(4):1457–1470

Parkins RN, Fessler RR, Boyd WK (1972) Stress corrosion cracking of carbon steel in carbonate solutions. Corrosion 28(8):313–320

Parkins RN, Wearmouth WR, Dean GP (1973) Role of stress in the stress corrosion cracking of a Mg-Al Alloy. Corrosion 29(6):251–260

Sedriks AJ (1996) Corrosion of stainless steels, 2nd edn. Wiley, New York

Sinigaglia D, Re G, Pedeferri P (1979) Cedimento per fatica e ambientale dei materiali metallici. CLUP, Milano, Italy (in italian)

Staehle RW (1977) Predictions and experimental verification of the slip dissolution model for stress corrosion cracking of low strength alloys. In: Staehle RW, Hochmann J, McCright RD, Slater JE (eds) NACE-5 Stress corrosion cracking and hydrogen embrittlement of iron base alloys. NACE International, Houston

Chapter 14
Hydrogen-Induced Damage

> *If a hydrogen atom is as small as a golf ball, a hydrogen molecule is as big as a basketball.*

Abstract Hydrogen induced damage (HID) can occur at high temperature (HT-HID) and at low temperature, (LT-HID). Hydrogen attack, affects steels operating at temperatures typically above 400 °C in high pressure hydrogen atmosphere. The interaction of atomic hydrogen and metals at low temperature occurs in different way. Atomic hydrogen is produced during electroplating processes (as chrome plating, galvanizing and phosphating), chemical and electrochemical pickling treatments, in welding if the humidity of consumables is too high, or by the cathodic process in corrosive fluids: in this last case, so called cathodic poisons, as H_2S, inhibit molecular hydrogen formation and promote atomic hydrogen diffusion into the metal. Once entered the metal, atomic hydrogen interacts with the metal structure and may produce a "damage" of various forms, such as delayed fracture, HIC (hydrogen induced cracking) and blistering, hydrogen embrittlement (HE). All of these forms of damage are discussed in this chapter.

Fig. 14.1 Case study at the PoliLaPP Corrosion Museum of Politecnico di Milano

© Springer Nature Switzerland AG 2018
P. Pedeferri, *Corrosion Science and Engineering*, Engineering Materials,
https://doi.org/10.1007/978-3-319-97625-9_14

14.1 Hydrogen Induced Damage

Hydrogen-metal interactions can occur at high temperature, indicated as HT-HID (High Temperature Hydrogen-Induced Damage) and at low temperature, LT-HID (Low Temperature Hydrogen-Induced Damage), whether temperature is above or below 200 °C. An example of hydrogen embrittlement (Lt-HID) is reported in Fig. 14.1.

There are other damage mechanisms induced by hydrogen in metals, when hydrogen reacts with hydride-forming metals or with carbon to form hydrides or methane, respectively. Typical hydride-forming metals are titanium and zirconium. In this case, hydrides cause the brittle fracture of the metal along the hydride-matrix interface.

On carbon steels, hydrogen reacts at high temperature (T > 200 °C) with carbon to give methane. This phenomenon occurs mainly at the surface of carbon steels, causing decarburization; if it occurs internally, methane is trapped like hydrogen in hydrogen induced cracking (HIC, see in the following), then creating internal cracks and blisters, reducing fracture toughness, fatigue and creep resistance. This type of damage is often called *hydrogen attack*. A scheme with classification of HID is reported in Fig. 14.2.

14.1.1 Adsorption, Dissolution and Trapping

Atomic hydrogen is the smallest atomic element, because composed by a proton and an electron, only. Accordingly, its size is so small that it can dissolve and diffuse in metals: both phenomena depend on the crystalline structure of metal lattice, and precisely on lattice structure and cell size.

Atomic hydrogen interacts with metals as schematically shown in Fig. 14.3. First of all, hydrogen atoms adsorb easily on the metal surface reaching very high

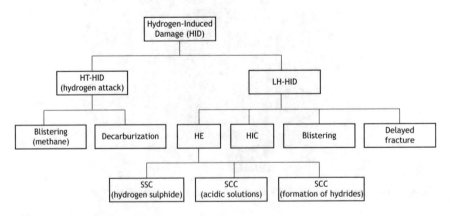

Fig. 14.2 Classification of hydrogen-induced damage (HID)

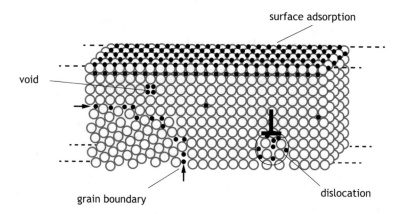

Fig. 14.3 Schematic representation of interaction of hydrogen atoms with metal lattice. Adapted from Pundt and Kircheim (2006)

concentration on the surface, in practice one hydrogen atom for each cell, i.e., one hydrogen atom for about 4 atoms of the metal. For example, for iron such concentration is about 4500 ppm ($10^6/224$).

In *bcc* structures (atomic packing factor 0.68) hydrogen atoms can occupy octahedral and tetrahedral sites as shown in Fig. 14.4, then producing a cell distortion. Therefore, maximum solubility would be 1 atom per cell (i.e., one hydrogen for two atoms of iron, which corresponds to about 900 ppm); this concentration cannot be obtained because lattice should deform strongly, so the practical maximum hydrogen content in steel is about thousand times lower, around 1 ppm (higher content up to 20 ppm can be measured when hydrogen is forced in and screened, for instance by plating the surface with coatings made of *fcc* or *hcp* metals, Cu, Ni, Cd and Zn).

Although compact structures, as *fcc* and *hcp*, have a packing efficiency 0.74, exhibit an atomic hydrogen solubility of about one order of magnitude higher than *bcc*, because *fcc* lattice has a large single octahedral interstitial site plus two small

Fig. 14.4 Interstitial sites in bcc structure where atomic hydrogen can store

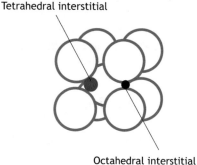

Fig. 14.5 Example of hydrogen storage at edge dislocations where lattice is distorted

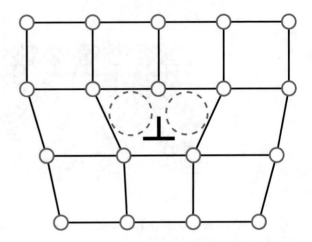

tetrahedral interstitial sites per cell. This is favourable to accommodate hydrogen without any excess energy requirement.

In *bcc* structures, in addition to interstitial sites, there are *hydrogen traps* inside crystals and grains, where hydrogen atoms store. To make it simple, hydrogen traps can be classified by two types: *reversible traps* and *irreversible* or *permanent traps*. Figure 14.3 shows most common reversible traps, precisely: dislocations, grain boundaries, vacancies. These traps are called reversible because hydrogen easily penetrates steel, when forced in, and easily leaves it by stopping hydrogen production and by heating. Figure 14.5 shows how dislocations can host hydrogen atom where the lattice is distorted.

Irreversible traps are coarse micro-voids, consisting of interfaces at inclusions, especially manganese sulphide, MnS (Mn(II)), as shown in Fig. 14.6. To remove this hydrogen, it is necessary to heat steel above 400 °C. As reported below, if trap size is big, hydrogen atoms form a hydrogen molecule, then making it impossible to recover, causing HIC and blistering (Fig. 14.7).

For iron and steels, Table 14.1 reports the types of traps for atomic hydrogen and the associated binding energy and the degassing temperature, which are strongly interrelated, to give an idea of the strength of such traps. For example, the hydrogen entrapped in the matrix or adsorbed on the metal surface is easily stripped out at low temperature when exposed to an atmosphere not containing hydrogen (in practice to the open air). As a further example, atomic hydrogen stored at grain boundaries is less bound than that at dislocations hence more easily depleted.

14.1.2 Diffusion

As said, atomic hydrogen has a different solubility which depends on crystalline structure. Once it is dissolved in the surface layer of a metal, hydrogen can diffuse

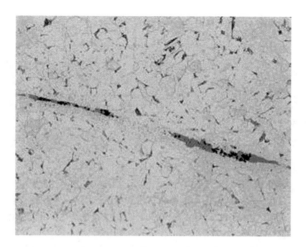

Fig. 14.6 Elongated MnS inclusions in carbon steel

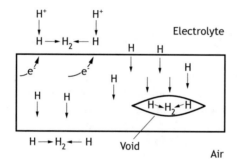

Fig. 14.7 Schematic mechanism of blistering and HIC formation on MnS inclusion traps

Table 14.1 Properties of reversible traps existing in iron and steel (average values from various literature data)

Metals	Traps	Binding energy (kJ/mol)	Degassing temperature (°C)
Fe	Matrix	7	25
	Grain boundaries	17	110
	Dislocations	20–26	200–215
	Microvoids	35–48	305
	Carbides interfaces	97	725
Carbon steel (0.47% C)	Microvoids	35–48	480
AISI 4340	Dislocations	20–26	270
	Microvoids	35–48	340
	MnS inclusions	72	495

Table 14.2 Diffusion coefficient of hydrogen in different materials at room temperature (average values from various references)

Material	D (m²/s)	Notes
Iron (carbon steel)	2.5×10^{-10}	Hydrogen solubility in the lattice is about 1 ppm
Ferritic stainless steel	10^{-11}	
Austenitic stainless steel	2.15×10^{-16}	Hydrogen solubility in the lattice is about 20 ppm
Martensitic stainless steel	2×10^{-13}	
Duplex stainless steel	10^{-13} to 10^{-14}	Depending on the ferrite/austenite ratio

into the inner volume flowing from the zones at higher concentration to the ones at lower concentration. Also the diffusivity within the lattice depends of the crystalline structure. It is worth noting that solubility and diffusion are in opposition. Diffusion is the prevailing mechanism which drives the interaction between hydrogen and metals and explains the HID occurrence. Table 14.2 reports the diffusion coefficient of atomic hydrogen in metals used in industry.

14.1.3   Atomic Hydrogen Produced by a Cathodic Process

During electroplating processes (as chrome plating or electrogalvanizing), phosphating, chemical and electrochemical pickling treatments, cathodic protection or in case of corrosion in acidic solutions, on metal surface the process of hydrogen ions reduction can take place; based on a simplified mechanism, this process follows two stages through the intermediate formation of atomic hydrogen:

$$2H^+ + 2e^- \rightarrow 2H \tag{14.1a}$$

$$2H \rightarrow H_2 \tag{14.1b}$$

Since each stage occurs with its own rate, generally different from each other, two situations apply, whether the second step is faster or slower than the first one:

- *Faster second step*. As soon as atomic hydrogen is produced, molecular hydrogen forms and evolves into solution, therefore, atomic hydrogen has no time to enter the metal; to give a rough idea, 99% of produced atomic hydrogen combines to give molecular hydrogen and less than 1% is available for entering the metal
- *Slower second step*. This happens when some chemical species, called *cathodic poisons*, which inhibit molecular hydrogen formation, are present. Cathodic poisons include: sulphur, arsenic, antimony in the metallic phase, and cyanides, sulphides and organic substances containing sulphur in the electrolyte. It follows that only some of atomic hydrogen recombines forming molecular hydrogen,

Table 14.3 Percentage of hydrogen entering the steel in the presence of different sulphide content in solution

$[S^{2-}]$ in solution (ppm)	p_{H_2S} in the gas in equilibrium with the solution (bar)	% of atomic hydrogen penetrating the steel
0.0035		1
0.035		16
0.35	0.0001 (=0.0014 psi)	33
1	0.0003 (=0.004 psi)	40
13	0.0036 (=0.05 psi)	60
100	0.028 (=0.39 psi)	75
1000	0.28 (=3.9 psi)	90
10,000	2.8 (=39 psi)	100

Extrapolated equation: $H_{ab}\% \cong 40 + 16 \cdot \log (H_2S, ppm)$
Adapted from Hudson et al. (1968)

while atomic hydrogen concentrates at metal surface so that it can enter the metal. For instance, again to give an estimate, only 10% of produced atomic hydrogen combines, while the remaining enters the metal. Table 14.3 reports the percentage of hydrogen entering the steel in the presence of different sulphide content in solution.

Oxide films formed on the metallic surface are barriers for H absorption and are hindering the hydrogen passage through the interface. When a cathodic poison is present, hydrogen flow rate inside the metal increases and can reach a maximum when cathodic overvoltage meets a critical value of approximately 100 mV. This condition is encountered when sulphides and cyanides are present and corresponds to maximum hydrogen solubility in metals (for instance, of the order of 10^{-1} mol/dm^3 for iron, about 12 ppm, at 25 °C).

Measurement of Hydrogen Diffusion in Metals
Figure 14.8 shows an experimental apparatus to measure diffusion rate of atomic hydrogen in metals (Devanathan 1962). A thin strip separates two compartments: the first one (I) contains hydrochloric acid and the second one (II) contains an alkaline sodium sulphate solution, which passivates the steel surface. The strip surface of compartment (I) works as cathode by means of an auxiliary anode, in order to produce atomic hydrogen on the metal surface: some atomic hydrogen enters the metal and diffuses in it, some others combine to give molecular hydrogen, which escapes through the solution.

Simultaneously, in compartment (II) the strip surface works as anode by means of an auxiliary cathode. By keeping with a potentiostat the anodic potential below oxygen evolution potential, the anodic reaction possibly occurring is oxidation of atomic hydrogen that reached the metal surface by

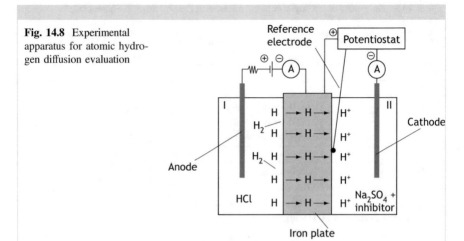

Fig. 14.8 Experimental apparatus for atomic hydrogen diffusion evaluation

diffusion. Therefore, the anodic current measures directly the diffusion rate of hydrogen in the metal, which is governed by Fick's law:

$$\frac{i}{F} = J = D\frac{C_1 - C_2}{s} \qquad (14.a)$$

where J is hydrogen flow (mol/m²s), i is circulating current density (A/m²), D is diffusion coefficient (m²/s), C_1 and C_2 are hydrogen concentrations on strip surfaces for compartment I and II respectively, s is strip thickness (m) and F is Faraday constant (96,485 C). After an initial transient, a steady state is reached when $C_2 = 0$, i.e., all hydrogen that diffuses through the strip is immediately oxidized, then relationship (14.a) becomes:

$$\frac{i}{F} = D\frac{C_1}{s} \qquad (14.b)$$

which allows the calculation of the diffusion coefficient, D, by the measurement of current density, i.

The diffusion coefficient of hydrogen in iron (*bcc* structure) has same order of magnitude of the one of ions in aqueous solutions ($D = 6.25 \times 10^{-9}$ m²/s at 25 °C); in other metals, for example austenitic steels (*fcc* structure), diffusion occurs more slowly and diffusion coefficient is three orders of magnitude lower.

Testing has shown that hydrogen diffusion occurs through interstitial crossing in crystal lattice, therefore the diffusion coefficient depends on temperature only. Accordingly, it increases with temperature, while is practically independent from either imperfections on grain boundaries or structure, whether poly or monocrystalline. In the presence of tensile stresses, hydrogen diffusion significantly increases, while it decreases if compressive stresses apply: this behaviour is due to changes of hydrogen solubility and not changes in diffusion coefficient.

14.1.4 Decomposition and Solubility of Hydrogen at High Temperature

In the presence of hydrogen containing gas at high pressure and at temperature above 200 °C, iron catalyses the split of molecular hydrogen, $H_2 \rightarrow 2H$, with atomic hydrogen dissolution in iron lattice. Atomic hydrogen solubility varies with temperature: maximum solubility is in molten iron, about 30 ppm, then decreasing below 0.1 ppm at ambient temperature. According to Sievert's law, the solubility of atomic hydrogen in *bcc* iron can be expressed as follows:

$$\ln[H]_{bcc} = 1.628 + \frac{1}{2}\ln(p_{H_2}) - \frac{1418}{T} \qquad (14.2)$$

where $[H]_{bcc}$ is atomic hydrogen concentration in ppm, p_{H_2} is hydrogen partial pressure in bar and T is absolute temperature.

Atomic hydrogen can derive also from the thermal decomposition of water coming from e.g., air humidity in contact with melted iron during steel production or the humidity possibly contained in weld flux or welding electrode coating.

14.2 HT-HID or Hydrogen Attack

High temperature hydrogen-induced damage, also called *hydrogen attack*, affects steels operating at elevated temperatures (typically above 400 °C) in high pressure hydrogen atmosphere, as in refineries, petrochemical, ammonia production plants and other chemical facilities and, possibly, high pressure steam boilers. Hydrogen attack is one of the major problems in refineries, where hydrogen and hydrocarbon streams are handled at up to 20 MPa and approximately 250 °C.

It is the result of reaction between dissolved atomic hydrogen and carbon of carbides present in steel to form methane ($C + 4H \rightarrow CH_4$). This reaction can occur either on the steel surface or inside steel: the former leads to skin decarburisation, then to a decrease in hardness and an increase in ductility near the surface, the latter forms blisters containing methane, at grain boundaries and/or at precipitate interfaces, with reduction of local carbon content and strength resistance as well as formation of fissures and cracks. Internal decarburisation, and in

(a)   **(b)**

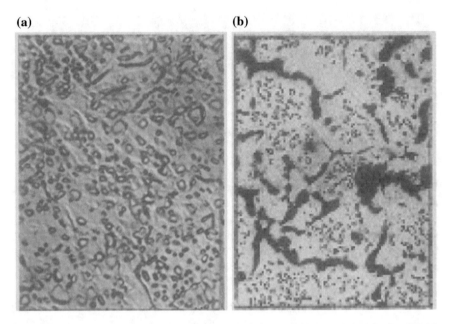

Fig. 14.9 Micrographs of carbon steel C-1095 (0.95% C, 0.40% Mn). **a** original, **b** after 500 °C; 1000 bar; 1 h exposure (from Thygeson 1964)

particular the formation of methane and consequent development of voids, can lead to substantial deterioration of mechanical properties of steels due to loss of carbides and formation of voids (Fig. 14.9).

Main factors influencing HT-HID are hydrogen partial pressure, temperature, exposure time and steel composition. The presence of elements forming stable carbides such as Cr, Mo and V is very important: steels with chromium more than 5%, and austenitic stainless steels, do not suffer this attack. Cr, Mo, W, V, Ti and Nb—i.e., carbide forming elements—are used in steels to improve resistance. Industry experience indicates that post-weld heat treatment of Cr-Mo steel is beneficial in resisting hydrogen attack in so-called hydrogen service.

In 1949, Nelson gathered and rationalised a number of experimental observations on different steels. Since that, API 941 Nelson curves (Fig. 14.10) are a universally used guidance for carbon and low alloy steel selection and has been updated a number of times. Nevertheless, today's trend is the combination of Nelson curves and risk-based inspection approach, as recommended by American Petroleum Institute.

Hydrogen Sources
Hydrogen is adsorbed as atom. Most typical occurrence is during steelmaking (production of forged steel), pickling/etching treatments, electrochemical plating, and primarily in acidic corrosion. During steel forging, hydrogen

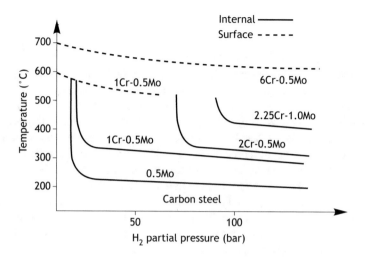

Fig. 14.10 Nelson curves for different grade of carbon and low alloy steels

comes from atmospheric moisture by water splitting and dissolves at high temperature in *fcc* austenitic phase (γ-iron). After cooling, as hydrogen solubility in *bcc* (α-iron) decreases by one order of magnitude, hydrogen atoms remain trapped in inclusions, micropores and lattice interstitial sites, often causing embrittling airline cracks (flakes). Similarly, atomic hydrogen is produced in welding operation with humid consumable welding rods.

In pickling, electrochemical plating and acidic-related corrosion processes, hydrogen atoms can also be directly absorbed by cathodic charging, which is enhanced by substances, called cathodic poisons, such as arsenic, antimony, sulphur, selenium, tellurium, and cyanide ions. These poisons stop hydrogen atoms from forming molecular hydrogen, H_2, then in atomic form it readily diffuses into the metal. The most common cathodic poison is hydrogen sulphide, which is often present in the petroleum industry, during drilling, completion and production of oil and gas wells.

14.3 LT-HID

The interaction of atomic hydrogen and metals at low temperature is different from the one at high temperature in all ways: mechanism, occurrence time and type of damage. It should be added that consequences of LT-HID are generally more severe and uncontrolled than hydrogen attack at high temperature.

14.3.1   Delayed Fracture

Ferritic steels suffer hydrogen-induced damage, when hydrogen content exceeds a critical value, through a drastic reduction of plasticity, while elasticity (i.e., Young modulus or elastic modulus) remains unchanged. For high strength steels with a tensile strength, σ_R, exceeding 1500 MPa, the critical hydrogen content is 1 ppm or less, while it is about 5–10 ppm for steels with lower strength, in the range 500–1000 MPa. The effect of dissolved hydrogen may appear mysterious, since the loss of ductility cannot be observed through neither impact test nor tensile test on smooth specimens even at low temperature; instead, it occurs in tests at low deformation rate and with a pre-existing notch or crack. This behaviour is therefore the opposite of the conventional one, where high strain rates and low temperatures favour ductile-to-brittle transition. This is due to hydrogen diffusion as controlling factor: hydrogen diffuses slowly towards micro-voids that form also slowly as steel deforms above ambient temperature. In the presence of notches, hydrogen accumulates at notch tip, where plasticization occurs, then mechanical strength decreases. This is called *delayed fracture* or *static fatigue* that leads to a fracture after times as longer as applied stress and hydrogen content are lower (Fig. 14.11).

Figure 14.12 shows the influence of hydrogen content on delayed fracture of a high yield strength steel: time-to-failure, at constant applied stress (or the opposite, applied stress at constant time-to-failure), increases as hydrogen content in steel decreases. Fracture does not take place below a stress threshold, which depends on hydrogen concentration, yield strength (i.e., microstructure), degree of deformation and temperature.

Fig. 14.11 Hydrogen delayed fracture caused by acid pickling before galvanizing

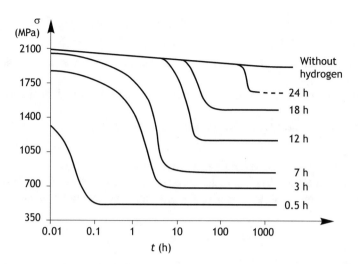

Fig. 14.12 Delayed fracture curves for ferritic low alloy steel, AISI 4340 grade, with different hydrogen concentrations obtained by heating hydrogen saturated steel at 150 °C for various times (from Barth 1969)

14.3.2 HIC and Blistering

When atomic hydrogen, while diffusing within the steel lattice, crosses a micro-void it is trapped; as soon as another hydrogen atom arrives, the two combine to molecular hydrogen with an extraordinary increase in pressure (two golf balls to give a basketball size: about 600 times!). Local pressure increases so highly that steel cannot withstand it, so cracking or deformation occurs. It has been estimated that molecular hydrogen in blisters can reach a pressure as high as 10^4 bar, therefore, if a hydrogen bubble forms around an elongated MnS inclusion that has a crack-like shape, the very high internal pressure causes at the inclusion tip an increase of stress in the metallic matrix that easily overtakes the tensile strength of the material causing its failure by ductile tearing.

Applying a very simplified approach, considering Mariotte equation:

$$\sigma = \frac{P \cdot \phi}{4s} \qquad (14.3)$$

where σ is the stress generated by the hydrogen pressure, P, created in a blister, ϕ is void diameter and s layer thickness, the tensile stress produced in a layer 0.001 m thick around a spherical void of 1 mm in diameter is about 250 MPa, which is a value higher than the yield strength of bcc iron, i.e., of mild steel.

When cracking occurs, the phenomenon is called *hydrogen-induced cracking*, HIC, when blisters form, the phenomenon is called *blistering*. Examples of blistering are shown in Fig. 14.13. When atomic hydrogen forms molecular hydrogen inside the metal in a microvoid, there is an equilibrium between:

- Concentration of hydrogen adsorbed on the metal surface, i.e., its degree of coverage, θ_H
- Concentration of dissolved hydrogen in the metal $C_{H,S}$, which is proportional to θ_H
- Pressure of molecular hydrogen inside microvoid, P_{H_2}.

The above parameters depend on the overvoltage, η_H, of the hydrogen evolution process which takes place on metal surface. When the slowest stage is the combination of atomic hydrogen to give the molecule, the following relation holds:

$$P_{H_2} = P_{H_2}^0 \cdot e^{\frac{2F\eta_{H_2}}{RT}} \tag{14.4}$$

where $P_{H_2}^0$ is the external pressure of hydrogen, often taken as 0.1 MPa, P_{H_2} is the pressure of molecular hydrogen inside the microvoid and the other parameters have the usual meaning. As cathodic overvoltage increases, similarly internal pressure increases. For instance, with a cathodic polarization of 0.1 V, internal pressure increases of about 30 times or 1000 times if polarization would be about 0.2 V.

For carbon steel plates used for pressure vessels and pipelines, the most affecting inclusions are manganese sulphide, MnS, type II, because their form is flat as obtained during hot rolling and are parallel to the rolling direction, hence providing an easy trap with a crack-like shape for diffusing atomic hydrogen. Trapped hydrogen forms small cracks, parallel to the plate. This phenomenon, called

Fig. 14.13 Example of blisters

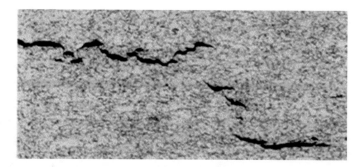

Fig. 14.14 Stepwise cracking (HIC attack) in carbon steel

Hydrogen Induced Cracking (HIC), depends on the amount of hydrogen available for diffusion into the steel and on time. For example, for a design life of 20 years, the H_2S partial pressure to produce an atomic hydrogen flow sufficient for HIC occurrence on susceptible steels is 0.1 bar (corresponding to a concentration in the aqueous phase of about 400 ppm). Higher partial pressures reduce the time more than proportionally (for instance, at 1 bar the same steel shows evidence of HIC in about hundreds of hours and blisters in a couple of years).

Figure 14.14 shows typical *step-wise-cracking* (SWC) occurring in a carbon steel plate manufactured by austempering heat treatment, consisting of lamination at temperature around 850 °C where the steel structure is austenitic.

14.3.3 HE Mechanism

As said, HE occurs if the following conditions are met:

- Sufficient atomic hydrogen in the metal
- Presence of a tensile stress
- Susceptible steel
- Temperature range that supports hydrogen transport (−50 to 150 °C) in steel.

In non-hydride forming elements, three mechanisms have been considered (Lynch 2007):

- HELP, Hydrogen-Enhanced Localized Plasticity
- HEDE, Hydrogen-Enhanced De-Cohesion
- AIDE, Adsorption-Induced Dislocation Emission.

It is worth noting that while cracking mechanisms are similar, the rate controlling processes are very different. A combination of these three mechanisms occurs in most cases. The most dominant mechanism will be dependent upon variables such as strength, microstructure, slip-mode, stress intensity factor, and temperature, thus affecting the fracture path and fracture surface appearance.

Furthermore, as practical evidence which has not yet a complete explanation, for a given hydrogen content, steels appear in general more susceptible as strength resistance increases and the tendency to embrittlement increases as strain rate decreases. HE susceptibility decreases as temperature increases; this behaviour prevails at room temperature, and disappears almost entirely in steels above 200 °C, as dissolved hydrogen escapes out of steel.

Hydrogen-Enhanced Localized Plasticity (HELP) . This mechanism is based on the presence of solute hydrogen ahead of cracks, specifically in hydrogen atmospheres around both mobile dislocations and obstacles to dislocation movements. By this mechanism, the hydrogen atmospheres distort when mobile dislocations approach obstacles, meaning that the repulsion by obstacles is decreased. Since hydrogen accumulation is localized near crack-tips, deformation is localized and facilitated near crack-tips, resulting in an overall lower strain for fracture.

Hydrogen-Enhanced De-cohesion (HEDE) . This mechanism is based on the weakening of iron-iron intermetallic bonds at or near crack tips due to a decrease in the electronic charge density due to the presence of hydrogen in the crystal lattice in interstitial sites, then favouring an easy tensile separation of the atoms. Fracture surfaces should appear basically featureless with a few cleavage steps and tear ridges separating de-cohered regions.

Adsorption Induced Dislocation Emission (AIDE) . This mechanism is based on hydrogen-induced weakening of interatomic bonds, but with crack growth occurring by localized slip. It has been proposed that adsorbed hydrogen weakens substrate interatomic bonds and thereby facilitates the emission of dislocations from the crack tips. There is also substantial dislocation emission ahead of the crack tip, resulting in the formation of voids around particles or at slip band intersections. This behaviour means that crack propagation occurs due to dislocation emission from crack tips also with a contribution from the void formation ahead of the crack tip.

14.3.4   Failure Mode

The failure mode caused by HE, unlike SCC does not branch and varies from brittle cleavage or quasi-cleavage fracture (i.e., with very little plastic deformation) to intergranular, as shown in Fig. 14.15. To explain this behaviour, Beachem (1972)

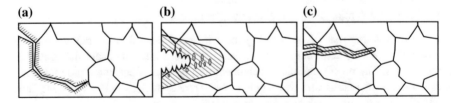

Fig. 14.15 Simplified schemes of HE fracture morphology. Adapted from Beachem (1972)

proposed that cracks can develop by both transgranular and intergranular paths as function of the stress intensity factor, K_I:

- High K_I (greater than 100 MPa√m) generates microvoids coalescence (Fig. 14.15a)
- Intermediate K_I leads to transgranular fracture by a quasi-cleavage mechanism (Fig. 14.15b)
- Low K_I (lower than 20 MPa√m) generates intergranular fracture (Fig. 14.15c); this is typical for high strengh steels, with K_{IC} much lower than 50 MPa√m.

14.3.5 HE by Hydrides

There are other damage mechanisms induced by hydrogen in metals, when hydrogen reacts with hydride-forming metals as titanium and zirconium. In this case, hydrides cause brittle fracture of the metal along the hydride-matrix interface.

It is worth illustrating the case of Ti. Titanium oxide is an excellent barrier to hydrogen intrusion. However, when hydrogen is produced by cathodic processes in galvanic coupling (typically with Al or Zn galvanic anodes or by impressed current CP or in strong acids), hydrogen atoms can enter titanium through its oxide. At temperature below 80 °C hydrogen diffusion is very slow, so hydrogen remains practically on the surface at a concentration of several thousand ppm, causing a surface hydriding that has little effect on mechanical properties.

Hydrogen solubility in α-Ti is 20–150 ppm at ambient temperature; hydrogen in excess forms titanium hydrides, TiH/TiH_2, which precipitate close to the metal surface and embrittle the metal, causing crack propagation by repeated formation and rupture of sub-surface hydrides. Ti hydrides appear as dark, acicular, needle-like shaped. Thermodynamic conditions necessary for hydrogen evolution depend, primarily, on pH and potential. For example, in seawater at 25 °C, hydrogen starts evolving below −0.7 V SCE, but experience has shown that TiH_2 formation requires potentials below −1.0 V SCE. An acceptable limit in cathodic protection design, considered sufficiently conservative to avoid Ti embrittlement, is −0.75 V SCE.

In α-β and in β-Ti, hydrogen solubility and diffusivity are much greater, due to the more favourable *bcc* lattice of the β-phase. At low hydrogen pressures, β-Ti grades are generally not susceptible to HE, due to the possibility of accommodating larger quantities of hydrogen; yet, at high temperature they may experience an increase in the ductile-to-brittle transition temperature, thus switching to brittle behaviour even above room temperature. Conversely, α-β Ti alloys are particularly prone to HE damaging by hydride formation, as hydrogen diffuses in the metal bulk through β grains until it reaches the a lattice, causing the precipitation of hydrides at α-β grain boundaries.

14.3.6 Sulphide Stress Cracking (SSC)

In the presence of hydrogen sulphide, H_2S, which is a strong poison for atomic hydrogen combination to give molecular hydrogen, in acidic oxygen-free solutions, HE is called *Sulphide Stress Cracking* (SSC) and the occurrence condition is called *sour condition*, as typical nowadays in oil and gas activities. To establish if sour condition applies, the standard NACE MR0175-ISO 15156 "Petroleum and natural gas industries—Materials for use in H_2S-containing environments in oil and gas production" is used worldwide. Based on that, sour conditions are determined by H_2S partial pressure, in situ pH and temperature. See Chap. 24 for more details.

As far as temperature is concerned, SSC is not an issue above 65 °C; as said above, more severe condition for HE and therefore for SSC is at room temperature. Atomic hydrogen is produced from the cathodic process in acidic oxygen-free solution, therefore the lower the pH the more atomic hydrogen is produced. For pH > 6.5 production of atomic hydrogen stops, so SSC does not take place. If actual pH is not known (i.e., not measured in separated brine), it can be estimated from CO_2 partial pressure (for instance, Eq. 8.7) or by using specific nomograms reported in the mentioned standard.

14.4 Prevention of LT-HID

The primary method to prevent LT-HID in all forms is to avoid the formation of atomic hydrogen at the surface of metals. When produced in a corrosion process, prevention or limitation of hydrogen production is achieved by removing or excluding the presence of cathodic poisons. For example, in acidic pickling both measures are adopted: the use of inhibitors to reduce corrosion rate, hence hydrogen production, and the elimination of poisons, as sulphides and cyanides. Similarly, in galvanic processes atomic hydrogen production is reduced by regulating the operating conditions, in particular the current density. In welding processes, HE prevention is done by adopting consumables or fillers free from moisture, which would be the hydrogen source. Another way is the use of screens to hydrogen diffusion, like impermeable coatings with *fcc* structure, as austenitic steels and nickel, or even thick rubber.

When hydrogen production cannot be avoided, resisting metals must be selected. In general, HE susceptibility increases significantly as mechanical strength increases, hence proper heat treatments have to be considered.

In many cases, reversible hydrogen is eliminated by heating steel above 150 °C for a time (in h) proportional to the square of thickness (in cm) (as rule of thumb, 2 h per ½ inch at temperatures between 150 and 200 °C). Treating time increases when zinc and cadmium plating, with little permeability to hydrogen, is present. A positive complimentary effect is the mechanical properties recovery.

When selecting ferritic steels, in principle highly susceptible to HE, the addition of nickel and molybdenum is beneficial while addition of chromium and molybdenum is detrimental.

14.4.1 Prevention of HIC and Blistering

As soon as H_2S partial pressure in separated gas exceeds 0.03 bar (about ten times the threshold used to classify sour conditions for risk of SSC), the prevention of HIC follows two strategies:

- Reduction of hydrogen production rate by the use of corrosion inhibitors (in practice, time increases proportionally with inhibitor efficiency, for instance 90% efficiency increases time-to-failure by ten times)
- Use of HIC resistant steels, by reducing or preventing the formation of manganese sulphide either by the addition of elements more reactive with sulphur than manganese, such as calcium and caesium (so-called rare earth treated steels) that form hard sulphides, which do not deform during hot rolling, or by lowering the sulphur content in steel below 20 ppm.

14.4.2 Materials for Sour Service

Metals susceptible to HE and specifically to SSC are characterized by a *bcc* crystalline structure, which is a susceptible microstructure. In general, *fcc* lattice metals do not suffer HE. There are case studies showing SCC behaviour of austenitic alloys at temperature exceeding 150 °C in the presence of high chloride content and high H_2S partial pressure; most likely, the cracking mechanism is primarily a chloride-induced SCC with a mixed anodic and cathodic embrittlement mechanism.

Based on experience, gathered since early 1950s by NACE International, ferritic steels resist SSC if the microstructure obtained by proper heat treatment shows a hardness, HRC scale, lower than 22 (equivalent to HV 248). This limit is increased for Ni, Mo containing low alloy steels up to HRC 26. As general rule of thumb, with same tensile strength, the most resistant ferritic steels are quenched and tempered, followed by bainitic and then normalized ones (ferrite and pearlite microstructure); conversely, cold drawing steels are the most susceptible.

Both NACE MR0175 and ISO 15156 list classes and proprietary alloys suitable for SSC service, i.e., resisting HE in sour service conditions. For more details, refer to Chap. 24.

14.5   Applicable Standards

- API 941 Steels for Hydrogen Service at Elevated Temperatures and Pressures, American Petroleum Institute, Dallas, TX.
- ISO 2626, Copper, Hydrogen embrittlement test, International Standard Organization, Geneva, Switzerland.
- ISO 7539, Part 11—Stress corrosion cracking. Part 11: Guidelines for testing the resistance of metals and alloys to hydrogen embrittlement and hydrogen-assisted cracking, International Standard Organization, Geneva, Switzerland.
- ISO 15156, Petroleum, petrochemical and natural gas industries—Materials for use in H_2S-containing environments in oil and gas production. Part 1: General principles for selection of cracking-resistant materials; Part 2: Cracking-resistant carbon and low alloy steels, and the use of cast irons; Part 3: Cracking-resistant CRAs (corrosion-resistant alloys) and other alloys, International Standard Organization, Geneva, Switzerland.
- ISO 17081, Method of measurement of hydrogen permeation and determination of hydrogen uptake and transport in metals by an electrochemical technique, International Standard Organization, Geneva, Switzerland.
- NACE MR 0175, Sulphide stress cracking metallic material for oil field equipment, NACE International, Houston, TX.
- NACE TM 0284, Evaluation of Pipeline and Pressure Vessel Steels for Resistance to Hydrogen-Induced Cracking, NACE international, Houston, TX.

14.6   Questions and Exercises

14.1 Design the cell and specify operating conditions for the measurement of the diffusion coefficient of hydrogen in metals.

14.2 Demonstrate that the maximum hydrogen solubility in iron at 25 °C, which is of the order of 10^{-1} mol/dm^3, corresponds to about 12 ppm.

14.3 List the hydrogen traps existing in iron and steel and classify them in reversible and irreversible. Is there a relationship between the trend of the binding energy and degassing temperature? Why?

14.4 Explain the trend of the values of the hydrogen diffusion coefficient reported in the Table 14.2 on the basis of crystal lattice and type of traps.

14.5 Which is the effect of an anodic or cathodic polarisation on the susceptibility of steels to hydrogen embrittlement (HE)?

14.6 Explain the difference between the mechanism of high temperature hydrogen-induced damage (HT-HID) and Hydrogen embrittlement (HE).

14.7 Explain the trend of the delayed fracture curves for ferritic low alloy steel saturated with hydrogen and exposed to high temperature for different times (Fig. 14.12): which is the main factor that explain the effect of the time?

14.8 Explain the effect of temperature and strain rate on hydrogen embrittlement susceptibility of steels; compare the effect of temperature on hydrogen embrittlement and SCC by slip dissolution mechanism.

14.9 Why α-β and β titanium are more susceptible to HE by hydrides than α-Ti?

14.10 List the methods for the prevention of LT-HID caused by: pickling, welding, formation of cathodic hydrogen (corrosion reactions or cathodic protection); for each method explain briefly which mechanism is exploited.

Bibliography

Barth CF, Steigerwald EA, Troiano AR (1969) Hydrogen permeability and delayed failure of polarized martenstic steels. Corrosion 25(9):353–358

Beachem CD (1972) A new model for hydrogen-assisted cracking (hydrogen "embrittlement"). Metall Mater Trans B 3(2):441–455

Devanathan MAV, Stachurski Z (1962) The adsorption and diffusion of electrolytic hydrogen in palladium. Proc R Soc A 270:90

Flis J (ed) (1991), Corrosion of metals and hydrogen-related phenomena. Elsevier, Amsterdam, Nederland, PWN—Polish Scientific Publishers, Warszawa, Poland

Hochmann J, Staehle Rw, McCrigth RD, Slater JE (eds) (1977) Stress corrosion cracking and hydrogen embrittlement of iron base alloys. NACE International, Houston

Hudson PE, Snavely Jr ES, Paune JS, Fiel LD, Hackerman N (1968) Corrosion. *NACE, 24*, 7

Lynch SP (2007) Progress towards understanding mechanisms of hydrogen embrittlement and stress corrosion cracking. In: NACE corrosion conference, Paper n. 07493, NACE International, Houston, TX, pp 1–55

Oriani RA (1970) The diffusion and trapping of hydrogen in steel. Acta Metall 18:147–157

Oriani RA, Hirth JP, Smialowski M (1985) Hydrogen degradation of ferrous alloys. Noyes Publications, Park Ridge

Pundt A, Kirchheim R (2006) Hydrogen in metals: microstructural aspects. Ann Rev Mater Res 36:555–608

Thygeson JR, Molstad MC (1964) High pressure hydrogen attack of steel. J Chem Eng Data 9:2

Chapter 15
Intergranular and Selective Corrosion

> *The most part of piping was stainless steel, and you know*
> *that stainless steel is a great material but does not allow,*
> *I mean does not yield if cold but if warmed up it's not*
> *so much anymore stainless.*
>
> Primo Levi, The Wrench

Abstract Metals consist of micrometric size crystalline grains. The border between these grains, called grain boundary, is a peculiar and delicate region, due to a distorted crystallographic structure and possible segregation of impurities and second phases. These characteristics of non-equilibrium make grain boundaries particularly reactive, and weaker in terms of corrosion resistance, so that in some cases a localized corrosion, called intergranular corrosion, can occur. This corrosion-type attack is very severe because it leads to grain detachment, then to a reduced mechanical resistance, despite the negligible metal consumption; in some environments and in the presence of tensile stresses it triggers stress corrosion cracking. In this chapter the most common intergranular corrosion forms are described, including stainless steel sensitization, knife-line attack, exfoliation of aluminium, and selective corrosion of brass and cast iron.

Fig. 15.1 Case study at the PoliLaPP Corrosion Museum of Politecnico di Milano

© Springer Nature Switzerland AG 2018

P. Pedeferri, *Corrosion Science and Engineering*, Engineering Materials,
https://doi.org/10.1007/978-3-319-97625-9_15

15.1   Impurities and Segregations

Sometimes intergranular attack is caused by the presence of specific alloying elements or impurities. This occurs on tin containing small content of aluminium in hydrochloric acid; on copper when the concentration of arsenic exceeds 0.5%; on silver alloyed with about 2% of gold, and others. In the past, cast Zn-Al alloy containing impurities of lead exposed to steam or warm marine atmospheres, frequently showed some spectacular effects as bulges or cracks due to intergranular corrosion.

Some concerns deal with intergranular attacks that are caused by phase precipitation at grain boundaries. This is the case of aluminium alloys strengthened by precipitation hardening, by which intermetallic compounds, often of sub-microscopic size, form at grain boundaries. For example, in 5000 (Al-Mg) alloy series, intermetallic Mg_2Al_8 precipitates which is more reactive than the aluminium matrix, and therefore corrodes selectively. Similarly, in 7000 (Al-Mg-Zn) series alloys, $MgZn_2$ precipitates: as it is less noble than the matrix, it is selectively attacked. In high-strength alloys, series 2000 and 7000, containing copper, $CuAl_2$ precipitates then depleting the surrounding adjacent areas from copper: this leads to a galvanic effect between the noble precipitate (cathode) and the surrounding matrix (anode) with an intergranular attack at the matrix side.

However, the most important case of intergranular corrosion involves austenitic stainless steels (one example is given in Fig. 15.1).

15.2   Sensitization of Stainless Steels

Austenitic stainless steels are supplied by steelworks as *stabilized*, which means that carbon, with maximum content of 0.08%, is dissolved in the metal matrix. This condition is achieved by a heat treatment, called *solubilization*, which consists of a thermal cycle as follows: maintenance at 1050 °C for 1 h per thickness of 1 inch (25.4 mm) in which carbides dissolve and carbon solubilizes in the metal matrix, then followed by a rapid cooling to avoid carbides to form again. Without further heat treatments these steels do not suffer intergranular corrosion; instead, if brought and maintained for some time in a temperature range approximately between 500 and 850 °C, chromium carbides, of $Cr_{23}C_6$ type, form and precipitate at grain boundaries. Since chromium carbide is enriched in chromium (chromium to carbon ratio is 16:1 by weight) this precipitation of chromium carbides at grain boundaries depletes chromium in the near matrix from 18%—as typical of stainless steels of most common use—to even below 12% as Fig. 15.2 shows, then jeopardizing passivity build up. This process, called *sensitization*, is the prerequisite for intergranular corrosion to occur as soon as sensitized stainless steels are exposed to a mildly aggressive or strongly oxidizing environment.

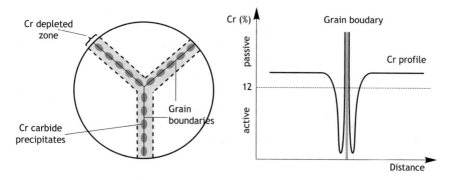

Fig. 15.2 Schematic representation of chromium carbides precipitation and % Cr profile at grain boundaries

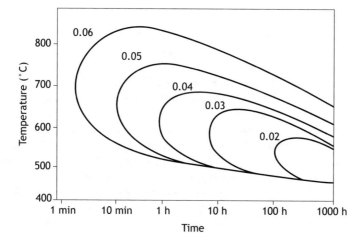

Fig. 15.3 Time of sensitization of an AISI 304 steel with changes in temperature and carbon content

The sensitization tendency of austenitic stainless steels depends on composition, in particular, the time required decreases as carbon content increases (Fig. 15.3). Also other alloying elements, such as nickel, molybdenum or nitrogen, influence the time of sensitization, but to a lesser extent than carbon. For example, a few seconds at 600 °C are sufficient for sensitization when carbon content is 0.08% and higher, while times longer than an hour are necessary for carbon content below 0.03%.

Also ferritic stainless steels can be sensitized through a different and more complex mechanism; because in this case the solubility of carbon and nitrogen in the ferritic matrix is much lower, sensitization time is shorter. The critical temperature interval for both chromium carbides and chromium nitrides is 500–900 °C; however, because chromium diffusivity in the ferritic matrix is much higher than that in austenite, the permanence of ferritic steels at temperatures between 700 and 900 °C, while causing carbides and nitrides separation, does not give rise to sensitization because the high chromium diffusivity favours homogeneous chromium redistribution. The critical temperature range for the sensitization of ferritic steels is therefore limited between 500 and 700 °C; in practice, sensitization occurs once a permanence at temperatures above 900 °C (where most of carbides and nitrates are dissolved) is followed by a rapid cooling to the range of 500–700 °C.

As far as other materials are concerned, austenitic-ferritic steels have good resistance to sensitization, while nickel super-alloys do not.

15.3   Corrosion Rate

Figure 15.4 shows an intergranular attack of austenitic stainless steel with 0.06% C, sensitized at 600 °C for several hours.

A schematic representation of the electrochemical mechanism of intergranular corrosion is shown in Fig. 15.5 in which reference is made to two curves, relating to the anodic behaviour of an austenitic stainless steel with a chromium content equal to that of the core of the grain (18% Cr) and of a steel with a chromium content equal to that of the sensitized grain boundary (10% Cr). The weakness of the grain boundary is reflected by the increase in activity and transpassivity intervals and the reduction of passivity interval. A sensitized steel is subject to intergranular corrosion in environments where the cathodic process brings the potential close to E_2 or E_4 (grain core is passive and grain boundary is active); if the potential is more noble or less noble than E_2 or E_4 (for example, close to E_1 or E_5) generalized and uniform corrosion takes place; finally for potential between E_4 and E_2 (for example close to E_3) there is no corrosion, being both grain core and grain boundary passive. In short, environments promoting intergranular corrosion are those with weak

Fig. 15.4 Intergranular corrosion of an austenitic stainless steel AISI 304 with 0.06% of C, sensitized at 600 °C for several hours

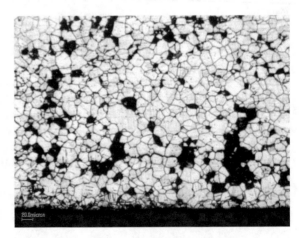

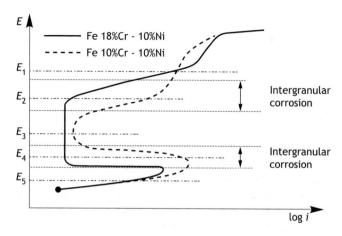

Fig. 15.5 Electrochemical conditions for intergranular corrosion in sensitized alloys

oxidizing power (E_4) or strong oxidizing power (E_2). Typical hazardous environments include: nitric and sulphuric acid, sulpho-nitric, sulpho-acetic, nitric-hydrofluoric and nitric-lactic acid mixtures, pickling solutions, organic acids such as lactic and acetic acids. The presence of ions characterized by two valences, for example iron (Fe^{2+}, Fe^{3+}) and copper (Cu^+, Cu^{2+}), are of remarkable importance, because the oxidized-to-reduced activity ratio determines the cathodic process potential, hence conditions for intergranular attack occurrence.

15.4 Prevention of Intergranular Corrosion

Sensitized austenitic stainless steels can be recovered by repeating the solubilisation heat treatment; in practice, this operation is rather difficult to carry out on final assembled structures or equipment, because of the risk of permanent deformation and induction of internal stress.

There are two main ways to prevent intergranular corrosion: by decreasing the carbon content and by the addition of stabilizing elements which form stable carbides. By the first one, which is the most followed, carbon content is reduced below 0.03%: AISI 304L or 316L (L means low carbon) are the most known commercial grades. These low carbon grades have the drawback to exhibit lower mechanical strength. By the second one, a typical German tradition, an element forming dispersed carbides in the lattice is added, with the aim to subtract carbon from the matrix and to avoid chromium carbides precipitation: these elements are titanium and niobium added by a content 5 and 10 times greater than carbon content, respectively. These modified stainless steels are called *stabilized*; typical grades are AISI 321, titanium stabilized, and AISI 349, niobium stabilized, as variant of classic AISI 304, and AISI 316Ti and AISI 316Nb as variants of classic AISI 316.

15.5   Weld Decay

Welding is without doubt the main cause of sensitization of stainless steels and may give rise to different types of attacks: the most important one is called *weld decay*.

To understand where and why weld decay localizes, let's consider a head-to-head welded joint. At the centre of the weld bead, called molten zone, the microstructure is a wrought-like structure as result of a mixture of base metal and weld metal, after complete fusion and solidification. Aside, there is a region, more or less extended, according to the heat input, generally of the order of the thickness of the joint, in which the base material undergoes one or more thermal cycles (in relation to the number of weld passes) with a temperature profile shown in Fig. 15.6. This zone, which is called the *heat affected zone*, comprises the sensitized steel that is between two zones where temperature or residence time are not critical for sensitization.

The exact location of the sensitized zone depends primarily on steel composition and all those factors that govern the thermal gradient during welding time such as thickness of steel, welding procedure, number and velocity of passes, heat input per pass, any pre and post welding heating. As rule of thumb, the sensitized area in austenitic stainless steels is one to two centimetres from the weld bead and even less in ferritic stainless steels.

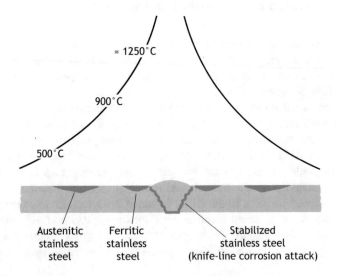

Fig. 15.6 Thermal gradient and location of the corrosion areas for austenitic steels and sensitized ferritic and austenitic steels

15.5.1 Knife-Line Attack

Stabilized stainless steels need a specific comment. Welding processes do not sensitize stabilized steels because titanium or niobium have sequestered carbon. However, near the melted zone, in a very narrow area, temperature is so high to dissolve also carbides of stabilizing elements, then releasing carbon which can combine with chromium, once steel is rapidly cooled within the critical temperature range of 500–800 °C for a sufficient time. In some oxidizing environments, such as those based on nitric acid, this particular sensitization gives rise to an attack limited to a few crystalline grains parallel to the weld bead and penetrating steel thickness up to cut off the weld: for this reason, this attack is called *knife-line attack* (Fig. 15.7).

15.6 Intergranular Corrosion of Nickel Alloys

Also nickel alloys can be sensitized, then undergoing intergranular attack. It is worth mentioning the case of Hastelloy C employed in annealed solution condition for its excellent resistance to oxidizing environments. It suffers intense intergranular corrosion if it is sensitized by heating in the range of 500–700 °C: accordingly, it is used only after solubilization heat treatment in the range 1150–1250 °C, hence welds are not accepted without post-welding solubilization heat treatment.

Figure 15.8 shows the intergranular corrosion of Ni-Cr alloy-20 (standard composition is Cr 20, Ni 25, Mo 4.5, Cu 1.5; Fe balance) in a 50% sulphuric acid solution containing chlorides at 50 °C. Sometimes corrosion occurs near welds

Fig. 15.7 Knife-line corrosion attack on an AISI 321 stainless steel plate welded with an AISI 304L stainless steel (nitric acidic solution)

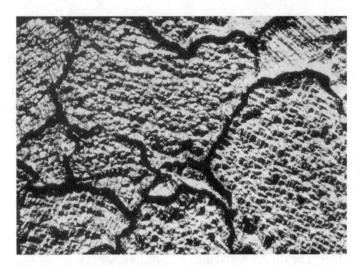

Fig. 15.8 Intergranular corrosion of nickel alloy

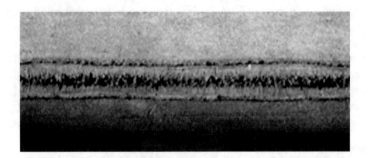

Fig. 15.9 Intergranular corrosion and knife attack on nickel alloy

where knife-kind attack can also take place, as shown in Fig. 15.9 for a nickel-based alloy. Corrosion at welds has twofold causes: micro-inhomogeneity in dendritic structure and knife attack due to sensitization in heat affected zone.

Since nickel alloys are often used at high temperature (for example, in nuclear industry for superheater tubes) sensitization is possible also during operating.

15.7   Intergranular Corrosion Without Sensitization

Non-sensitized stainless steels are susceptible to intense intergranular attack when exposed to particularly aggressive environments containing a strongly oxidizing ion, for example boiling nitric acid solutions and high temperature aqueous solutions. Susceptibility is determined by the presence of some impurities, in particular

Fig. 15.10 Intergranular
attack on titanium in
nitric-hydrofluoric acid
mixture

Fig. 15.11 Silver coin Incusa
(sixth century B.C.)

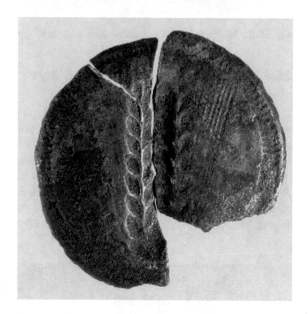

phosphorus; instead, a high content of silicon is beneficial. Today, silicon-rich ELI
(Extra Low Interstitial) stainless steels are not subject to this form of attack.

Intergranular attack without sensitization is also observed on other metals. For
example, Fig. 15.10 shows the intergranular attack on titanium in nitric-
hydrofluoric acid mixture. Figures 15.11 and 15.12 are related to a silver coin
'Incusa' of the VI century BC of a Greek colony in Magna Graecia (Metaponto,
Basilicata, Italy) affected by intergranular attack that has led to its embrittlement. In
this case, it is possible that the attack was favoured by the presence of internal
stresses originating from the drawing process of the coin which was not followed by
an annealing.

Fig. 15.12 Micrograph of intergranular attack on coin of Fig. 15.11

15.8 Exfoliation of Aluminium Alloys

It is a special type of intergranular corrosion typical of aluminium alloys which proceeds through preferential intergranular paths, usually parallel to the direction of extrusion or rolling. The corrosion product that forms has a greater volume than the volume of the parent metal. The increased volume forces the layers apart, and causes the metal to exfoliate or delaminate. It is also called *lamellar corrosion*. Al-Cu-Mg, Al-Zn and Al-Zn-Mg alloys are typically susceptible to this type of attack, which is enhanced if a galvanic action with noble metals is present. Figure 15.13 shows an example of this form of corrosion obtained in a testing.

Fig. 15.13 Exfoliation on aluminium plate exposed to atmosphere

In addition it is worth mentioning that some kind of exfoliation corrosion is sometime found in forged steel characterized by excessive internal growth of oxide, which has a volume some seven times that of the steel.

15.9 Intergranular Corrosion Tests

An important quality control test is the check of susceptibility of intergranular corrosion of stainless steels and nickel-based alloys, if sensitized because of incorrect heat-treatment or improper welding. Aim of the test is revealing of chromium-depleted areas caused by the precipitations of carbides and sigma phase. Three tests are used.

Huey Test, ASTM A262—Practice C. The specimen is immersed in a boiling 65% solution of nitric acid for five periods, each of 48 h. Corrosion rate is calculated from weight loss measurements. If there is no sensitization, i.e., grain is homogeneously passivated, nitric acid maintains passivity and weight loss refers to some oxide dissolution, only. When sensitized, intergranular attack occurs on chromium-depleted zones, which are not passive so that galvanic corrosion occurs between these zones and passive surrounding ones. The cathodic reaction is hydrogen evolution (reduction of nitrate anion does not contribute to the cathodic process), therefore corrosion rate is given by the cathodic current density of hydrogen evolution reaction multiplied by the surface area ratio which normally exceeds 10. Hence corrosion rate is often of the order of tens mm/y which is easily measured by weight loss.

Strauss Test, ASTM A262—Practice E. The specimen is immersed in a boiling $Cu/CuSO_4$—16% sulfuric acid solution. After exposure for 1 h, the specimen is 180° bent over a rod with diameter equal to the specimen thickness and visually examined: no cracks are allowed. The cathodic process is copper ion reduction and to a smaller extent hydrogen evolution. Corrosion rate, much higher than in Huey test, is difficult to predict.

Streicher Test, ASTM A262—Practice B. The specimen is immersed in a boiling $Fe_2(SO_4)_3$—50% sulfuric acid solution for 24–120 h. As in Huey test, corrosion rate is calculated from weight loss measurements. The cathodic process is reduction of ferric to ferrous ion and to a smaller extent hydrogen evolution. Corrosion rate is more difficult to predict than in Huey test, however it can be assumed that corrosion rate is the same or higher.

15.10 Selective Corrosion of an Alloying Element

Selective attack of a constituent of an alloy is fairly common: the most important example is brass dezincification, but other copper alloys show selective etching of aluminium, manganese, nickel and cobalt. In copper-silver alloys, copper is selectively attacked; in gold-silver and lead-tin alloys, silver and tin are attacked, respectively. As a rule of thumb, the more reactive element corrodes while the more noble or passive does not.

Selective corrosion often affects a surface layer only without producing structural damages; in some cases it can cause an indirect damage, for example, when it causes the leaching of the less noble metal, there can be the setup of galvanic corrosion on closer metals.

15.10.1   Dezincification of Brass

Selective etching of zinc in brass is easily diagnosed by the change of colour toward copper appearance (Fig. 15.14). It affects both alpha and alpha-beta brass types and is produced mainly in stagnant or slow-flowing solutions, that is, conditions favouring deposit formation.

This attack occurs as either spread or localized, the latter preferably in neutral or weakly alkaline solutions on brass with low zinc content (20–30%), while the former on brass with high zinc content, for example on Muntz metal (60% Cu, 40% Zn). In biphasic brass, at least in a first stage, the zinc-rich phase is attacked preferentially.

The corrosion mechanism involves in some cases a selective corrosion of zinc, or dissolution of both zinc and copper, followed by a re-deposition of copper in a typical spongy form.

Brasses with high copper content (>85% Cu) do not suffer dezincification attack. A strong increase in the dezincification resistance of alpha brass is obtained by adding small contents (0.02–0.06%) of arsenic, antimony or phosphorus; instead, no elements to inhibit dezincification on alpha-beta or beta brass types are known: in this case, inhibitors are used.

Fig. 15.14 Selective etching of zinc in brass

15.10.2   Cast Iron Graphitization

In saline environments, in certain types of water and even in weakly acidic solutions or in soils, the so-called graphitization of grey cast irons can occur; it consists of a selective etching of iron so the content of graphite on the surface increases. As the attack grows, graphite remains with no change of size and shape of the component, instead mechanical properties drop down. Unlike grey cast iron, thite, spheroidal and malleable cast iron do not suffer graphitization (Fig. 15.15).

15.11   Applicable Standards

- ASTM A 262, Standard practices for detecting susceptibility to intergranular attack in austenitic stainless steels, American Society for Testing of Materials, West Conshohocken, PA.
- ASTM A 763, Standard practices for detecting susceptibility to intergranular attack in ferritic stainless steels, American Society for Testing of Materials, West Conshohocken, PA.
- ASTM G 28, Standard test methods of detecting susceptibility to intergranular corrosion in wrought, nickel-rich, chromium-bearing alloys, American Society for Testing of Materials, West Conshohocken, PA.
- ASTM G 67, Standard test method for determining the susceptibility to intergranular corrosion of 5xxx series aluminium alloys by mass loss after exposure

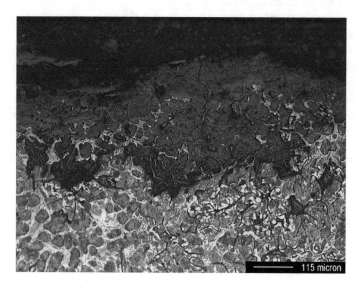

Fig. 15.15 Graphitization of cast iron pipe buried in the 1930s

to nitric acid (NAMLT test), American Society for Testing of Materials, West Conshohocken, PA.

- ASTM G 110, Standard practice for evaluating intergranular corrosion resistance of heat treatable aluminium alloys by immersion in sodium chloride + hydrogen peroxide solution, American Society for Testing of Materials, West Conshohocken, PA.
- ISO 3651-1, Determination of resistance to intergranular corrosion of stainless steel. Part 1: austenitic and ferritic-austenitic (duplex) stainless steel. Corrosion test in nitric acid medium by measurements of mass loss (Huey Test), International Standard Organization, Geneva, Switzerland.
- ISO 3651-2, Determination of resistance to intergranular corrosion of stainless steel. Part 1: ferritic, austenitic and ferritic-austenitic (duplex) stainless steel. Corrosion test in media containing sulfuric acid (Strauss Test), International Standard Organization, Geneva, Switzerland.
- ISO 9400, Nickel based alloys. Determination of resistance to intergranular corrosion, International Standard Organization, Geneva, Switzerland.
- ISO 11846, Corrosion of metals and alloys—Determination of resistance to intergranular corrosion of solution heat- treatable aluminium alloys, International Standard Organization, Geneva, Switzerland.

15.12   Questions and Exercises

15.1   Why do grain boundaries become sensitive to the corrosion attack in case of precipitation of chromium carbides?

15.2   Discuss the effect of temperature, time and carbon content on sensitization of austenitic stainless steels.

15.3   Why isn't the intergranular attack prevented in ferritic stainless steel even if carbon is reduced to 0.3%? What is the temperature range in which it can occur?

15.4   Which information does the time-temperature sensitization curve provide? Determine the sensitization time for an AISI 304 stainless steel with 0.05% carbon content during a heat treatment at 650 °C. Consider the time-temperature sensitization curve of Fig. 15.3.

15.5   What is a stabilized steel? How can the use of a stabilized steel prevent sensitization of stainless steels?

15.6   Why do stabilized stainless steels undergo knife-line attack?

15.7   Consider a head-to-head welded joint. Where is the sensitized zone located with respect to the molten zone? Consider: (a) a ferritic stainless steel; (b) an austenitic stainless steel; (c) a stabilized stainless steel.

15.8   Discuss the two approaches used to prevent intergranular corrosion of stainless steels.

15.9 A sensitized austenitic stainless steel suffers intergranular corrosion in a strong oxidizing environment. Discuss the electrochemical behaviour of the metal by using Evans diagram.

15.10 Dezincification of brass is a typical example of selective corrosion of an alloying element. Discuss the corrosion mechanism. What are the methods to prevent dezincification of alpha brass and beta brass?

Bibliography

Cihal V (1984) Intergranular corrosion of steels and alloys. In: Materials science monographs, vol 18. Elsevier, Amsterdam

Fontana M (1986) Corrosion engineering, 3rd edn. McGraw-Hill, New York

Shreir LL, Jarman RA, Burstein GT (1994) Corrosion. Butterworth-Heinemann, London

Steigerwald RF (1978) Intergranular corrosion of stainless alloys, ASTM special publication N 656. American Society for Testing of Materials, Philadelphia

Chapter 16
Erosion-Corrosion and Fretting

In the cycling return of many solar years a ring, furiously
worn on finger, thins inside, a water drop falling continuously
carves the stone, curved ferrous plough wears hidden in soil,
on streets people feet have consumed the pavement, and on
doors bronze statues show worn by the frequent right
hand touch of those who greet when passing through.

Lucretius, De rerum natura, I, 311–318

Abstract This chapter presents the forms of corrosion related to the contact of a metallic surface with something moving on it, be it a fluid or another material. In the former case, erosion-corrosion phenomena may onset due to the rapid flow of a fluid on a metal, which combines corrosion with physical-mechanical interactions as turbulence, cavitation or impingement of particles on its surface. On the other hand, if a solid body slides on a metal surface, typically in the form of cyclic micrometric slips such as those created by vibration, fretting corrosion establishes, causing a range of damages from simple loss of brightness to the formation of craters that then trigger fatigue cracks.

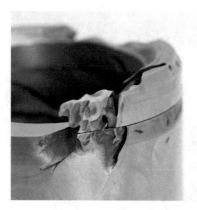

Fig. 16.1 Case study at the PoliLaPP Corrosion Museum of Politecnico di Milano

© Springer Nature Switzerland AG 2018
P. Pedeferri, *Corrosion Science and Engineering*, Engineering Materials,
https://doi.org/10.1007/978-3-319-97625-9_16

16.1   Erosion-Corrosion Forms

Erosion-corrosion is defined as the conjoint action of erosion and corrosion and consists of the progressive loss of material from a solid surface due to corrosion and to the mechanical interaction between the surface and a flowing, single or multiphase fluid. It includes erosion of protective films in a turbulent fluid, solid and liquid particles impingement corrosion, and cavitation corrosion. Otherwise, flow-induced corrosion is defined as the increased corrosion resulting from increased fluid turbulence intensity and mass transfer because of the flow of a fluid over a surface, without mechanical interaction (Uhlig 2000).

An erosion-corrosion attack is the consequence of a continuous local damage of the protective film by the mechanical action of an aggressive environment, which is followed by the oxidation of bare areas, sometimes also accelerated by the galvanic action exerted by the surrounding passive areas, acting as cathode. Figures 16.1, 16.2, 16.3, 16.4, 16.5 and 16.6 show some examples of this type of attack.

Fig. 16.2 Corrosion-erosion on a choke valve made of AISI 304 in oil production wellhead

Fig. 16.3 Corrosion-erosion on carbon steel in naphthenic acid containing oil

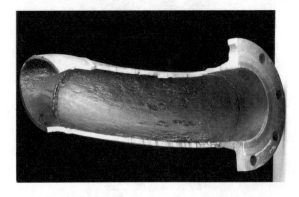

Fig. 16.4 Corrosion by turbulence on a carbon steel pipe in the presence of CO_2, at a cross section reduction

Fig. 16.5 Corrosion-erosion on top of a reactor made of stainless steel (AISI 304 grade), caused by liquid droplets in turbulent water vapour

16.1.1 Corrosion by Turbulence

Metals exposed to flowing liquids are potentially subject to such attack, especially pump impellers, turbine blades, agitators, tube inlets in heat exchangers and all conditions where there is a sudden change of hydraulic regime for the presence of obstructions of any kind (for example weld beads, gaskets, valves) or due to sharp changes of fluid direction (Fig. 16.3) or of cross section area (Fig. 16.4).

Corrosion by turbulence is closely related to surface conditions; surface defects often originate and localize the attack, which, once started, contributes to further increase the turbulence locally, then accelerating the attack. The morphology of the

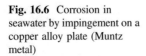

Fig. 16.6 Corrosion in seawater by impingement on a copper alloy plate (Muntz metal)

attack, different in each case, is always defined by hydrodynamics as shown in Fig. 16.5: there are smooth grooves, hydro-dynamically profiled, flaming-like or wavy streaks, sometimes reminding desert dunes.

The onset of corrosion-erosion depends on the synergy between the mechanical action exerted by the fluid and corrosion resistance properties of the metal. The former, i.e., mechanical action, depends on the kinetic energy of the liquid, which goes with the square power of the fluid velocity, and the latter with the metal surface hardness. An empirical relationship proposed by API (American Petroleum Institute) for a two-phase flow system and often extended to one phase fluids as water streams, provides a link between fluid kinetic energy and material property as follows (API-RP 14-E):

$$v_f = \frac{C}{\sqrt{\gamma_f}} \tag{16.1}$$

where C is a constant depending on the metal, v_f (m/s) and γ_f (kg/m^3) are fluid critical velocity and gas/liquid mixture density at flowing pressure and temperature, respectively. Industry experience indicates that for solids-free fluids constant C used in design phases for metal selection is 40 for copper, 60 for copper-nickel 70/30, 120 for carbon steels and 500 for stainless steels; these values are generally considered conservative. When fluid velocities are particularly high, hard and corrosion resistant coatings such as stellites (i.e., cobalt-chromium alloys designed for wear resistance) and ceramic coatings are used.

Copper alloys suffer corrosion-erosion easily (low constant C). Attacks are favoured by the presence of gas bubbles crushing against metal walls (called *impingement attack*) as shown in Fig. 16.6, so the protective film is damaged. The attack is characterized by sharp edges and, inside tubes, appears as craters oriented in direction opposite to the flow, showing a typical horseshoe shape, also called *horseshoe attack* (Figs. 16.7 and 16.8). If suspended solids are present (gravel,

Fig. 16.7 Horseshoe-attack due to erosion-corrosion in a copper tube

Fig. 16.8 Typical shape of horseshoe-attack

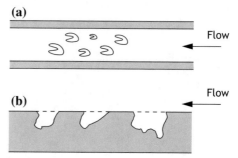

sand, silt, catalyser dust, coating debris, shell debris) the mechanical action increases and the API equation (Eq. 16.1) is not valid anymore (not conservative); abrasion is enhanced and localizes at elbow extrados, as depicted in Fig. 16.9.

Other zones particularly affected by corrosion-erosion are those where suspended solids are trapped in geometric recesses; for example, deep corrosion-erosion attacks are produced on pumps body near seals where sand particles are entrapped because of the lack of appropriate conductors for continuous removal. Also zones where solids accumulate are affected by corrosion-erosion, for instance at the bottom of vertical pipes where accumulated solids are in constant agitation (Fig. 16.10).

16.1.2 Cavitation Corrosion

Cavitation corrosion is caused by the formation and collapse of vapour bubbles in a liquid in contact with a metal surface. Cavitation removes protective surface scales by the implosion of gas bubbles in a fluid. This attack occurs when fluid pressure

Fig. 16.9 Extrados zone of a
curve with more intense
abrasive action

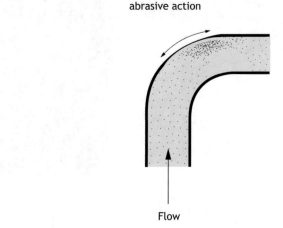

More intense
abrasive action

Flow

Fig. 16.10 Foot of a vertical
tube, where abrasives
accumulate

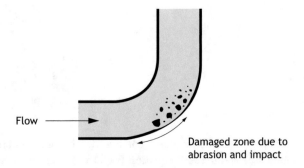

Flow

Damaged zone due to
abrasion and impact

locally drops below the vapour tension of the transported liquid, especially in high
elevation parts of large equipment or on pump impellers, or in presence of vibrations (for example, inside cooling sheaths of cylinders in diesel engines). The
mechanism is as follows: a pressure drop produces gas bubbles which rapidly
collapse where pressure becomes again normal, then generating violent shockwaves, with a mechanical damage or also a permanent deformation or even fatigue
of the metal. In aggressive environments, cavitation can result in typical localized
attacks, characterized by numerous deep craters looking like a spongy appearance
(Fig. 16.11). Calculations have shown that implosions produce shock waves with
pressures approaching 60 ksi (about 410 MPa).

The severity of the attack depends on either mechanical factors or fluid
aggressiveness; passivating substances and the presence of gaseous bubbles, present
in the fluid or locally produced by cathodic processes of hydrogen evolution, have a
beneficial effect because they dampen the shockwaves effect.

Fig. 16.11 Cavitation on the
impeller of a pump in
stainless steel AISI 304

16.1.3   Metal Affecting Properties

The corrosion resistance of metals to turbulence, abrasion or cavitation depends
primarily on metal hardness, so that films formed on soft metals are easily removed
together with the metal itself. For example, equipment or pipes made of lead can
resist dilute sulphuric acid exposure for years in stagnant conditions, while only few
days under flow.

Naturally, metal hardness does not influence a pure corrosion process, while it
does influence the resistance to erosion; however, hardening processes, as alloying
or heat treatments, improve erosion resistance if the resulting structure is homogeneous: on the contrary, inhomogeneous structures lead to a decrease in resistance.
For instance, among inexpensive materials, silicon-based cast iron, which is a high
hardness homogeneous solid solution containing 14.5% silicon, is certainly the one
which better resists corrosion-erosion and abrasion, then widely used in chemical
industry, despite its poor workability and impact resistance.

More than hardness, self-healing properties of surface films are the key
parameter which determines the resistance to abrasion and corrosion-erosion. As a
rule of thumb, thick films, such as those formed on copper alloys, are easily
removed and difficult to repair because too thick; conversely, thin films, such as
those formed on stainless steels, are more resistant also because of their easier
self-repairing ability. Often, the combination of mechanical action and galvanic
coupling provokes corrosion-erosion also in environments which are safe for the
involved metals separately. A weakening of the protective film is usually observed
when the metal is cathodically polarized without achieving immunity condition.

16.1.4 Environment Affecting Properties

The nature and mechanical strength of protective film as well as its ability to re-build depend on both the environment and the metal. The presence of oxidizing or passivating species is beneficial for film consolidation; for example, addition of chromates to cooling water of diesel engines prevents cavitation attacks. On the other hand, corrosion rate on bare zones depends on the environment: if it is not aggressive, the local mechanical breakdown of surface film is not followed by any corrosion attack. For example, copper alloys in oxygen-free seawater practically do not suffer corrosion-erosion on damaged surfaces: for this reason, aluminium brass tubes resist higher water velocity in a desalter than in a heat exchanger, because in the former water is deaerated.

Figure 16.12 summarizes the influence of the fluid velocity for different industry used metals and alloys.

16.1.5 Prevention

Prevention methods, which are different in each case, comply with all considerations carried out so far; therefore, they are based on:

- An appropriate choice of materials in relation to flowing conditions
- Protection, at least of most stressed areas, by thick coatings, such as rubber or ebonite or solvent-free epoxy coatings

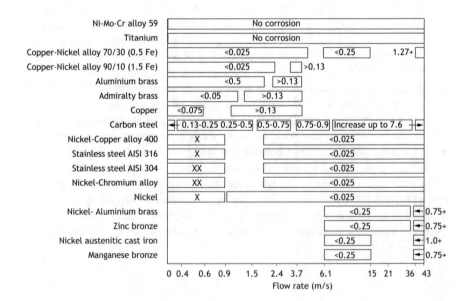

Fig. 16.12 Corrosion-erosion chart (inside boxes: corrosion rate corresponding to the specific condition, in mm/y) (NiDi 1987)

- A design intended to eliminate all possible causes of turbulence or cavitation, by suitable choice of shapes and sizes and characteristics of equipment (e.g., pumps should be chosen with a value of NPSH—*Net Positive Suction Head*—lower than that of the system operating condition), fluid flow rate and pressure conditions
- Environment control as far as particles in suspension are concerned.

16.2 Fretting Corrosion

Fretting corrosion is a form of damage that occurs between two metallic surfaces, subject to cyclic slips with an amplitude not higher than a few tens of microns, generally caused by vibrations. Sometimes, this form of attack only shows a loss of brightness, other times the formation of craters filled and surrounded by corrosion products as fine powder of metal oxide, typically red rust in case of steel or black debris in case of aluminium; these craters often trigger fatigue cracks. The formation of oxide powder can cause malfunctioning because of loss of tolerance limits or galling effects.

Fretting corrosion occurs in machinery and structures subject to vibrations such as:

- In aeronautics and railways related structures
- Between wires of metallic ropes
- In chemical equipment (plate heat exchangers where vibrations are caused by fluid flow or pulsing action of pumps; in safety valves)
- In electrical contacts, then causing change of contact resistance, as in telephone junction boxes
- In the core of nuclear reactors due to the flow of cooling fluid.

16.2.1 Mechanism

Two mechanisms have been proposed to explain fretting corrosion depending on the sequence of mechanical action and chemical oxidation. According to the first mechanism, the mechanical action is the first step, consisting of a micro welding of contact points which are subject to very high pressures due to their limited contact surface, then followed by the mechanical break of such bridges with production of metal debris, which are immediately oxidized for the high temperatures reached due to the friction. The second hypothesis starts from the observation that the majority of metals is covered by an oxide film, so the slip between the two surfaces produces the destruction of the oxide film at contact points forming oxide powder and leaving bare and most likely work hardened areas which are particularly reactive, hence re-oxidizing immediately, and repeating a new destructive cycle.

According to both theories, the produced debris act as abrasive agents, hence increasing the damage. It is proved by practice that both mechanisms apply, depending on the environmental conditions: in inert environments or under vacuum, where metals cannot oxidize, fretting occurs according to the first mechanism, while the second occurs in oxidizing environments. Indeed, the presence of oxygen accelerates the fretting attack of many metals, in particular of ferrous alloys.

16.2.2 Main Factors

Fretting corrosion occurs when a small cycling slip, generally of shear type, exists between two metallic surfaces. However, there are conditions where a fluctuating load is perpendicular to the surfaces, as in the case one of the metallic objects is a sphere or a cylinder; in these cases, the process is called *hammering* and the damage is extremely severe.

Fretting damage increases as applied load, number of cycles and slip amplitude increase.

As far as the environment is concerned, fretting corrosion of steel surfaces is almost identical in air or in oxygen; instead, in vacuum or in oxygen-free protective atmosphere, for example nitrogen or helium, the fretting damage is much lower. Humidity is very important in oxygen containing environments, while it has a very modest effect in protective atmospheres.

Among metal-related factors, hardness is the most important. As rule of thumb, as surface hardness increases, fretting damage is less severe. A further important factor is the oxide hardness compared to that of the metal: the higher the oxide hardness the higher the grinding effect on the metal surface.

Surface treatments such as carburizing, nitriding, shot peening, which increase surface hardness, should reduce the wear damage; instead, experimental data assessed by weight loss testing showed that these treatments have little influence, while confirming an increase in fretting induced fatigue resistance.

16.2.3 Fretting Corrosion Fatigue

Fretting represents a fatigue initiation stage when shear stresses cause micro-cracks exceeding about 100 μm in length. The stress level is generally lower, often by a factor of 3–6, than that needed for smooth non-corroded specimens. Typically, the fatigue crack initiation zone is between the corroded area, where slip takes place, and the adjacent non-corroded one. In fact, fretting micro-cracks propagate diagonally to the surface to originate a true fatigue crack which proceeds perpendicular to the stress direction. Sometimes, two or more micro-cracks from opposite sides of the affected zone merge and cause metal spall off.

The effect of fretting on corrosion fatigue of metals is similar to the one of a severe notch; indeed, metals with high sensitivity to notches are also particularly sensitive to fretting corrosion. It should also be noted that fretting corrosion, in regimes characterized by low loads and high number of cycles, reduces fatigue strength more than in case of the presence of notches.

The most dangerous conditions seem to be those characterized by small slip amplitude, between 7 and 15 µm, that is, unfortunately, the most common ones in engineering applications. It appears that slips with higher amplitude, which show a higher wear rate, are able to eliminate larger micro-cracks and, conversely, to promote the development of numerous smaller cracks which hardly reach the critical size to trigger fatigue cracks.

Finally, it is observed that fatigue does not occur when a considerable quantity of powder is produced, because powder works as a lubricant and also eliminates micro-cracks through its abrasive action.

16.2.4 Prevention

Prevention of fretting corrosion is based on the following measures:

- Elimination of slips by design
- When slip is unavoidable, reduction of stress concentration on affected areas
- Use of lubricants for a twofold aim: hindering oxygen access and reducing friction coefficient
- Choice of suitable surface treatments.

16.2.5 Lubricants

Good lubrication conditions help reduce fretting. Lubricants are liquid and, for not easily accessible parts, solid fats. Liquid lubricants are made of hydrocarbons (mineral oils) with the addition of additives; solids ones (fats) consist of hydrocarbons thickened with soap, clay, carbon black, silica gel and additives. Lubricants which are generally considered not corrosive can become aggressive as a result of deterioration processes. There are two ways in which the lubricants deteriorate in service: by contamination with substances such as dust, wear products, fuels, combustion products, water, and through physical-chemical modifications due to oxidation which leads mainly to the formation of acidic substances.

To improve the properties of lubricants and additives, inhibitors are added. It should be paid particular attention to the additives which have the function of binding the lubricant to the metal material. Generally these substances are carboxylic acids, alcohols, sulfonic acids, which contain functional groups that react with the metal surface locally forming an extremely thin layer. If the operating conditions are very harsh, these layers are continuously formed and destroyed, resulting in a corrosion itself.

Among solid lubricant coatings for applications on steel there are sulphur rich layers (sulphide process) and molybdenum sulphide, MoS_2.

16.3   Applicable Standards

- API RP-14E, Recommended practice for design and installation of offshore production platform piping systems, American Petroleum Institute, Northwest Washington, DC.
- ASTM D 2809, Standard test method for cavitation corrosion and erosion-corrosion characteristics of aluminium pumps with engine coolants, American Society for Testing of Materials, West Conshohocken, PA.
- ASTM E 2789, Standard guide for fretting fatigue testing, American Society for Testing of Materials, West Conshohocken, PA.
- ASTM G 32, Standard test method for cavitation erosion using vibratory apparatus, American Society for Testing of Materials, West Conshohocken, PA.
- ASTM G 73, Standard test method for liquid impingement erosion using rotating apparatus, American Society for Testing of Materials, West Conshohocken, PA.
- ASTM SPT 1425, Fretting fatigue: advances in basic understanding and applications, American Society for Testing of Materials, West Conshohocken, PA.
- ISO 16203, Corrosion of metals and alloys. Guidelines for the selection of methods for particle-free erosion corrosion testing in flowing liquids, International Standard Organization, Geneva, Switzerland.

16.4   Questions and Exercises

16.1   API equation provides a relation between fluid critical velocity and its mass density. What does the constant C represent? (Hint: consider the mathematical equation by an energetic approach).

16.2   Based on API equation for erosion corrosion assessment, calculate the fluid critical velocity of seawater (mass density 1100 kg/m^3) for a copper-nickel 70/30.

16.3   Based on API equation, indicate the minimum diameter of different spools used in a loop for the following fluids: seawater; oil; oil + gas; brine +

oil + gas. The fluid density is 1100–880–1.5 kg/m^3 for seawater, brine, oil + gas (in standard condition), respectively. Composition is GOR (gas-oil-ratio) 1000 m^3/m^3, WC (water cut) 10%. Flow rate is 0.1 m^3/s. [Hint: density of multiphase fluids is the weighted average].

16.4   Why are copper alloys less resistant than stainless steels to erosion-corrosion?

16.5   A heat exchanger consists of an AISI 316L stainless steel water box, an AISI 316 shell, a Cu-Ni tube-sheet and pure Cu tubes. Cooling fluid into the pipes is seawater. After 2 years operation, a leakage of copper tubes was detected. Eddy current analysis revealed a corrosion attack on the internal surface of some tubes. Seawater total flow rate is 104 m^3/h. Pipes cross section is 1 m^2. Which is the cause of corrosion? Is pure copper a correct material selection? Does the use of Cu-Ni solve the problem?

16.6   Discuss the effect of the presence of oxidizing or passivating species on erosion corrosion.

16.7   Describe the mechanism by which cavitation occurs.

16.8   Cavitation occurs on pump impellers. Explain why and how it is more risky when pumping seawater than gasoil.

16.9   Discuss the two main mechanisms proposed to explain fretting corrosion.

16.10 Try to explain why and how the addition of molybdenum sulphide to lubricants is beneficial to reduce fretting damage. How can phosphatizing the surface reduce the fretting hazard?

16.11 Discuss the difference between the expected fretting damage in a metallic rope when used in air with an incorrect lubrication and in full immersion in seawater. Explain why and how cathodic protection could be beneficial.

16.12 What is the effect of metal hardness on fretting corrosion resistance?

16.13 Discuss the effect of fretting on corrosion fatigue.

16.14 Water jet is used to cut many metallic alloys. Explain the principle on which it is based. Do you think that corrosion prevails on erosion?

Bibliography

Nickel Development Institute, NiDI (1987) Nickel stainless steel for marine environments, natural waters and brine. Guideline for selection, series 11 003, London, UK

Shreir LL, Jarman RA, Burstein GT (1994) Corrosion. Butterworth-Heinemann, London

Waterhouse RB (1972) Fretting corrosion. Pergamon Press, New York

Winston Revie R (2000) Uhlig's corrosion handbook, 2nd edn. Wiley, London

Chapter 17
Corrosion Prevention by Coatings

A tin coating applied to the buccal makes it more pleasant to the taste and halts the virus of rust and, wonderful to say, the weight does not increase.

Pliny, Nat.Hist., 34.160

Abstract This chapter deals with the method of corrosion prevention based on the use of a physical barrier to separate the metal surface from the environment. This barrier consists of a coat that can be classified, as often generally agreed, into four types, namely: metallic coatings, conversion coatings, inorganic coatings and paints. In this Chapter, only corrosion protection related coatings are considered, although other characteristics, such as mechanical strength, hardness, wear resistance, appearance, and electrical, optical or thermal properties are also of remarkable importance.

Fig. 17.1 Case study at the PoliLaPP Corrosion Museum of Politecnico di Milano

© Springer Nature Switzerland AG 2018
P. Pedeferri, *Corrosion Science and Engineering*, Engineering Materials,
https://doi.org/10.1007/978-3-319-97625-9_17

17.1 Metallic Coatings

The use of metallic coatings is one of the most common methods adopted for corrosion prevention. Metallic coatings are usually named according to the metal used, for example, copper, zinc, nickel, gold coatings. This nomenclature is rather generic as any coating may have a crystalline or amorphous structure, a fibrous morphology, or columnar, yet coarse or fine, lamellar or strip banded; furthermore, they present a variable chemical composition through small percentages of metals and/or impurities as hydrogen, oxygen, carbon, sulphur.

17.1.1 Coating Defects

From a corrosion point of view, continuity, uniformity as well as the absence of porosity are of particular importance for the coating performance. In fact, when a coating is continuous and non-porous, the base metal is thoroughly covered, so corrosion takes place on the coating metal as it would happen on a solid plate (Fig. 17.1).

If the metallic coating is porous, galvanic couples form between the metal of the coating and the base metal on uncovered areas; the result of galvanic coupling can lead to a cathodic protection effect exerted on the base metal or, conversely, to an accelerated anodic dissolution of it, depending on the practical nobility of the two metals, as discussed in Chap. 10.

Coating defects are primarily produced in the application phase. The coating metal may not deposit or adhere to the base, or result porous due to poor surface preparation or inappropriate pre-treatment or incorrect application conditions—for instance, in case of galvanic deposits, excessive operating current densities. For example, a coating applied to cast iron may be non-adherent due to graphite powder left on the surface by a previous pickling, as well as on titanium a metallic coating shows poor adhesion because of a titanium oxide film, which forms immediately after pickling. On the other hand, the technique used can set up a high porosity, as for example in the case of sprayed coatings.

Other defects arise during manufacturing or processing (cutting, drilling, welding) or during operating events as mechanical damage, wear or corrosion attacks.

17.1.2 Cathodic Coatings

Let's consider first the case study of a metallic coating made of a metal nobler than the base, hence with a cathodic behaviour against the base metal that is anodic. As Fig. 17.2a shows for the case of a copper coating applied on carbon steel, the coating metal, i.e., copper, is nobler than the base, i.e., carbon steel, so that

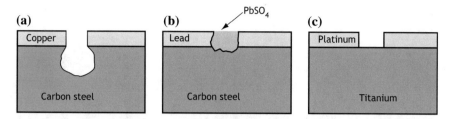

Fig. 17.2 Effect of porosity in a cathodic-type coating

corrosion takes place on the latter metal in coating defects, favoured by the small overvoltage of the anodic process (active metal dissolution). These considerations can be extended to the use of coating metals such as nickel, silver, lead, all nobler than carbon steel, and materials with active-passive behaviour such as chromium, titanium and stainless steels, having a practical noble behaviour. For the estimation of corrosion rate, reference is made to Chap. 10.

Metallic coatings made of passive metals are generally less effective in accelerating corrosion rate than active noble metals (for instance, stainless steel versus copper) because the overvoltage associated with oxygen reduction is much higher. Furthermore, if the passive film is an insulator as in the case of anodized aluminium, the galvanic effects stops.

Noble metallic coatings can be used when corrosion products seal defects and pores of the coating in which corrosion initially takes place, then impeding corrosion to proceed. For example, this occurs to porous lead coatings on carbon steel operating in industrial environments, where the coating porosities are filled with lead sulphate as initial corrosion product (Fig. 17.2b); or when exposed to hard fresh water where, due to the alkalinity produced on the surface by the cathodic reaction, a mixture of carbonates and iron corrosion products forms.

In atmospheric corrosion or in low conductivity solutions, the influence of porosity is less important because the galvanic effect is modest or even negligible, as the ohmic drop between cathodic and anodic areas is high and the geometry of the condensate film makes the surface area ratio unitary.

If the base is an active-passive metal, corrosion conditions are different from what previously described. For example, a coating made of platinum or gold on titanium, which is passive, has no dangerous consequences (Fig. 17.2c); the same applies to stainless steels in chloride-free environments.

17.1.3 Anodic Coatings

If the coating metal is less noble than the base, any porosity or defect of coating is not intrinsically harmful because the coating metal itself works as anode, also ensuring a cathodic protection effect (Fig. 17.3a).

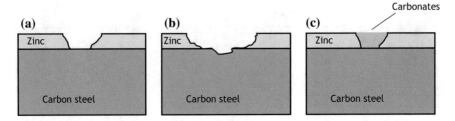

Fig. 17.3 Effect of porosity on an anodic-type coating: **a** porosity; **b** coating defect larger than a critical diameter; **c** sealing effect by corrosion products

Typical coatings of this type for carbon steel are zinc, aluminium and, for specific environments, tin. Industrially, zinc coatings are often referred to as galvanized goods. The critical size of a coating defect which allows a sufficient protection effect depends on conductivity and geometry of the aggressive environment. For example, for outdoor galvanized carbon steel, the critical defect size is slightly more than 1 mm in diameter (the electrolyte is condensed water, i.e., distilled water); for larger size, evidence of a corrosion attack, i.e., a rusting point, appears in the middle (Fig. 17.3b). In brackish water or seawater the critical defect size increases by at least one order of magnitude. Also for anodic-type coatings, sealing effects can arise, as for example on galvanized steel exposed to hard fresh waters by the precipitation of carbonates triggered by the alkalinisation on cathodic areas, that is, on the carbon steel surface (Fig. 17.3c). Anodic coatings can lead to hydrogen evolution as cathodic reaction on the base metal, therefore, if the latter is susceptible to hydrogen embrittlement, the use of these coatings has to be carefully analysed case by case.

17.1.4   Multilayer Coatings

A single metal layer is not always possible or convenient as metallic coating, while a multiple layer can solve smartly corrosion concern.[1]

The presence of defects in the coating causes consequences similar to those already described for single-layer systems. However, as corrosion penetrates and inner layers arise, the electrochemical system changes accordingly. For example, in the multilayer depicted in Fig. 17.4, composed of—from external to internal— chromium, nickel and copper on the base carbon steel (i.e., iron), copper and nickel are initially anodic versus chromium and become cathodic versus iron, once the electrolyte reaches the inner layer, enhancing galvanic corrosion on the latter.

---

[1]For some applications, for example for electronic components, multi-layered coatings are used for functional requirements and not for corrosion prevention.

Fig. 17.4 Corrosion damage in a multilayer Cr-Ni-Cu-Fe coating due to a coating defect

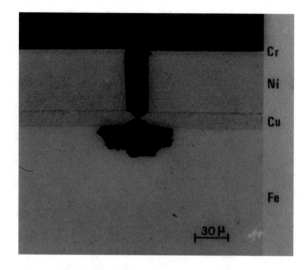

Nickel-chrome coatings. A good example that illustrates the behaviour of multilayer coatings is the so-called *decorative chrome plating*, largely used in the past in automotive industry (chromate bumpers and slips of cars). Despite their name, these coatings are not a monolayer of chromium, instead they are one or more layers of nickel for a total thickness of some tens of μm, covered by a thin layer of chromium, 0.25–0.8 μm thick, to give shiny appearance and resistance to abrasion. Only chromium is not used because it would be cathodic to iron, then accelerating galvanic corrosion beneath coating defects and the formation of rust that would be difficult to remove by washing, so consumers would not accept it essentially for aesthetic reasons. Unfortunately, it is not possible to obtain thin, defect-free chromium films, because even in the presence of little deformation they suffer micro-cracking, then exposing the iron beneath.

The presence of a nickel layer does not change the electrochemical scenario because nickel is less noble than chromium, so nickel corrodes as iron does in correspondence of chromium defects, but nickel corrosion products are almost invisible and easily washed out.

A layer of nickel below chromium could therefore appear as the correct solution, because nickel layers can be obtained almost without defects and they do not crack in operating due to its ductility. However, this solution may appear weak because the nickel layer is still stimulated to corrode at chromium film defects; luckily, the passive chromium surface is a bad catalyst for oxygen reduction, so the corrosion process is under cathodic control. This leads to the final result that even a high porosity chromium layer reduces the penetration rate on nickel. In fact, the effective cathodic surface does not change with the defectiveness of chromium, so the available oxygen amount is constant as well as the total amount of dissolved nickel, therefore the nickel corrosion rate decreases as chromium porosity increases (Fig. 17.5a).

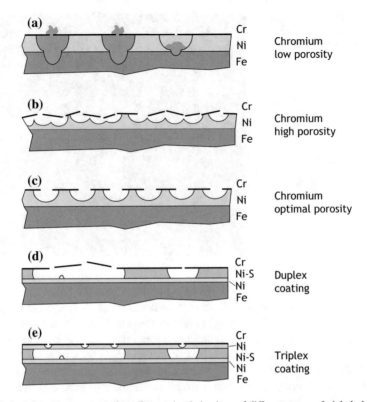

Fig. 17.5 Schematic representation of corrosion behaviour of different types of nickel-chromium deposit

A promising solution could be the use of a high porosity chromium film with distributed micro-cracks that spread the corrosion process on the whole surface (Fig. 17.5b). These high porosity coatings can be obtained by depositing on the nickel layer a thin chromium film (0.25 μm) enriched with small particles of alumina, or a highly micro-cracked chromium layer, with high thickness (0.8 μm) that ensures high stresses and therefore film micro-cracking.

For each thickness value of the nickel film there is a critical value of the coating defect density that maximizes the time required for nickel consumption (Fig. 17.5c). Such value can be found assuming that defects in the chromium film are pores, and that corrosion of nickel proceeds by forming hemispherical cavities; on this basis, pores spacing in the chromium film should be twice the nickel layer thickness. These coatings are typically used in environments with moderate aggressiveness.

In polluted environments, such as those in contact with the bumper of cars, these nickel-chromium coatings show a too short duration, therefore a different philosophy is adopted by switching to so-called *duplex coatings*. These consist of two layers of nickel: one, non-porous and ductile, on steel and a second one, sulphur-enriched, less noble, defective and brittle, working as galvanic (sacrificial) anode (Fig. 17.5d) on top of the first. These duplex coatings are also used to prevent corrosion in different environments (for example in chemical reactors) by acting as an efficient and reliable barrier. This solution is also recommended when small quantities of corrosion products are permitted. These coatings last proportionally to the thickness of the sacrificial layer.

In the specific case of bumpers, duplex coatings are not always working properly. In fact, as corrosion of the external nickel layer proceeds, adhesion of the chromium layer decreases up to delaminate (Fig. 17.5d). To strengthen the coating a three-layer nickel coating is used (Fig. 17.5e), where the third layer is again nickel. It is curious to note that once the best coating to protect bumpers was eventually developed, chromate steel bumpers were abandoned.

17.1.5 Methods for Obtaining Metallic Coatings

Metallic coatings are obtained by different methods:

- *Mechanical.* By hot rolling (cladding), by jacketing or by spraying a molten metal
- *Physical at high temperature.* By immersion in molten metal (zinc, lead, aluminium, tin and tin-lead alloys); by diffusion of metals from powders (zinc, aluminium, chromium or silicon); by welding (for example stainless steel). These coatings show high adhesion to the base metal, due to the formation of alloys as result of a metal diffusion at the interface
- *Physical at low temperature.* A physical method at low-temperature is evaporation under vacuum. The adhesion of coatings is merely mechanical
- *Chemical at high or low temperature.* Among the former, chromium, silicon and aluminium coatings are obtained from reduction or dissociation of the respective chloride through chemical vapour deposition technique; or from molten salts by displacement or electrodeposition. These coatings are characterized by a good adhesion based on the formation of alloys at the interface. Low temperature methods are electrodeposition, displacement and chemical reduction.

In the following, some details are provided.

Sprayed coatings. The metallic coating is obtained by spraying molten metals or alloys directly on the metal surface. It is used particularly for zinc and aluminium and their alloys. By employing a plasma flame it is also possible to deposit high-melting metals, oxides (Al_2O_3), compounds (carbides, nitrides, borides), which are materials with high mechanical characteristics (in particular wear resistance) and corrosion resistance. Sprayed coatings have special features. First of all, adhesion to the base metal is poor being purely mechanical; an increase in roughness is generally beneficial. The porosity of these coatings is high, since they are obtained by the impact of small drops of liquid metal covered by a layer of oxide; this leads to the conclusion that only anodic coatings can provide good protection. Furthermore, because of their porosity they can be easily painted, obtaining the sealing of pores. The greatest advantage of this method is that the size of the component is not a limit, either considerable or tiny such as riveting or bolting.

Immersion coatings. This is the oldest, easiest and often cheapest method. The metallic coating is obtained by dipping the goods, previously degreased, deoxidized and dried, in a bath of molten metal or alloy for a short period. In practice, this method is used to coat carbon steel with low melting metals, such as zinc, lead and their alloys, tin and aluminium. The cross section of a metallic coating obtained by immersion shows a range of compositions, varying from the base metal through gradual alloying up to the composition of the metallic coating. The formation of these alloys, often consisting of intermetallic compounds, is the necessary condition for obtaining a good adhesion. For example, iron and copper, which do not form alloys with lead, cannot be covered directly with it, because of lack of wetting conditions; to obtain adherent coatings, an intermediate layer is used to bind the base metal to the final coating, as for example tin and antimony as intermediates for iron and lead. The main advantages of this technique are simplicity and high production speed. There is a limitation for complex geometries for which it may be difficult to obtain uniform thickness. The presence of small holes or cracks should be avoided. This method has some disadvantages, such as the decrease of ductility. The process to obtain zinc coating by immersion is known as galvanizing or hot-dip galvanizing and is the most world widespread.

Chemical reduction. This process exploits a redox reaction which reduces the ions of the metal to form the metallic coating by a suitable reducing agent. The reduction reaction of metallic ions occurs, for kinetic reasons, preferentially or exclusively on the surfaces to be coated and not within the solution. The metal deposition is obtained on a variety of materials, from metals to non-conducting ones such as ceramics, polymers, glass. The most used reducing agents are: sodium hypophosphite to deposit nickel, cobalt and their alloys, gold, silver, palladium (with simultaneous deposition of phosphorus that is derived from the decomposition of hypophosphite) and formaldehyde to deposit copper. For nickel, cobalt and precious metals also boranes are used, which lead to a co-deposition of nickel-boron alloy, and hydrazine. Coatings obtained by this method, either with microcrystalline or amorphous structure, show low porosity, high hardness, almost uniform thickness regardless the shape, therefore particularly suitable for profiled surfaces as threads, holes, valves. The nickel coating, called *chemical* or *electroless nickel*, which is in practice the most popular obtained by this method, can be hardened by a precipitation heat treatment because of the presence of phosphorus or boron, reaching hardness values comparable with those of the chromium coating, making the coating highly resistant to wear.

Electrodeposition. This is the most used method for obtaining metallic coatings, and is generally performed in aqueous solutions. Since in aqueous solutions hydrogen evolution can take place in competition with metal deposition, metals less noble than zinc are excluded: in those cases molten salts or non-aqueous solutions have to be used. Common electrodeposited coatings are: gold, silver, copper, lead, tin, zinc, cadmium, chromium, nickel, iron, cobalt, platinum, rhodium and others, either as pure metals or alloys. The obtained thickness is regulated and varies from fractions to hundreds of micrometres. Physical and mechanical properties, such as brightness, porosity, hardness or mechanical strength, can be controlled and varied over a wide range by changing nature and composition of solution and setting current density, temperature and stirring. The most used solutions are based on salts of the metal to be deposited as sulphates, cyanides, pyrophosphates, fluoroborates and sulphamates, with a complex composition due the addition of substances with various purposes, as:

- Substances with buffer effect, for pH control at the cathodic boundary layer where the hydrogen evolution process produces alkalinity
- Complexants
- Substances increasing conductivity
- Catalysts
- Substances increasing cathodic overvoltage to increase the throwing power for obtaining more uniform deposits
- Depassivating substances to have a uniform attack of the anodes
- Levellers or brighteners to get levelled or bright deposits
- Wetting agents or surfactants to get rid of hydrogen bubbles which would shield the cathode then impeding the metal deposition.

These substances, especially organic ones, can dope the deposit and modify chemical-physical or mechanical properties as hardness, corrosion resistance or electrical conductivity. Sometimes, neutralizing agents are added; for example, brighteners or levellers can promote tensile or rarely compressive stresses within the deposit, accordingly other substances are added to balance opposite effects. Current density influences many properties as appearance, uniformity, porosity, mechanical and chemical-physical characteristics; therefore, it is maintained in proper ranges of values, generally increasing as temperature, concentration of metal ions and agitation increase. Outside those validated ranges (i.e., if current density is too high or too low) non-uniform coatings, very porous or even spongy, often polluted with oxides or hydroxides, and poorly adherent are obtained. Among aspects worth of discussion, those dealing with microstructure, adhesion, mechanical and more generally physical-chemical properties, homogeneity (i.e., throwing power) and alloying should be remembered. In the following, only throwing power is mentioned, because impacting also on the more general problem of current distribution in galvanic cells as established in galvanic corrosion and cathodic protection.

Throwing power and uniformity of metallic coatings. In an electrolytic cell of cylindrical geometry, where electrodes are the two parallel bases, perpendicular to the cylinder axis, current lines are parallel and perpendicular to the electrodes, and current density is uniform, so that a homogeneous coating is obtained. For all other geometries, conditions are different. In practice, current density is never uniform on the cathode because current lines concentrate where facilitated, for instance where anode-to-cathode distance decreases or on edges, and their density is lower where screens, recesses and cavities are present. As a result, metallic coating thickness is not uniform. As discussed in Chap. 9, current distribution is primary when the overvoltage is negligible on both electrodes, accordingly it follows the Ohm's law and depends on geometry, only; conversely, when overvoltage prevails, current distribution becomes secondary and depends also on resistivity. In primary distribution conditions, either current density or coating thickness depends on the anode-to-cathode distance. When the secondary distribution applies, ohmic drop in solution and overvoltage on electrodes lead to a compensation which in practice makes the current distribution more uniform, which means that throwing power is high. For instance, this is the case of cyanide-based salts and more in general of complex salts used as electrolyte. Various tricks are used to achieve uniform current distribution even in low throwing power conditions. For instance, by making round corners and edges; or shaping the anode as the cathode; or by placing insulating screens, or auxiliary electrodes (cathodes) called *current thieves* as parasite electrodes.

17.1.6   Zinc Coatings

Zinc coatings, called galvanizing, are widely used for protecting carbon steel in atmospherically exposed structures. Their anodic behaviour reduces the requirements regarding uniformity and porosity, facilitating the application and reducing costs. Their performance is a function of environmental conditions and their lasting is proportional to the thickness. Typical corrosion rates are the following:

- 0.2–3 µm/year in rural areas
- 0.5–8 µm/year in marine environment
- 2–16 µm/year in urban and industrial environment.

The corrosion rate of zinc is 10–30 times lower than the one of carbon steel according to environmental conditions; in water, zinc behaviour is good in the pH range 7–12.

As temperature increases the difference in nobility between iron and zinc decreases. In fresh aerated waters at temperatures around 40–60 °C, zinc and iron undergo a polarity reversal which makes zinc cathodic versus carbon steel, then the beneficial protection effect in correspondence of coating porosity or defects is lost. This behaviour derives from the formation of a conductive zinc oxide which allows oxygen reduction, while the normal zinc hydroxide does not, being a good insulator.

Galvanizing is obtained through two techniques: by immersion in a bath of molten zinc or by electrodeposition. By the hot immersion method, high thicknesses are obtained, in the order of 100 μm, while by electrodeposition typical thicknesses are in the range 2–5 μm. Because corrosion duration depends on the zinc layer thickness, electro-galvanizing is used only as primer for painting systems, as is typical in automotive industry.

17.1.7 Tin Coatings

Among metallic coatings for laminated products, tinplate is the most manufactured. It is produced by electrodeposition. The first output from the electrolytic cell is a matt, porous product which has to be heat treated at the tin melting temperature around 250 °C to make it brilliant, more compact and more adherent to the base.

Tin is an amphoteric metal, corrosion resistant at pH close to neutrality. As low melting metal, the overvoltage of the hydrogen evolution process is very high, so corrosion rate in non-oxidizing acidic environments is very low.

The use of *tinplate* for food cans derives from two fortunate circumstances:

- Tin salts are not toxic, although human tolerance is limited to very low contents
- Tin exposed to food substances becomes anodic versus iron, then giving protection effects, because organic anions present in food form complexes of stannous ions and help passivate iron, leading to this final result: tin potential decreases and that of iron increases.

As a result, iron does not corrode due to the cathodic protection effect given by tin and tin does not corrode due to the high overvoltage for the hydrogen evolution process. Conversely, without that beneficial mechanism, the usual nobility sets up, therefore iron rapidly corrodes at tin coating defects and because of extremely unfavourable cathodic-to-anodic area ratio with two injuries on the cans: a rapid perforation and an evident bulging due to hydrogen formation.

17.1.8 Nickel Coatings

Nickel coatings are widely used for a variety of features, such as mechanical resistance, hardness, ductility, as well as corrosion resistance to industrial atmosphere, waters, alkaline solution, basic molten salts, and solvents; it is also used for decoration purposes.

Decorative coatings exposed to atmosphere, typically 10–50 μm thick, are covered by a thin layer of chromium, about 0.3–0.8 μm thick, to prevent the lack of gloss; this also increases the resistance to wear. Used in chemical plants with thickness as high as 250 μm, nickel coatings are obtained by electroplating as well as by chemical reduction (known as *electroless nickel deposition*) when equipment has complicated shape.

As far as electroless nickel is concerned, which consists of nickel-phosphorus based coating, its use is typically in car parts, machine parts, electronic parts, office automation apparatus parts, precision product and other applications. Thickness varies from 25 to 75 μm. The content of phosphorus in the film is in the order of 6–13%. The altered composition gives to the coatings higher mechanical characteristics, wear resistance and hardness, especially after annealing. Corrosion resistance is improved in the as-deposited state, especially to organic acids and caustic environments, due to the formation of a protective, amorphous glassy film, where the absence of grain boundaries eliminates the risk of SCC.

Since many years, nickel coatings have been prohibited for personal goods as glasses and jewellery because it causes allergic reactions.

17.1.9 Chromium Coatings

Chromium coatings are bright, hard, brittle, highly tensioned, porous (if thin) and cracked (if thick). They can be divided in two main types: thin coatings for decorative purposes and thick ones (also called *hard chrome*) for corrosion prevention. Decorative chromium coatings, about one micron thick, are generally deposited on a layer of nickel; hard coatings are much thicker, about 200–300 μm thick, and are used to resist wear and corrosion even at high temperature.

Chromium plating is more recent than others like copper, nickel or gold plating, which appeared in the nineteenth century, while chromating dates back to 1920s. It is interesting to note that chromating is obtained in an unexpected condition: the electro-deposit is obtained from chromic acid and not from chromium salts, provided the presence of a catalyst which is usually a small percentage of sulphuric acid, or hydrofluoric acid and fluosilicic acid. Since chromium plating baths have poor throwing power, it is difficult to deposit chromium on objects with complicated shape; furthermore, current efficiency zeros below a critical current density value, while above it efficiency grows as current density increases.

17.1.10 Copper Coatings

Copper plating finds less applications than nickel and chromium, although today high throwing power baths are available, able to give perfectly polished and uniform deposits. The reason is that copper plating is rarely used for decorative purposes, because of easy oxidation unless protected with lacquers or special surface treatments. On the contrary, there are a lot of industrial applications. It is used on cast zinc before nickel plating, which occurs in acidic baths (pH = 4) where zinc would dissolve, jeopardizing the adhesion: a previous copper plating in an alkaline bath allows to overcome the problem. A second example is the use of copper on plastic objects before nickel plating, with the purpose to absorb the differential thermal expansion between plastic and metal.

17.1.11 Precious Metals

In the past, the deposition of precious metals was limited to gilding and platinizing for decorative purpose and protection of electrical contacts. Nowadays, gold deposits are widely used in electronic industry to avoid oxidation of contacts and because of good thermal and electrical conductivity.

Among other precious metals, rhodium should be remembered, which is very hard, wear-resistant, resistant to acids, even *aqua regia*, used to produce mirrors with very uniform reflectivity and resisting to scratching. Coatings made of oxides of precious metals such as ruthenium and iridium, called mixed-metal-oxides (MMO) are used on titanium since 1970s in industrial electrochemistry and for anodes employed in cathodic protection applications, because such oxides show low overvoltage and good resistance to aggressive environmental conditions (high acidity and chlorine evolution) while ensuring good conductivity.

17.2 Conversion Coatings

These types of coatings are formed on the surface of various metals as the result of chemical or electrochemical reactions after immersion in suitable solutions. Most popular processes for obtaining conversion coatings are phosphatizing, chromating and anodic oxidation.

17.2.1 Phosphate Coatings

Phosphate coatings are obtained on carbon and galvanized steels by immersion in acidic phosphate solutions that form on the metal surface a porous, crystalline phosphate layer strongly adherent to the metal. This layer, which works by itself as a mild corrosion prevention, is an excellent and economical primer for anchoring paints or as temporary protection of semi-finished products. It was extensively used in automotive industry and presently it is widely adopted for indoor goods as metallic scaffolds and household machines.

Manganese and zinc phosphates are employed, which—as well as iron—may form the following relevant phosphates:

- Primary phosphates of $M(H_2PO_4)_2$ type, very soluble in water
- Secondary phosphates, $MHPO_4$, slightly insoluble
- Tertiary phosphates, $M_3(PO_4)_2$, insoluble in phosphoric acid solutions.

In acidic solutions with pH between 2 and 4, the above phosphates are in hydrolytic equilibrium according to the following reactions:

$$M(H_2PO_4)_2 \rightarrow MHPO_4 + H_3PO_4 \tag{17.1}$$

$$3MHPO_4 \rightarrow M_3(PO_4)_2 + H_3PO_4 \tag{17.2}$$

As soon as the metal, for example iron or zinc, is exposed to the solution, the first stage is a corrosion reaction, such as the following for iron:

$$Fe + 2H_3PO_4 \rightarrow Fe(H_2PO_4)_2 + H_2 \tag{17.3}$$

with a local reduction of acidity on the metal surface, which pushes reactions (17.1) and (17.2) rightward, causing the separation of the tertiary phosphate. This happens in the narrow range of the free acid to phosphate ratio of 0.12–0.15: in other words, this ratio must be sufficiently low to allow the precipitation of phosphates but not too low to have also precipitation inside the solution. Hydrogen evolution is the slow stage of the process and tends to shield the metal surface; to avoid these drawbacks some oxidizing substances, such as nitrites, nitrates, chlorates and peroxides are added, to oxidize hydrogen into water. Depending on the applications, composition, thickness and morphology can vary.

Conversion coatings can be used as primers for painting cycles as composed of the phosphates of the three metals, Fe, Mn and Zn, to obtain a weight of 10–30 g/m^2. Phosphatizing obtained by spraying is often preferred to the one by immersion, especially for temporary protection.

17.2.2 Chromate Filming

In the past, chromate filming treatment was employed for finishing aluminium, zinc, cadmium and magnesium (but not for steel), either as corrosion prevention or as primer for painting cycles. The former is based on the presence of hexavalent chromium as strong passivating agent; however, due to its toxicity the treatment has been banned. Alternative treatments based on non-toxic trivalent chromium are much less effective.

Chromate-based coatings are obtained by immersion for a few minutes in an acidic chromate solution, usually sodium bichromate, $Na_2Cr_2O_7$, containing depassivating additives that co-precipitate in the coating as chlorides, fluorides and also phosphates. Composition and pH of chromate filming solution vary according to the metal. A chromate coating is generally yellow or even transparent, amorphous, thin (a fraction of a micron thick) and with semiconductor characteristics.

It is believed that the mechanism of film formation entails, initially, the dissolution of substrate surface (metal or oxide), then the partial reduction of hexavalent chromium and simultaneous separation on the metal surface of a film consisting of chromium chromate, $Cr_2(CrO_4)_3$, mixed with oxides of the treated metal and other species that are present in solution such as phosphates.

17.2.3 Anodic Oxidation

Anodic oxidation is an electrochemical process carried out in order to thicken the oxide film naturally present on metals, with the aim to improve the corrosion resistance or the aesthetic look or to obtain oxide films with insulating properties.

Anodic oxidation of aluminium. Aluminium anodizing is certainly the most widely used electrochemical process of anodic oxidation for the production of aluminium goods, known as anodized. Anodizing improves corrosion resistance and gives a good pre-treatment for further painting especially after chromate filming has been banned. Aluminium anodizing goes through three stages:

- Surface preparation
- Anodizing, possibly followed by a colouring treatment
- Sealing.

Surface preparation aims to remove all contaminants, in particular iron inclusions, which often remain after mechanical processing.

Table 17.1 summarizes the parameters relating to the three most used processes for aluminium anodizing. The properties of oxide films depend on bath composition and operating conditions (i.e., temperature and current density). Oxide thickness ranges from a few microns to about 50 μm as depending on time and current applied. Figure 17.6 shows the aluminium oxide structure, which is composed of an inner *barrier layer* covered by a *porous layer*, with a typical porous honeycomb structure, obtained in less than one hour with applied voltage of 20–60 V; these films are used typically as anticorrosive coating. For other applications, such as for electrolytic capacitors, other solutions are used, for example, boric acid or ammonium phosphate solutions which allow to obtain oxide films with only the barrier layer structure.

Table 17.1 Characteristics of the three most frequently used processes for aluminium anodizing

Acid (% by weight)	T (°C)	Cell voltage (V)	Current density (A/m^2)	Thickness (μm)
Sulphuric 10–15%	15–25	10–20	100–300	3–50
Chromic 3–10%	30–40	30–50	30–40	2–8
Oxalic 3–8%	20–40	30–60	100–300	10–60

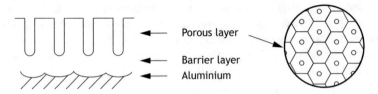

Fig. 17.6 Schematic representation of the barrier layer and of the porous oxide film produced by anodic oxidation of aluminium

To eliminate the porosity of the porous layer, a sealing process is necessary, which consists of the hydration of aluminium oxide which seals the pores by swelling. In the past this treatment was done by immersion in boiling water for about 5–20 min or exposure to steam at 150 °C; today cold-sealing catalysed processes are used, for example by immersion in a nickel fluoride solution. During this operation, some substances can be conveniently entrapped into pores, such as corrosion inhibitors, or pigments to colour the film.

Anodic colouring of titanium. A few years after Volta's pile invention, the Italian physicist Leopoldo Nobili (1784–1835) discovered that by anodizing some metals, such as lead and tin, amazing interference colours could be obtained: he called this electrochemical technique *metallocromia*. Today this anodic colouring technique is applied to titanium and other so-called valve metals. Stainless steels can also be oxidised to obtain a colouring effect, but the treatment is a chemical one, based on the immersion of the metal in a suitable chemical bath.

On titanium, anodic oxidation gives rise to beautiful interference colours, as Pietro Pedeferri obtained for artistic decorations and as new painting technique. By varying operating conditions and bath composition, amorphous or crystalline films are obtained. The latter, in the form of rutile or anatase crystal structures, can exhibit interesting antibacterial, photocatalytic and superhydrophilic properties.

In many solutions, with the application of very high cell voltage for anodic oxidation, an electrical sparking occurs which perforates the anodic film (*anodic spark deposition*), giving brown or grey coloured oxides, several microns thick. It is interesting to note that the oxide film can incorporate ions present in the used solutions as sodium, potassium, calcium, phosphorus and others. As shown in Fig. 17.7, a typical morphology of such anodic films is observed at the microscopic scale, showing the presence of craters that sparks form.

Figure 17.7 shows another interesting structure consisting of oxide *nanotubes* obtained in fluoride ions containing solutions, which are characterized by a huge surface, tens or hundreds of times the nominal one.

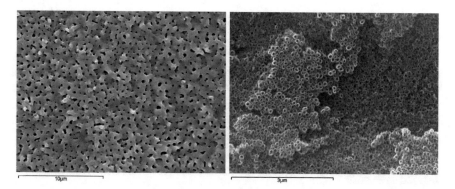

Fig. 17.7 Micro and nanostructure of films of titanium dioxide obtained by anodic oxidation: left, anodic spark deposition; right, nanotubes

17.3   Other Inorganic Coatings

17.3.1   Hot Enamels

Metals as cast iron, copper, aluminium and mostly carbon steel can be coated with enamels applied at high temperature (glass lining) which result very adherent to the metal surface, resistant to abrasion and corrosion especially in acidic environments (except hydrofluoric acid) and produced with pleasant finishing and stable colours.

Enamels are obtained from a mixture of silica, titanium oxides, feldspars, clay and melting additives as borax, fluorosilicates, nitrates and carbonates of lithium and potassium, melted at a temperature between 1000 and 1400 °C to obtain alkaline aluminium borosilicate compounds, then rapidly cooled in water and successively grinded. This powder, added with coloured pigments, is sprayed on the metal surface and heated in oven at 750–850 °C for some tens of minutes.

Chemical resistance in acidic environments increases with the addition of silica, titanium oxide and zirconia and decreases with the addition of boron oxides.

To ensure a good adhesion to the metal surface, which is enhanced by nickel and cobalt oxides, and a good resistance to abrasion despite the possible increase in brittleness, the enamel must have a thermal expansion coefficient compatible with the one of the metal, which is normally higher.

Since chemical resistance grows as thermal expansion coefficient decreases, acid resistant silica-rich coatings have very low coefficient and consequently poor resistance to thermal shock. To minimize the risk of cracking due to thermal shocks, multilayer enamels are used: the inner one with low expansion coefficient, the external one with greater resistance to corrosion.

17.3.2 Thick Cementitious Coatings

Due to their alkalinity (pH $\approx$ 13) cementitious mortars applied on carbon steel offer a perfect protection, as proved by successful long lasting results offered in the last century by reinforced concrete structures. Cementitious coatings have been used for internal protection of pipelines for water mains, applied by a centrifugation process, for submerged marine pipelines, where concrete has also a weighting function, and for the external protection of oil and gas well casings.

The behaviour of steel embedded in concrete is illustrated in Chap. 23.

17.3.3 Thick Corrosion Resistant Coatings

To resist very high aggressive corrosion conditions, such as those found inside chemical reactors, thick layers of corrosion resistant materials are used: glassy materials, cured inorganic silicates, ceramics, sulphur, graphite and carbon, often applied in the form of tiles or bricks stuck with suitable adhesives. To absorb the inevitable differential expansion strains, membranes made of resins, asphalt or synthetic elastomers are often interposed.

Glassy and ceramic materials are resistant in all environments, except in hydrofluoric and caustic ones. For their use, their properties must be remembered: insulating, brittle, with poor mechanical strength, and sensitive to thermal shock. Graphite and carbon are conductive with a good resistance to rapid temperature changes. Sulphur bricks which stuck each other at temperatures above 120 °C show good resistance in oxidizing environments and in hydrofluoric acid.

Photocatalytic and Superhydrophilic Properties of Titanium Dioxide.
A photocatalytic reaction is a chemical process occurring on a surface hit by a radiation because of the presence of a photocatalytic agent. Anatase and rutile titanium oxides are the best photocatalytic substrates for oxidation or reduction of many organic and inorganic substances. Anatase is better than rutile for degradation of pollutants; yet, both titanium oxides need UV radiation to activate, therefore they are proposed for the sunlight-activated degradation of air pollutants or inside UV reactors for water purification.

Other applications deal with the degradation of inorganic compounds, removal of metal ions from water, photovoltaic cells, antifouling, self-cleaning and antibacterial surfaces, which have found interesting applications in civil constructions.

Anatase shows interesting self-cleaning capacity through its photoactivated properties. In detail, anatase is normally slightly hydrophilic as proved by a drop of water on its surface, which remains stably of semi-spherical shape. As soon as the surface is irradiated with UV light the drop of water spreads on the surface because anatase becomes highly hydrophilic, due to the formation of hydroxyl groups. By this process, the wettability of the surface increases to allow the formation of a continuous water film instead of drops, inducing a tiding and antifogging effect.

17.4 Paintings

Paints are heterogeneous systems formed by solids dispersed in a film-forming component, said *binder*, whose viscosity is regulated by adding suitable *solvents* and thinners, then applied to a metal surface to form in time a solid and adherent film. This chapter deals with anti-corrosion paints only, then regardless those formulated for other purposes as decoration or signalling.

The use of organic coatings based on oils, greases, waxes and bitumen for protective aims dates back to antiquity, however, for the protection of steel, practical applications started in England in the eighteenth century employing linseed oil, pigmented with red lead oxide. The extension of its use to resist aggressive chemicals as acids, bases, saline solutions and hot corrosive fumes is more recent; for example, chlorine-rubber and vinyl paints were introduced in the late 1920s and epoxy and polyurethane resins only after World War II.

17.4.1 Components

Paints are composed of two parts: a liquid one and a solid one. The former contains the binder, the solvent and additives as plasticizers and auxiliaries. The latter is a mixture of pigments and fillers.

Binders. Once based on natural oils and resins, now based on synthetic polymers, the binder is the constituent of the paintings that gives body to the paint film, whose filming properties derive primarily from solvent and diluent evaporation—so-called physical filming—or from chemical reaction, or crosslinking. Some of the best-known physical filming binders are:

- *Vinyl resins*: highly used as anti-corrosion paints due to their high chemical resistance, elasticity and ease of application. They are mostly composed of copolymers of polyvinyl acetate and polyvinyl chloride. Other polymeric and carboxylic radicals can be introduced in the polymer chain to improve adhesion to the metal surface. The chemical resistance of these polymers is excellent in contact with chemicals, seawater, electrolytes, as well as with industrial and marine atmospheres

- *Chlorinated resins*: they are obtained by chlorination of natural or synthetic polyolephins and polydiolefins, as chloro-caucciu and chlorinated polyethylene or polypropylene. The presence of a high percentage of chlorine in the molecule makes these paints less flammable. They are highly resistant to a wide range of environments, but their brittle character requires the use of plasticizers
- *Bitumen (coal tar) and asphalt*: the former is obtained as by-product of crude oil distillation, then having a very variable composition; asphalts are natural products found in quite big reservoirs. These hydrocarbons dissolved in a solvent give low-cost and thick coatings (0.1–1 mm), largely used for waterproofing membranes of buried structures.

Crosslinking filming is obtained by a chemical reaction which links polymer chains to form a three-dimensional structures. Binders of this type are:

- *Epoxy resins*: their chains contain reactive compounds, hydroxyl and epoxy groups, which react with a crosslinking agent, for instance a polyamide, to form highly cross-linked structures with exceptional chemical resistance. When the crosslinking reaction occurs at room temperature, as when using polyamides in suitable percentages, the paint is of two-component type; conversely, when crosslinking occurs at high temperatures, typically with amines or phenolic resins as curing agents, paints are provided ready to use as a mono-component system. Epoxy resins are formulated in a wide variety of compositions to obtain different properties for various applications. The main types are: (1) *solvent-free epoxy* resins, neither porous nor subject to shrinkage because the solvent is eliminated, then also showing the advantage of being less dangerous in terms of flammability and harmfulness; (2) *epoxy systems with solvents*, which are the usual epoxy paints, particularly resistant to highly aggressive marine or industrial atmospheres; (3) modified systems that consist of epoxy resins to which other resins are added to improve resistance to aggressive environments: epoxy-tar, with excellent corrosion and chemical resistance; epoxy-vinyl, which have excellent adhesion to the support and excellent mechanical properties; epoxy-phenolic, having excellent water resistance; epoxy-silicones, very resistant to high temperatures
- *Phenolic resins*: offer excellent resistance to aggressive chemicals as diluted acids and weak alkalis and seawater due to the presence of unreactive bonds. Their mechanical properties are rather low; besides, they are hard, rigid with low adhesion property. For this reason they are modified with oleoresins or with alkyd resins. Curing is obtained in air or in oven
- *Polyurethane resins* are obtained by reaction of a polyester or a polyether containing reactive hydroxyl groups, with an isocyanate group-containing compound which acts as curing agent. The latter is obtained by condensation of an aromatic poly-isocyanic derivative with a polyfunctional hydroxyl compound: by this way a pre-polymer of poly-isocyanic derivative is obtained, avoiding harmful effects, as volatility and toxicity. Polyurethane resins are used to prepare anti-corrosion paints, with excellent resistance to chemicals,

industrial atmospheres and abrasion. They are bright glossy and stable to sunlight exposure, then used as final coats.

For other binders, the curing agent is oxygen, as for example:

- *Linseed and similar vegetable oils*: although no longer in use, it is interesting to review their mechanism. The binder consisted of a mixture of glycerol esters with unsaturated fatty acids. Once applied as a film, these oils polymerize producing a crosslinking process triggered by atmospheric oxygen. To accelerate this curing process, special treatments were used: for example, for linseed oil a boiling process with addition of oxides or soaps of cobalt, manganese, lead or zinc, or heating by blowing air or an inert gas
- *Alkyds*: they are widely used for the reasonable cost, good aesthetic features, good elasticity and adhesion on non-sandblasted surfaces, good performance in non-polluted atmospheres; conversely, they are not suitable for aggressive environments, especially if alkaline. They are high molecular weight polyesters obtained from reaction of poly-alcohols with fatty acid mixtures. As far as fatty acids are concerned, there is a wide variety of available substances; the choice depends on the properties required for the resin. Fatty acids are generally added in the form of vegetable oils (linseed, coconut, soy oils), which are tri-esters of glycerol and fatty acids. Alkyd resins are classified based on the percentage of oil: long oil (55–65% oil), medium oil (45–55%) and short oil (45%) resins. Depending on the percentage and type of oil, dry and non-drying resins are obtained. The first, long oil and in some cases also medium oil, are characterized by the presence of a high percentage of double bonds, which react with atmospheric oxygen, then bridging polymer chains. Non-drying alkyl resins include most of medium oil and all short oil resins, containing a few number of double bonds: curing is obtained by increasing the temperature. To improve finishing aspects, as for cars or household machines, low percentages of amine resins, as urea, formaldehyde and/or melamine-formaldehyde are added
- *Silicone binders*: these resins have a –Si–O–Si–O– structure, with some organic side groups to improve their solubility. Curing takes place by action of atmospheric humidity by a hydrolysis process of poly-silicates, which eliminates organic groups as volatile compounds. The –O–Si– bond, similar to that in glass or in quartz, is stable to oxidation processes. For this reason, silicone resins have excellent resistance to high temperatures (normally up to 250 °C and, if pigmented with aluminium, up to about 500 °C). This type of binder is also used for the preparation of zinc paints, which exhibit excellent resistance to many near-neutral pH environments.

Solvents and thinners. They are fundamental constituents of paints to facilitate their application and allow the formation of the film. Usually, they consist of volatile organic substances, as aliphatic and aromatic hydrocarbons, alcohols, esters, ketones and, for so-called green formulations, water. The most requested property is an easy application, to help avoid formation of film defects and drops and as long as possible to give the polymer the best mechanical properties. Once the paint film has been applied, solvents and thinners must evaporate rapidly and thoroughly. To meet all of these requirements, a mixture of components is used, paying attention to prevent variations of solvent composition during the evaporation process which can jeopardize the integrity and resistance properties of the film paint. Solvents and thinners are dangerous to health and flammable; however, fire hazard disappears after paint drying.

Plasticizers. For crosslinking polymers, paint properties depend on type and amount of the curing agent; conversely, for thermoplastic polymers film properties are improved, by modifying the chemical composition of the polymer, by the addition of plasticizers, consisting of polymers or monomers which interact with polymer chains to improve their flexibility and consequently film plasticity.

Other constituents. So-called auxiliary and filler constituents are substances that optimize film performance, such as dryers, suspensive wetting agents, antioxidants and resin stabilizers.

Pigments and additives. Pigments and additives are solid constituents sparsely dispersed in the binder. Pigments usually indicate synthetic substances, while additives derive from natural minerals after grinding, washing and thermal treatments. Additives are selected to resist the chemical action of exposed environments; for example, for coatings exposed to acidic environments, a low content of carbonates is allowed. Pigments are classified as active or inert. Active ones have the primary function of stopping the anodic or cathodic process, accordingly they are used for the formulation of primers; conversely, inert pigments aim to reduce the permeability of the film and improve adhesion properties. Typical inert additives are metal oxides (titanium oxide, iron oxide, alumina, silica, and zinc oxide) and some inorganic salts as carbonates and sulphates or complex silicates (mica, talc, kaolin). Their function is twofold: to improve paint thickness and physical and mechanical properties such as hardness, abrasion resistance and flexibility, and furtherly to increase chemical resistance and decrease film permeability. For paints applied to structures exposed to intense solar action, among other pigment types, aluminium flakes are used to form a low porosity fish skin-like scale barrier, enabling to reflect solar radiations and to resist the action of ultraviolet rays as well as of heat.

To classify the added pigments, a parameter called *volumetric concentration of pigments*, VCP, is used which is the pigment-to-dry paint volume ratio. There is a maximum value, called critical VCP (CVCP), for which paint properties change drastically once exceeded. A high VCP is required for some paints to facilitate the absorption of water needed for the migration of active metal ions toward the metal surface, however, it should not be too high because also aggressive ions such as sulphates and chlorides may be transported.

Figure 17.8 shows the classification of anti-corrosion paintings on the basis of VCP-to-CVCP ratio. As this ratio increases, paint protection mechanism switches from barrier-based to active-type paints (see also Sect. 17.4.3).

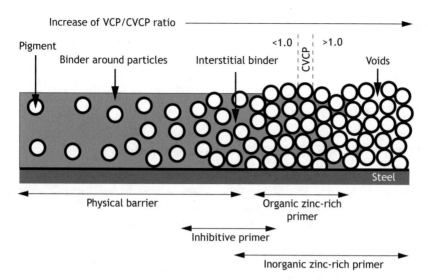

Fig. 17.8 Volumetric fraction of pigment in paint for different classes of anticorrosive coatings

17.4.2 Corrosion Under Paintings

Until World War II, although the anti-corrosive action of some pigments—in particular red lead oxide—was known, it was opinion that paints acted only as a physical barrier with the purpose to prevent the penetration of water and oxygen from an aggressive environment.

For the truth, physical filming paints are never an efficient barrier because water and oxygen can diffuse through by a quantity higher than the one needed for the corrosion process of the bare metal, hence the anticorrosive action is more complex than that of a simple barrier. In practice, only in dry atmospheres the reduction of water diffusion determines the stop of corrosion processes under paint, while in wet atmospheres water supply is always in excess. Instead, oxygen diffusion can become the controlling stage for high thickness paints.

Besides by water and oxygen diffusion, corrosion rate is also controlled by the electrical resistance of the paint, the so-called *ohmic control*, which is effective if paint conductivity is low. This parameter depends first of all on the amount of water absorbed by the paint, which in turn depends on the chemical-physical properties of paint binders and pigments; for example, bituminous-based paints have little tendency to absorb water. In addition, also the salt content of the exposed solution determines paint conductivity.

When ionic groups form in the binder as a result of hydrolysis caused by the absorbed water, the paint may become an anionic or cationic semipermeable membrane, then selective for ion migration (positive or negative). This behaviour setups in many paints a saline concentration gradient, which determines a beneficial effect when the ionic concentration decreases toward inner layers; the presence on the metal surface of corrosion products or oxides not removed before painting, which retain moisture, vanishes that beneficial effect.

17.4.3 Protective Action of Paints

The protective action of paints follows two mechanisms:

- *Barrier effect*: which reduces oxygen and water diffusion
- *Active effect*: when specific pigments, adsorbed onto the metal surface, exert an active protective action by changing the electrochemical condition (promoting passivity or increasing overvoltage or setting up cathodic protection conditions).

The barrier effect is adopted when exposure electrolytes, as acids, bases or saline solutions, react with active pigments, making their use unsuitable. For atmospherically exposed components, active based paints are preferred; however, usual painting cycles, composed of different coats, take advantage of both mechanisms.

Barrier effect. An efficient barrier effect is obtained by acting on composition, chemical-physical characteristics of components and thickness. The first action deals with the molecular structure of the binder, which must ensure a dense film as obtained by thermosetting polymers with a high degree of crosslinking or partially crystalline thermoplastic ones; both these polymers are hydrophobic and have a strong resistance to diffusion of water, oxygen and ions. Among thermoplastic binders, high molecular weight polymers, such as fluoropolymers and polyvinyl-chloride, are particularly attractive; as regards thermosetting polymers, epoxy, urethane, polyester and vinyl esters are used. Water absorption and oxygen permeability are also influenced by the nature, shape, content and size of pigments such as mica or aluminium of flake-shape type. Other influencing parameters are the composition and quantity of solvents and the method of application in combination with the filming process. In general, it is known that the lower the amount of solvent used, the lower the permeability of paint films. In addition, some solvent may remain trapped even in small quantities in the hardened film, then favouring water absorption.

Thickness. The barrier effect increases with thickness: in practice, to reduce paint permeability enough to withstand aggressive environments, a minimum thickness of 200–300 μm is necessary. Ordinary paints reach a maximum film thickness of 30–35 μm per coat because of the filming process, which is triggered by atmospheric oxygen, starts from the outer surface and proceeds inward hardly; accordingly, higher thicknesses require multiple coats. Conversely, thermosetting paints develop a fully hardened film regardless the thickness, allowing the application of a single high thickness coat. Thickness homogeneity on the entire metal surface is of great importance in determining the barrier effectiveness. Many causes affect local film thickness variations: first of all the substrate roughness profile, which depends on surface preparation (for example, sandblasting may offer a peak type profile instead of a rounded one). With rough profiles, small quantities of paint produce uneven films. Secondly, the design complexity of the structure to be painted due to the presence of cracks, sharp edges, recesses. Typical defects are in these cases: brush strokes, crinkles, orange skin, spraying fumes, glides on vertical surfaces or micro slit with peak emersion.

Non-stick coatings. A further improvement to coatings with barrier effect is to produce a non-stick surface, i.e., coating the metal with a polymers with very low surface free energy. This involves the use of fluorinated polyols as precursors, which can be applied either directly to the metal surface or as topcoat of urethane or epoxy primers, and then cured to obtain a fluoropolymer, whose fluorinated groups demonstrate the non-stick function. Such coatings are particularly appreciated in industrial, marine and automotive applications where having a clean surface is a primary requirement, for instance to prevent soiling or fouling which may compromise the functionality of components.

Paints with active primers: cathodic protection mechanism. The most important cathodic pigment for steel protection is zinc. Relevant primers, called *zinc rich primers*, consist of zinc powder with particle size between 3 and 20 μm. To give cathodic protection effect, the zinc content should exceed 90% by weight to ensure an efficient electrical connection between zinc grains and steel (Fig. 17.9).

Attempts to use aluminium or magnesium dust instead of zinc failed because these metals do not give electrical continuity, being the oxide which covers each particle an insulator: manganese would show a similar behaviour as zinc, but for economic reasons it has found little application.

Two types of zinc rich primers are used:

- *Organic zinc-rich primers*: the binder is typically based on epoxy, urethane, styrene or chloro-caucciu polymers, as many other paints; the zinc powder content is as minimum 85% by weight in dry film. A thin paint layer covers each zinc particle for a twofold function: sticking particles to each other and to the

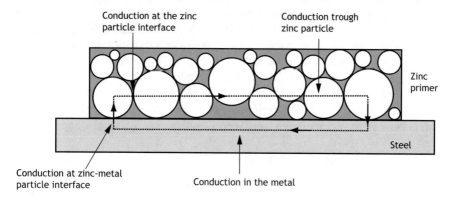

Fig. 17.9 Electric conduction in an organic zinc paint

steel surface, and ensuring electrical continuity, which requires it to be very thin (Fig. 17.9)

- *Inorganic zinc-rich primers*: zinc content is up to 95% by weight. The binder is an inorganic silicate that forms a hard conductive film, more adherent and porous than the polymeric binder. These inorganic zinc rich primers provide a higher protection than organic ones, but require a more stringent surface preparation.

These primers provide a protection degree comparable to galvanizing and, unlike other paints, they do not suffer from filiform or under paint corrosion starting from gaps or scratches. Protection conditions continue also after corrosion of zinc particles, although the electrochemical action fades, because zinc corrosion products, as hydroxides and carbonates, seal the pores giving an efficient barrier effect (see also Sect. 17.1).

Paints with active primers: inhibition mechanism. The aim of these pigments is the formation or the strengthening of a passivating protective film on the metal surface, according to the mechanisms of passivating inhibitors, through the establishment of alkaline environment or by passivating mechanism (see Sect. 17.2.3). Others, as silicates and phosphates, form a barrier film on the metal surface. Chromates, which are excellent passivating inhibitors, have been banned because carcinogenic; before that, also red lead was prohibited because of its toxicity, causing saturnism.

Self-healing coatings. Self-healing coatings can be included among coatings with an active function. Mechanisms of protection vary from the above described passivating effect, by inhibition or cathodic protection, to the actual sealing of defects and holes that may form in the coating by a shape recovery effect, or by the release of swelling compounds or of chemically reactive molecules that react with the matrix, sealing it. The active components are generally encapsulated, so that they exert the sealing or inhibiting action only in presence of a stimulus—be it mechanical, thermal, or a pH variation—in order to reduce risks of deactivation or leaching, which is why they are also referred to as smart coatings.

Although the matrix incorporating the active component is generally organic, mostly epoxy, in some cases oxide matrices (e.g., SiO_2, ZrO_2, TiO_2) are used, applied either by sol-gel or by chemical vapour deposition. As active principles, chromium compounds were initially used, being the formation of chromium oxide the mechanism of self-repair. Due to the restrictions on the use of Cr(VI), other systems were then investigated, and currently span from natural oils (for instance, linseed oil or tung oil) to synthetic organic or inorganic compounds (e.g., alkyl ammonium salts, alkoxysilanes, mercaptobenzothiazole, hydroquinone, MoO_4^{2-}, Ce^{3+}).

Few commercial products are currently available, with current applications in industrial and marine sectors, especially as above water heavy duty corrosion protection technique; field tests indicated a threefold up to six-fold coating lifetime extension.

17.4.4 Paint Film Properties

A painting can last for the entire design life exploiting either the barrier effect or the inhibition mechanism, provided that the film can show:

- A good chemical resistance to withstand environmental exposure conditions
- A good adherence to the metal surface, to avoid either spontaneous or mechanically caused macro defects as peeling off, to ensure total protection. In the case of poor adherence, under paint corrosion attack occurs starting from micro defects as pinholes
- A mechanical hardness to withstand hits, abrasion and erosion, and elasticity as well as flexibility to avoid cracking when stressed
- Suitable rheological properties, as per nature and composition of the film, in particular binder, solvent and curing components, which influence the application method (brush, spray or immersion) and the filming process. A paint is thixotropic when it shows the ability to change viscosity while stirred, being then suitable for brush or spray applications because once spread it becomes viscous, hence avoiding dripping.

The above requirements are often in conflict; for example, thermoplastic polymers should have high molecular weight to resist mechanical impact but the ease of application worsens.

17.4.5 Painting Cycles

One single paint coat can hardly show all required features, hence the so-called painting cycle, composed of multiple coats, is used. A painting cycle consists basically of:

- A primer or first coat in contact with the metal surface
- An intermediate coat
- A final or finishing coat, directly in contact with the environment.

The cycle is called homogeneous when there is only one binder, conversely it is called heterogeneous. The primer has a twofold aim: to stick to the metal surface and provide the active pigment. The finishing coat must resist the environmental aggressiveness. The intermediate coat has the purpose to thicken the paint and reduce permeability while using less expensive components than primer and finishing coats.

Figure 17.10 shows the effect of a wrong painting cycle where no primer was used.

Primer. The primer, as first coat, must ensure the maximum adherence to the metal surface. Adherence mechanism can be a simple mechanical anchoring enhanced by the roughness of the metal surface, generally obtained by sandblasting, and in

Fig. 17.10 Corrosion under painting of a gate in Venice due to the absence of the primer

addition chemical-physical interactions between the binder and the metal. Adhesion properties refer to the binder wetting capability, which varies greatly with type of binder and relevant modifications; for example, for chloride-acetate type vinyl resins, the insertion of a few hydroxyl or carboxylic groups greatly improves wetting, while for pure vinyl chloride acetate copolymers wetting is very low. Adhesion is modified by the presence of pigments, fillers, plasticizers and additives. It also depends on the nature of the metal: for example, it can be entirely different for carbon steel and galvanized steel, often worse for the latter. Adherence is strongly influenced by the cleanness of the metal surface to an extent that depends on the type of binder. For instance, an oleo-phenolic binder does not require a particularly accurate cleaning, unlike an epoxy based binder that requires a white sandblasted metal surface. For a primer, adhesion is more important than chemical resistance or appearance, which may be opaque; conversely, the feasibility of the over coating application is important. In this regard, the opaque appearance is generally associated with poor binder content in the paint that facilitates the adhesion of the subsequent coat, which is always problematic in the case of ther-mosetting binders. It is worth mentioning the so-called *shop-primers*, for example based on epoxy-phenols, used in factory for a temporary protection of maximum 1 year, before final assembling. Of course, shop-primers must be easily coated with the widest spectrum of paintings to avoid restriction in the selection of intermediate and finishing coats.

Finishing coat. Aesthetic aspects are of primary importance for the formulation of the finishing coat, hence chemical resistance comes first rather than adherence. Inert binders and chemically resistant pigments with a high volumetric concentration are privileged to obtain maximum waterproofness. This leads to a film with glossy or semi-glossy appearance. For example, in contact with aggressive chemical environments, vinyl epoxy, chloro-caucciu or polyurethane resins should be chosen.

Instead, alkyd resins, oleo-resinous binders, and generally all binders containing reactive ester groups or reactive pigments have to be avoided. Sometimes, finishing coats may require special features besides those of chemical resistance and, of course, in addition to aesthetic ones. For example, in moderate polluted atmospheres aluminium containing binders are used, which also give protection against solar and ultraviolet rays as well as effective protection against heat (for example, they are used for coastal tanks of petroleum products). For the same reasons, a urethane finishing is used on epoxy cycles. For marine structures such as ship hulls, the finishing coat must also have anti-fouling properties to prevent or limit marine fouling growth. Fillers used for this aim are: copper oxide or organometallic compounds containing arsenic, lead or zinc.

Intermediate coat. The aim of the intermediate coat is to give thickness and to increase waterproofness of the cycle. Provided a good compatibility and proper adhesion with primer and finishing coats, cost-effective materials are generally used.

17.4.6 Pre-treatments

Before painting or performing any other surface treatment as metal deposition, the surface must be accurately cleaned by removing oxides, corrosion products, fat, oil and dirt of various origin and nature, and subjected to a further roughening to increase adhesion.

Such treatments vary with type of metal as well as nature of paints. For example, an alkyd linseed oil paint requires much less accurate treatment than epoxy or urethane based ones; an inorganic zinc-rich primer requires higher surface preparation than an organic zinc-rich primer. Similarly, surface treatment in galvanic deposition is different whether the solution is acidic or alkaline: the latter tolerates higher amount of adsorbed organic substances than acidic ones.

The choice of a suitable pre-treatment depends on a variety of parameters. In general, reference is made to international agreed standards promoted by recognized organizations, such as in USA the SSPC (Society for Protective Coatings, formerly the Steel Structures Painting Council) with its "Surface Preparation Specifications"; in Europe, CEN-ISO and the Swedish Standard Association, which reports widely used photographic standards as a practical guide.

Degreasing with solvents and detergents. The aim of this practice is cleaning from fats, oils, dirt, especially animal fat, by means of solvents, emulsions or detergents. To remove dust particles that remain stuck on the metal surface even after degreasing a mechanical action is necessary, as produced by impingement made by liquids or gas bubbles in electrochemical treatments or by ultrasounds. Degreasing agents are classified as: organic solvents, emulsions, biphasic detergents and alkaline detergents.

Pickling. It consists of dipping the metal in strong inorganic acids to remove oxides that form on the surface during manufacturing at high temperature or by a corrosion attack. This surface treatment is quite ideal for the application of coatings. Common acids used for treating steels are: sulphuric, hydrochloric, phosphoric, nitric, hydrofluoric or a mixture of them at a temperature higher than the ambient one. The most suitable acid as well as the concentration recommended for the treatment depend on steel composition and nature of oxide. It is good practice to add proper corrosion inhibitors to reduce metal wastage during the treatment as well as hydrogen evolution, then reducing risks of hydrogen embrittlement and entrainment of acid in vapours (so-called acidic fumes). Pickling is followed by rinsing in neutralizing solutions and drying. For hot laminated carbon and low alloy steels sulphuric acid, as solutions of 1–50%, is used; also the more reactive hydrochloric acid is used—at room temperature, for being volatile. For titanium and stainless steels, a mixture of nitric and hydrofluoric acids is used. It is worth remembering that pickling requires treatment of exhaust solutions before sewage disposal and precautions for health.

Pickling of hot rolled steels. After hot rolling, steels are covered by an oxide scale, the rolled scale, also called calamine. If rolling has been carried out at a temperature above 575 °C, the scale consists of three layers:

- Internal layer in contact with the base metal, made of wüstite, FeO
- Intermediate layer consisting of magnetite, Fe_3O_4
- External layer, of hematite, Fe_2O_3.

The ease of pickling relies primarily on the inner layer composition and on the porosity of the outer layers. Both composition and porosity depend essentially on the transformations of oxides occurring during cooling below 575 °C: a rapid cooling does not affect the wüstite layer and pickling becomes more difficult; similarly, pickling is also difficult when the scale forms at temperatures below 575 ° C because no wüstite forms.

Conversely, a slow cooling from 575 °C partially decomposes wüstite into magnetite and metallic iron dust, so that pickling is favoured because acid penetrates through the pores and easily detaches wüstite, also thanks to the development of hydrogen produced by the corrosion of iron dust (Fig. 17.11).

Fig. 17.11 Illustration of pickling action on carbon steel sheets, laminated at high temperature (>575 °C), and then cooled slowly

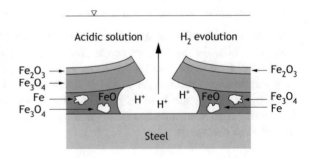

Sandblasting. Sandblasting is the best technique for preparing steel surfaces. It consists of spraying at a high-speed abrasive powder onto the metal surface, then removing hot rolled scale and rust until obtaining a so-called *white metal surface*. Silica sand, whether natural or obtained by grinding, is now generally discarded as grinding medium as it causes silicosis, and replaced with other abrasives as corundum, cast iron and steel granules, whose use is restricted to continuous cycle production plants, being very expensive. It is still preferred if old paints of very high hardness have to be removed from the surface and if steel, just released from production, is completely or nearly coated with very adherent rolling scales. Sandblasting can be performed dry or wet; in the latter, abrasive particles are dispersed in a water stream containing a corrosion inhibitor. Dry sandblasting is normally preferred due to the shorter working time and because the steel surface remains dry, then allowing a long storage time before further painting.

The degree of blasting (not to be confused with the sandblasting profile that indicates surface roughness) defines the percentage of removed scales, rust and old paint. In current practice, the following sandblasting grades are defined, according to ISO 8504:

- White metal (grade SA 3)
- Almost white metal (grade SA 2½)
- Commercial (grade SA 1)
- Brushing or coarse sandblasting.

17.4.7 Paint Application

The good performance of a painting depends on:

- Appropriate painting cycle to withstand the aggressive environment
- Proper surface preparation
- Correct application.

To select proper painting cycles, including surface preparation, reference is made to specialized literature and standards. Excluding the description of application systems (brushing, rolling, spraying, with or without air, electrostatic, by immersion, by electrophoresis) for which reference is made to specialized handbooks, in the following, only the atmospheric and environmental conditions for the correct application are considered.

Abnormal conditions of temperature and relative humidity or the presence of wind must be avoided. As far as temperature is concerned, too low temperatures (less than 7–10 °C) or an excessive cooling due to rapid solvent evaporation would jeopardize the final result because curing stops or is altered, and once temperature drops below the dew point, water condensation forms.

Also high temperatures (above 35 °C) of both ambient and metal surface are harmful. In this case solvent evaporation is too fast, hence increasing porosity and contrasting paint relaxation. Moreover, the higher the temperature, the lower the pot-life time, or the maximum the time from the mixing of components. Painting work should not be carried out when relative humidity exceeds 85%, and on rainy days. This depends on the fact that non-polymerized binders tend to absorb water physically or chemically, resulting in decreased adhesion and chemical resistance.

About the application time schedule, it is essential that the primer is applied immediately after surface preparation, in particular after sandblasting, in order to prevent the oxidation of the metallic surface that is very reactive after sandblasting, which would compromise adhesion and the effectiveness of the paint. Also the elapsed time between the applications of successive coats is limited. It is a stringent requirement for two-component paints to avoid lack of adhesion, and less rigid for solvent-based paints, although also in this case it should not be too long to avoid total dryness and surface contamination by dust.

17.4.8   Painting Maintenance

To ensure the design life of a paint, preventative maintenance work, consisting of periodical retouching or partial overlays, at a frequency determined on a cost effective and exposure conditions basis, as well as programmed inspections are necessary and usually planned. These interventions, which represent the cheapest way to extend the painting life, require special attention for ensuring good adhesion between old and new coats. For paints of thermoplastic binder type, since they are soluble in solvents, maintenance work is possible at any time; conversely, with thermosetting paints adhesion between old and new coat is difficult and always poor, especially when crosslinking is tight, since any solvent is ineffective. In these cases, it is necessary to completely remove old paint and corrosion products before applying the new paint.

17.4.9   Threats at the Workplace

As final paragraph of this chapter, a mention of health threats for personnel involved in all painting-related activities, from production plants to field applications, is mandatory. In short, risks deal with:

- Surface preparation by sandblasting because of the presence of dust (risk of silicosis with siliceous sand), noise and vibrations
- Degreasing operation with risk of inhalation of solvent vapours (in particular for chlorine containing and aromatic compounds)
- Chemical and electrochemical treatments, risk associated to handling of acids, bases or poisoning solutions (for example cyanide) and exposure to poisoning metal ions (lead, zinc and chromate)

- Metal hot dipping, risk of inhalation of metallic vapours
- Painting application for risk of inhalation or contact with solvents, pigments, dusty charges, toxic curing agents.

17.5 Applicable Standards

- ASTM D62, Standard Test Methods for Holiday Detection in Pipeline Coatings —West Conshohocken, Pa.: American Society for testing of Materials
- ASTM D714, Standard Test Method for Evaluating Degree of Blistering of Paints, West Conshohocken, Pa.: American Society for testing of Materials
- ASTM D1653, Standard Test Methods for Water Vapor Transmission of Organic Coating Films—West Conshohocken, Pa.: American Society for testing of Materials
- ASTM D3359, Standard Test Methods for Rating Adhesion by Tape Test— West Conshohocken, Pa.: American Society for testing of Materials
- ASTM D4541, Standard Test Method for Pull-Off Strength of Coatings Using Portable Adhesion Testers—West Conshohocken, Pa.: American Society for testing of Materials
- ASTM D5162, Standard Test Method for Discontinuity (Holiday) Testing of Nonconductive Protective Coating on Metallic Substrates—West Conshohocken, Pa.: American Society for testing of Materials
- ASTM F941, Standard Practise for Inspection of Marine Surface Preparation and Coating Application—West Conshohocken, Pa.: American Society for testing of Materials
- ISO 8504, Preparation of steel substrates before application of paints and related products. Surface preparation methods, International Organization for Standardization, CH-1211 Geneva 20
- ISO 10309, Metallic coatings—Porosity tests—Ferroxyl test, International Organization for Standardization, CH-1211 Geneva 20
- ISO 21809, Petroleum and natural gas industries—External coatings for buried or submerged pipelines used in pipeline transportation systems, International Organization for Standardization, CH-1211 Geneva 20

17.6 Questions and Exercises

Metallic coatings

17.1 Grounding systems are typically made of: bare copper, galvanized steel and copper-coated steel. Illustrate (suggest) the process for obtaining galvanized and copper-coated earthing rods. Explain the advantages or drawbacks of these goods compared to the solid copper rods.

17.2 Explain the advantages of chromating the galvanized steel as finishing or surface preparation for further painting.

17.3 Al-Zn coatings obtained by hot spraying are used, also on submerged surfaces, as bare. Discuss why and give a reason for being a bad active primer.

Conversion coatings

17.4 Explain why anodizing of titanium cannot be adopted as surface preparation for painting.

17.5 Phosphatizing is sometime used as surface preparation for painting. Explain why it is used for indoor application and not recommended for marine applications.

17.6 Anodized aluminium if often painted. Discuss about the surface preparation needed.

Painting

17.7 Calculate the water and oxygen diffusion rate through a paint film if corrosion rate is the same as in atmospherically exposed bare metals. These values represent the threshold for which the barrier effect becomes beneficial. Assuming that the minimum paint thickness for reaching the threshold value calculated above is 100 μm, and the dependence with the paint thickness is parabolic, suggest a thickness for a solvent based paint for an effective barrier effect.

17.8 Based on the results of exercise 17.1, predict the minimum thickness of an epoxy-based paint if water and oxygen diffusion rates for a paint thickness of 100 μm is 10 and 30 mg/m^2 d, respectively.

17.9 Explain which is the mechanism by which red lead oxide and zinc chromate would work. Try to compare the two mechanisms by a representation on the Evans diagram. Estimate the minimum primer thickness required for a duration of 10 y, assuming that VCP is unitary.

. Explain the mechanism by which zinc-rich primers work. Show the difference with a primer based on phosphatizing treatment.

17.11 Try to estimate the consumption rate of a primer based on an inorganic zinc-rich and a hydroxyl zinc chromate.

17.12 Give all justifications you find reasonable on why the adopted painting cycle world-wide used for offshore structures, either atmospherically or submerged, typically consists on: sandblasting, inorganic zinc-rich primer, epoxy based for the intermediate coat and polyurethane for the final one.

17.13 An acid pickling treatment is carried out at about 60 °C in 20% sulphuric acid solution. Suggest the maximum dipping time if no corrosion inhibitor is added.

17.14 With reference to the previous exercise, suggest an inhibitor for the acid pickling treatment and the associated dipping time.

17.15   The Golden Gate Bridge is continuously maintained, with a period of about two years (that is the time required for the team involved to run from one side to the opposite). Suggest a painting cycle and the nature of the most suitable binder taking into account the environment.

17.16   Explain the difference between a painting cycle for a galvanized steel and the one for anodized aluminium. Explain the influence of the total coating thickness.

Bibliography

Munger CG (2014) Corrosion prevention by protective coatings, 3rd edn. NACE International, Huston

Parkinson R (2001) Properties and applications of electroless nickel. Nickel Development Institute

Roberge P (2008) Protective coatings, corrosion engineering—principles and practice. McGraw Hill, New York, pp 587–662

Stankiewicz A, Barker MB (2016) Development of self-healing coatings for corrosion protection on metallic structures. Smart Mater Struct 25:084013

Shreir LL, Jarman RA, Burstein GT (1994) Corrosion. Butterworth-Heinemann, London

Chapter 18
Environmental Control

Boil the amurca down to one-half of its volume and use
it as a polish for any kind of copper vessel, [...],
Each object will have a lustre and will be protected from rust.
Catone, De agri cult., 98, 108, 99

Abstract An effective and widely used corrosion prevention method is obtained through a modification, often called conditioning, of the aggressive environment. The rationale of the approach is to influence cathodic or anodic processes or both, in order to decrease the corrosion rate below an acceptable limit to meet the design requirement about design life and safety. The main actions, illustrated in the following, are based on the modification of pH, the reduction of oxygen content to slow down the cathodic process, the use of biocides to control microbiological activity, or on the injection of specific chemicals, as scaling agents and corrosion inhibitors, to slow down both anodic and cathodic processes. In particular, for corrosion inhibitors, the mechanism (anodic, cathodic, mixed) is described and the effectiveness is discussed.

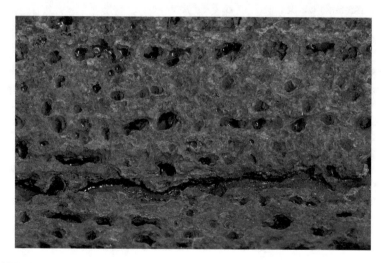

Fig. 18.1 Case study at the PoliLaPP Corrosion Museum of Politecnico di Milano

© Springer Nature Switzerland AG 2018
P. Pedeferri, *Corrosion Science and Engineering*, Engineering Materials,
https://doi.org/10.1007/978-3-319-97625-9_18

18.1   pH Control

To inhibit the hydrogen evolution as main cathodic process in slightly acidic solutions, as industrial waters, the pH is increased. This pH modification is also beneficial on the anodic process of many metals as a high pH favours passivation. More common pH corrections are:

- Alkalization of boiler water to annihilate the hydrogen evolution and induce the formation of a magnetite film on carbon steel
- pH variations to promote scaling in waters
- Addition of alkaline substances to avoid acidic condensates.

The latter treatment refers to the conditioning of steam for the presence of acidic substances such as CO_2 and HCl, which are harmless in dry steam while give very aggressive conditions in condensates. Addition of neutralizing agents, such as ammonia or amines by direct injection in the steam, is adopted for example, in distillation columns of crude oil.

In waters, injection of NaOH, ammonia based salts or amines is adopted.

18.2   Oxygen Control

The presence of oxygen is the main cause of corrosion in waters and natural environments (Fig. 18.1). In particular, the corrosion rate is proportional to the amount of oxygen dissolved (see Chap. 8); accordingly, an easy and effective corrosion control is the oxygen removal regardless the need to establish passive conditions.

Two methods are adopted for oxygen removal:

- Physical
- Chemical.

The principle of the first method is based on the temperature and pressure dependence of oxygen solubility: as the temperature increases or pressure decreases, oxygen solubility zeros. Therefore, oxygen is almost totally removed by heating up the water close to boiling temperature or in vacuum columns, or even by gas stripping. Typically, a physical treatment reduces the oxygen content to less than 1 ppm.

A residual oxygen content as low as 0.015 ppm, as required for feeding water of low or medium pressure boilers, can be achieved in industrial plants only combining the physical method with a chemical process.

Chemicals most frequently used for oxygen removal are *sodium bi-sulphite* and *amines* (hydrazine, hydroxylamine) according to the following reactions:

$$2Na_2HSO_3 + O_2 \rightarrow 2Na_2SO_4 + 2H^+ \tag{18.1}$$

$$N_2H_4 + O_2 \rightarrow 2H_2O + N_2 \tag{18.2}$$

Sodium bi-sulphate is effective up to 60 °C, while hydrazine only reacts sensibly with oxygen at temperatures above 140 °C. Compared to sodium sulphate, hydrazine has the advantage of not increasing either the acidity or the salinity.

In the case of Na_2HSO_3, attention should be paid to the scavenger dosage; as a rule of thumb, for each ppm of oxygen about 8 ppm of sodium bi-sulphite should be added. A higher dosage will reduce the pH of the solution, making the electrolyte more acidic.

On the other hand, hydrazine can partially decompose to give ammonia, which increases alkalinity, then facilitating the formation of the protective magnetite film:

$$N_2H_4 \rightarrow 4NH_3 + N_2 \tag{18.3}$$

The mechanism involving hydrazine as passivating agent is effective if its concentration exceed 100 ppm. Due to toxicity problems, hydrazine is partially replaced by less toxic compounds such as hydroxylamine.

18.3 Corrosion Inhibitors

Corrosion inhibitors are substances added in small quantities to the aggressive environments in order to slow down and practically zero the corrosion rate without changing the environment composition. There is a variety of substances having these characteristics.

A corrosion inhibitor can work in various ways: through the modification of the metal surface as a result of a physical adsorption or a chemical interaction, by increasing the overvoltage of cathodic or anodic reactions, by preventing oxygen from reaching on the metal surface, and by favouring passivation conditions.

An inhibitor must be compatible with the environment where it is injected. Characteristics such as pH, the nature of the metal, dissolved ions, temperature, the presence of microorganism, aeration and flowing conditions can affect the performance of a candidate corrosion inhibitor. In particular, pH plays an important role because it influences the anodic and cathodic reaction. If pH enables passivation, the inhibitor helps strengthen the oxide existing film. Conversely, when an oxide film cannot form, the inhibitor builds a protective film by adsorption at the surface. Inhibitors for acidic environments are generally not effective in neutral or alkaline solutions, and vice versa. The electrical charge, present on the metal surface and the nature of the inhibitor are also important factors as they control the adsorption of the inhibitor onto the metal.

The use of inhibitors, although this constitutes an excellent and tested method to prevent corrosion, can exhibit considerable side effects such as fluid contamination, environmental pollution and damage to subsidiary parts. For example, some inhibitors, injected to protect a part of a plant can be aggressive to others.

Table 18.1 presents some inhibitors that have been used with success in typical environments, and Table 18.2 summarizes most common inhibitors used for the corrosion protection of some metals.

Table 18.1 Typical inhibitors used in industry

Environment		Inhibitor	Metal	Dosage
Waters	Drinking water	Polyphosphate	Fe, Zn, Cu, Al	5–10 ppm
		Silicate	Fe	10–20 ppm
		$Ca(OH)_2$	Fe, Zn, Cu	10 ppm
	Cooling System	$Ca(HCO_3)_2$	Fe	10 ppm
		Chromates	Fe, Zn, Cu	0.1%
		$NaNO_2$	Fe	300–500 ppm
		Polyphosphates	Fe	10–20 ppm
		Silicates	Fe	20–40 ppm
	Boiler	NaH_2PO_4	Fe, Zn, Cu	10 ppm
		Polyphosphates	Fe	10 ppm
	Engine cooling circuit	Na_2CrO_4	Fe, Pb, Cu, Zn	0.1–1.0%
		Nitrites	Fe	300–500 ppm
	Brine and seawater	Na_2SiO_3	Fe	0.01%
			Zn	10 ppm
		Amines quaternaries	Fe	10–25 ppm
		$NaNO_2$	Fe	0.5–3%
		$NaNO_2 + NaH_2PO_4$	Fe	0.5% + 10 ppm
Pickling solution	Sulphuric acid	Phenylthiourea Orthotoluenthiourea Mercaptans Sulphides	Fe	0.003–0.01%
	Hydrochloric acid	Organic compounds with N and N rings	Fe	100–200 ppm
		Hexamethylenetetramine	Fe	5%
	Nitric acid	Amine, thiourea	Fe	
		Aromatic amines	Cu	
Petroleum industry	Extraction	Various amines	Fe	100–500 ppm
	Refining	Imidazoline and derivatives	Fe	100–1000 ppm
Concrete	Marine De-icing salts	$CaNO_2$	Fe	10 kg/L
		Amines Alkanol-amines Carboxylate salts	Fe	1% versus cement weight

Table 18.2 Typical corrosion inhibitors for metals in different environments

Metal	Environment	Inhibitor
Carbon steel	Citric acid	Cadmium salts
	H_2SO_4 diluted	Aromatic amines
	H_3PO_4 concentrated	Dodecyl-amine 0.01–0.5%
	Brine containing oxygen	Methyl, ethyl or propyl-dithiocarbamates 0.001–3%
	Mixtures glycol ethylene-water	Tri-sodium phosphate 0.025% Phosphates or alkaline borates
	NaCl 0.5%—neutral pH	Sodium nitrite 0.2%
	Brine containing sulphides	Formaldehyde
	Water	Benzoic acid, polyphosphate, silicate
	Hydrocarbons and water	Sodium nitrite
	Concrete	Calcium nitrite Amines or carboxylates
Aluminium	HCl 1N	Phenilacridina, Naphthoquinone, Acridine, Thiourea
	HNO_3 10%	Hexamethylenetetramine 0.1%
	H_3PO_4 20%	Sodium chromate 0.5%
	H_2SO_4 concentrated	Sodium chromate 5%
	Chlorinated water	Sodium silicate
	Seawater	Amyl stearate 0.3%
	Na_2CO_3 1%	Sodium silicate 0.2%
	Na_2S	Sodium metasilicate 1%
	Mixtures glycol ethylene-water	Nitrite or molybdate of sodium
	Chlorinated aromatic	Nitroclorobenzene 0.1–2%
	Commercial ethanol	Carbonates, Lactates, Acetates or alkaline borates
Cadmium (on steel)	Mixtures glycol ethylene-water	Fluophosphate sodium 1%
Magnesium	Alcohols	Alkali sulphides
	Tri-chloro-ethylene	Formamide 0.05%
	Water	Potassium dichromate 1%
Lead	Neutral solutions	Sodium benzoate
Copper and brass	H_2SO_4 diluted	Benzilthiocianate
	Mixtures glycol ethylene-water	Borates or alkaline phosphates Mercaptobenzothiazole Benzotriazole
	Neutral solutions	Mercaptobenzothiazole Benzotriazole 0.2–0.3%

(continued)

Table 18.2 (continued)

Metal	Environment	Inhibitor
Tin (steel)	Alkaline soaps	Sodium nitrite 0.1%
	NaCl 0.05%	Sodium nitrite 0.2%
Titanium	HCl	Oxidizing Agents
	H_2SO_4	Oxidizing Agents
Zinc (galvanized steel)	Water	Calcium and zinc metaphosphates 15 ppm

The main use of inhibitors deals with:

- Neutral or lightly alkaline waters and solutions with a pH between 5 and 9
- Acidic environments such as in pickling processes, chemical cleaning and descaling
- Hydrocarbon production and refining
- Concrete that is expected to be contaminated by chlorides.

18.3.1   Classification of Inhibitors

The usual inhibitors are classified based on:

- The chemical nature: organic or inorganic inhibitors
- The electrochemical mechanism: anodic, cathodic or mixed inhibitors
- The application: inhibitors for feeding water of boiler, for pickling, for descaling, for packaging, for concrete, for hydrocarbons, for chloride containing environments and others
- The effectiveness, which can always be beneficial (*safe* inhibitors), or detrimental where the corrosion rate sometimes increases (*unsafe* or *hazardous* inhibitors) because of an insufficient dosage.

Hereinafter, inhibitors will be classified according to their electrochemical mechanism as summarized in Fig. 18.2 by distinguishing among *cathodic*, *anodic* (oxidizing or non-oxidizing) and *mixed* inhibitors:

- Cathodic inhibitors increase cathodic overvoltage and thus lowering the corrosion potential (Fig. 18.2a)
- Anodic inhibitors, instead, increase anodic overvoltage and therefore an ennoblement of corrosion potential is expected (Fig. 18.2b)
- Mixed inhibitors, which act on both anodic and cathodic processes, can give rise to a slight change of potential (Fig. 18.2c).

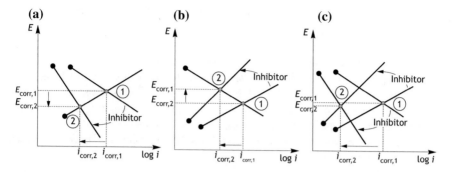

Fig. 18.2 Electrochemical mechanism of inhibition: **a** Cathodic; **b** anodic; **c** mixed. Point 1: Operating point without inhibitor. Point 2: Operating point in the presence of an inhibitor

In all cases, inhibitors cause a reduction of the corrosion rate from $i_{corr,1}$ to $i_{corr,2}$ (Fig. 18.2).

18.3.2 Cathodic Inhibitors

Acidic solutions. When cathodic reaction is hydrogen evolution, inhibitors are those substances that increase the cathodic overvoltage. Typical substances are some inorganic compounds, as salts of arsenic, antimony, bismuth, sulphuric or hydrogen halide ions, and many organic substances. The latter sometimes can also act also on the anodic process. When cathodic inhibitors stop the recombination of atomic hydrogen, the risk of hydrogen embrittlement increases on high-strength steels (see Chap. 14).

   Neutral or slightly alkaline solutions. Effective inhibitors for these environments include deoxidizing substances, pH correctors and substances that give rise to the formation of hydroxides or scaling on a metal surface, such as for instance salts of zinc, magnesium, manganese, nickel and calcium that give hydroxides or carbonates, which start forming on cathodic surfaces due to the locally pH increase. Since anodic and cathodic surfaces overlap in uniform corrosion, scales formed on the cathodic side interact with anodic reaction products to form a compact mixture of iron and calcium carbonates.

18.3.3 Anodic Inhibitors

These inhibitors increase the anodic overvoltage by promoting the formation of protective or passivating films. They can be oxidant (or passivating) or non-oxidant.

Non-oxidizing anodic inhibitors. These inhibitors modify the anodic overvoltage according to the electrochemical behaviour of the metal. On active metal, the inhibitor promotes the passivity. On a passive metal, the inhibitor strengthens the passivity and reduces the primary passivation current density. Typical non-oxidant anodic inhibitors are hydroxides, borates, phosphates, polyphosphates, silicates and benzoates. These inhibitors only work in the presence of oxidizing species as shown in Fig. 18.3, since the operating point does not move from point 1. A more noble cathodic process is mandatory to reach a passivity condition (point 2).

Oxidizing (or passivating) anodic inhibitors. These inhibitors have a twofold action as illustrated in Fig. 18.4:

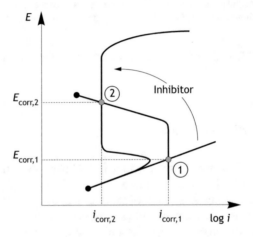

Fig. 18.3 Mechanism of action of an anodic inhibitor non-oxidizing

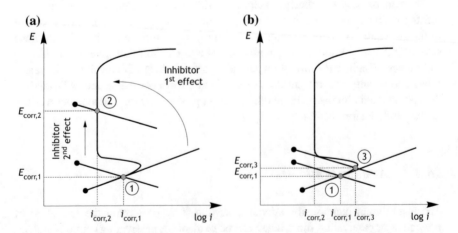

Fig. 18.4 a Electrochemical effects of a passivating anodic inhibitor; **b** electrochemical effects of a unsafe inhibitor in proper concentration (point 2) and inadequate dosage (point 3)

- Modify the anodic curve
- Provide a more noble cathodic process in order to bring the metal to passive condition.

Chromates, nitrites, permanganate, molybdates, tungstates and, obviously, oxygen, are inhibitors of this type. Because of their toxicity, some of them are banned as chromates; in many cases, it may be useful to employ mixtures of passivating inhibitors to also reduce the critical passivation current.

Oxidizing anodic inhibitors work by setting up passivity conditions, then reaching point 2 of Fig. 18.4b, provided a threshold is exceeded, because, conversely, corrosion may accelerate by reaching point 3. For this reason, these inhibitors are called *unsafe*. Instead, inhibitors are *safe* when even a low dosage does not accelerate the corrosion rate.

Therefore, the use of these inhibitors may be dangerous when the environment composition changes due to a local increase of de-passivating agents such as chlorides, or when passivity conditions are not reached due to locally insufficient inhibitor concentration. These circumstances typically happen in dead spaces and low flow conditions where the inhibitor is not renewed.

Once passivity conditions are reached, the consumption rate of the inhibitor is of the order of the passivity current density, thus depending on temperature, pH and chloride content, which are the same factors affecting passive conditions. The consumption of an effective inhibitor is very low; conversely, consumption is high when an inhibitor works badly.

18.3.4 Mixed Inhibitors

Adsorption inhibitors, for vapour phase or packaging, work by a mixed mechanism. Figure 18.5 shows the chemical structures of the most used organic substances. **Absorption inhibitors**. Normally used in acidic environments, for example in pickling, they consist of:

- Substances that contain elements of the 5th and 6th group of the periodic table, as N, P, As, S and O, with pairs of free electrons
- Organic compounds with double or triple bonds
- High molecular weight compounds such as proteins and polysaccharides, which should not be considered strictly mixed inhibitors.

Due to their functional groups, these inhibitors are absorbed to the metal surface where they form a protective monomolecular layer.

Volatile corrosion inhibitors. Volatile corrosion inhibitor (VCI), also known as vapour phase inhibitors, are organic substances characterized by high vapour pressure, typically between 10^{-2} and 10^{-7} mmHg, so they easily vaporize and condense on a metal surface in a close environment. They are employed to prevent

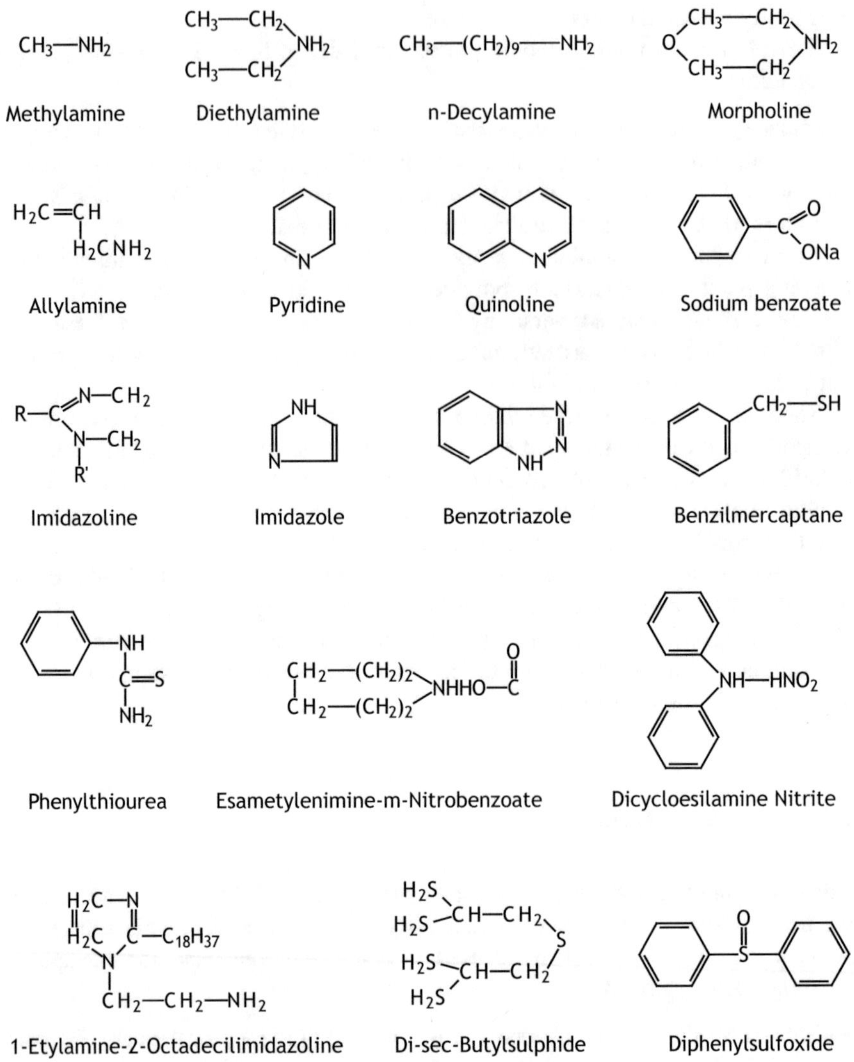

CH₃—NH₂
Methylamine

Diethylamine

CH₃—(CH₂)₉—NH₂
n-Decylamine

Morpholine

Allylamine

Pyridine

Quinoline

Sodium benzoate

Imidazoline

Imidazole

Benzotriazole

Benzilmercaptane

Phenylthiourea

Esametylenimine-m-Nitrobenzoate

Dicycloesilamine Nitrite

1-Etylamine-2-Octadecilimidazoline

Di-sec-Butylsulphide

Diphenylsulfoxide

Fig. 18.5 Organic substances frequently used as corrosion inhibitors

atmospheric corrosion in indoor spaces, for example within plastic containers, during storage or shipment, as well as for packaging equipment. VCIs are impregnated in polymeric films, in cardboard sheets or in papers, then wrapped around the metallic component to be protected. They are also inserted in polymeric boxes and sponges, placed in the environment where the metals are stored.

VCIs offer an important advantage over conventional inhibitors since they can spontaneously penetrate into inaccessible crevices, gaps and slots, reach surfaces of complex-shaped articles that are difficult to coat, and get adsorbed on the surface from the vapor phase.

It has been found that the most effective VCIs are made up of a weak volatile base and a weak volatile acid. Amine nitrite and carboxylate are susceptible to hydrolysis and provide inhibition as follow:

$$R_2NH_2NO_2 \rightarrow R_2NH_2^+ + NO_2^- \tag{18.4}$$

18.3.5   Inhibitor Adsorption Mechanism

Organic corrosion inhibitors work by adsorbing on the metal surface. This process is governed by two factors: the residual electrical charge at the metal surface and the chemical structure of the inhibitor. The former plays an important role: it produces an electrical field that modifies the orientation of the solvent molecules, and attracts (or repels) the ions present in the solution. The interactions are of physical nature, hence calling this type of adsorption *physisorption*. The presence of certain functional groups in the structure of a substance indicates whether a substance is more favoured to adsorb onto a substrate: the presence of polar groups such as $-NH_2$, $-OH$ or $-COOH$ contributes to the formation of bonds with the surface, involving interactions of chemical nature. This type of adsorption is denominated *chemisorption*.

Physisorption. Physical adsorption is due to the electrostatic attraction between the inhibitor and the electrically charged surface of the metal. The surface charge, ϕ, is defined by the difference between the free corrosion potential of the metal, E_{corr}, and its zero charge potential (ZCP). At ZCP the net charge on the electrode is zero:

$$\phi = E_{corr} - ZCP \tag{18.5}$$

This potential is fundamental for adsorption because it controls the kind of species adsorbed onto the surface: as this potential becomes positive, the adsorption of anions is favoured and as it becomes more negative, the adsorption of cations is favoured. It also controls the electrostatic interaction of the metal with dipoles in adsorbed neutral molecules, and hence the orientation of the dipoles and the adsorbed molecules.

As an expression of the charge present on the surface, the ϕ-potential is dependent on the nature of the metal, on the cathodic process and on the chemical nature of the adsorbed substance.

Chemisorption. It is probably the most important type of interaction between the metal surfaces and an inhibitor molecule. The adsorbed species is in contact with the metal surface, and a bond is formed between them, almost as strong as in stoichiometric compounds. Electron transfer between the metal and the inhibitor forms this bond through a coordinate link. In the metal, this process is favoured by the presence of vacant electron orbitals of low energy, as observed in transition metals. Conversely, in the adsorbed species the electron transfer is favoured by the presence of relatively loosely bound electrons, such as may be found in anions

(–COO⁻, –O⁻), and neutral organic molecules containing lone pair electrons on the donor atom (–NH$_2$, –OH) or π-electron systems with multiple bonds or aromatic rings.

These groups, either reactive or active, are the sites used by the organic molecules to adsorb onto the metal. The strength of the coordinate bond between the metal and the reactive groups depends upon the electron density on the donor atom of the functional group and the polarizability of the group. Therefore, increasing the polarizability by adding electron-donor groups to the inhibitor molecule provokes an increase in the strength of the coordinating bond, which raises the corrosion inhibitorefficiency. However, not only the ability to affect the electron density is important in the adsorption of an organic inhibitor: the molecular area, molecular weight and molecular configuration also have a great effect on its corrosion inhibition efficiency. For example, an increase in the hydrocarbon chain may increase the inhibition efficiency due to steric effects and the hydrophobic nature of the chain. Under other circumstances, an increment in the hydrocarbon chain can produce opposite effects.

18.3.6   Adsorption Isotherm

An adsorption isotherm describes the relationship between the coverage of an interface with an adsorbed species (the amount adsorbed) and the pressure of gas or concentration of the species in solution; in electrochemical reactions, the coverage will depend also on the potential difference at the interface. The isotherm is typically used to verify the performance of organic adsorbent-type inhibitor.

Usually three types of isotherms are considered to describe this relationship: Langmuir, Freundlich and Temkin isotherms.

In order to apply the *Langmuir isotherm* to an adsorption process, the following assumptions must be taken into account: there is no lateral interaction between the adsorbed species; the surface is smooth and defect free; and eventually the surface can be saturated by the adsorbate. It is described by the following equation:

$$K_{ads} \cdot C = \frac{\theta}{1 - \theta} \tag{18.6}$$

where K_{ads} is the equilibrium constant for the adsorption process, C is the bulk concentration of the adsorbate, and θ is the fractional surface coverage.

The *Freundlich isotherm* is the most common isotherm equation due to its simplicity and its ability to fit a variety of adsorption data. It is based on four assumptions: all of the adsorption sites are equivalent and each site can only accommodate one molecule; the surface is energetically homogeneous and adsorbed molecules do not interact; there are no phase transitions; at the maximum adsorption, only a monolayer is formed. Adsorption only occurs on localized sites on the surface, not with other adsorbates. These assumptions are seldom all met:

there are always imperfections on the surface, adsorbed molecules are not necessarily inert, and the mechanism is clearly not the same for the very first molecules to adsorb to a surface as for the last.

The Freundlich isotherm can be expressed as:

$$K_{ads} \cdot C = \left(\frac{\theta}{1 - \theta}\right) \cdot e^{f \cdot \theta} \qquad (18.6)$$

where f is molecular interaction constant. The Langmuir isotherm can be seen as a special case of the Freundlich isotherm, when the interaction constant, f, is equal to zero, i.e. there are no interactions between the adsorbate molecules. If f is negative, lateral interactions must be of attractive character, while positive values indicate repulsive lateral interactions.

The *Temkin isotherm* makes the same assumptions as the Freundlich isotherm for the presence of heterogeneities on the surface and interaction between the adsorbed molecules. It is defined by the following equation:

$$K_{ads} \cdot C = e^{f \cdot \theta} \qquad (18.7)$$

where f is the molecular interaction constant. This isotherm can be considered a special case of the Freundlich isotherm as well.

Once adopted the proper isotherm, the equilibrium constant for the adsorption process, K_{ads}, can be used to define the standard free energy of the adsorption as follow:

$$\Delta G_{ads} = -RT \cdot \ln(55.5 \cdot K_{ads}) \qquad (18.8)$$

where R is the ideal gas constant, T is temperature (in K) and 55.5 is the molar concentration of water in solution. Chemisorption generally requires a standard free energy in the order of -100 kJ/mol, while lower values are typical of a physical adsorption.

18.3.7 Inhibitor Effectiveness

Inhibitor effectiveness depends on a variety of factors such as nature, concentration and chemistry of the inhibitor as well as nature and surface condition of the metal, and composition and operating conditions of the aggressive environment, as temperature and velocity; further factors as those affecting the inhibitor availability to the metal surface (agitation, structure geometry and the presence of cracks).

The effectiveness of a corrosion inhibitor, ξ_{inib}, is the reduction of the corrosion rate as follows:

$$\xi_{inib} = 100 \cdot \frac{C_{rate}^0 - C_{rate-inhib}}{C_{rate}^0} \qquad (18.5)$$

where C_{rate}^0 and $C_{rate-inhib}$ are corrosion rates in the absence and in the presence of an inhibitor, respectively.

Concentration. Corrosion inhibitors work properly if the dosage exceeds a minimum, typically lower for oxidant inhibitors and higher for the others, which often also act as pH correctors. The minimum dosage depends on surface conditions (smooth and clean surfaces are better than rough surfaces or those covered by corrosion products or scales), as well as on the environment composition (inhibitor concentration should be increased when contrasting species are present).

Obviously, minimum concentration must be exceeded on the metal surface and for the entire operating life by continuous or alternating refurbishing, in order to compensate for any chemical and mechanical losses. The start-up is critical since inhibitor consumption can be high for the film formation or by reaction with contaminants. Accordingly, in the start-up phase, the rule of thumb is to maintain the highest concentration especially for the anodic inhibitor.

In stagnant environments and in the presence of scales or porous deposits, inhibitor concentration on a metal surface may be insufficient due to a lack of supply: it follows that in these circumstances, dosage and dispersion efficiency must be significantly increased.

As far as the influence of dosage is concerned, Fig. 18.6 schematically demonstrates the influence on the corrosion rate of a *safe* and an *unsafe* inhibitor: if the inhibitor concentration is less than the minimum effective one, there is a risk of an increase in the corrosion rate only in the case of unsafe inhibitors.

The nature of metal. Most inhibitors are suitable only for a specific metal. For example, benzotriazole and mercaptobenzothiazole are specific for copper and its

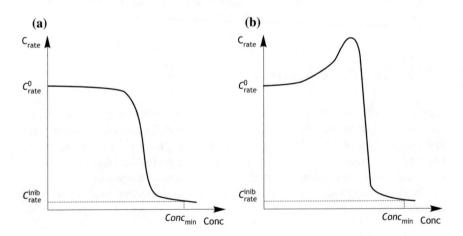

Fig. 18.6 Concentration effect in the case of a safe inhibitors (**a**) and an unsafe inhibitor (**b**)

alloys, benzoates for steels, fluorides for magnesium, silicates for iron, zinc and magnesium, and polyphosphates for steel and zinc. Conversely, chromates show a general inhibiting action in neutral or slightly alkaline environments. Table 18.3 summarizes the degree of effectiveness of the commonly used inhibitors in industry. In the case of equipment made with different metals, proper mixtures of inhibitors are adopted. For example, in cooling circuits in cars, copper alloys, steel, cast iron and aluminium are present, hence the cooling liquid, composed of ethylene glycol and water in a ratio of 1 to 3, is added of phosphates, benzoates, nitrites and borates to protect iron and aluminium, and mercaptobenzothiazole or benzotriazole to protect copper-made components. It is worth noting that if there is a lack of inhibitor for copper, aluminium suffers the greatest damage because of a displacement reaction of the copper with aluminium (see Chap. 7), thus causing localised attacks. **Environmental condition.** Inhibitors may lead to adverse effects when used in unfavourable conditions. We already discussed the danger of oxidant inhibitors used in environments containing de-passivating species, such as chlorides. Similar considerations apply to their use in critical pH ranges. For example, chromates works in the pH range 8–8.5; inhibitors for neutral or alkaline solutions become ineffective below pH 5 for nitrites, 6 and 7.2 for benzoates and phosphates, respectively. Inhibitor dosage normally increases as the temperature increases. For some inhibitors, effectiveness fades if a critical temperature is exceeded.

18.4 Biocides

Biocides are chemicals used to control microbiological-related activity, which can influence the corrosion behaviour as MIC. Typically they can be divided into two categories:

- Non-oxidizing agents, as formaldehyde, glutaraldehyde, acroleine, quaternary ammonium salts, bisthiocyanate, isothiazolines and dodecylguanidine hydrochloride
- Oxidizing agents, as chlorine and its derivate, chlorine dioxide, hydrogen peroxide, ozone and per-acetic acid.

Non-oxidising biocides for industrial use are effective in controlling legionella, slime-forming bacteria, sulphate reducing bacteria (SRB), algae and most of the aerobic bacteria. They are able to penetrate deposits to access sessile bacteria. They are compatible with many systems and with other chemicals, above all oxygen scavengers.

Oxidising biocides are effective in controlling microbiological activity in commercial cleaning, environmental hygiene, food and pharmaceutical industry, disinfection industrial and process water treatment activities. Among oxidizing agents, chlorination is the most used treatment in food industry, pharmaceutic industry and in public building, such as hospital of schools, where legionella and gallionella must be eliminated. The source is typically sodium hypochlorite or chlorine

Table 18.3 Effectiveness of various inhibitors in solutions close to neutrality with respect to different metallic materials

Metal	Chromates[a]	Nitrites	Benzoates	Borates	Phosphates	Silicates	Tannin
Mild steel	Effective	Effective	Effective	Effective	Effective	Partially effective	Partially effective
Cast iron	Effective	Effective	Ineffective	Variable	Effective	Partially effective	Partially effective
Zinc and galvanized steel	Effective	Ineffective	Ineffective	Effective	–	Partially effective	Partially effective
Copper and its alloys	Effective	Partially effective	Partially effective	Effective	Effective	Partially effective	Partially effective
Aluminium and its alloys	Effective	Partially effective	Partially effective	Variable	Variable	Partially effective	Partially effective
Lead and lead alloys by welding	–	Aggressive	Effective	–	–	Partially effective	Partially effective

[a]Banned in some European countries due to their toxicity

dioxide, rarely chlorine gas. Since the storage of such biocides is harmful, hypochlorite is often produced on site, combining an HCl and $NaClO_2$, or by direct electrolysis on site. The treatment is effective to reduce bacteria to less than 10 colonies/mL if the residual free chlorine content is about 0.5–1.0 ppm. An under-treatment may contribute to the increase of bacteria colonies. An over-treatment is not recommended; in fact, chlorine is a highly corrosive gas and a strong oxidiser that may initiate corrosion on passive metals (pitting and crevice corrosion). Moreover, the excess of chlorine may react with oxygen scavengers, if present. Chlorine is not persistent: chlorination should be continuous to prevent microbiological problems downstream. Chlorination does not reduce the effect of SRB.

Although all biocides are efficient on the planktonic population (i.e., microorganisms in the water phase), few are efficient in the case of microorganisms in a biofilm, so that an increase in the dosage is needed. Besides, the use of biocides is more and more limited by environmental legislation due to their toxicity to higher organisms. Table 18.4 highlights applications, advantages and drawbacks of the most commercially used biocides.

18.5 Questions and Exercises

18.1 What are the effects of a pH increase on the corrosion behaviour of a carbon steel water boiler? [Hint: consider the Pourbaix diagram of the iron-water system. Is an oxygen removal still necessary?

18.2 Write the oxygen removal chemical reactions by sodium bi-sulphite and hydrazine. What are the main difference between these oxygen scavengers?

18.3 In a water injection plant, oxygen content must be reduced to control carbon steel corrosion. A physical treatment (a stripping column) is performed before the use of oxygen scavengers. Which is in your opinion the reason? Is it advisable to carry out firstly the chemical treatment?

18.4 The oxygen content of a room temperature process water is 3 mg/L. Which is the oxygen scavenger dosage necessary to reduce oxygen content to 0.05 ppm?

18.5 Discuss the basic difference between anodic and cathodic inhibitors in terms of variation of free corrosion potential.

18.6 Discuss the difference between non-oxidizing anodic inhibitors and oxidizing (or passivating) anodic inhibitors. Provide some examples of inhibitors of the two groups.

18.7 What is the difference between safe and unsafe inhibitors?

18.8 Due to a pump out of service, for 2 weeks the inhibitor dosage was lower than the recommended one. Is there any risk on the corrosion of the pipe?

18.9 Discuss the inhibition mechanism of volatile corrosion inhibitors (VCI). How do they work?

Table 18.4 Available biocides

Biocide	Reaction	Application	Advantages	Drawbacks
Oxydizing biocides				
Chlorine	$Cl_2 + H_2O \rightarrow$ $HCl + HOCl$ $HOCl \rightarrow H^+ + OCl^-$	Freshwater pH 6–8	- Economical - Monitoring is simple - Broad spectrum activity	- Operator hazard - Ineffective for biofilm - Ineffective for SRB
Sodium hypochlorite	$NaOCl \rightarrow Na^+ + OCl^-$	Freshwater pH 6–8	- Economical - Monitoring is simple - Broad spectrum activity	- Expensive
Chlorine dioxide	$2HCl + NaOCl \rightarrow$ $Cl_{2,aq} + H_2O + NaCl$ $Cl_{2,aq} + 2NaClO_2 \rightarrow$ $2NaCl + 2\,ClO_2$	Downhole cleaning and injection Wide pH range	- Good oxidizing agent - Remove organic materials biomass - Dissolves FeS	- Expensive - Toxic - Special equipment is required for generation
Chloroamine	$Cl_2 + H_2O \rightarrow$ $HOCl + HCl$ $NH_3 + HOCl \rightarrow$ $NH_2Cl + H_2O$	Petrochemical plant	- More effective than chlorine - Low toxicity - Reduced corrosiveness	- Ammonia injection is required - More expensive than chlorine
Hydrogen Peroxide 35% Solution	$H_2O_2 \rightarrow H^+ + OOH-$	Freshwater pH 6–8	- Safe - Non-foaming - Decomposition products are completely safe	- Contact with catalytic metal will accelerate its decomposition - Contact with concentrated acid, alkalis reducing agents, cause spontaneous decomposition
Peracetic Acid		Food industry Disinfection of medical supplies	- Non-foaming - Broad spectrum activity	- Contact with catalytic metal will accelerate its decomposition - Contact with concentrated acid, alkalis reducing agents, cause spontaneous decomposition
Bromine 1-bromo-3-chloro 5-5 dimethylhydantoin	In water it releases HOBr and HOCl	Freshwater pH 6–8	- More effective than chlorine - Broad spectrum activity	- Expensive

(continued)

Table 18.4 (continued)

Biocide	Reaction	Application	Advantages	Drawbacks
Ozone	$O_3 + OH- \rightarrow$ $O_2 + HO_2\cdot{}^-$	Freshwater pH 6–8	– Broad spectrum activity	– Low stability
Non oxidizing biocides				
Formaldehyde	37% aqueous solution	Petrochemical plants	– Economical	– Carcinogen – High dosage required – React with oxygen scavenger
Glutaraldehyde	Used in a blend with other biocide	Petrochemical plants	– Broad spectrum activity – Insensitive to sulfide	– React with ammonia and oxygen scavenger
Acroleine	Effective biocide and sulphide scavenger	Petrochemical plants	– Broad spectrum activity	– Difficult to handle – Highly toxic – React with ammonia and oxygen scavenger
Halogenated compounds 2-bromo-2 nitropropane 1,3 diol		Widely used in drilling and produced fluid	– Low toxicity – Broad spectrum activity	– Unstable at high pH
Quaternary ammonium compounds		Widely used in O&G	– Broad spectrum activity – Persistence – Do not interact with other chemicals	– Form foams – No tolerance for brine – Slow reaction
Quaternary phosphonium salts		Injection waters	– Stable – Broad spectrum activity – Tolerance for sulfide	

18.10   What are the differences between physisorption and chemisorption? What are the factors that govern them?

18.11   Which information provide an adsorption isotherm? What are the main differences between Langmuir and Freundlich isotherms?

18.12   Corrosion rate of carbon steel in acidic solution, pH 4.5, is reduced by adding amines as cathodic inhibitor. Calculate the inhibitor efficiency to decrease corrosion rate from 1.1 to 0.01 mm/year.

18.13   What is the main drawback related to the chlorination over-treatment of water?

18.14   NaClO is added to water to control bacteria proliferation in an AISI 304 stainless steel tank in a food plant used to clean apples. Due to problem in the software controlling NaClO dosage, for one week a double dosage was added to the water. Could the stainless steel tank suffer corrosion?

18.15   The cold-water plant in a hospital is treated with chlorine mainly to kill Gallionella bacteria. As a rule of thumb, operators suggest to dose chlorine in order to have at the end of the piping a maximum chloride content lower than 0.1 mg/L. Which is the main reason of this limit?

Bibliography

Fontana M (1986) Corrosion engineering, 3rd edn. McGraw-Hill, New York, NY
Kuznetsov YI (1996) Organic inhibitors of corrosion of metals. Plenum Press, New York, USA
Sastri S (1998) Corrosion inhibitors: principles and applications. Wiley, Chichester, UK
Shreir LL, Jarman RA, Burstein GT (1994) Corrosion. Butterworth-Heinemann, London, UK

Chapter 19
Cathodic and Anodic Protection

> *Electrical current [exchanged between metal and electrolyte]*
> *modifies exceptionally the oxidation state.*
> A. Volta, *Letter to Editors of Britannic Library of Geneva*,
> March 18th 1802

Abstract As Volta already observed in early 1800s, a current that is exchanged between a metal surface and the electrolyte to which it is exposed modifies the behaviour of the metal, increasing the corrosion rate or reducing it. Some years later, Davy put in practice such concept demonstrating the possible implications, which, after about a dormant century, started to have effective industrial use: with Davy, cathodic protection was born. A century and a half from Volta's observations, Edeleanu showed that Hickling potentiostat could control corrosion by making an anodic polarization and Riggs made the first commercial application of anodic protection: hence, anodic protection was also born. Both techniques are set up by establishing a cathodic or anodic current that is exchanged with the electrolyte, which means supplying or taking electrons. The following chapter deals with these techniques from a theoretical and practical point of view: mechanism, protection criteria, protection potential and current density, applications, monitoring.

Fig. 19.1 Case study at the PoliLaPP Corrosion Museum of Politecnico di Milano

© Springer Nature Switzerland AG 2018
P. Pedeferri, *Corrosion Science and Engineering*, Engineering Materials,
https://doi.org/10.1007/978-3-319-97625-9_19

19.1 Cathodic Protection (CP)

Cathodic protection is an electrochemical corrosion control technique used to prevent the corrosion of metal structures, by lowering their potential by means of a cathodic direct current supplied by an anodic system, such as the one reported in Fig. 19.1. Some examples are:

- Underground facilities (gas pipelines, pipelines, aqueducts, tanks)
- Marine exposed structures (harbours, offshore platforms, ship hulls)
- Internal of heat exchangers or equipment often in contact with seawater
- Chloride-contaminated reinforced concrete structures.

As shown in Fig. 19.2, it is carried out by supplying electrons through a current circulation between an electrode (anode) in contact with the environment and the structure (cathode).

The mode by which the current circulation is performed defines the two types of cathodic protection: galvanic (or sacrificial) anodes (GACP) and impressed current (ICCP).

Figure 19.2a shows the GACP obtained by galvanic coupling with a less noble metal; for example, aluminium and zinc are used for the protection of steel in seawater, magnesium in soils and in freshwater; iron is usually used to protect copper alloys or stainless steels. Figure 19.2b shows the electrical scheme of ICCP by the use of DC feeding systems with the negative pole connected to the structure and the positive to an anode, which can be soluble, as iron, or inert as silicon-cast iron, graphite or platinized and activated titanium.

19.1.1 Protection Potential

CP reduces or stops corrosion by two distinct effects as the result of potential lowering, namely:

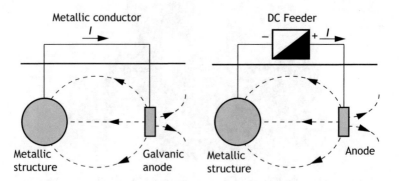

Fig. 19.2 Cathode protection principle: **a** by galvanic anodes; **b** by impressed current

- Thermodynamic effect which reduces or cancels the driving voltage for the corrosion process, ΔE
- Kinetic effect which increases reaction electrical resistances.

19.1.2 Thermodynamic Effect

When potential is brought below the equilibrium potential, $E < E_{eq}$, the tendency of a metal to oxidize stops or instead reverses toward reduction of metal ions, if they are present. This condition is called *thermodynamic immunity*. Table 19.1 shows the equilibrium potential of different metals in contact with solutions containing their ions at a concentration of 10^{-6} mol/L, which can be assumed as potentials for immunity protection.

If potential decreases below the free corrosion potential, E_{corr}, without zeroing the driving voltage (i.e., $E_{eq} < E < E_{corr}$), corrosion reduces rather than stops; from an engineering point of view, provided that corrosion decreases below an agreed threshold (10 µm/year), this condition, called *quasi-immunity* is acceptable as shown in Fig. 19.3a. For example, quasi-immunity of steel immersed in seawater or buried in soil, is achieved if potential is below −0.85 V CSE in aerated condition or below −0.95 V CSE in anaerobic condition, when SRB are present.

19.1.3 Kinetic Effect

Rather than the reduction of driving voltage, kinetic effects associated to the potential lowering derive from the increase of the corrosion reaction resistance when the metal undergoes or strengthens passive conditions (Fig. 19.3b), like stainless steels in neutral solution or carbon steel in alkalies, for example in concrete. This condition is defined as *CP by passivity*.

Table 19.1 Immunity potential as equilibrium potential, E_{eq}

Metals	Equilibrium potential as per Pourbaix (V SHE)		
	pH = 0	pH = 7	pH = 14
Silver	+0.44	+0.44	+0.32
Copper	+0.12	+0.12	−0.38
Lead	−0.31	−0.31	−0.74
Iron	−0.62	−0.62	−0.92
Aluminium	−1.78	−1.78	−2.06

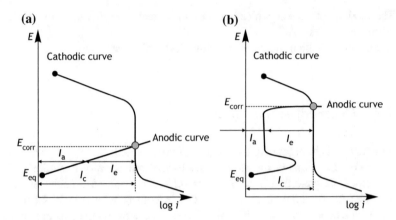

Fig. 19.3 Electrochemical conditions of CP for a metal **a** active (quasi-immunity condition) and **b** active-passive (passivity)

19.1.4 Protection Criteria and Overprotection

The protection potential, E_{prot}, is a potential value lower than, or equal to one at which an acceptable corrosion rate is reached for the metal considered. Some practical values are listed in Table 19.2, which assure immunity or quasi immunity or passive conditions.

If potential is reduced to more negative values, hydrogen evolution becomes the predominant cathodic reaction and this condition is defined as *overprotection*. For example, for carbon steel in aerated soil, overprotection is reached when the potential is more negative than −1.2 V CSE, as reported on ISO 15589-1.

Overprotection has to be avoided, for economic reasons first, because for example a 100 mV lowering of potential increases by one order of magnitude the cathodic current according to Tafel law of hydrogen evolution reaction, hence the cost of CP. Secondly, overprotection may have two negative effects due to the local hydrogen evolution:

Table 19.2 Protection potentials for most commonly used metals in soil and seawater

Metals	Soil	Seawater	
	V CSE	V Ag/AgCl	V Zn
Carbon steel:			
– Aerobic	−0.85	−0.80	+0.25
– Anaerobic	−0.95	−0.90	+0.15
– In concrete	−0.75	−0.70	+0.35
Copper	−0.30	−0.20	+0.70
Lead	−0.50	−0.45	+0.50
Zinc	−1.00	−1.05	0
Aluminium	−0.80	−0.90	+0.15
Stainless steel	−0.40	−0.50	+0.55

- Disbonding of coating
- Hydrogen embrittlement of susceptible steels.

19.1.5   Protection Current Density

Protection conditions are achieved when an adequate cathodic current density is supplied as Fig. 19.4 shows. It is called *protection current density* and corresponds to the electrons taken by all cathodic processes occurring at the protection potential. By applying Kirchhoff's law, the cathodic external current, I_e, is given by:

$$I_e = I_c - I_a \tag{19.1}$$

where I_a is the residual anodic current and I_c is the cathodic current. Anodic current zeros when the external current equals the cathodic current, then giving the protection current, I_{prot}:

$$I_e = I_c = I_{prot} \tag{19.2}$$

The protection current density, I_{prot}, is the ratio between the protection current, I_{prot}, and the protected surface. It depends on environmental conditions in the same way as the cathodic processes.

In natural environments, as soil and waters, the protection current density for carbon steel equals the oxygen limiting current density, hence it depends on dissolved oxygen content, temperature and turbulence. For example, for steels in seawater, the protection current density ranges from 1 to 20 mA/m^2 in seamud; 50–70 mA/m^2 in warm, stagnant water and 100–200 mA/m^2 in cold and agitated water, reaching 1000 mA/m^2 for propellers zone in ships because of maximum agitation and oxygenation. Table 19.3 shows the current density values used in industrial applications.

In natural waters, protection current decreases with time because a calcareous deposit forms after the alkalinization produced by the cathodic reactions. In general,

Fig. 19.4 Principle of cathodic protection from the electrical viewpoint

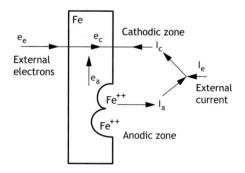

Table 19.3 Protection current density adopted in typical industrial applications

Environment	Protection current density (mA/m^2)
Bare steel	
Soil	
– Neutral aerated	20–150
– Water saturated	2–20
– Hot metal surface	30–60
Concrete	
– Atmospherically exposed	2–20
– Water saturated	0.2–2
Fresh water	
– Cold	30–160
– Warm	50–160
Seawater	50–600
Seamud	20–33
Acidic electrolytes	50–1500
Coated steel	
Soil	0.01–1
Seawater	0.1–10

to reduce the protection current density, coatings are applied to limit the exchanged current at the coating pinholes and defects then reaching a small percentage of the exposed surface, typically less than 1%.

If the applied current density exceeds the protection values as reported above, potential is reduced to more negative values, and overprotection conditions may occur.

19.1.6   Anodic Reactions

On anodes, different reactions take place depending on anode type and environment as follows.

Galvanic anodes. The anode reaction is the metal dissolution, typically zinc, magnesium, aluminium and iron dissolution, followed by the hydrolysis reaction which produces a local acidification:

$$M \rightarrow M^{z+} + z\,e^{-} \tag{19.3}$$

$$M^{z+} + z\,H_2O \rightarrow M(OH)_z + z\,H^{+} \tag{19.4}$$

This acidity increase plays an important role in avoiding passivation of anodes, especially in the case of aluminium anodes. The chemical composition of the anodes is very important. In the case of zinc anodes, the presence of some elements, as iron with a content in excess of 50 ppm, can lead to passivation of the anodes or may reduce its efficiency; in the case of aluminium anodes, iron or copper impurities are harmful; conversely, the presence of other elements, such as indium, mercury or tin, is necessary to keep anodes active.

Inert anodes. Anodes used in ICCP systems are in general insoluble. Therefore, the anodic reaction is oxygen evolution and, in the presence of chlorides, chlorine evolution, according to the following reactions:

$$2H_2O \rightarrow O_2 + 4H^+ + 4e^- \tag{19.5}$$

$$2Cl^- \rightarrow Cl_2 + 2e^- \tag{19.6}$$

Chlorine evolution prevails in the case of seawater applications if $i_a > 10$ A/m^2. Oxygen evolution causes a local acidification which greatly increases the aggressiveness against the anode material.

19.1.7 Coatings and Scales

The application of an insulating coating reduces significantly the protection current density, since only pores, pinholes, defects and damages can exchange the current. Accordingly, the protection current density for a coated surface is given by the following expression:

$$i_{prot} = i_0(1 - \xi) \tag{19.7}$$

where i_0 is the protection current density of the bare metal and ξ is the coating efficiency (or unitary coated surface fraction).

Coating efficiency varies over time; for example, for buried pipelines can reach 90% after 10–20 years; in seawater coating degradation might be even more accelerated as for shiphulls for which efficiency can shift from 99.9 to 99% in the first year exposure.

In seawater, cathodic protection causes the formation of a scale consisting of a mix of calcium carbonate and magnesium hydroxide, called *calcareous deposit*; it reduces favorably of about one order of magnitude the protection current density as an effective barrier to the oxygen diffusion. Parameters affecting the quality of the calcareous deposit are seawater composition (for instance, it hardly forms in low salinity seas, as Caspian and Baltic seas), current density and mechanical actions (abrasion and vibration). Once protection current is interrupted, the calcareous deposit slowly starts to dissolve.

19.1.8   Current Distribution

The protection current must be uniformly distributed over the cathodic surface to obtain even protection conditions. Conversely, non-uniform distribution can lead to either over-protection or under-protection, with risks in the former case of hydrogen embrittlement on high strength steels and disbonding of coatings and in the latter case of some residual corrosion.

For structures with complex geometry, it may be difficult to achieve a good current distribution, especially when high protection currents are required with high resistivity electrolyte and small anode-to-cathode distance or linear structure as pipelines. In the latter case, a considerable ohmic drop throughout the metal takes place, which establishes so-called *potential attenuation*. Conversely, in the case of reinforced concrete structures there is a strong ohmic drop in the concrete cover although applied current densities are moderate. Finally, for offshore and marine structures, ohmic drop is negligible despite the very high protection current densities.

The usual adopted design procedures are based on empirical criteria regardless the problem of electric field distribution. Nowadays, it has become a common practice to verify for complex structures, as nodes of platforms and heat exchangers, the current and potential distribution by the resolution of the electric field through modelling (FEM and BEM). As discussed in Chap. 9, the analytical resolution of the electric field, regulated by the Laplace equation, is possible only for very simple geometries and with drastic simplification, for example, homogeneous electrolyte and negligible overvoltage or linearly increasing with current. For some distributions, reference is made to Table 9.1.

Conditions for the application of modelling
The most commonly used numerical model is the finite elements method, developed some decades ago, especially in the field of structural engineering to solve problems of strain and deformation analysis.

The method has been extended to solve a wide variety of physical, field-related problems connected to a flow of some magnitudes, such as heat, mass or current under the action of a gradient of temperature, pressure or electrical potential, respectively. There are issues regarding heat transmissions, diffusion through porous media, hydrodynamics, electrical and magnetic fields, and finally CP. The common equation is the almost harmonic Poisson-Laplace equation. In the case of CP, the potential field in the electrolyte follows the Laplace equation as follows:

$$\nabla^2 E = 0 \tag{19.8}$$

By using the finite element method, the resolution of (19.8) is addressed by dividing the global system into a finite number of elements, each of which interacts with contiguous ones crossing the points of its outline, called nodes. The values of the potential function, E, in the nodes are unknown to the discretized problem and are obtained by introducing the following boundary conditions: on electrodes, $i = f(\eta)$ and E_m = constant (equipotential electrodes) or $dE_m = r\,I_x\,dx$ (when an ohmic drop occurs in the metal); on insulating surfaces, $i = 0$. Instead, curiously, Leopoldo Nobili in 1835 gave a coloured solution of the Laplace equation (see Chap. 9).

19.2 CP Applications

CP is typically applied to structures exposed to natural environments and reinforced concrete ones.

Galvanic anodes are used in high conductivity environments, such as seawater, and in some cases when small currents are required even in low conductivity environments, such as soil and in the cathodic prevention of reinforced concrete.

Impressed current systems are required in high resistivity environments such as soil and concrete and are preferred for the protection of extended structures when a limited number of anodes is mandatory. A significant advantage is that the system has a great flexibility of operation, which can vary, allowing an easy regulation of the current delivered.

Table 19.4 summarizes benefits and limitations of the two applications.

Table 19.4 Advantages and drawbacks of GACP and ICCP systems

	GACP	ICCP
Advantages	• Feeding power not required • Current adjustment not required • Easy and almost inexpensive installation • No stray current arising • No maintenance costs required • Almost uniform current distribution • Additional areas not required around facilities	• Voltages and currents can vary • High current output if necessary • A single groundbed can protect large surface structure • Suitable for high resistivity environments • Effective for protecting bare or badly coated structures
Drawbacks	• Low driving voltage • Low current output • Expensive installation after commissioning • Bare or badly coated structures require many anodes • Unsuitable in high resistivity environments	• Stray current problems • Subject to vandalism • Maintenance necessary • Need of feeding • Operating cost • Overprotection risks • Risks of failure of cables and connections • Additional areas required around facilities

19.2.1   Galvanic Anodes Cathodic Protection Systems (GACP)

CP with galvanic anodes is mainly used to protect marine structures as offshore platforms and submerged pipelines, internal surfaces of equipment: from domestic boilers to chemical or petrochemical industry heat exchangers or condensers of power plants, shiphulls and internal of doublewall tankerships and many more. In principle, GACP can be made with any metal less noble than the one to be protected; in practice, for carbon steel structures, only aluminium, magnesium and zinc alloys are used.

Two parameters characterize an anode: working potential and current capacity. The former determines the current output of an anode and therefore the minimum number needed to establish the protection current; the current capacity, i.e., the charge by weight, determines the anode consumption, hence the minimum weight to ensure the protection during the lifetime.

Anode capacity. A galvanic anode while working dissolves through its electrode reaction ($M \rightarrow M^{z+} + ze^-$, where M is the anode metal and z is its valence). By Faraday law, the product of electrochemical equivalent and charge gives the *theoretical anode consumption*, usually expressed in kg/A year. In practice, true anode consumption is higher because the anode corrodes by its own as it occurs in galvanic corrosion; for zinc and aluminium anodic alloys the efficiency is about 95%, for magnesium alloys is as low as 50%. Often *theoretical anode capacity* is used, defined as the inverse of theoretical consumption, hence expressed in A h/kg. For ease of calculations, the product between theoretical capacity, A h/kg, and theoretical consumption, kg/A year, is 8760. Table 19.5 shows the consumption and theoretical abilities of aluminium, zinc and magnesium.

Table 19.5 Electrochemical properties of Mg, Zn and Al anodic alloys

Parameters	Mg	Zn	Al
Atomic mass	24.32	65.38	26.97
Equivalent mass	12.16	32.69	8.99
Specific gravity a 20 °C (g/cm^3)	1.74	7.14	2.70
Anode theoretical consumption – kg/A year – dm^3/A year	3.98 2.3	10.69 1.5	2.94 1.1
Anode theoretical capacity (A h/kg)	2200	820	2980
Anode working potential	−1.55/−1.75 (V CSE)	−1.00 (V SSC)	−1.05 (V SSC)
Anode efficiency (%)	50	95	90–95

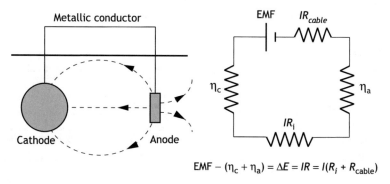

Fig. 19.5 Equivalent electrical circuit for galvanic anode systems

Anode output. With reference to the equivalent electric circuit depicted in Fig. 19.5, the current supplied by an anode, I, is given by the Ohm's law:

$$I = \Delta E / R \tag{19.9}$$

where ΔE is the driving voltage, i.e., the energy available to overcome ohmic drops, and R is the total electrical resistance of the circuit, expressed as:

$$R = R_i + R_{cables} \tag{19.10}$$

where R_i is the ohmic resistance of the electrolyte and R_{cables} is the resistance of the metallic circuit, most often negligible unless in the case of big remote anodes. The electrolyte resistance can be reduced to the anode resistance, R_a, which is calculated by empirical equations as function of anode size and electrolyte resistivity (Table 19.6). In order to decrease the anode resistance, elongated anodes are used, as slender anodes for offshore applications and rod type anodes for soil applications; in soil, a low resistivity backfill is also used. The driving voltage, ΔE, is obtained by subtracting the overvoltage from the electromotive force, hence it is the difference between the working potential of the anode, E_a, and the protection potential, E_{prot}. For example, for steel structures, driving voltage, ΔE, is 250 mV for zinc anodes, 300 mV for aluminium and 800 mV for magnesium.

Anode selection. Anodes are selected on the basis of the metal-environment couple. Table 19.7 summarizes the types of anode for water and soil, suggested in practice as function of environment resistivity. This rule of thumb derives from the evidence that once fixed the anode size, the current output becomes a function of the environment resistivity only. For soil applications, it is possible to lower the resistivity by using a proper backfill, which also can help decrease overvoltage and prevent anode passivation. Typical backfill composition is a mixture of gypsum, bentonite and sodium sulphate in the weight ratio of 70:20:10. Gypsum ($CaSO_4$) has the main function to keep the anode active so to favour a uniform dissolution,

Table 19.6 Equations used for the calculation of the anode resistance

Type		Notes
Slender anodes		
Dwight	$Ra = \frac{\rho}{2\pi L}\left(\ln\frac{4L}{r} - 1\right)$	Most used for offshore slender anodes
Peters	$Ra = \frac{\rho}{2\pi L}\ln\frac{2L}{r}$	Little used
Shepard & Graeser	$Ra = 6.2\frac{\rho}{2\pi L} \cong \frac{\rho}{L}$	Valid for $L/2r \geq 64$
Bracelet anodes		
Naval Research Laboratory	$Ra = \frac{1.66\,\rho}{(S_a)^{0.727}}$	When $S_a > 500$ cm^2 is equivalent to Mac Coy's. Little used
Mac Coy	$R_a = \frac{\rho}{\pi \cdot \sqrt{S_a}}$	Most used for bracelet anodes
Flat anodes		
BKL	$R_a = \frac{1.5\,\rho}{\pi\,(L + 0.8B + 0.5s)}$	For anodes with $L/r < 10$
Lloyds	$Ra = \frac{\rho}{2L'}$	For anodes applied on flat surfaces
Suspended anodes		
Lloyds	$Ra = \frac{\rho}{4L'}$	For thin, flat anodes

ρ = electrolyte resistivity ($\Omega.m$); L = anode length (m); r = equivalent radius (m)
S_a = anodic surface (m^2); B = width (m); s = thickness (m);
$L' = (a + b)/2$ (m) where a and b are the sizes of flat anodes, $b < 2a$

Table 19.7 Galvanic anodes suitable for water and soil

Anode	Resistivity (Ω m)	
	Water	Soil (plus backfill)
Aluminium	Up to 1.5	Not used
Zinc	Up to 5	Up to 15
Magnesium low potential (-1.5 V CSE)	All	Up to 40
Magnesium high potential (-1.7 V CSE)	All	Up to 60

bentonite to reduce resistivity by absorbing moisture and sodium sulphate (Na_2SO_4) to increase electrolyte conductivity.

In particular applications, the choice of the anode material should comply with safety criteria. For example, magnesium and to a lesser extent aluminium, can cause sparks with a rusted steel surface due to a reaction with iron oxide if freely dropped, which may ignite a deflagration of hydrocarbon vapors. For this reason, within storage tanks or tankers used to transport flammable products, only the use of zinc anodes is permitted without restriction, while magnesium anodes are always excluded and for aluminium it is necessary to observe specific restrictions as maximum height from the floor.

From the economic point of view, aluminium alloys are the cheapest ones, as the cost to produce the same charge, taking into account the practical consumption and average unitary cost follows an approximate ratio 100:300:750 for Al, Zn and Mg respectively.

Anode number calculation. The design of CP involves, after the choice of the anode type, the calculation of weight and number of anodes and eventually their location on the structure. Among a variety of possible solutions as number, weight and duration of anodes, the adopted one is generally the cheapest solution. A serious error can be made if the number of anodes is simply calculated dividing the total weight of the anodes (calculated on the basis of the lifetime) by the weight of a single anode selected by a catalog without verifying its output capacity. In this case, in fact, the total anode mass is sufficient for the expected duration of the protection, but may not be able to provide the current required for the protection, which depends on anode surface area rather than weight. In other words, it is necessary to optimize the weight-to-surface area ratio of the anodes to ensure both protection and duration.

Procedure steps for determining the anode number are based on the following calculations:

- Protection current (total surface area times the protection current density)
- Anode weight (current times consumption times duration)
- Anode resistance of a selected anode (first choice) by applying empirical equations reported in Table 19.6 (resistivity is the one of the electrolyte surrounding the anode)
- Current output of an anode (driving voltage divided by anode resistance)
- Number of anodes (total weight divided by the weight of an anode)
- Minimum anode output (total current divided by the number of anodes)
- Compliance between anode output and minimum current requirement
- Acceptance of the solution or repetition with a second anode choice.

19.2.2 Impressed Current Cathodic Protection Systems (ICCP)

CP by impressed current is primarily applied to buried and reinforced concrete structures and also to water tanks, heat exchangers, desalinators and ships. For these systems, a direct current is supplied by a DC feeder through an anode immersed or embedded in the environment. Figure 19.6 illustrates the equivalent electrical circuit of an ICCP system.

Anode materials. Table 19.8 reports compositions, consumption rate and working conditions of main used anodes of ICCP systems. The anodic reaction occurring on anode surface depends on both anode material and environment. For example, for carbon steel anodes, the anode reaction is iron dissolution; for inert or so-called insoluble anodes, for example platinized or MMO (mixed metal oxide) activated titanium, graphite and others, the anodic reaction is oxygen evolution or chlorine

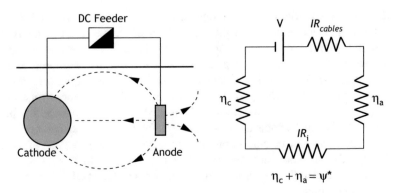

Fig. 19.6 Equivalent electrical circuit for ICCP

evolution or both depending on the electrolyte, soil or seawater, and anodic current density.

Feeding voltage. The first step for the design of an ICCP system is the calculation of the minimum feeding voltage, V_{min}, required. From electrical and electrochemical considerations, it is the so-called cell voltage given by:

$$V_{min} = I R_{tot} + \psi^*$$
(19.11)

where I is the total design protection current; R_{tot} is the total resistance of the circuit (calculated or given) and ψ^* is the thermodynamic and kinetic contribution of electrode reactions, which is the minimum feeding tension out of the ohmic drop. It is negligible for dissolving iron anodes and 2–3 V for inert anodes.

Groundbeds. In seawater and inside equipment, the anodes work without backfill as directly exposed to the electrolyte. Anode shape can be easily adapted as for the shape of the cathodic surfaces and the mechanical requirements for the installation. In soil applications, the need for a low anode resistance prevails on other requirements. As a rule of thumb, resistance is often less than 2 Ω. Furthermore, for safety reasons, the feeding voltage should not exceed 50 V in compliance with standards. To achieve this goal a backfill is used to reduce the anode resistance. Typical groundbeds for soil applications are of three types: horizontal, shallow vertical, deep vertical.

Calculation scheme. The design of an ICCP system deals with type and sizes of groundbed and feeder requirements. Often, adopted technical solutions are determined by external input, for example in urban areas vertical deep groundbeds are mandatory for lack of suitable large areas or lack of usage permits.

The general calculation procedure consists of the following steps:

- Calculation of the protection current (product of the cathodic surface area times the protection current density)

Table 19.8 Properties and characteristics of the main insoluble anodes

Material	Description—composition	Density Mg/m³	Resistivity Ω m 10^{-8}	Consumption		Anode current density (A/m²)			
				UF	kg/A year	Soil	Fresh water	Seawater	Mud
Iron	Scrap steel	7.1–7.8	12–55	0.5	10	5	NR	(−)	(−)
Graphite	Graphite impregnated with oil, resins, wax	1.6	800–1500	0.6	1	2.5–10 (*)	2.5	20	2.5
Silicon iron	Cast iron 14% Si, 0.75% Mn, 0.95 C	6.8–7.0	72	0.5	1	10	10	NR	(−)
	Cast iron 14% Si, 4.5% Cr, 0.75% Mn, 0.95 C	7.0	72	0.5	0.5	10	10	15	(−)
Lead	Solid lead with 2% Ag	11.3	25	0.6	0.1	(−)	(−)	32–65	(−)
	Solid lead with 1% Ag, 6% Sb	10.9	25	0.6	0.2	(−)	(−)	50–200	(−)
Magnetite	Non stoichiometric Fe_3O_4	5.2	33,000	0.4	0.05	20	20	60	20
Pt (on Ti)	Platinized titanium	4.5	48.2	0.85	10-4	100(*)	150	500	(−)
Pt (on Nb)	Platinized niobium	8.5	15.2		10-4	100(*)	150	500	(−)
Plastic support anodes	Polyolephyne plastics with conductive fillers (carbon)				1	0.5	0.5	(−)	(−)
Activated titanium	Titanium activated by mixed noble metal oxides (Ir, Rh)	4.5	48.2	0.95	10-4	50–100 (*)	150	600	100

UF utilisation factor; (−) not recommended; (*) with calcined carbon coke backfill

- Choice of groundbed and calculation of anodic resistance, R_a, and cable resistance, R_{cable}
- Calculation of the total resistance, R_{tot} ($R_{tot} = R_a + R_{cable}$)
- Calculation of feeding voltage $V_{min} = I R_{tot} + \psi^*$, where ψ^* is thermodynamic and kinetic constant, and I is the maximum protection current
- Check of anode life.

19.2.3 CP Monitoring

CP monitoring includes all operations aimed to check, directly or indirectly, the protection condition of a structure, which is related to local conditions rather than to global parameters. Therefore, protection conditions are not monitored by the total protection current output or by the potential measured in a remote location; instead, protection on a point of a structure is reached if the current density, at that point, is equal to or greater than the protection value.

Measurement of potential. The measurement of potential is the criterion universally used to verify CP. The measurement follows the scheme shown in Fig. 19.7; a reference electrode is placed in contact with the environment (as soil, water, concrete) and connected to the negative pole of an high impedance voltmeter (at least greater than 10 MΩ for most used reference electrodes[1]), while the structure is connected to the positive one. This simple and easy measurement needs however a correct interpretation, as discussed in the following.

Interpretation of potential measurements. When measuring the potential by a portable reference electrode (as shown in Fig. 19.7), there is an ohmic drop contribution, IR, due to the current flowing in the electrolyte (protection current, stray current, etc.). This contribution alters the potential reading, hence has to be eliminated. The measured value, E, is the sum of three contributions, as shown in Fig. 19.8:

$$E = E_{eq} + \eta + IR \tag{19.12}$$

[1] The current flowing during measurement is given by the ratio of the potential difference and the circuit resistance which practically coincides with the internal impedance of the voltmeter. If, for example, the reference electrode has a surface area of 10 cm^2 and the potential difference is 1 V with a voltmeter impedance of 1 MΩ, current density exchanged on the reference electrode surface is 1 mA/m^2 which is less than the exchange current density of active metals like copper, zinc and iron, but is greater by at least one order of magnitude than the passivity current density of Ti. Therefore, when using a small size MMO activated Ti as reference electrode, to avoid its polarisation the voltmeter impedance must be greater than 10 MΩ for short measurements and higher than 1 GΩ for continuous monitoring.

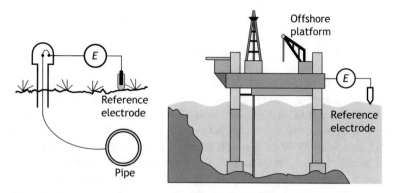

Fig. 19.7 Measurement of the potential of buried and immersed structures

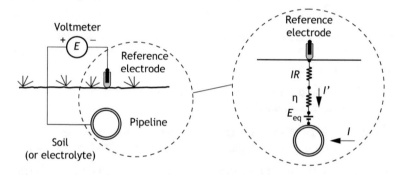

Fig. 19.8 Measurement of the potential of structures and the meaning of the measure

Fig. 19.9 Schematic indication of overvoltage contribution and meaning of potential measurement

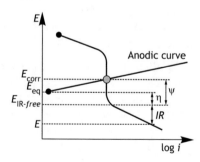

where E_{eq} is equilibrium potential defined by Nernst equation, η is overvoltage related to the exchanged current measured with respect to the equilibrium potential and IR is ohmic drop in the electrolyte. The sum $(E_{eq} + \eta)$ is the so-called IR-free potential, $E_{IR\text{-free}}$, also referred to as "true potential", which can be expressed through the free corrosion potential, as shown in Fig. 19.9:

$$E_{IR-free} = E - IR = (E_{eq} + \eta) = (E_{corr} + \psi) \qquad (19.13)$$

where ψ is the overvoltage measured with respect to the free corrosion potential, E_{corr}.

The ohmic contribution, IR, is the spurious term that contaminates the measurement of potential. According to the second Ohm's law, it depends on the position of the reference electrode with respect to the monitored structure, the resistivity of the electrolyte, and the flowing current density, i. Hence, the measured potential is a function of the position of the reference electrode: it decreases when the reference electrode is placed close to the structure, therefore, the simplest technique to minimize it, consists in placing the reference electrode as close as possible to the structure. Resistivity plays an important role: for example in seawater, where resistivity is low, even if current is high, the ohmic drop is often negligible. On the contrary, in concrete and in soils, especially in the most resistive ones, the ohmic drop is not negligible at all, even if a small current circulates.

Elimination of ohmic drop contribution from potential measurement
The two ways to eliminate the ohmic drop from potential measurement are based on the following actions:

- Reduction of the distance reference electrode—structure
- Elimination of the circulating current.

In the first case, local (fixed) reference electrode or potential probes are used. The elimination of the circulating current is achieved by the ON-OFF method.

Local reference electrode. If the reference electrode is placed close to the structure, the IR drop contribution is reduced or even eliminated. In practice, this is done in seawater, for example on offshore platforms, where it is easy to place the reference electrode close to the jacket at the desired point. For buried structures, it is necessary to make use of fixed reference electrodes buried very close to the structure, up to a maximum of 0.2 m.

While this method is efficient for bare structures, it is questionable for coated structures, since to know the IR-free potential at a coating holiday, the reference electrode must be as close as possible to it. In other words, by simply installing a reference electrode close to the coated structure, the potential measured is not representative of the lower protection level established at holidays, unless the reference electrode is, by chance, close to the biggest one.

Potential probe. To overcome this problem, an artificial defect, called *corrosion coupon*, is used in combination with a close reference electrode. This device is called *potential probe*. The coupon is connected to the structure to simulate the presence of a coating defect. Coupon surface is in the range 1–10 cm^2, depending on the insulation resistance of the pipe to which it is connected. Indeed, the coupon shall be representative of a defect in the pipe coating or of any buried structure. The potential probe represents an elegant and brilliant method to measure the protection

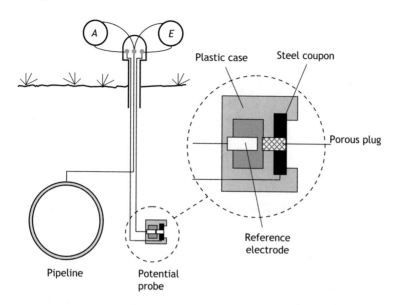

Fig. 19.10 Fixed probe with built-in reference electrode

potential of the coupon because ohmic drop is eliminated physically by placing the reference electrode in close proximity to the coupon ($E = E_{\text{IR-free}}$). Figure 19.10 shows a potential probe, which consists of a coupon embedded with an incorporated reference electrode.

ON-OFF technique. This method is based on the experimental evidence that when the protection current is interrupted, the *IR* drop disappears in a very short time, on the order of 10^{-6} s. Therefore, by recording the potential reading shortly after current interruption, the *IR* drop is no longer included. This technique is not valid in presence of currents different from the protection current, as stray currents.

Obviously, by interrupting the protection current, overvoltage is also eliminated but over a much longer period than *IR* drop, ranging from milliseconds for activation overvoltage (related to hydrogen evolution reaction) to seconds or even days for concentration overvoltage (related to oxygen diffusion). The latter is the prevailing one in CP. Figure 19.11 reports a typical potential recording obtained by means of a high frequency acquisition voltmeter (at least 10 Hz sampling frequency) where the so-called OFF potential, E_{off}, is the potential value recorded immediately (within 1 s according to international standards) after current interruption. Nevertheless, some typical situations should be carefully considered:

- *Overprotection conditions.* When measurement is taken on coated structures in overprotection conditions, that is with hydrogen evolution, E_{off} potential does not correspond to the *IR*-free potential, since the activation overvoltage of hydrogen evolution disappears very rapidly, in less than 1 ms, so that $E_{\text{off}} > E_{\text{IR-free}}$. Therefore, by taking the OFF reading 0.1–1 s after current interruption,

Fig. 19.11 Example of potential recording in ON-OFF technique

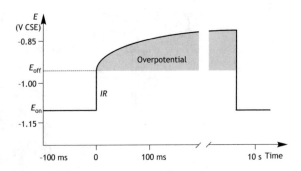

error can be as high as 200–300 mV and overprotection condition can not be recognized;

- *Coated structures.* When coating holidays of various sizes are present, two problems arise. The first depends on the different polarisation level achieved by each defect according to its size: smaller defects are more polarised than larger ones. The resulting E_{off} reading is difficult to interpret and can not be associated to an isolated defect. The second problem is related to the fact that once the protection current is interrupted, the more polarized holidays (the smaller ones) become anodes, releasing some current to the less polarized holidays, acting as cathodes. The result is the circulation of a so-called *equalizing current*, which makes an apparent lower protection on the small holidays and an overestimated protection on the larger ones;
- *Grounded structure.* If the ON-OFF technique is applied on a grounded structure (for examples on a carbon steel pipeline with a copper-made earthing), when current is OFF, galvanic corrosion between the two metals occurs and the measured potential is typically the potential of the wider surface, specifically the grounding system. For this reason, the ON-OFF technique can not be applied on grounded structures. To overcome this problem, the OFF reading should be carried out on a buried corrosion coupon, without interrupting pipeline protection, simply by disconnecting it from the pipeline. The drawback of this device is that an *IR* drop contribution, due to the CP current of the pipeline, may still be present.

Close Interval Potential Profile. In case of a coated pipeline in cathodic protection condition, the monitoring of CP, as well as the localisation of coating defects, can be worked out by examining the potential profile along the pipeline, as schematically illustrated in Fig. 19.12. To obtain the profile, an electrical connection with the pipeline is required and the measurement must be synchronised with the operator's movement. The ON potential, E_{on}, measured before the interruption of the protection current by a portable reference electrode, may provide useful information. However, because of *IR* drop, the E_{off} potential profile has to be used. Two distinct effects are recorded in correspondence with defects: a peak towards more positive potential values of both E_{on} and E_{off} (the polarisation is lower and therefore the potential is more positive) and a decrease in ohmic drop measured as the difference ($E_{on} - E_{off}$).

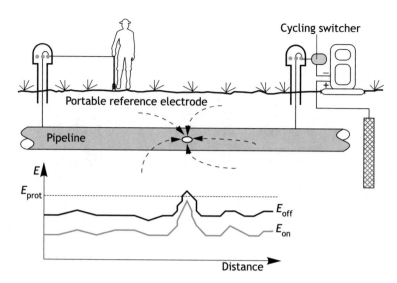

Fig. 19.12 On-off potential profile along a pipeline with coating defects

CP of Marine Structures

The exploitation of offshore oil and gas fields developed special structures as platforms and submerged wellheads with their ancillary components like pipeline and risers. The most impressive of these structures are the platforms. Used to locate new deposits or to exploit them, they are fixed or movable, laying on the bottom by gravity or fixed with piling, designed to operate in deep waters, depths exceeding 150 m, or semi-submersible or floating and maintained in position with appropriate anchoring systems. Construction materials are carbon-manganese steel for tubular structures and reinforced or pre-compressed concrete for those operating by gravity. A few meters above the waves area, all the equipment and crushing materials, sludge deposits and various pumping systems, the power plant and the housing module are installed on the work platform and sometimes a plant for a first crude treatment. The collateral structures used for the production, transport and temporary storage of crude oil are also of great importance, and must also be cathodically protected. To name a few: floating tanks, mooring buoys, jets, pipes that carry crude oil on the surface (risers) or on the bottom (seals), drill pipes, casings. On the other hand, due to the need to exploit reservoirs in deep water, early production systems are in use, including delicate parts, such as well test or collecting groups (manifolds and so-called umbilicals) often made of non-carbon material, e.g. stainless steel and monel, which are cathodically protected.

CP is the corrosion protection technique for immersed zones.

ICCP is generally cheaper than GACP, nevertheless is used in practice only for retrofitting purposes. Some concerns derived from the past experience of ICCP which faced various failures of cable-anode connections and premature anode consumption, most likely because anodic current density was too high. GACP is mandatory when no energy is available; nevertheless it is mostly used because more reliable, assuring good current distribution and no need for maintenance. Drawbacks of the method are high investment cost, high weight and wave induced mechanical load especially for large size anodes; furthermore, operating conditions are unchangeable. Anodes normally used are aluminium-based.

Calcareous deposit

Before discussing the CP design in seawater, it is important to draw attention to the formation of calcareous deposits promoted by the application of CP, especially to its importance for a fruitful protection condition. As mentioned in Sect. 19.1.5, the cathodic reaction causes on the cathodic surface an increase in pH. For a protection current density equal to the oxygen limiting current density, the measured pH is about 10.8–11, and for higher current density where also hydrogen evolution takes place, pH reaches 11.5. This pH increase moves the calcium carbonate-bicarbonate equilibrium to the precipitation of calcium carbonate, $CaCO_3$, as soon as the equilibrium pH, around 8.7, is exceeded. In addition, magnesium hydroxide precipitates when pH exceeds 10.

Protection potential

The exploitation of offshore oil and gas reservoirs started in the Gulf of Mexico after World War II (more than 1000 platforms were present in 1960) followed by the North Sea. The analysis of failures occurred in the 1950s and 1960s indicated two main reasons: plastic collapse due to overloading by action of hurricanes on structures weakened by generalized corrosion and generalized corrosion itself.

From this experience, a design philosophy developed and extended to all other seas that consisted of a mechanical design to withstand the maximum loads induced by so-called 100-year storm and the application of CP. The experience demonstrated that a protection potential of −0.8 V Ag/AgCl was sufficient to reduce generalized corrosion; indeed, it was proved that even potential in the range −0.7 to −0.75 V Ag/AgCl could stop corrosion efficiently, extending the life from 3 to 5 years without CP to more than 30–40 years.

In practice, more negative potentials are recommended, on one hand to allow the formation of beneficial calcareous deposits, and on the other hand also to prevent MIC by SRB particularly at sea bottom. Hence, the recommended protection potential is −0.9 V Ag/AgCl.

Protection current

By adopting a protection potential of –0.8 V Ag/AgCl, the cathodic process is oxygen reduction under diffusion control, then approximating the oxygen limiting current density. The build-up of the calcareous deposit reduces the oxygen availability with time and accordingly the protection current density. For example, based on local conditions, adopted protection current density is 65 mA/m^2 in Gulf of Mexico, 85 mA/m^2 in Persian Gulf and or West Coasts of United States, West Africa and Australia, 120 mA/m^2 in Mediterranean Sea and 250 mA/m^2 in the North Sea.

Typical protection current density values are given in Table 19.9.

Values in Table 19.9 or from standards are the initial ones, then after the so-called polarization period, lasting about a few months where calcareous deposit forms, protection current density drops by about 30–50%. Calcareous deposit starts forming near galvanic anodes where current density is higher, then spreading far away; according to this mechanism, an initial uneven current distribution helps the polarization, hence galvanic anodes are more suitable than impressed current by remote anodes.

Specific conditions affect protection current density, as vibrations or hot surfaces, for example the so-called hot risers; for the latter, because strong convection sets up, the oxygen limiting diffusion current density increases by two or three times.

Finally, once again, it must be emphasised the importance of the build-up of the calcareous deposit, which is affected by: hydrodynamic conditions, such as strong turbulence associated with the abrasive action of suspended solids; the first period of polarisation; the quality of calcareous deposit: low porosity, high thickness, good adhesion and mechanical strength to resist turbulence and abrasive actions. High initial current densities assure the best characteristics.

Current and Potential Distribution

Current output

As previously stated in Sect. 19.2.1, the current output depends on anode size and seawater resistivity. As the latter is function of temperature, output varies considerably from tropical to Arctic regions, as well as with seasons and depth. For example, from the Persian Gulf to Gulf of Mexico and to the North Sea (where the benefits of the Gulf Stream are present) to even colder seas,

Table 19.9 Protection current density for marine structures

Environment	Protection current density (mA/m^2)
Seamud	20–33
Stagnant seawater	50–110
Typical condition	60–140
Stormy seawater	130–550
Hot pipelines	120–600

for example in Alaska, water resistivity increases from 0.2 Ω m up to 0.5 Ω m; hence anode output decreases proportionally.

However, because in colder seas, with higher resistivity, the protection density is also the highest, ohmic drop is not anymore localized on the anode only. For example, in North sea with resistivity 0.3 Ω m or more, protection current density exceeds 150 mA/m^2, hence ohmic drop within 1 m in cathodic region may also be 50 mV: consequently, driving voltage reduces by about 20% and also current output.

Potential distribution

Once checked that the CP system can deliver the protection current, potential distribution must be verified throughout the structure. As complexity and size of the structure increase, the difficulty of this task also increases, because for tubular jacket-like structures like platforms, neither few remote external anodes nor insulating screens are suitable; instead, anodes inside the structure are necessary.

Furthermore, potential distribution needs to take into account specific local conditions, as for example, complex nodes for which ohmic drop is not negligible, so that potential ennobles by more than 100 mV compared to the one far away from them. Besides nodes, under-protection conditions may occur at numerous local situations as shadowed areas, recesses and shielded zones.

Overprotection

In GACP systems, anodes are installed preferably far from surface and not directly in contact with the structures or on a shield. This reduces high anode consumption, as well known in applications other than those of off-shore structures, for example tanker tanks.

In ICCP systems risk of overprotection is much higher due to the higher driving voltage. Adopted solutions are similar to those for galvanic anodes, as distance from structure and use of shields. The availability of a variety of size and shape of anodes in combination with their number helps avoid overprotection conditions.

Current demand versus current availability

Anode current output and current requirement or protection current must balance. For ICCP systems, this condition is easily achieved because current can be changed manually or automatically; instead for GACP, in spite of an intrinsic self-regulating current output capacity, the problem exists and is difficult to solve if not properly considered in the design phase. In fact, current output variation with time has to be forecast in design, by changing shape, number and size of anodes, for example to obtain a high initial output to facilitate the calcareous deposit buildup. In the past bi-anodes, zinc cored with a magnesium shell were proposed for the same purpose.

Cyclic variations of protection current occur as environmental conditions change (turbulence, temperatures, oxygen content) due to tides and seasons; by damaging the calcareous deposit, storms provoke an increase of protection current up to the maximum initial values.

Finally, protection current may increase with time simply because the surface of the structure increases due to additional auxiliary structures as well casings installed after platform installation.

A critical situation to be investigated in design is the one at the middle of design lifetime when a depolarization occurs after an exceptional violent storm, because there is the risk that anodes are no longer able to repolarize the structure by rebuilding the calcareous deposit. Obviously, this is a critical situation because the structure corrodes and anodes consume rapidly.

19.3   Anodic Protection (AP)

Anodic protection was first experienced by C. Edeleanu in 1954. It is applied to active-passive metals for bringing the potential inside their passivity interval by means of an anodic polarization, obtained, as shown in Fig. 19.13, by an electrical circuit where the structure is the anode connected to the positive pole of a DC

Fig. 19.13 Schematic representation of the anodic protection of a stainless steel tank containing sulphuric acid

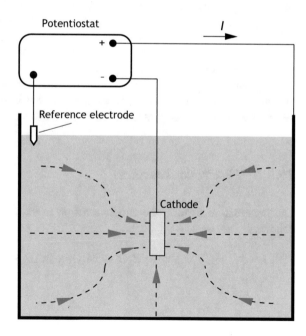

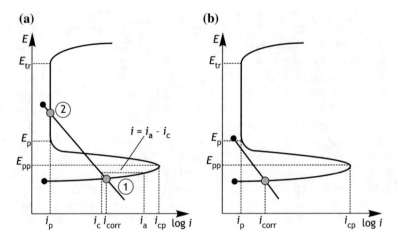

Fig. 19.14 Anodic protection principle: **a** the metal remains passive after the anode current is interrupted; **b** the metal returns under active conditions if the protection is interrupted

feeder and a cathode, connected to the negative pole, is immersed in the solution. The potential is maintained in the passivity interval by a potentiostatic control.

Figure 19.14a illustrates the principle of the method. For instance, with reference to a stainless steel in sulphuric acid, point ① represents the initial corrosion condition as active behaviour. Starting from E_{corr} (i.e., point ①), to achieve an anodic polarization an increasing external current ($i_e = i_a - i_c$) has to be supplied, which initially causes an increase in the corrosion rate, expressed by i_a; then, once reached the primary passivation potential, E_{pp}, the external current decreases sharply because passive conditions are achieved (point ②). To reach this condition, the external current must exceed the critical passivation current, i_{cp}. The potential should be fixed to a value neither too high to avoid transpassivity conditions, nor pitting potential if chlorides are present.

In general, there may be two cases: the first where passive conditions, once established, persist even when the external current is interrupted (Fig. 19.14a); a second case in which the protective action terminates as current stops (Fig. 19.14b).

19.3.1   Electrode Reactions

On the structure surface, the anodic reaction is the formation of passive film, as follows:

$$xM + yH_2O \rightarrow M_xO_y + 2yH^+ + 2ye^- \tag{19.14}$$

with possible competing reaction of oxygen evolution:

$$2H_2O \rightarrow O_2 + 4H^+ + 4e^- \tag{19.15}$$

On the cathode in neutral or alkaline environments, cathodic reaction is the reduction of oxygen, while in acidic solutions, as in most of applications, hydrogen evolution prevails.

In neutral or alkaline electrolytes, anodic reactions cause a slight acidification on the metal surface; this effect is exploited in seawater for titanium components, typically on submarines, because the decrease in pH inhibits the fouling growth.

19.3.2   AP Applications

The anodic protection has found various applications for reducing the corrosion rate of passivating metals in highly aggressive electrolytes (acids, bases and saline solutions) as shown in Table 19.10.

The most known application deals with the storage of concentrated sulphuric acid in carbon steel tanks: AP reduces corrosion rate by two orders of magnitude as well as iron ions contamination and the amount of hydrogen developed, hence reducing the related hazards.

Again in concentrated sulphuric acid, AP is applied as prevention method of corrosion-erosion of AISI 316L in heat exchanger tubes. Without AP, this attack is particularly severe (with penetration rates of several mm/year) because the passive film, enriched with iron, chromium and nickel sulphates, is mechanically unstable. Conversely, AP converts that film into oxides that have better mechanical resistance, hence able to withstand the turbulent regime even at high temperatures.

19.3.3   AP Versus Active-Passive Metals

The anodic characteristic, $E - \log i$, of the active-passive metal determines whether AP is possible, easy and effective, through the following parameters:

Table 19.10 Typical anodic protection applications

Metal	Electrolyte
Carbon steel	Oleum, sulphuric acid, phosphoric acid, ammonia in aqueous solution, aqueous ammonium nitrate solutions, aqueous ammonia and ammonium nitrate solutions, ammonium nitrate and urea
Stainless steel	Sulphuric acid, sulphuric and nitric acid mixtures, phosphoric acid, aqueous ammonium nitrate solutions, aluminium sulphate aqueous solutions, caustic soda, sulphonic acids, oxalic acid, sulphamic acid
Titanium	Sodium chloride solutions, rayon industry, chlorides of Cr^{2+} and Cr^{3+}

- Critical passivation current density, i_{cp}, which determines the maximum current needed (easiness, feasibility)
- Passivity current density, i_p, which measures the residual metal dissolution rate (effectiveness)
- Passivity interval, e.g. the difference between the primary passive potential and the transpassive potential, which defines upper and lower limits of potential range (possibility, feasibility, easiness).

Titanium, for example, passivates more easily than steel in the vast majority of environments, because both i_{cp} and i_p are small. Similarly, austenitic stainless steels, particularly those containing molybdenum, are better than ferritic ones because of low i_{cp}.

Both the passivity current density, i_p, and the critical passivation current density, i_{cp}, depend on both metal and environment nature, composition and temperature (Table 19.11).

Table 19.11 Operating variables for anodic protection of different materials in different environmental conditions

Metal	Environment	Temperature (°C)	Critical current density, i_{cp} (mA/m^2)	Passivity current density, i_p (mA/m^2)
AISI 304	H$_3$PO$_4$ (115%)	24	0.15	1.5×10^{-3}
	H$_3$PO$_4$	82	0.31	1.5×10^{-3}
	H$_3$PO$_4$	177	650	22
	HNO$_3$ (80%)	24	25	0.31
	HNO$_3$	82	120	3.1
	H$_2$SO$_4$ (67%)	24	50×10^3	1
	H$_2$SO$_4$	82	46	2.9
	NaOH (50%)	25	9	4.4×10^{-3}
AISI 316	H$_3$PO$_4$ (75–80%)	104	–	140
	H$_3$PO$_4$	121	–	350
	H$_3$PO$_4$	135	–	440
	H$_2$SO$_4$ (67%)	24	5×10^3	1
	H$_2$SO$_4$	66	40×10^3	3
	H$_2$SO$_4$	93	110×10^3	9
Carbon steel	Oleum	25	1.1	4.4×10^4
	H$_2$SO$_4$ (96%)	27	–	11
	H$_2$SO$_4$	49	–	120
	H$_2$SO$_4$	93	–	1.1×10^6
Alloy 20	H$_2$SO$_4$ (50%)	120	–	10
Titanium	H$_2$SO$_4$ (40%)	60	200	0.2

Often, the presence of contaminants and impurities is important; for example, a small content of chlorides in concentrated sulphuric acid can change by more than one order of magnitude i_{cp} and i_p for either carbon steel or stainless steels.

The upper limit of passivity range depends on metals; for example for titanium, it is always in the order of several volts although it varies with environmental conditions; for carbon steel in concentrated sulphuric acid, it varies from about 2 to 5 V when the concentration of acid increases from 67 to 96%.

The passivity interval of carbon steel and stainless steels does not generally exceed 1 V and shortens considerably, especially at high temperatures or when depassivating species such as chlorides are present.

19.3.4 Throwing Power of AP

The throwing power of AP is generally higher than that of CP, at least once met the passive conditions, for the following reasons:

- Currents are small
- Electrolyte conductivity is high
- The resistance of passive films is high.

Table 19.12 shows the passivity current density, i_p, and film resistivity in typical applications. It appears that there is a relationship: as one grows the other decreases.

The throwing power of AP is so high that passivation is easily achieved even for equipment with complex geometry, with uniform distribution of potential, although primary distribution would be strongly uneven. For example, for carbon steel pipes carrying concentrated sulphuric acid (although passivity current density, i_p, is high and oxide resistivity is low) cathode distance from tube inlet with diameters 25, 50 and 150 mm can be 3, 5 and 9 m, respectively.

For opposite reasons, the throwing power is not good in the initial phase especially when critical passivation current density, i_{cp}, is high and in the presence of interstices, where the achievement of AP may be critical. For example, comparing laminated against wrought stainless steel plate (AISI 304 vs. CF-8 type) the

Table 19.12 Passivity current density and resistivity of passive film in 67% sulphuric acid applications

Metal	Passivity current density, i_p (mA/m^2)	Oxide film resistivity (kΩ m)
Carbon Steel	150	0.26
AISI 304	2	5
AISI 310	1.5	21
AISI 316	1	180
Titanium	0.8	180
Alloy 20	0.3	460

potential and current distribution inside a crevice by imposing different external potentials would show strongly different behaviour because i_{cp} is different:

- The AISI 304 laminated plate, with low i_{cp}, passivates perfectly inside crevice even with external potential +0.2 V SCE
- CF-8, with high i_{cp}, does neither passivate completely under the same conditions, nor with an external potential as high as +0.8 V SCE, that is, the maximum potential before falling into transpassivity.

This is because the throwing power is low before current exceeds i_{cp} inside the interstice, as two out of three necessary conditions are absent: high current density and low oxide film resistance. In conclusion, in the presence of crevices, cracks or interstices, AP is driven, besides the geometry and the external potential, by the value of the critical passivation current density, i_{cp}. Accordingly, for metal/environment couples with very high i_{cp} AP works satisfactorily only in the absence of cracks or after passivation pretreatments carried out in suitable environments.

19.3.5   Potentiostatic Feeding

To apply AP a potentiostat is necessary. As mentioned in Chap. 29, the potentiostat is an electronic device able to maintain a prefixed potential, by regulating automatically the circulating current between the structure and a counter electrode (the cathode). The potential, E, is measured by a reference electrode and compared with the prefixed protection potential. The feeder uses the difference to vary dynamically the current until the potential scatter zeros (Fig. 19.15).

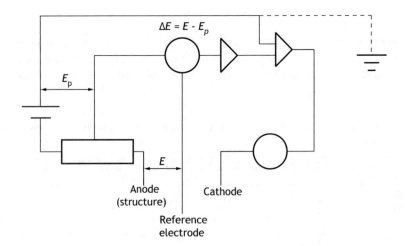

Fig. 19.15 Operating scheme of a potentiostat

Power of the feeder. To set up the protection, a current proportional to i_{cp} is necessary while the one to maintain passivity conditions is proportional to i_p. These two currents differ by orders of magnitude. For example, based on data shown in Table 19.11, AP of a stainless steel AISI 304 tank containing concentrated sulphuric acid (67%) requires 5 A/m^2 to reach passive conditions, while only 1 mA/m^2 to maintain it. To avoid immobilizing an expensive high power feeder not necessary once AP conditions are met, different strategies are used:

- Pre-passivation treatments by employing low i_{cp} electrolytes (for example, using phosphoric acid which shows i_{cp} two or three orders of magnitude lower than sulphuric acid)
- Addition of anodic inhibitors to reduce i_{cp}
- Lowering the temperature
- Step-by-step passivation obtained by a progressive partial filling operation.

Generally, the time needed to reach passive conditions is very short, in the order of minutes; however in a few cases, longer times are required as for cast iron tank containing boiling concentrated sulphuric acid (75–96% at 310 °C). In this case, passivation starts with acid at 60 °C, then increasing temperature by steps of 20 °C per hour. For example, with a tank surface of 20 m^2, the initial current is 130 A, after 24 h as target temperature is reached, current drops to 30 A, then to 5 A after a week and finally to 3 A after one year.

Consequences of feeding current interruption. By interrupting the current, the system goes back to free corrosion in active condition. This is not so harmful if AP has the aim to reduce metal ion pollution when the metal itself resists enough corrosion (for instance, in concentrated sulphuric acid tank applications); conversely, consequences can be catastrophic when free corrosion conditions match with high corrosion rates (Fig. 19.14b): in this case, other prevention methods than AP are necessary.

19.3.6 CP-AP Comparison

Table 19.13 reports and compares the main features of the two techniques. It is worth mentioning that an imperfect CP is almost never more harmful than its absence, instead an insufficient AP can result in disastrous attacks, even more serious than without its application. Only a correct AP leads to a strong reduction of corrosion rate.

It is also interesting to compare CP by passivity and CPrev and AP. For all of them, passive conditions are set up: the first two by supplying a cathodic current, while in the latter by an anodic current.

Table 19.13 Comparison between AP and CP

	Anodic protection	Cathodic protection
Metal	Only active-passive Only bare	All Bare or coated Attention to amphoteric and less noble
Environment	From low to high aggressive	Low to moderate aggressive
Throwing power	Relatively high	Generally low Good with coating or low protection current density
Protection current	Comparable to passivity current Depends on metal	Depends on environment Determined by the coating efficiency
Design parameters	Determined in lab	From practice Empirical
Cost per plant unit	Medium	Low
Cost of operating	Very low	Low to medium
Partial protection conditions	Harmful	Always beneficial

19.4 Applicable Standards

- ASTM B 843—Standard Specification for Magnesium Alloy Anodes for Cathodic Protection.
- ASTM F 1182—Standard Specification for Anodes, Sacrificial Zinc Alloy.
- DNV RP B401—Recommended practise for Cathodic Protection design in seawater.
- EN 12495—Cathodic protection for fixed steel offshore structures.
- EN 12496—Galvanic anodes for cathodic protection in seawater.
- EN 12499—Internal cathodic protection of metallic structures.
- EN 12696—Cathodic protection of steel in concrete.
- EN 12954—Cathodic protection of buried or immersed metallic structures—General principles and application for pipelines.
- EN 13509—Cathodic protection measurement techniques.
- EN 14505—Cathodic protection of complex structures.
- EN 15280—Evaluation of a.c. corrosion likelihood of buried pipelines applicable to cathodically protected pipelines.
- ISO 12473—General principles of cathodic protection in seawater, International Organization for Standardization, CH-1211 Geneva 20
- ISO 15257—Cathodic protection—Competence levels of cathodic protection persons—Basis for certification scheme, International Organization for Standardization, CH-1211 Geneva 20.

- ISO 15589-1—Petroleum, petrochemical and natural gas industries—Cathodic protection of pipeline systems—Part 1: On-land pipelines, International Organization for Standardization, CH-1211 Geneva 20.
- NACE SP0169, Recommended Practice: Control of External Corrosion on Underground or Submerged Metallic Piping Systems, NACE Int., Houston, TX.
- NACE SP0675, Recommended Practice: Control of Corrosion on Offshore Steel Pipelines, NACE Int., Houston, TX
- NACE SP0176, Control of Corrosion on Steel Fixed Offshore Platforms Associated with Petroleum Production, NACE Int., Houston, TX.
- NORSOK M-503, Cathodic Protection, Lysaker, Norway.

19.5 Questions and Exercises

19.1 Discuss by means of Evans diagram the three following cathodic protection criteria: cathodic protection by immunity, by quasi-immunity and by passivity.

19.2 What is the protection criterion used for carbon steel in aerated soil? What is the recommended potential range? Can the protection criterion by immunity be applied to aluminium in pitting condition? Why?

19.3 What are the drawbacks of overprotection condition of a carbon steel coated pipeline?

19.4 The cathodic protection potential of carbon steel in aerated soil is −0.85 V CSE. Estimate the residual corrosion rate at the protection potential considering that in free corrosion condition (−0.65 V CSE), the corrosion current density is about 0.1 A/m^2. Consider the Tafel slope of steel dissolution reaction equal to 0.1 V/decade.

19.5 What are the main parameters that affect cathodic protection current density in seawater? Estimate the cathodic protection current density in the following conditions: (a) seamud; (b) stagnant seawater; (c) cold stormy seawater; (d) propeller zone in ships.

19.6 Demonstrate that the driving voltage of a cathodic protection employing a Zn-based galvanic anode to protect a steel structure is 250 mV. Explain why it is independent of the electrolyte.

19.7 What are the main constituents of an anode backfill of a galvanic anode for cathodic protection application in soil? How does the backfill affect anode current output?

19.8 What are the electrochemical reactions on an insoluble anode for ICCP system in seawater? Which is the prevalent reaction? Discuss the effect on anodic current density.

19.9 Demonstrate that the term $\Psi*$ which is the thermodynamic and kinetic contribution of electrode reactions in an ICCP system is about 0 for steel anodes and 2–3 V for inert ones. Show that for seawater applications it is lower than in soil application.

19.10 Consider an ICCP system of the external bottom of a carbon steel tank in soil, connected to a copper grounding network. A deep vertical anode is used to supply the protection current. Which cathodic current would you design for carbon steel bottom tank (10 m diameter) and for the grounding network (8 copper nets, 1.5 m length and 30 mm diameter)? Which is the power supply of the DC feeder you suggest?

19.11 Explain why a high impedance voltmeter is used to make accurate measurement of pipe-to-soil potential. Why shall the internal impedance be higher if a MMO-Ti reference electrode is used rather than a CSE reference electrode?

19.12 Discuss the main drawbacks of the use of the ON-OFF technique for cathodic protection assessment.

19.13 Consider the following potential measurements of a buried carbon steel pipe: (a) $E = -1.35$ V CSE: measured by a potential probe with internal reference electrode; (b) $E = -1.75$ V CSE measured by a portable CSE reference electrode in ON condition; (c) $E = -1.10$ V CSE: measured by a portable CSE reference electrode in OFF condition. Is the pipe in over-protection condition? Calculate ohmic drop contribution in ON-OFF technique.

19.14 Estimate the cathodic protection current density of carbon steel in neutral fresh water and in acidic solution (pH 3) to polarize the metal to -1.2 V CSE. Oxygen limiting current density is 0.1 A/m^2 and Tafel slope of hydrogen evolution 0.12 V/decade.

19.15 Determine the design parameters for anodic protection of an AISI-304 tank to store 10 m^3 of sulphuric acid 65% at 50 °C.

Humphry Davy, Alessandro Volta and the Cathodic Protection[2]
The Priority of Humphry Davy
In 1823, the British Admiralty commissioned Sir Humphry Davy to investigate the corrosion of copper sheathing of the hulls of wooden ships of that time. The following year, before the Royal Society of which he was President, he read a paper in which he announced the discovery of the method to control metal corrosion nowadays called «cathodic protection».

«In the Bakerian Lecture for 1806, I have advanced the hypothesis [..] that the chemical attractions may be exalted, modified or destroyed by changes in electrical state of bodies; that substances will only combine when they are in different electrical states; and that by bringing a body naturally positive artificially into a negative state its usual powers of combination are altogether destroyed. [..]

---

[2]This paper is extracted from Pietro Pedeferri, *Humphry Davy, Alessandro Volta and the cathodic protection*, Istituto Lombardo: Accademia di Scienze e Lettere, 2001.

It was in reasoning upon this general hypothesis likewise, that I was led to the discovery. Copper is a metal only weakly positive in the electro-chemical scale; and accordingly with my ideas it would only act upon sea water when in the positive state; and, consequently if it could be rendered slightly negative the corroding action of sea water upon it would be null. [..] But how was this could be to be effected? I at first thought of using a Voltaic battery; but this could hardly be applicable in practise. I next thought of the contact of the zinc, tin or iron. I resolved to try some experiments on the subject. I began with an extreme case. I rendered sea water slightly acidulous by sulphuric acid, and I plunged into it a polished piece of copper, to which a piece of tin was soldered equal to about one twentieth of the surface of copper. Examined after three days the copper remain perfectly clean whilst the tin was rapidly corroded: though in the comparative experiment, when copper alone and the same fluid mixture was used, there was a considerable corrosion of copper».
[1]

Meanwhile Davy had advised the Admiralty to utilise the new idea. Cast iron and zinc plates were installed on three copper-sheeted ships in record time. The iron anodes proved to work and halted the corrosive process but, unfortunately, unforeseen side effects, such as the development of marine fouling organisms on copper no longer hindered by the toxic action of its corrosion products, made it unfeasible: fouling reduced the speed of ships under sail and the Admiralty decided against the new protection technique. The method's failure greatly vexed Davy. He was, as he said, *«burned out»*.

Nearly a century passed before the technique proposed by Davy would regain credibility and be utilised with the name of cathodic protection to prevent corrosion of buried structures, oil and gas pipelines, offshore platforms, marine reinforced concrete constructions, ships, chemical reactors or simply to prolong the service life of the electric water heaters in our homes.

The world of cathodic protection recognises that Davy first applied the cathodic protection in 1824 and also first understood the principle which it is based on. On the occasion of the second centennial celebration of invention of the pile, I am happy to prove that the second priority belongs to Alessandro Volta. Until now this fact has been overlooked.

The Priority of Alessandro Volta

Alessandro Volta in a paper to the editors of the Swiss magazine «Bibliothècque Britannique» in 1802 [2]—and *«lu par l'auteur a la Societé de phisique e d'histoire naturelle de Geneve dan la séance du 27e ventose»* (March 18th) of that year—wrote (Fig. 19.16):

L'oxydation est en partie indépendante de l'action galvanique, ou pour mieux dire électrique; car elle est l'effet chimique ou ordinaire de tel ou tel fluide, sur tel ou tel métal; elle en dépend aussi en partie, en tant que le courant électrique modifie singulièrement cette oxydation, en l'augmentant beaucoup dans le métal d'où le courant sort pour passer dans l'eau ou tout autre liquide oxydant, et en le diminuant ou supprimant tout à fait dans le métal où le courant électrique entre, et où le gaz hydrogène se développe. Ainsi donc, le courant électrique exerce une action oxydante et une désoxydante, suivant qu'il passe d'un métal dans un liquide o du liquide dans le métal.

In these few lines, Volta makes three crucial statements:

- First of all, he explains that the oxidation rate of a metal which exchanges current with a solution depends in part on the corrosive processes that take place on the metal and in part on the current exchanged (*«Oxidation is partially independent of galvanic or better electric action: because it is the usual chemical effect of one or the other agent on this or that metal; it is in part dependent on this action because the electric current modifies in a singular manner this oxidation»*)
- Second, he observes that the oxidation rate increases when the current leaves the metal and passes into solution («The oxidation increases very much on the metal from which the current goes out for entering in water or any other oxidising liquid»). This occurs, for example, when a metal immersed in an aqueous solution is coupled with a more noble one, and in general when, as we say today, the metal is polarised in the anodic direction.
- Finally, he points out that the rate of oxidation decreases if the current passes from the solution to the metal even to the point of extinction where hydrogen is evolved. (*«The oxidation decreases or entirely suppresses it on metal in to which current enters and on which hydrogen is evolved».*) This occurs, for example, to a metal coupled with a less noble or when, as we say today, the metal is cathodically polarised or when cathodic protection is applied to it.

First Point

As far as the first point is concerned, no one before Volta had ever described what happens to a corroding metal when it exchanges galvanic current with the solution in which it is immersed. Nobody had before explained the corrosion phenomena and the corrosion rate in term of 'local corrosion' and external polarisation so clearly and correctly.[3]

---

[3]Piontelli wrote: «Volta's attitude against the 'chemical theories' must be considered remembering that the chemical phenomena considered at this epoch very often were 'local processes' at the electrodes». [3] According to Piontelli many of the theories of the pile given after Volta where affected by the confusion between local chemical phenomena at the metal surface/solution and the chemical reaction connected with the current.

Obviously Volta, like Davy few years later, did not realise that corrosion was an electrochemical phenomenon even if he stated that galvanic current can stop corrosion. The belief that the mechanism of corrosion is electrochemical will be expressed by de la Rive only in 1830. It is worth underlining how the concepts expressed by Volta in 1802 will have their complete clarification more than one hundred years after, in the period between the two world wars. In particular, the confusion between local 'chemical' phenomena and the ones produced by galvanic current will end only in 1938 when Hoar, Mears and Brown gave the basic electrochemical theory of cathodic protection and Wagner and Traud the one of mixed potentials.

Second Point

With regards to the second point, observations had already been made, even before the invention of the pile, showing how the corrosion rate of a metal increases when it is in contact with a more noble metal. This type of corrosion, which today we call galvanic or bimetallic, had firstly been described by Giovanni Fabbroni in 1792 in a paper given to the Accademia dei Georgofili of Florence in 1792 but printed only in 1799 [3].[4]

Immediately following the invention of the pile, there was a proliferation of observations on the fact that corrosion increased in metals coupled with a positive pole during the functioning of the pile (Nicholson, Davy, Wollaston). Volta himself in a letter dated September 22nd, 1800 to Landriani [4] writes that already in April 1800 he had shown to Brugnatelli not only the development of hydrogen and oxygen but also the calcination of metals and particularly of zinc in the areas in contact with water. But until 1802 everybody thought that the corrosion which took place on the metal *«from which the current goes out for entering in water or any other oxidising liquid»* was due only to the current exchanged, while this was a part of the oxidation process, as correctly stated by Volta.

---

[4]Fabbroni wrote: «J'avois observé aussi que l'alliage employé a la soudure des plaques de cuivre qui couvrent le toit mobile de l'Observatoire de Florence, s'étoit promtement altéré, chargé manifestement en oxide blanc à son conctact extreme avec ce métal. J'avoix appris [..], en Angleterre, que le clous de fer dont on se servoit autrefois pour assujettir le feuilles de cuivre employées au doublage des vaisseaux, les rongeoient tellement par leur conctact que bientôt le trou étoit dilaté, jusqu'à surpasser la tête de clou qui le retenoit. Il me parut qui il n'en falloit pas davantage pour reconnoître que les métaux exerçoient dans set cas un action réciproque, et que c'etoit elle qu'on devoit attribuer le cause des phénomènes qui s'opéroient par leur réunion ou contact». With observations like these Fabbroni laid the foundations both of the chemical theory of galvanism and of the electrochemical theory of corrosion.

Third point

Finally with reference to the third point that regards the effects of cathodic polarisation, Volta was the very first to state that if the metal is cathodically polarised, the oxidation rate of a metal decreases even to the point of extinction (Pourbaix after 150 years would have said *immunity*). Volta was thus the first to understand the principles underlying cathodic protection and to describe its effects including side effects such as the development of hydrogen. Today every specialist in cathodic protection knows that when steel—or any other more noble metal like tin, nickel or copper—is polarised to the point that hydrogen develops on its surface, corrosion stops. Volta had already grasped this point in 1802.

Conclusion

No doubt exists that Davy in 1824 was the first to apply cathodic protection and that Volta in 1802 was the first to express the basic principles of this technique.

References

[1] Davy H (1824) Phil. Transactions, 114, part 1.
[2] Volta A (1923) Bib. Brit., 274 e 339, 1802. Opere di Volta, Ed. Nazionale Tomo II, 151, Milan, Italy.
[3] Piontelli R (1961) *On the theory of electrochemical systems*, C. R. 3e Reunion CITCE, Milan, Italy, 357.
[4] Fabbroni G (1799) Journal de Physique, 49, 348.

Fig. 19.16 Excerpt from the letter «Aux rédacteurs de la Bibliothèque Britannique» (March 18th, ▶ 1802), in which Volta illustrates the principles of cathodic protection. The letter is kept in the «Cartellario Voltiano» at «Istituto Lombardo di Scienze e Lettere» in Milan, Italy: «...oxidation is partially independent of galvanic or better electric action: because it is the usual chemical effect of one or the other agent on this or that metal; it is in part dependent on this action because the electric current modifies in a singular manner this oxidation. The oxidation increases very much on the metal from which the current goes out for entering in water or any other oxidising liquid. The oxidation decreases or entirely suppresses it on metal in to which current enters and on which hydrogen is evolved. The electric current produces an oxidant or a reducing action as it passes from the metal into the liquid or from the liquid into the metal...»

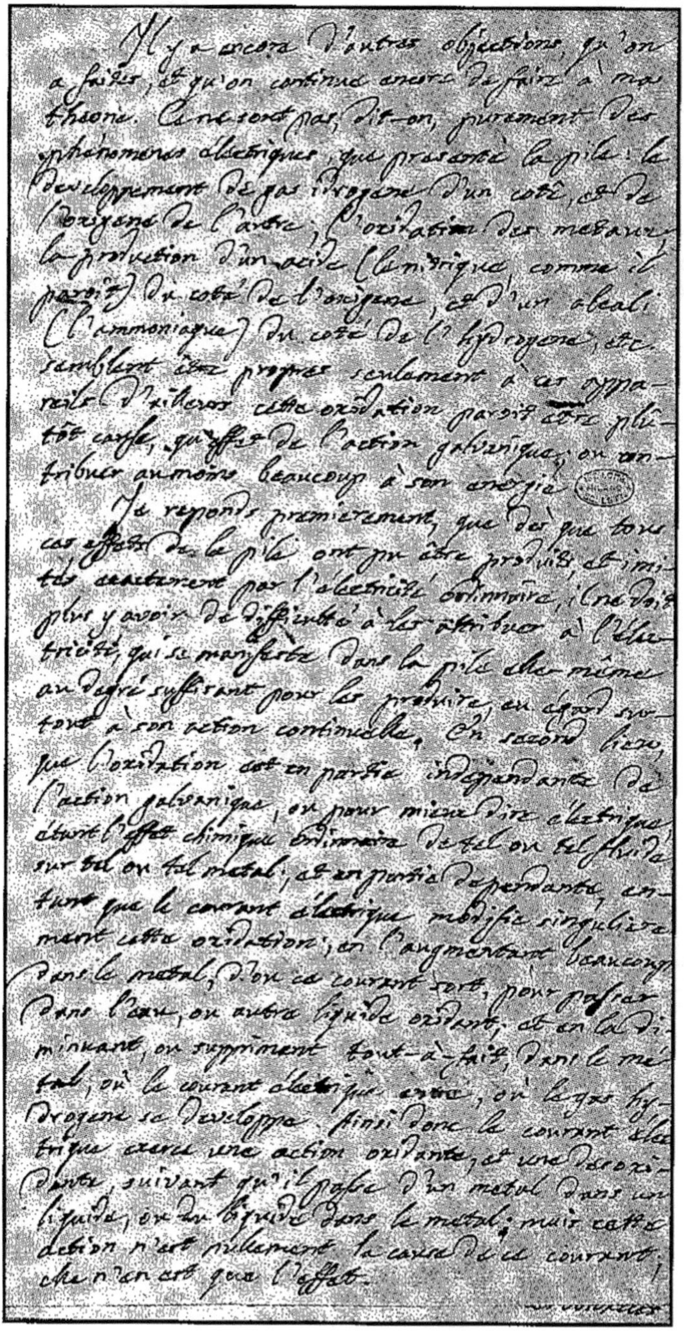

Bibliography

Bianchetti R (2001) Peabody's control of pipeline corrosion, 2nd edn. Nace International, Houston
Ashworth V (1982) Cathodic protection theory and practice. The Present Status. In: 1st international conference, Coventry, UK, 28–30 April 1982
Ashworth V (1989) Cathodic protection theory and practice. In: 2nd international conference, Stratfordon-Avon, UK, 24–28 June 1989
Lazzari L, Pedeferri P (1981) Protezione catodica. CLUP, Milano, Italy
Lazzari L, Pedeferri P (2006) Cathodic protection. Polipress, Milan, Italy
Morgan JH (1992) Cathodic protection. Nace International, Houston
Peabody AW (1967) Control of pipeline corrosion. Nace International, Houston
von Baeckman W, Schwenk W, Prinz W (1997) Handbook of cathodic corrosion protection. Theory and practice of electrochemical protection processes, 3rd edn. Gulf Publishing-Nace International, Houston

Chapter 20
Corrosion in Waters

*Water is smooth and spreads softly, but can corrode rocks
and also destroy rigid and hard things.
This is another paradox: what is soft is strong.*

Lao-Tsu (600 b.c.).

Abstract All natural and industrial waters are corrosive toward common construction metals if a cathodic process can occur. Oxygen reduction is the most typical cathodic process, for about 95% of dealings, then followed by slightly acidic conditions, the presence of oxidizing species as chlorine and more rarely by bacteria. An important factor that reduces water corrosiveness is the tendency to form protective scales. The main damages due to corrosion are alteration of water quality, especially for drinking water requirements, reduction of components service life, due to wall perforation or other localized corrosion forms, and obstruction inside small pipes due to the high volume of corrosion products. The performance of most used metals are reported with reference to their use in freshwater, brackish and seawater.

Fig. 20.1 Case study at the PoliLaPP Corrosion Museum of Politecnico di Milano

© Springer Nature Switzerland AG 2018
P. Pedeferri, *Corrosion Science and Engineering*, Engineering Materials,
https://doi.org/10.1007/978-3-319-97625-9_20

20.1   Types of Water

Waters differs according to the chemical composition and physical properties, as resistivity and pH. The most known types of water are:

- *Distilled* or *demineralized water*, with a very low amount of salts, and a resistivity higher than 20,000 Ω m
- *Freshwater*, characterized by a low salinity level, less than 2 g/L. According to its origin, freshwater is divided in rainwater, surface water and groundwater:
  - *Rainwater* is the product of condensation of water vapour in atmosphere and, in theory, should be chemically pure water. However, in industrial and densely populated areas, rainfall dissolves a series of gases, in particular nitrogen and sulphur oxides, which provide, together with CO_2, an acidic character to rain. Varying amounts of solids, organic substances and even bacteria are also often present
  - *Surface waters* (rivers, lakes, ponds) contain highly variable concentrations of organic matter and minerals due to the contact with atmosphere and soil and because of urban and industrial sewage. Salt content is function of geology of soil and is usually between 50 and 1000 ppm; in Italian rivers it varies between 150 and 300 ppm
  - *Groundwater* percolating through the soil (water table, artesian springs) undergoes a natural filtration process that removes suspended solids and bacteria and dissolves the minerals composing soil and rocks, in a variable amount that depends on the chemical nature. For example, in CO_2 containing water in contact with gypsum ($CaSO_4 \cdot 2H_2O$) or limestone ($CaCO_3$ and $MgCO_3$) there is an enrichment with Ca^{2+} and SO_4^{2-} or Ca^{2+}, Mg^{2+} and CO_3^{2-}, respectively
- *Potable (drinking) water* is the most important freshwater from the corrosion viewpoint. Its origin can be groundwater (as in Italy) or obtained by chemical-biological treatments of surface waters. Basic requirements for potable water are the absence of bacteria, limitations of the content of chemical species dangerous to health, maximum level of ions (chlorides and sulphates lower than 250 ppm, sodium lower than 200 ppm, resistivity higher than 4 Ω m, total dissolved salts lower than 1.5 g/L. The European Directive 98/83/EEC lists all the requirements for a water to be considered for human consumption
- *Industrial waters*, characterized by a wide range of chemical composition, with a salinity varying from few ppm to some g/L. Some of the industrial waters are treated with oxygen scavengers to reduce oxygen content, with biocides to kill aggressive bacteria and with other compounds to control composition and above all pH
- *Brackish water and seawater*. When salinity exceeds 2 g/L water is defined as brackish. Seawater is the most abundant "brackish" water, with an average salinity of 34–36 g/L

- *Formation water* is associated with oil or multiphase hydrocarbons, which are extracted from the reservoir. It is characterized by a very high ionic content, typically from 3 to 4 times that of seawater, and a slightly acidic pH. Formation waters are practically oxygen free, but rich in CO_2 and H_2S: the main cathodic process is hydrogen evolution. Corrosiveness of formation water is described in Chap. 24.

20.2 Factors Influencing Corrosion Likelihood

According to the international standard EN 12502, corrosion of metals in water is influenced by the following main factors: the cathodic reactant, mainly dissolved oxygen, which is the responsible for corrosion rate, water hardness and scaling power, pH, temperature, resistivity, presence of bacteria. On the other side, at a fixed water aggressiveness, metal composition plays an important role in defining type of corrosion and corrosion rate. In order to minimize corrosion damages, special care should be taken during design, construction, commissioning and operation.

20.2.1 Oxygen Content

Oxygen dissolved in water is the main cathodic reagent. It is always present in natural waters in contact with the atmosphere: its solubility depends on temperature, salinity and fluid dynamic regime. Oxygen content in seawater (salinity about 35 g/L) varies from 9 mg/L at 0 °C, to 6 mg/L at 30 °C, 3 mg/L at 60 °C, to zero at 100 °C (at the pressure of 1 bar). As salinity increases, solubility decreases until zero above 150 g/L; in the Dead Sea, which is salt saturated (more than 200 g/L), there is no dissolved oxygen and therefore there is no life and no corrosion for iron. In natural waters, photosynthesis and fouling may determine local conditions of either over- or under-saturation or even anaerobic; the absence of oxygen, which would be ideal to impede corrosion, instead favours microbiologically influenced corrosion. Dissolved oxygen increases as water velocity increases, reaching a maximum in turbulent condition.

20.2.2 Water Hardness

Water hardness is a relevant parameter, as it expresses the content of Ca^{2+} and Mg^{2+} ions, i.e. ions that contribute to the formation of calcareous deposits. There are three types of hardness: total, temporary and permanent. Total hardness expresses the

Table 20.1 Classification of
water as a function of
hardness

Hardness (°F)	Classification
0–7	Soft
7–15	Slightly hard
15–22	Moderately hard
22–35	Hard
>35	Very hard

total content of Ca^{2+} and Mg^{2+}; permanent hardness is that persisting after boiling and filtering water; the temporary is the difference between the first two. Hardness is usually measured in mg/L of equivalent $CaCO_3$ (accounting for both Ca and Mg contents). For example, if the sum of Ca^{2+} and Mg^{2+} equivalents in 1 L of water is 1.5, water hardness is $1.5 \cdot 50 = 75$ mg/L, where 50 is $CaCO_3$ equivalent weight (MW/z = 100/2). Natural waters have hardness from a few mg/L to hundreds of mg/L. Hardness is often measured in French degrees (°F) where 1 °F corresponds to 10 mg/L of equivalent $CaCO_3$. Table 20.1 shows the classification of fresh waters according to hardness.

20.2.3   Scaling Tendency

Depending on composition and temperature, freshwater can form a scale made of calcium carbonate on the metal surface (Fig. 20.1). This scaling tendency is governed by the chemical equilibrium of carbonate/bicarbonate precipitation, because calcium carbonate is insoluble (it is marble) while calcium bicarbonate is highly soluble.

Once formed, the calcareous deposit reduces the oxygen diffusion proportionally to its thickness and porosity. These properties are summarized through two indexes, whose meaning is reported in the box, namely:

- Langelier Saturation Index, *LSI*
- Ryznar Stability Index, *RSI*.

If the Langelier index is positive (corresponding Ryznar index lower than 6), i.e. water has a pH higher than that of saturation, the water will have a tendency to deposit calcium carbonate, then is said "scaling". If the Langelier index is negative (corresponding Ryznar index higher than 6), i.e., pH is lower than the saturation one, water will have a tendency to dissolve any deposits present and is therefore defined as "aggressive". Figure 20.2 shows the map of scaling and non-scaling water as a function of water hardness (expressed as equivalent mg/L of $CaCO_3$ and pH). Scaling waters, with high values of Langelier index, are typically less corrosive. An aggressive water is corrosive only if a cathodic reactant, i.e., oxygen, is present.

Fig. 20.2 Scaling and non-scaling water as a function of hardness and pH

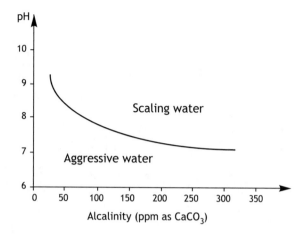

Alcalinity (ppm as $CaCO_3$)

Ryznar and Langelier Indexes

Wilfred Langelier[1] in 1936 proposed and developed an index, called Langelier Saturation Index (*LSI*). It is based on the equilibrium of precipitation of calcium carbonate: $Ca(HCO_3)_2 \leftrightarrow HCO_3^- + H^+ + CaCO_3$ that is a function of pH which is controlled by the amount of CO_2. The Langelier index is the positive or negative offset of the water pH from the equilibrium pH, called saturation pH, pH_s. Hence: *LSI*= pH − pH_s.

The saturation pH, pH_s, is calculated based on the equilibrium condition according to the equation:

$$pH_s = \log K_2 - \log K_{ps} - \log[Ca^{2+}] - \log A$$

where: K_2 = second dissociation constant of H_2CO_3; K_{ps} = solubility product of $CaCO_3$; $[Ca^{2+}]$ = concentration of calcium ions in g · ions/L; A = total alkalinity (alkM) in eq/L. The following relationships are used, depending on temperature:

T < 25 °C

$pH_s = 12.65 - 0.0142(1.8T + 32) - \log[Ca^{2+}] - \log(alkM) + 0.1\log(TDS)$

T > 25 °C

$pH_s = 12.27 - 0.00915(1.8T + 32) - \log[Ca^{2+}] - \log(alkM) + 0.1\log(TDS)$

---

[1]Wilfred F. Langelier (1886–1981) was an American chemist. From 1909 to 1916 he worked for the Illinois State Water Survey. In 1916, Langelier accepted an assistant professorship in Sanitary Engineering at the University of California, Berkeley, where he remained until his retirement in 1955.

where T is °C; $[Ca^{2+}]$ in mg/L; alk_M in mg/L as $CaCO_3$; TDS is total dissolved solids in mg/L.

If the index is positive, the water is scaling. If the index is negative, the water will have a tendency to dissolve scales if present and is said aggressive.

In 1944 the Ryznar Stability Index (*RSI*) was introduced. It is defined by the equation:

$$RSI = 2pH_s - pH$$

Ryznar index 6.5 roughly equals Langelier index 0. This index provides a qualitative and semi-quantitative assessment of either scaling or aggressive tendency of water: when smaller than 6, water has a scaling tendency and, conversely, for an index greater than 6 it is aggressive.

Alkalinity M and P

The alkalinity of water defines its ability to neutralize acidic solutions. Hydroxide (OH^-), carbonate (CO_3^{2-}) and bicarbonate (HCO_3^-) ions which are in equilibrium through the reaction:

$$HCO_3^- + OH^- = CO_3^{2-} + H_2O$$

define the alkalinity.

At pH greater than 8.3, only CO_3^{2-} is present. HCO_3^- exists only at lower pH.

Neutralization reactions are:

(1) $OH^- + H^+ = H_2O$
(2) $CO_3^{2-} + H^+ = HCO_3^-$
(3) $HCO_3^- + H^+ = CO_2 + H_2O$

Reaction (2) is complete at pH 8.3, while reaction (3) at pH 4.5.

Phenolphthalein and methyl orange dyes are used as indicators to check the completion of above reactions after titration of water.

Methyl orange alkalinity (AlkM) is the number of equivalents of strong acid needed to turn methyl orange colour (around pH 4.5).

Phenolphthalein alkalinity (AlkP) is the equivalents of strong acid needed to turn phenolphthalein (around pH 9).

The alkalinity of water gives information on its buffering power, i.e. the ability to receive additions of acidic substances without changing pH significantly.

Acidity P
The acidity of water (named AcidP) is measured by titration with an alkaline solution to turn phenolphthalein. It measures the amount of free CO_2 present. This parameter is used to determine the aggressiveness toward steel and zinc.

20.2.4 Water Resistivity

Resistivity has a strong influence on localized corrosion, either galvanic or by differential aeration; it depends on total dissolved salinity (*TDS*) and temperature: as rule of thumb, the following relation can be used:

$$\rho(\Omega\,m) \cong \frac{30}{(1 + 0.02 \cdot \Delta T) \cdot TDS} \tag{20.1}$$

where *TDS* is in g/L and ΔT is *T*-25 °C.

20.2.5 Bacteria

Bacteria are often present in water, but not all of them cause corrosion. The determination of the presence and quantity of bacteria is generally performed with observations under an optical microscope or through biological cultures. The most dangerous bacteria with regard to corrosion are the so-called sulphate-reducing bacteria (SRB), which develop under anaerobic conditions in the presence of sulphate ions. The corrosive attack is characterized by the formation on the metal of a black deposit of sulphide-containing corrosion products.

20.2.6 Other Cathodic Reactant

In oxygen-free water, other possible cathodic reactions are the reduction of chlorine to chlorides, hydrogen gas evolution in acidic environments, reduction of highly oxidizing compounds (Fe^{3+} ions, H_2O_2, O_3, …). The presence of chromates, permanganates, molybdates or nitrites is dangerous only in the presence of chlorides, which may trigger localized corrosion.

20.3 Uniform Corrosion Rate Evaluation

In the presence of oxygen and chlorine, the maximum uniform corrosion rate of carbon steel, C_{rate}, expressed in μm/year, is given by the following equation:

$$C_{rate} \cong 12 \cdot 2^{\frac{T-25}{25}} \cdot \{[O_2] + 0.14 \cdot [Cl_2]\} \cdot (1 + \sqrt{v}) \qquad (20.2)$$

where $[O_2]$ and $[Cl_2]$ are oxygen and chlorine contents in mg/L, T is temperature in °C and v is water velocity in m/s.

As discussed in Chap. 8, corrosion rate equals oxygen availability in the case of uniform corrosion or is proportional to it in case of localized attack, as in galvanic and differential aeration corrosion mechanisms. Factors influencing corrosion rate in freshwaters are:

- Oxygen availability
- Scaling tendency with precipitation on steel surface of a calcareous deposit which works as a barrier to oxygen diffusion
- Resistivity that mainly affects localized corrosion.

For *non scaling freshwaters*, defined by a negative value of Langelier Index, $LSI < 0$, corrosion rate is given by (chlorine not included):

$$C_{rate,LSI<0} \cong 12 \cdot [O_2] \cdot (1 + \sqrt{v}) \cdot 2^{\frac{T-25}{25}} \qquad (20.3)$$

For *scaling fresh waters*, $LSI > 0$, corrosion rate depends also on the permeability or porosity, p, defined as fraction of bare metal surface (i.e., $1 - p$ is scaling efficiency):

$$C_{rate,LSI>0} \cong 12 \cdot [O_2] \cdot 2^{\frac{T-25}{25}} \cdot p \qquad (20.4)$$

p ranges between 0 when scaling efficiency is unitary, i.e., no oxygen diffusion is possible, to 1 when there is no scale formation. When a calcareous deposit is present, water velocity has a negligible influence, because the layer diffusion thickness in stagnant conditions practically coincides with the scale thickness.

Porosity, p, is a function of saturation index, LSI. As proposed by Lazzari (2017), the following empirical equation can be used:

$$p \cong \frac{\left[10 - (LSI)^2\right]}{10} \qquad (20.5)$$

Therefore, corrosion rate (mm/year) in aerated fresh water with scaling tendency is given by:

$$C_{rate,LSI > 0} \cong 1.2 \cdot [O_2] \cdot 2^{\frac{T-25}{25}} \cdot \left[10 - (LSI)^2\right] \tag{20.6}$$

hence it results as function of oxygen content, temperature and saturation index, *LSI*.

In the presence of a non-homogeneous scale, differential aeration may occurs. The maximum corrosion rate is given by the macrocell current between anodic corroding area and surrounding cathodic area, as seen in Chap. 9:

$$C_{rate,MC > 0} \cong 32 \cdot \sqrt{[O_2, ppm] \cdot 2^{\frac{T-25}{25}} \cdot \left[10 - (LSI)^2\right] \cdot TDS(g/l) \cdot (1 + 0.02\Delta T)}$$

$$\tag{20.7}$$

where parameters and symbols are known.

> **Even Small Changes Have an Influence**
> Iron in distilled (or demineralised) water corrodes with formation of typical brownish rust. The addition of 1 g/L of sodium chloride increases the corrosion rate; corrosion rate increases even more if 1 g/L ferric chloride is added. The addition of 2 g/L of sulphuric acid causes corrosion with hydrogen evolution; the addition of 2 g/L of sodium bisulphite slows down corrosion rate and a black film of magnetite forms on iron surface. The addition of 1 g/L of caustic soda halts corrosion and preserves on iron surface its natural colour; 1 g/L of sodium chromate causes the same effect, but, if 1 g/L of NaCl is also added, localized attack takes place. The addition of 0.2 g/L of permanganate reduces corrosion rate, while 2 g/L stops corrosion; by adding 0.3 g/L of hydrogen peroxide corrosion rate reduces and stops when 3 g/L is added. When calcium or magnesium salts are added to give the same composition of tap water—if you prefer, bring tap water directly—iron corrodes if water is stagnant and does not if stirred.
>
> (from M. Pourbaix, "Lectures on Electrochemical Corrosion", Plenum Press, New York, 1973)

20.4 Metals for Freshwater

Pipelines used to transport freshwater are made with cast iron, reinforced concrete, carbon steel, galvanized steel, polyethylene and PVC. The choice of material depends not only on economic considerations, but also on engineering factors, as diameter, pressure and flow rate. For example, pre-stressed cylinder concrete pipes (PCCP) are used for large diameter, low pressure and little change in load; cast iron for pressures not exceeding 20 bars. For high operating pressure, steel and galvanized steel are used. In civil works and in in-house installations, copper, copper alloys, mainly brass, galvanized steel and stainless steel are used.

For each class of metals, standard EN 12502 highlights the influencing factors related to corrosion issues.

20.4.1 Steel and Cast Iron

Ferrous materials in contact with aerated water are susceptible to both uniform and localized corrosion. In neutral, oxygen-free waters, corrosion is almost negligible. Oxygen concentration thresholds are 0.1 and 0.02 ppm in cold and hot water, respectively.

Corrosion rate decreases in time as a stable scale starts forming composed of corrosion products and calcareous deposit due to the local increase in alkalinity. Oxygen diffusion is controlled by the scale porosity and thickness. Figure 20.3 shows schematically the typical trend of corrosion rate in aerated scaling fresh water.

If the forming scale is porous, not adherent, localised corrosion takes place. This is the most frequent corrosion form on steels and cast irons in water distribution and storage systems. It occurs in the form of galvanic coupling, tubercles or shallow pits, differential aeration, when surface deposits are only partially protective or insufficient.[2] It happens mainly at the welds and heat affected zones in waters with high levels of chlorides and sulfates and low total alkalinity. The rationalization of the experience gained in Europe in the field of drinking water has led to the identification of criteria for estimating the probability of corrosion based on the characteristics of drinking water (Table 20.2).

Localised corrosion can be enhanced by microbial activity, which can take place in anaerobic conditions, especially under debris.

Carbon steel is also used for industrial waters such as, for example, in boilers, where water pH and oxygen content are adjusted to reduce corrosion rate, or in cooling waters formed by the condensation of steam from process gas containing small amounts of acids or organic acids. In both cases, pH adjustment should be carried out, completed by addition of corrosion inhibitors, in particular zinc salts in the case of cooling water, film-forming inhibitors in the case of acidic condensates process, or oxygen scavengers in the case of boilers. These treatments substantially decrease the aggressiveness of the liquid phase and therefore allow to use carbon steel even with good reliability.

---

[2]When water velocity is high, protective deposits are removed mechanically and corrosion morphology is corrosion-erosion. The erosion resistance of surface deposits is lower in slightly aerated water, and corrosion-erosion occurs in waters with low oxygen content. The areas of removal of deposits behave as anodic areas and corrosion rate is accelerated by the effect of turbulence induced in the area of the attack.

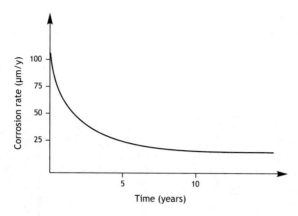

Fig. 20.3 Corrosion rate trend in time

Table 20.2 Conditions for acceptable corrosion rate of carbon and galvanized steel

Parameter—index	Carbon steel	Galvanized steel
pH	>7	>7
AlkM (mEq/L)	>2	>2
Acidity P (mEq/L)	<0.05	<0.7
$[Ca^{2+}]$ (mg/L)	>20	>20
O_2 (mg/L)	<0.02	<0.1
$HCO_3^-/(SO_4^{2-} + Cl^-)$ (mEq/L)	>1.5	>1
$(2SO_4^{2-} + Cl^-)/AlkM$ (mEq/L)	<5	< 1

The effect of pH on carbon steel corrosion rate is clearly depicted in Fig. 20.4. In almost neutral condition (5 < pH < 10) corrosion rate is constant and it depends on oxygen availability. At more acidic pH, the main cathodic reaction is hydrogen evolution, no corrosion products deposit form and the corrosion rate increases

Fig. 20.4 Carbon steel corrosion rate in aerated fresh water as a function of pH

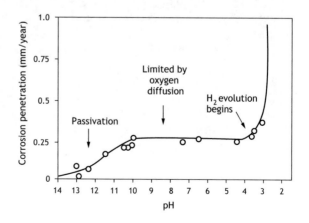

exponentially at reducing pH. In alkaline condition, pH > 10, carbon steel passivates according to Pourbaix diagram. Consequently, corrosion rate reduces to negligible values, lower than 1 μm/year. Only the presence of activating agents, chloride ions for instance, may promote severe localised corrosion in the form of pitting.

Water Treatment for Boilers

Before World War II, cooling water in power plants underwent a physical deoxygenation followed by a chemical treatment with sodium bisulphite and sodium hydroxide. At that time, carbon steel plates were riveted and not welded, then suffering frequent caustic cracking. Around the 1950s, the availability of ion exchange resins for deionization allowed the use of ultra-pure water which was deoxygenated, as in the past, with added a mixture of phosphates (so-called *coordinated phosphate treatment*). Above 100 °C steel reacts with water to form magnetite according to the reaction:

$$3Fe + 4H_2O \rightarrow Fe_3O_4 + 4H_2$$

In the 1970s, as operating conditions became more severe, a new treatment was introduced called AVT (*All Volatile Treatment*) consisting of physical de-aeration followed by hydrazine treatment as oxygen scavenger and injection of ammonia as alkalizer. Hydrazine reacts at temperatures above 100 °C to give nitrogen:

$$N_2H_4 + O_2 \rightarrow N_2 + H_2O$$

In the temperature range 150–200 °C hydrazine tends to decompose into NH_3 and N_2.

Unfortunately, the layer of magnetite, which forms with this treatment, tends to grow hindering heat transfer, so periodic acid cleaning is necessary. The choice of pH between 10 and 10.5 allows to limit these drawbacks. Currently, for its toxicity, hydrazine is replaced by other oxygen scavengers, which, however, may decompose giving rise to carbon dioxide or organic acids.

At the end of the 1970s a new treatment was introduced in parallel to the previous one, called *Combined Water Treatment* (CWT) with dosages of 100 ppb of oxygen and addition of ammonia to bring pH in the range 8.3–8.5. By this treatment, large magnetite crystals, covered with finer hematite crystals, form on the steel surface and grow more slowly, then reducing acid cleaning frequency.

(from P. V. Scolari, *Corrosion in power plants*, AIM Conference, Milan, Italy, 2006).

20.4.2   Galvanized Steel

Hot-dip galvanized steel pipes are used in particular in civil buildings as the resistance to corrosion of zinc is higher than steel. The thickness of zinc and the corrosion rate determine the operational life. The lower corrosion rate of zinc depends on the formation of protective corrosion products that is favoured by the presence of bicarbonates, phosphates, silicates, aluminium ions and organic molecules. Attention should be paid to the temperature because above 50 °C a dangerous polarity inversion between steel and zinc may take place, for which zinc becomes cathodic.

When different metals are present, in particular copper, copper alloys and stainless steel, zinc coating suffers galvanic corrosion in correspondence of junctions; the severity of the attack increases with water conductivity and pipe diameter. Moreover, zinc may locally corrode due to the deposition of copper.

Table 20.2 shows the characteristics of water that make the likelihood of corrosion of carbon steel and galvanized steel minimum, according to EN 12502.

20.4.3   Copper

Copper is the most used metal for water pipes in buildings, due to its high corrosion resistance. Uniform corrosion is rare and if occurring would give as side effect an unwanted release of copper ions in water. Uniform corrosion is prevented by the formation of protective corrosion products, as basic carbonate, $Cu_2(OH)_2CO_3$. Its solubility increases as pH decreases and concentration of carbonic and bi-carbonic ions increases. Uniform corrosion rate is negligible if the following requirements are fulfilled: AcidP < 1.5 mEq/L and AlkM > 1 mEq/L. The more typical corrosion morphology is pitting that happens in situations of altered state of tube surface, in particular for the presence of carbon particles originated during drawing as a result of cracking of lubricants. The weak point of pure copper is the very low resistance to corrosion-erosion. As soon as the water velocity exceed 2 m/s, copper pipes suffer severe localized attack at corrosion-erosion rates as high as 2–5 mm/year. To overcome the problem, copper alloys (brass, Cu–Ni) must be used: their critical velocity increase up to 3 m/s.

Brass with Zn content higher than 15% suffers dezincification.

Pitting on Copper Pipes for Drinking Water
Copper for drinking water pipes works satisfactorily, showing sometimes a few perforations within one year after installation. For pipes conveying cold water, pitting is classified according to two types: *type I*, where pit is triggered by the presence of carbonaceous particles present on the metal surface; *type II*, which occurs when a protective film does not form due to unfavourable environmental conditions.

Type I. Carbon debris, which trigger pitting, may have different origins: cracking decomposition of lubricants (oils or grease) used in drawing or of waxes, fats or oils present on surfaces subjected to welding, or decomposition of soldering fillers. Regulations are nowadays very strict on requirements for proper cleaning (chemical etching) after manufacturing to remove any contaminants.

Type II. Pitting is associated to the formation of copper protoxide, Cu_2O, which appears shiny after cleaning. It forms in the presence of a number of interrelated factors that are not easy to predict. For example: water composition, flow conditions, presence of deposits, for example corrosion products of iron, exposure (the bottom of pipes is more risky). As far as water composition is concerned, influencing parameters are: pH (as it increases, pitting susceptibility decreases), bicarbonates, sulphates and chlorides, which are beneficial, while conversely iron, sodium, manganese and aluminium ions have a detrimental effect; stagnant conditions are much worse than flowing ones. Since pitting takes place especially within a few months from operating start-up, initial conditions are particularly important (chemical composition and hydrodynamics). The influence of copper metallurgy seems poor, although literature reports that copper containing oxygen seems to be more susceptible than deoxidized copper and that the addition of small amounts (1%) of tin or aluminium is beneficial (Fig. 20.5).

Fig. 20.5 Localised corrosion on a copper pipe transporting fresh water

20.4.4 Stainless Steel

Stainless steels are used for transportation and distribution of water and as construction materials in food and pharmaceutic industry, where no kind of metallic contamination is allowed. Most common stainless steel types are austenitic (AISI 304 and AISI 316), the most used, and austenitic-ferritic ones (22Cr5Ni3Mo). Among austenitic types, the choice between AISI 304 and AISI 316 depends on working conditions, namely: level of chlorides, stagnant or flowing regime, the presence of cracks or deposits and the presence of welds (Fig. 20.6).

As described in Chaps. 11 and 12, in selecting the proper stainless steel, PREN index and chloride content must be primarily considered. As rule of thumb:

- Stainless steels with a PREN lower than 18 are recommended in the presence of low chloride content or under special conditions as discontinuous operation, absence of oxygen and other oxidants, cathodic protection or favourable galvanic coupling or at high pH as in concrete
- In flowing water, without cracks and welds, it is possible to use AISI 304 up to 200 ppm of chlorides
- Molybdenum containing stainless steels, as AISI 316 type, with PREN 24–28, can be used for brackish waters with chlorides content up to 1 g/L, not acidic, at temperature not exceeding 30–40 °C
- Stainless steels with PREN 35–40 or higher, resist pitting attack in seawater, provided there are no harmful galvanic coupling conditions as for instance with carbonaceous materials and titanium

Fig. 20.6 Corrosion of AISI 316 stainless steel in the presence of bacteria in water containing tank

- The best performance, even in presence of chlorine, is achieved using stainless steel with PREN greater than 45 (as superaustenitic steels or superduplex: alloys with 6% molybdenum)
- In case of water containing bacteria, especially of the oxidizing type gallionella and manganese oxidizing strain, attacks have occurred even at levels of chlorides lower than those reported.

Stainless steels for potable water and foodstuffs or for devices possibly in contact with food must be sufficiently inert to exclude any transfer of metals to water and food in quantities that endanger human health, change water composition unacceptably, or deteriorate its organoleptic characteristics. To comply with these requirements, steels must be tested at conditions in accordance with standards.

20.5   Brackish Water and Seawater

When salinity exceeds 2 g/L water is defined as *brackish*. Seawater is the most abundant "brackish" water, with an average total salinity of 34–36 g/L with a few exceptions such as Baltic Sea (about 9 g/L), Caspian Sea (average 12 g/L) and Dead Sea (practical at saturation, around 34% by weight, water density 1.240 g/L). Table 20.3 shows the synthetic seawater composition, according to ASTM D1141, with a salinity 35 g/L and density 1.023 g/L at 25 °C.

Oxygen (as well as nitrogen and carbon dioxide) dissolves from the atmosphere and in addition is produced and consumed by photosynthesis and microbiological processes, respectively. The amount of dissolved oxygen can change according to temperature, local turbulence and salinity, as already mentioned in Sect. 20.2.1. The high salinity of seawater influences the carbonate/bicarbonate equilibrium so strongly to determine the seawater as non-scaling water.

Table 20.3 Concentration of the main chemical species in seawater

Ion or molecule	Concentration	
	(mM/L)	(g/kg)
Na^+	468.5	10.77
K^+	10.21	0.399
Mg^{2+}	53.08	1.29
Ca^{2+}	10.28	0.412
Sr^{2+}	0.09	0.0079
Cl^-	545.9	19.354
Br^-	0.842	0.0673
F^-	0.068	0.0013
HCO_3^-	2.3	0.14
SO_4^{2-}	28.23	2.712
$B(OH)_3$	0.416	0.0257

Seawater has a slightly alkaline pH, around 8.3, and buffering properties (i.e., ability to maintain constant pH although small quantities of acids or bases are added) through a complex series of chemical equilibria between dissolved carbon dioxide and carbonates.

Another important factor is *fouling*, which consists of micro and macro organisms, growing on submerged structures and even inside plants. Fouling can be either beneficial or detrimental. In a few cases, it protects the metal beneath as an oxygen barrier; instead, most likely it creates anaerobic conditions, which allow MIC by sulphate-reducing bacteria.

The salinity of seawater is derived from chlorinity, which represents the total content of halogens (chlorides, iodides, bromides) expressed as weight of chlorides in a kilogram of water, or in parts per thousand (‰) obtained by titration with silver nitrate. The following empirical relationship is used:

$$\text{salinity}(‰) = 0.03 + 1.805 \times \text{chlorinity}(‰) \tag{20.8}$$

The content of ionic species in solution determines the electrical conductivity (σ), which is about 100–200 times higher than that of fresh water. The conductivity can be calculated as function of salinity (or chlorinity) and temperature as follows:

$$\sigma = 1/\rho = (0.15 + 0.005 \times T) \text{ chlorinity } (‰) \tag{20.9}$$

where σ is expressed in S/m, resistivity ρ in Ω m and chlorinity in g/L.

20.5.1 Corrosion Zones in Seawater

Among natural environments, seawater is the most corrosive towards carbon steel, due to high conductivity, oxygen availability and formation of porous corrosion products. Localized corrosion may occur on passive metals. Microbiologically influenced corrosion (MIC) is also a concern for localized corrosion attack promoted by SRB (i.e., sulphate-reducing bacteria).

For the classification of corrosion, hydrodynamics and oxygen availability condition determine four different zones as follows.

Atmospheric zone. It is the zone exposed to the atmosphere, free from seawater spray. Relative humidity and pollutants (chloride and sulphate) govern the intensity of the attack, so it varies with the geographical site. Affecting factors are direction and speed of wind, temperature, solar radiation, rainfall, pollution and dust. In this zone, corrosion takes place with the mechanism of atmospheric corrosion. Painting is the most used corrosion prevention method for carbon and low alloy steels.

Splash and tidal zone. It is the zone exposed to alternating immersion and emersion (the tidal width is determined by the geographical site) and includes the zone exposed to a continuous water spray. Corrosion rate is high, almost an order of magnitude higher than that in continuous immersion, due to the continuous wetting and the high oxygen availability. Corrosion prevention is achieved by the use of thick and strong protective coatings, either organic or metallic, as for example, copper-nickel alloys or fibre reinforced epoxy coatings.

Submerged zone. In this zone, the metallic structure is permanently immersed in seawater. The corrosion behaviour depends on temperature and turbulence, which are highly variable from site to site. In stagnant condition, corrosion rate decreases with time because corrosion products mixed with calcareous deposit (i.e., calcium carbonate) contribute to reduce oxygen availability, although not forming a tough scale. In summary, the following equations can be used to estimate uniform corrosion rate (μm/year):

$$C_{\text{rate,seawater,stagnant}} \cong \frac{12 \cdot [O_2]}{\sqrt{t}} \tag{20.10}$$

$$C_{\text{rate,seawater,flowing}} \cong 12 \cdot [O_2] \cdot \left(1 + \sqrt{v}\right) \tag{20.11}$$

where t is time (y) for $t > 1$. In stagnant conditions, affecting parameters are oxygen content, temperature and exposure time; yet, temperature is already accounted for in the variation of oxygen content in the different seas—for instance, the North Sea contains roughly 15–20 ppm O_2 (oversaturation) while this decreases to 8–10 ppm in tropical seas. In turbulent conditions, the main parameters are oxygen content and water velocity. In this zone, corrosion prevention is achieved by cathodic protection.

Mud zone. It is the zone placed below the seabed, where oxygen content is very low, almost close to zero, hence corrosion rate significantly reduces below 20 μm/year. Anaerobic condition may trigger SRB corrosion, with corrosion rate as high as 1 mm/year. Also in this zone cathodic protection is adopted.

Figure 20.7 shows the relative thickness loss for a carbon steel structure, as an offshore platform, operating in the different corrosion zones.

20.5.2   Materials for Seawater

Two main categories are used: corrosion resistant alloys (CRA) and active metals properly protected by cathodic protection and coatings, as described in dedicated chapters. Corrosion resistant alloys, in particular copper, copper alloys, nickel alloys and stainless steels are described in the following.

Copper and copper alloys. In principle, copper suffers oxygen-related corrosion, therefore it is expected to corrode in seawater; nevertheless, long experience has shown that copper and copper alloys resist corrosion in seawater and are considered

Fig. 20.7 Relative thickness loss for a carbon steel structure operating in the different corrosion zones (adapted from Humble 1949)

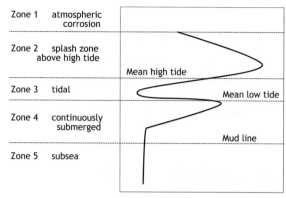

Zone 1	atmospheric corrosion
Zone 2	splash zone above high tide
Zone 3	tidal
Zone 4	continuously submerged
Zone 5	subsea

Mean high tide

Mean low tide

Mud line

Relative loss in metal thickness

as ideal for marine applications. The reason is that copper passivates in the presence of chlorides by forming cupric oxi-chloride, $Cu_2(OH)_2Cl_2$. This corrosion product leads to a passivation of copper and copper alloys, not as strong as stainless steel passivity, nevertheless sufficient to reduce corrosion rate to negligible values from an engineering point of view. Because of passivation and not passivity, this protective layer cannot resist corrosion-erosion; typically, when shear velocity exceeds 2 m/s corrosion erosion takes place. Copper-nickel alloys (typically, 90-10 and 70-30) and Ni-Al bronze are most used in applications when erosion-corrosion is a concern. For heat exchangers, brass (i.e., copper-zinc alloy) is often used.

Stainless steels. On the contrary to copper alloys, stainless steels suffer localized corrosion because of the presence of chlorides. *PREN* is the key for selection of the corrosion resistant stainless steels. Most often, additional preventative measures are adopted, typically cathodic prevention. For safe and reliable use, *PREN* must exceed 40, therefore the use of Mo-containing stainless steels becomes mandatory. Common stainless steels such as AISI 304 and AISI 316 are often successfully used provided cathodic protection is applied (more appropriately, cathodic prevention when applied since the installation). In seawater, crevice conditions should be avoided. For stress corrosion cracking and corrosion fatigue behaviour, reference is made to the dedicated chapters.

Nickel and nickel alloys. Nickel-based alloys offer excellent corrosion resistance to a wide range of corrosive media in energy, power, chemical and petrochemical industries, for applications in seawater and reducing electrolytes. They are also successfully used in nuclear submarines. Some commercially important nickel-copper alloys include so-called Monel as: Alloy 400 (66% Ni, 33% Cu), Alloy R-405, Alloy K-500, which combine formability, mechanical properties and high corrosion resistance, and so-called nickel-based super-alloys. These latter are employed in high temperature applications due to their high mechanical strength and oxidation resistance. Their composition is carefully balanced by additions of chromium, cobalt, aluminium, titanium and other elements. Hastelloy is the trade name of the most known super-alloy family, based on Ni-Mo and Ni-Mo-Cr alloys. Hastelloy B is known for its resistance to HCl. Hastelloy C resists active oxidizing agents such as wet chlorine, hypochlorite bleach, iron chloride and HNO_3. Hastelloy C-276 (17% Mo plus 3.7% W) resists seawater, pitting, stress corrosion, cracking and reducing atmospheres. Alloy 625 (9% Mo plus 3% Nb) offers high-temperature resistance as well as pitting and crevice corrosion resistance.

20.6 Applicable Standard

- ASTM D1141, Standard Practice for the Preparation of Substitute Ocean Water, ASTM International, 100 Barr Harbor Drive, PO Box C700, West Conshohocken, PA 19428-2959, United States.
- ASTM G52, Standard Practice for Exposing and Evaluating Metals and Alloys in Surface Seawater, ASTM International, 100 Barr Harbor Drive, PO Box C700, West Conshohocken, PA 19428-2959, United States.
- Directive 98/83/EEC of the European Community and the Council of 3 November 1998 on the quality of water intended for human consumption. (O. J. EC L 330, 05.12.1998, p. 32–54).
- EN 12502-1, Protection of metallic materials against corrosion—Guidance on the assessment of corrosion likelihood in water distribution and storage systems —Part 1: General, European Committee for Standardization, rue de Stassart, 36 B-1050 Brussels.
- EN 12502-2, Protection of metallic materials against corrosion—Guidance on the assessment of corrosion likelihood in water distribution and storage systems—Part 2: Influencing factors for copper and copper alloys, European Committee for Standardization, rue de Stassart, 36 B-1050 Brussels.
- EN 12502-3, Protection of metallic materials against corrosion—Guidance on the assessment of corrosion likelihood in water distribution and storage systems —Part 3: Influencing factors for hot dip galvanised ferrous materials, European Committee for Standardization, rue de Stassart, 36 B-1050 Brussels.
- EN 12502-4, Protection of metallic materials against corrosion—Guidance on the assessment of corrosion likelihood in water distribution and storage systems —Part 4: Influencing factors for stainless steels, European Committee for Standardization, rue de Stassart, 36 B-1050 Brussels.
- EN 12502-5, Protection of metallic materials against corrosion—Guidance on the assessment of corrosion likelihood in water distribution and storage systems —Part 5: Influencing factors for cast iron, unalloyed and low alloyed steels, European Committee for Standardization, rue de Stassart, 36 B-1050 Brussels.

20.7 Questions and Exercises

20.1 Discuss the effect of pH and water hardness on the scaling tendency. Which are the indexes used to define the scaling tendency? Make examples of water with different indexes.

20.2 What is the effect of water resistivity on corrosion rate of carbon steel in a non scaling freshwater? And in the presence of a non-homogeneous scale?

20.3 A carbon steel fire-system plant suffered localized corrosion after 4 year. The thickness of the pipe is 4 mm; pipe diameter is 2″. During visual inspection, a corroded area 1 cm^2 wide was found. Deposits were detected on the internal side of the pipe. Water composition is as follows: hardness 2 °F, oxygen 2 ppm, chlorides 50 ppm, sulphates 80 ppm, pH 7.5, T = 15–20 °C, conductivity 500 mS/cm. Make a corrosion assessment.

20.4 Estimate the corrosion rate of carbon steel in seawater in the following conditions: oxygen 6 mg/L, stagnant condition, laminar regime (v = 0.3 m/s), turbulent regime (v = 4 m/s).

20.5 Consider the following conditions:

- Stagnant fresh water, pH 6.5, T = 18 °C, 50 mg/L chlorides, 6 mg/L oxygen
- Deaerated fresh water, pH 6.5, T = 20 °C, water velocity 2 m/s, 500 mg/L chlorides
- Brackish water, 2 g/L chlorides, T = 40 °C; 5 mg/L oxygen
- Seawater.

For each condition, propose a stainless steel material selection based on PREN index.

20.6 An AISI 304 stainless steel tank (18–8 CrNi) contains stagnant seawater at 15 °C. In few months, corrosion has penetrated a 4 mm thick plate in the bottom. What is the cause of corrosion? How can corrosion prevention be improved?

20.7 The same steel (AISI 304) is used in a pipeline carrying water to which ferric chloride (FeCl$_3$) is added to such an extent that the free potential is about 600 mV SCE. The temperature is 50 °C. Explain what will happen. Propose alternative material selections.

20.8 A carbon steel platform is designed to work in the Adriatic Sea. Indicate the corrosion zones. For each section, list the influencing parameters, estimate the expected corrosion rate and suggest a possible protection technique.

20.9 The same carbon steel platform has to work in the Nordic Sea. Which are the main differences in the corrosion rate evaluation?

20.10 Explain why in the Dead Sea, aluminium and magnesium alloys corrode, while carbon steel does not.

20.11 What are the chemical treatments adopted to reduce the corrosion rate of carbon steel boilers?

20.12   According to Fig. 20.7, carbon steel corrosion rate in tidal zone is lower than in seamud zone. Explain the reason of this behaviour.

20.13   A water injection plant is used to inject at high pressure huge quantities of water, suitably treated, into the hydrocarbon reservoir in order to increase oil recovery in a petrochemical plant. In principle, it consists of the following five components: water supply pump (so-called lifting pump); a flow-line from supply well area to injection area (even some km long); a booster pump to increase pressure; a distribution system (manifold); injection wells.

- Please indicate corrosion-related problems for each plant unit, comparing the use of carbon steel and stainless steel, for the following three different waters: (a) low salinity water; (b) high salinity water, such as formation or brine separated from hydrocarbons ($TDS > 250$ g/L as NaCl); seawater (TDS 35 g/L as NaCl)
- Suggest treatments for the use of carbon steel for all plant units. Possible treatments are biocides, oxygen removal, inhibitors, filters, corrosion allowance. [Hint: separate the plant into homogeneous zone from a corrosion viewpoint, for instance aerated zones, de-aerated zones].

Giuseppe Bianchi

Giuseppe Bianchi (1919-96) graduated from Politecnico di Milano (Milan, Italy), first in chemical engineering and then in electrical engineering, and there began his teaching and scientific career in 1943. In 1959 he was called to the University of Milan where he held the chair of Electrochemistry for more than thirty years and gave life to the Institute of Electrochemistry and Metallurgy to make it one of the main European research centres in the field of electrochemistry and corrosion. He was a man of high culture and moral stature, a great researcher and a talented technician of anti-corrosion, indeed he was the first true Italian corrosionist. It is an aptitude that is always present in its activity, but becomes prominent since the 1980s when he first tackled the issue of reliability of plants in relation to the risk of corrosion, and then that of transferring the corrosion experience gained in the field to the emerging expert systems. He was also a great teacher. With his lessons he fascinated generations of students, researchers and technicians. His great teaching ability also transpires from his book of corrosion (written with Francesco Mazza) and his splendid monographs: from the one on cathodic protection awarded by the Ministry of Industry in 1954, to those on corrosion and protection in cooling circuits of thermal and nuclear power stations, published in early 1970s. In these works, the perfect knowledge of electrochemistry and of the behaviour of materials in use allowed him to rationalize very complex corrosion processes and to be able to predict and control them.

Bibliography

LaQue FL (1975) Marine corrosion. Causes and prevention. Wiley, New York

Lazzari L (2017) Engineering tools for corrosion. Design and diagnosis. European Federation of Corrosion (EFC) Series, vol 68. Woodhead Publishing, London

Pourbaix M (1973) Lectures on electrochemical corrosion. Plenum Press, New York

Humble HA (1949) Cathodic Protection of Steel Piling In Sea Water. Corrosion 5(9):292–302

Chapter 21
Corrosion in Soil

*I see that water, nay, I see that fire and air and earth, and all
their mixtures become corrupt, and but a little while endure.*
Dante, Paradise Canto 7

Abstract Soil can be defined as a complex agglomeration composed of an aqueous
solution with solid particles dispersed in, originated from the fragmentation of rock.
Its pores entrap either water or air as competitors: these situations determine different corrosion mechanisms, related to the presence (or absence) of oxygen. In this
Chapter, the corrosion forms in soil will be described, divided in three main groups:
oxygen-related corrosion (general and localized, differential aeration), microbiologically influenced corrosion (MIC) and stray current corrosion, by DC and AC.
For these last conditions, acceptance criteria of interference are highlighted.

Fig. 21.1 Case study at the PoliLaPP Corrosion Museum of Politecnico di Milano

© Springer Nature Switzerland AG 2018
P. Pedeferri, *Corrosion Science and Engineering*, Engineering Materials,
https://doi.org/10.1007/978-3-319-97625-9_21

21.1 Soil Classification

The main constituents of soils are coarse particles, sand, clay and silt; typically, an agricultural soil contains sand, clay and silt in comparable quantities. Table 21.1 lists the geological classification of soils, according to AASTHO Soil Classification System (US code). Dry soil is not corrosive, instead corrosion can take place in the presence of water. Water in soil may be available as:

- *Underground water*, with salinity varying from 80 to 1500 ppm
- *Meteoric (rain) water*, collected from atmospheric precipitation and available only for a short period unless retained by clayey soils where it may be held back longer
- *Capillary water*, that is, low salinity water retained in capillaries of clayey soil or lime.

Soil entraps either water or air in pores as competitors: dry soil is aerated (*aerobic conditions*) while wet soil is oxygen-free (*anaerobic conditions*) since the presence of water impedes oxygen diffusion through the pores. The capacity of a soil to retain water increases as the average particle size decreases; accordingly, the presence of coarse particles provides the soil with a high drainage capacity and therefore a low degree of water retention, as typical for sandy and coarse soils. On the contrary, lime and clay have a high capacity to retain water with poor draining capacity, therefore establishing anaerobic conditions. These two situations determine different corrosion mechanisms, which are related to the presence (or absence) of oxygen.

As regards the chemical composition of soil, key factors are the presence of soluble salts, mainly chlorides and sulphates, the presence of bicarbonates able to form calcium carbonate deposits, pH that is usually between 6.5 and 7 but, in extreme conditions, reaches 3 in acidic soils and 9.5 in alkaline soils. Temperature can vary by several tens of degrees above and below zero depending on the season and geographical location. When temperature drops below zero and the water contained in the pores ices, corrosion stops.

The corrosion forms in soil can be divided in three main groups in accordance with corrosion mechanisms involved: oxygen-related corrosion, microbiologically influenced corrosion, MIC, particularly by sulphate-reducing bacteria under anaerobic conditions, and stray current corrosion (Fig. 21.1).

Table 21.1 Classification of soils based on particles size (AASTHO soil classification system)

Class	Definition	Diameter of particles (average)
1	Stones–Gravel	20–2 mm
2	Sand	2–0.2 mm
3	Fine sand	0.2–0.02 mm
4	Silt	20–2 μm
5	Clay	<2 μm

21.2 Corrosion in Aerated Soils

Oxygen in soils as well as in water is the most important factor determining corrosion in the absence of bacterial activity. In certain acidic soils, even the hydrogen evolution reaction is important. The amount of oxygen reduced in the corrosion process is that arriving through diffusion, hence regulated by Fick's law (i.e., corrosion process is under cathodic control of oxygen transportation). The forms of the oxygen-related corrosion are *uniform corrosion* in homogeneous aerated soils with low conductivity and in acidic soils, *localized corrosion* with formation of pustules in the presence of chlorides and sulphates, *differential aeration corrosion* in the presence of non-uniform distribution of oxygen, corrosion by *galvanic coupling* due to the contact with noble metals. As discussed in previous chapters, corrosion rate equals oxygen availability in the case of uniform corrosion or is proportional to it in case of localized attack, as in galvanic and differential aeration mechanisms.

21.2.1 Uniform Corrosion

Although uniform corrosion is very rare in soils, except in some acidic and high resistivity soils, it is useful to consider its theoretical aspects, especially concerning oxygen related corrosion. In the presence of oxygen, the *maximum uniform corrosion rate* is given by the following equation:

$$C_{rate} \cong 12 \cdot 2^{\frac{T-25}{25}} \cdot [O_2] \qquad (21.1)$$

where C_{rate} is expressed in μm/year, $[O_2]$ is the oxygen content in mg/L ($\approx$ppm) and T is the temperature (°C). In soil, since turbulence conditions are absent, the oxygen limiting current density is only slightly variable, generally between 10 and 100 mA/m^2, and the corresponding uniform corrosion rate is limited in the range 10–150 μm/year. In practice, the corrosion rate differs from these values, as it is an order of magnitude lower due to the formation of protective corrosion products after the oxidation of ferrous ions to ferric ions.

NBS long term testing results. After extensive experimentation carried out in the United States, the National Bureau of Standards (NBS) through a statistical approach derived an equation that correlates the metal thickness loss (X, μm) of different metals to the period of exposure and soil resistivity:

$$X = k \cdot t^n \qquad (21.2)$$

where t is exposure time in years, and k and n are two constants that depend on metal, soil resistivity and aeration. Corrosion rate is worked out by derivation from Eq. 21.2:

$$C_{\text{rate}} = k' \cdot t^{n-1} \tag{21.3}$$

The value of n for carbon steel, extrapolated from the results obtained by NBS, ranges from 0.1 (good soil aeration) to 0.8 (poor soil aeration). In well-aerated soil, corrosion rate, although initially great, falls off rapidly with time because the precipitation of ferric hydroxide occurs in the presence of oxygen and this scale on the surface reduces corrosion rate. On the other hand, in poorly aerated soils, corrosion rate decreases slowly with time. As a general consideration, corrosion rate decreases with time, especially for low values of n. k values are inversely proportional to soil resistivity.

21.2.2 Localized Corrosion

The presence of a high content of chloride and sulphate ions influences the corrosion rate through the tendency to depassivate and form soluble corrosion products with scarce protective properties and promotes conditions for localized corrosion attacks.

In particular, the corrosion attack localizes beneath corrosion products or deposits, forming so-called *pustules* or *tubercles*, where oxygen can not be replaced, while the surrounding surface works as cathode, according to the *macrocell mechanism* described in Chap. 9. Inside the tubercle, which works as a pit (Fig. 21.2), the oxidation and hydrolysis reactions produce Fe_3O_4 and Fe_2O_3, increasing the acidity (the pH drops below 4) due to the poor diffusion towards the outer zones. At the same time, sulphates and chlorides enter inside the tubercle transported by the macrocell current, creating an aggressive local environment. If the cathodic area is sufficiently large, the corrosion rate may exceed a millimetre per year.

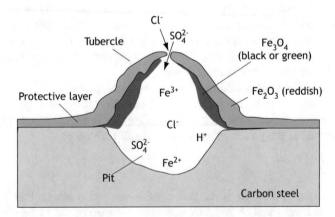

Fig. 21.2 Corrosion of carbon steel with formation of pustules due to the presence of chlorides and sulphates

According to laboratory data and extensive field experimentation (Jailloux 1989), the corrosion rate becomes high, that is, greater than 100 µm/year, when threshold values of 200 ppm of chlorides and 1000 ppm of sulphates are exceeded. With a chloride content less than 200 ppm, a sulphate content less than 1000 ppm, and a resistivity higher than 10 Ω m, the constants k and n of Eq. 21.2 are $k = 25$ and $n = 0.65$. When contents exceed the limits indicated, the constants k and n are calculated as follows:

$$k_{Cl} = 0.21 \cdot [Cl]^{0.86} \quad n_{Cl} = 1 \tag{21.4}$$

$$k_{SO_4} = 2.74 \cdot [SO_4]^{0.32} \quad n_{SO_4} = 0.65 \tag{21.5}$$

When both chlorides and sulphates are present, an additional property applies for k:

$$k = k_{Cl} + k_{SO_4} = \left[0.21 \cdot [Cl]^{0.86} + 2.74 \cdot [SO_4]^{0.32}\right] \quad n = 0.65 \tag{21.6}$$

21.2.3 Corrosion Index

As reported in Harris and Eyre (1994), DVGW (the German Gas and Water Works Engineer Association) developed an index in 1960s to evaluate soil corrosiveness. Trabanelli et al. (1972) and later other researchers (Elsener et al. 1988) revised the model. Soil corrosiveness is evaluated by the use of indices associated with corrosion factors such as resistivity, redox potential, pH, humidity, chlorides, sulphates and sulphides content (Table 21.2).

21.2.4 Differential Aeration Corrosion

Differential aeration corrosion takes place in all those conditions where a non-homogeneous oxygen distribution exists, as in the presence of corrosion products or deposit, which limits the oxygen diffusion.

In soil, differential aeration corrosion takes place when the soil surrounding the structure is not homogeneous and shows variable oxygen permeability. A typical case study is the presence of both a clayey layer and a sandy layer (Fig. 21.3). The metal surface in contact with clay, which is not permeable to oxygen, becomes anodic and metal dissolution takes place ($Fe \rightarrow Fe^{2+} + 2e^-$). On the other hand, the metal surface in contact with sand, which is highly permeable to oxygen, becomes cathodic and oxygen reduction takes place ($O_2 + H_2O + 4e^- \rightarrow 4OH^-$). Therefore, a macrocell sets up, where cathodic and anodic surface areas are separated. Corrosion rate is given by the cathodic process rate, equal to the oxygen diffusion

Table 21.2 Index for corrosion rate estimation

Resistivity		E_{redox}		pH		Humidity	
Ω m	Index	mV SHE	Index	pH	Index	% wt	Index
>120	0	>400	+2	>5	0	<20	0
120–50	−1	400–200	0	<5	−1	>20	−1
50–20	−2	200–0	−2				
<20	−4	<0	−4				

Chlorides		Sulphates		Sulphides	
ppm	Index	ppm	Index	ppm	Index
<100	0	<200	0	Absent	0
100–1000	−1	200–300	−1	<0.5	−2
>1000	−4	>300	−2	>0.5	−4

Index sum	Expected corrosion	Corrosion rate (mm/year)
0	Nil	<0.1
−1 to −8	Low	0.1–0.2
−8 to −10	Moderate	0.2–0.5
<-10	Severe	>0.5

limiting current density multiplied by the cathodic-to-anodic surface area ratio. For example, with an oxygen limiting current density of 100 mA/m^2 (possible in well-aerated zones) and a cathodic-to-anodic surface area ratio of 5, which is a common ratio based on a soil resistivity around 20 Ω m, the corrosion rate is about 0.5 mm/year.

Figure 21.4 shows the electrochemical interpretation of differential aeration corrosion of Fig. 21.3 by means of Evans diagram: in well-aerated soil, the cathodic reaction (i.e., oxygen reduction) forms corrosion products which passivate the steel, then ennobling the potential, while the anodic zone (oxygen-free) remains active. The established macrocell shows a maximum driving voltage of about 200 mV. Figure 21.5 shows a typical example of corrosion by differential aeration on a coated pipeline with a bituminous, i.e., rather porous coating.

21.2.5 Galvanic Corrosion

The presence of two metals showing different nobilities may lead to galvanic corrosion. The less noble metal, which has a lower potential in the natural environment, behaves as anode and the more noble one shows prevalent cathodic behaviour. The mechanism and influencing factors of corrosion due to galvanic coupling are described in Chap. 10. Galvanic corrosion is favoured in low resistivity soils: the lower the resistivity, the greater the corrosion effects. Although soil is an environment with medium to high resistivity, especially if compared to seawater, where bimetallic corrosion is dreaded, galvanic corrosion is not negligible

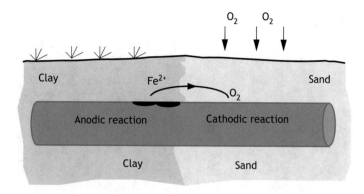

Fig. 21.3 Example of differential aeration corrosion

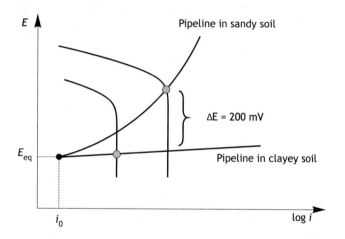

Fig. 21.4 Electrochemical interpretation by means of Evans diagram of differential aeration corrosion

because of an extended electric field geometry (typical cathodic-to-anodic surface area ratio is more than double that the one establishing by the differential aeration mechanism).

To predict galvanic corrosion, the free corrosion potential of concerned metals, as shown in Table 21.3, must be considered. Corrosion rate is high when a small anodic zone is coupled to an extensive cathodic one as it happens in the case of coupling with copper grounding net and buried steel structures. Figures 21.6 and 21.7 show some typical examples for tank bottoms and coated pipelines. So far, cathodic protection has demonstrated to be the best prevention method to stop corrosion. Other prevention measures, such as the application of a coating to the cathodic surface, can only reduce the corrosion rate, but never stops it.

Fig. 21.5 Corrosion by differential aeration on a pipe with a bituminous coating

Table 21.3 Potential series of metals in soil and concrete

Metals	Potentials (V CSE)
Stainless steel (passive)	≈0
Aluminium (passive)	≈0
Copper	−0.2
Steel in concrete	
– Active (carbonated concrete)	−0.55/−0.30
– Active (chloride contaminated concrete)	−0.55/−0.30
– Passive (aerated)	−0.25/+0.20
Carbon steel	
– Bare steel	−0.55
– Coated steel	−0.70/−0.55
Steel in concrete	
– Passive (oxygen free)	−0.80
Galvanised steel	−0.95/−0.80
Zinc anode	−1.04/−0.95
Magnesium anode	−1.70/−1.55

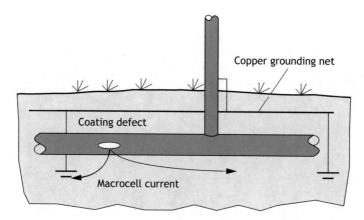

Fig. 21.6 Example of galvanic corrosion in soil between a carbon steel pipe (anode) and a copper grounding net (cathode)

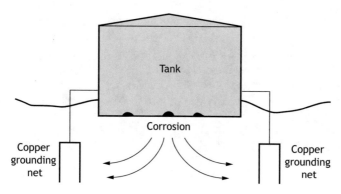

Fig. 21.7 Example of galvanic corrosion of the bottom of a carbon steel tank because of the coupling with a copper grounding net

Resistivity Measurement

The measurement of soil resistivity serves two purposes: evaluate soil corrosiveness and, in CP application, calculate the groundbed resistance. The Wenner method is the most widely used for measuring soil resistivity. It is also referred to as the four-pin method. The method is easy to use and to interpret when soil is assumed homogeneous in the zone under investigation. Measurements are taken with four electrodes, usually made of copper or stainless steel, aligned along the soil surface and equidistant by a length, d. A direct or half-wave rectified current, I, is circulated between the two external electrodes while the ohmic drop, ΔV, between the two internal electrodes is measured. In the case DC is used, the measurement is repeated, inverting the direction of the current. In homogeneous soil, the electric field generated by the circulating current has a semi-spherical shape, as schematically shown in the figure, and the ohmic drop, ΔV, is worked out by solving the electric field equation: $\Delta V = \rho \cdot I/(2\pi d)$, from which the resistivity is obtaibed.

If the soil is not homogeneous and, for example, made up of two parallel layers with different resistivity, the electric field is distorted with respect to a homogeneous condition. The resistivity value worked out from the equation indicated above, suitable for homogeneous layers, represents the so-called *apparent resistivity* and may vary considerably from the true value. In conclusion, the Wenner method is appealing because of its directness and simplicity but often provides an approximate resistivity value. Moreover, it does not provide information on soil stratigraphy.

To overcome the limits of the Wenner's method, Schlumberger proposed an improvement that consists of a series of measurements taken as in Wenner's method, with the difference that the external electrodes are spaced successively, while the spacing of internal electrodes is kept constant. Each measurement provides a value of the apparent resistivity, which in a simplified way may be considered equal to the average resistivity of layers crossed by the current. By increasing external electrode spacing, the depth the current may reach increases and the apparent resistivity measured at each spacing is closer to the resistivity value of the deepest, most influential soil layer (Fig. 21.8).

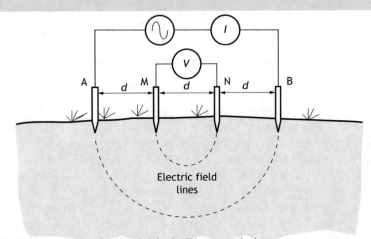

Fig. 21.8 Measurements of soil resistivity by Wenner method

21.2.6 *Effect of Soil Resistivity*

When considering the macrocell mechanism, as in differential aeration corrosion or galvanic coupling, the corrosion rate depends on macrocell circuit resistance and therefore on soil resistivity. A qualitative empirical model can be based on the empirical correlation with resistivity: the greater the resistivity, the smaller the macrocell current and therefore the lower the corrosion rate (Table 21.4). This model is often unreliable and conflicting when one considers that the low resistivity of a soil is due to its high water content, as in the case of clayey soil. However, a water-saturated soil does not allow oxygen diffusion, hence it should be concluded that despite its low resistivity, oxygen corrosion is nil.

Nevertheless, in the absence of oxygen, SRB corrosion may develop without a direct and clear relation to resistivity, although water is necessary for bacteria to thrive. Moreover, high resistivity denotes a dry soil, which permits high oxygen diffusion rate, thus favouring differential aeration corrosion and a high corrosion rate. Therefore, high resistivity soils such as those in desert zones may experience a

Table 21.4 Empirical correlation between soil corrosiveness and soil resistivity

Soil resistivity (Ω m)	Corrosiveness
<5	Very severe
5–10	Severe
10–30	Moderate
30–100	Light
100–250	Poor
>250	Negligible

Table 21.5 Influence of resistivity on corrosion in soil

Factors	Resistivity (Ω m)	
	Low $\rho < 10$	High $\rho > 100$
Water content	High	Low
Composition		
– silt/clay	High	Low
– sand/stones	Low	High
– salt content	High	Low
Water drainage	Low	High
Oxygen diffusion	Low	High
Anodic overvoltage	Low	High
Electrochemical control	Cathodic	Anodic
Free corrosion potential	Less noble	More noble
Tendency to uniform corrosion	High	Low
Tendency to differential aeration corrosion	Low	High
Likelihood to microbial corrosion	High	Low

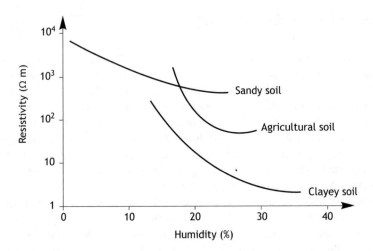

Fig. 21.9 Variation of resistivity with water content (schematic trend). This plot was presented by Mario Arpaia in a cathodic protection course held at Politecnico di Milano, Italy

high localised corrosion rate. Table 21.5 lists the factors that directly or indirectly influence resistivity. The relationship between resistivity and water content is schematically illustrated in Fig. 21.9.

21.3   Microbial Corrosion

Trabanelli et al. (1972) reported that about half of the corrosion attacks on buried pipelines can be attributed to bacterial and microbiological corrosion (MIC, *Microbiologically Influenced Corrosion*). It has been observed that this takes place mainly in neutral or alkaline soils, enriched in organic matter, and is absent in acidic or neutral soils with a high salt content. The most damaging family of these microorganisms is the sulphate-reducing bacteria (SRB) that develop and live in anaerobic environments (Figs. 21.10 and 21.11). The properties of clayey soils, which are neutral and oxygen-free, appear ideal for excluding corrosion from an electrochemical perspective, yet they actually provide conditions that most favour the growth of these bacteria and thus provide the worst outcome from a corrosion point of view. Literature data agree on a corrosion rate even above 1 mm/year. SRB are highly adaptable and are generally able to resist a temperature of 60 °C. Certain families of bacteria resist even up to 80 °C and a pressure of 400 bars, which are conditions encountered in oil and gas reservoirs. Although they are not active in aerobic conditions, they survive and are ready to resume growth as soon as micro and macro anaerobic environments are settled.

Chapter 7 reports a unified model of MIC for active and active-passive metals. For localized corrosion of active metals, typically carbon and low alloy steels in anaerobic environments, the cathodic process is the reduction of sulphate ion to

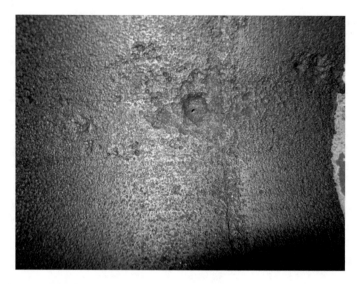

Fig. 21.10 Corrosion attack by sulphate-reducing bacteria on a carbon steel pipeline beneath a welding joint sleeve

sulphide. This cathodic reaction, although as noble as oxygen reduction in neutral solutions, cannot take place spontaneously, if not catalysed, because of its slow kinetics. The most known catalyser is given by SRB metabolism, as proposed by Von Wolzogen and Van der Vlugt (1934). Through an enzyme, the hydrogenase, bacteria make use of hydrogen produced at the cathode to reduce sulphate ions through the following reactions:

$$4Fe \rightarrow 4Fe^{2+} + 8e^- \qquad (21.7)$$

$$8H^+ + 8e^- \rightarrow 8H$$

$$SO_4^{2-} + 8H \rightarrow S^{2-} + 4H_2O$$

$$Fe^{2+} + S^{2-} \rightarrow FeS$$

$$3Fe^{2+} + 6OH^- \rightarrow 3Fe(OH)_2$$

The reduction of sulphate to sulphide occurs also in nature, in anaerobic, sulphate-enriched environments, where the oxidation reaction is carbon to carbon dioxide or methane; if SRB find metallic iron available, the thermodynamically preferred reaction becomes iron oxidation.

Redox potential measurements with a platinum electrode make it possible to establish if a soil is prone to the development of SRB (Table 21.6). In fact, once the pH is fixed, the redox potential measures the oxygen concentration, so that noble redox potentials are characteristic of aerobic conditions. In order to compare

Table 21.6 Relationship between redox potential and probability of corrosion

E_H (mV SHE)	E_H (mV CSE)	SRB corrosion probability
<100	< -200	High
100–200	−200 to −100	Moderate
200–400	−100 to 100	Limited
>400	>100	Nil

measurements of potentials obtained at different pH values, it is necessary to bring them to scale at pH 7 through the following equation:

$$E_H = E_p + E_R + 0.059 \cdot (pH - 7) \tag{21.8}$$

where E_p is the potential measured versus a reference electrode, E_R is the potential of the reference electrode versus standard hydrogen electrode (SHE) and pH is the measured soil pH. For example, if the measured potential, E_H, is −20 mV, the soil

Fig. 21.11 Pipeline for methane transportation (48″ in diameter) affected by SRB corrosion. The attack took place beneath disbonded polythene tape where cathodic protection could not penetrate, while providing conditions for bacterial growth

is prone to SRB corrosion. However, this does not mean that bacteria are present. On the other hand, soil with a measured potential, E_H, equal to +400 mV, therefore well aerated, can not sustain bacterial growth.

21.4 Corrosion by Stray Currents

Interference corrosion greatly worries owners of buried structures, because of the severe damage it causes. It occurs when a DC electric field influences a buried metallic structure, determining the onset of cathodic and anodic surface areas. The latter may suffer severe corrosion called *stray current corrosion*. Interference can be *stationary* and *non-stationary*.

Stationary interference takes place when the structure is immersed in a stationary electric field generated, for example, by a cathodic protection system, and the effect is greater as the structure is closer to the groundbed (GB). Figures 21.12 and 21.13 illustrate two general case studies; in the first case, the interfered pipeline crosses the protected one and the zone close to the GB tends to gather current from the soil (cathodic zone), which is released at the crossing point, causing corrosion (anodic zone). In the second one, the two pipelines are almost parallel and the current is released more extensively, typically in zones exposed to a soil with low resistivity. In both cases, if the interfered structure is provided with an integral coating, interference cannot take place, but when the coating has a number of faults, corrosion is very severe since current concentrates in them.

Non-stationary interference occurs when the electric field is variable, as in the typical case of stray currents dispersed by DC traction systems, illustrated in Fig. 21.14. Interference takes place only during the trains transit, and often, in spite of the limited duration, a few minutes, the effects may be severe due to high circulating current. The corrosion mechanism is simple: the DC traction system

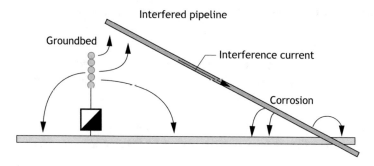

Fig. 21.12 Scheme of stationary interference between two crossing pipelines

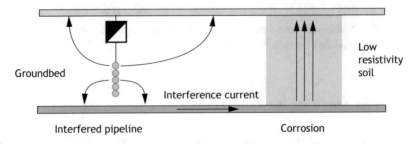

Fig. 21.13 Scheme of stationary interference between two almost parallel pipelines

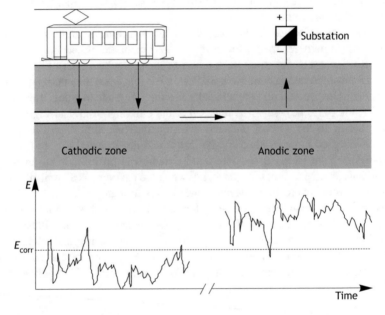

Fig. 21.14 Scheme of non-stationary interference caused by stray current dispersed by a DC transit system

has an aerial conductor as a positive and the track as a negative, so that the current return path is through both track and soil. If a pipeline is close to the track, interference takes place and corrosion occurs where the current leaves the structure near the substation.

In both cases, corrosion attacks are localized and very severe, with corrosion rates even higher than 1 mm/year, depending on current densities reached locally (Fig. 21.15).

Fig. 21.15 Corrosion by stray currents on a coated carbon steel pipe

21.4.1   Electrochemical Reactions on the Interfered Structure

Stray currents influence potentials of both anodic and cathodic zones: the former become more positive and the latter more negative than the free corrosion potential, as Fig. 21.14 shows. On cathodic zones, which receive current from soil, reactions are oxygen reduction first and hydrogen evolution when sufficiently negative potentials are reached, according to the following reactions:

$$O_2 + 2H_2O + 4e^- \rightarrow 4OH^- \tag{21.9}$$

$$2H_2O + 2e^- \rightarrow H_2 + 2OH^-$$

On anodic zones, which release current to the soil, the reaction is metal dissolution when the metal is active, for instance in case of active steel:

$$Fe \rightarrow Fe^{2+} + 2e^- \tag{21.10}$$

When the metal is passive, the anodic reaction depends on the type of metal and environment. For example, in alkaline media, such as pristine concrete where carbon steel is passive, for an initial period the anodic reaction is oxygen evolution, by the following reaction:

$$2H_2O \rightarrow O_2 + 2H^+ + 4e^- \tag{21.11}$$

Then, because acidity is produced, passivity may be destroyed, provoking metal dissolution. This also happens on stainless steel, which is rapidly depassivated, so the corrosion reaction is metal dissolution with same harmful corrosion effects as for carbon steel. According to Faraday law, the amount of metal that dissolves by reaction (21.10) is directly proportional to current and time. A flow of 1 A dissolves about 9 kg/year of iron and a current density of 1 A/m^2 produces a thickness loss at a rate of 1.17 mm/year.

21.4.2   Interference Current

Stationary interference. An evaluation of the interference current is obtained by solving the electric field equation. In practice, the interference current can be estimated from the balance of electrical tensions, as for example shown in Fig. 21.16 for stationary interference. Path 1 is the current path of cathodic protection, where the current leaves the anode and enters the pipeline through the soil. Path 2 is the one of the interference current and sums different contributions: (a) in soil from the anode to the interfered structure; (b) within the structure; (c) in soil from the interfered structure to the protected one. Ignoring the overvoltage at the anode, common to both paths, path 1 is characterised by an *IR* drop in soil and a cathodic polarisation contribution, Ψ_{c1}; path 2 includes *IR* drop in soil and structure, cathodic and anodic overvoltage on the interfered structure (Ψ_{c2} and Ψ_{a2}) and the cathodic polarisation contribution, Ψ_{c1*}. Ψ indicates the overvoltage with respect to the free corrosion potential ($\Psi = E - E_{corr}$) localised at anode (Ψ_a) and cathode (Ψ_c). The balance of electrical tensions is the following:

$$\rho \cdot \int_1 I_1 \frac{\partial L}{S} + \Psi_{c1} = \rho \cdot \int_2 I_2 \frac{\partial L}{S} + \Psi_{c2} + \rho_{str} \cdot \int_3 I_2 \frac{\partial L}{S} + \Psi_{a2}$$
$$+ \rho \cdot \int_4 I_2 \frac{\partial L}{S} + \Psi_{c1*} \tag{21.12}$$

By ignoring the structure ohmic drop and assuming constant cathodic overvoltage ($\Psi_{c1} \approx \Psi_{c1*}$, although it depends on the effective local current density), the electrical balance is:

$$\Delta E = \Psi_{c2} + \Psi_{a2} = \rho \left[\int_1 I_1 \frac{\partial L}{S} - \left(\int_2 I_2 \frac{\partial L}{S} + \int_4 I_2 \frac{\partial L}{S} \right) \right] \tag{21.13}$$

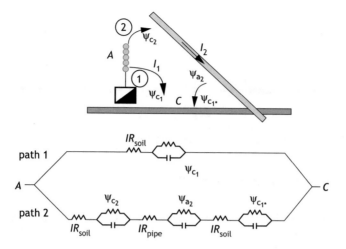

Fig. 21.16 Electrical scheme of stationary interference

where ρ is environment resistivity. ΔE is the resulting driving voltage imposed by the interference system, which drives the interference current and represents the "ohmic drop saving" in soil.

As general approach, the interference current density on the anodic zone (i_{int}), corresponding to the corrosion rate, is expressed by Tafel equation for active metals:

$$i_{int} = i_{corr} \cdot 10^{\frac{\Psi_a}{b}} \tag{21.14}$$

where b (V/decade) is anodic Tafel slope and i_{corr} (A/m^2) is corrosion rate in free corrosion condition.

Non-stationary interference. Cell balance for interference current evaluation is based on the electrical scheme shown in Fig. 21.17, where current I is the current that passes through the rail (estimated to be about 50% of the total current) and I^* is interference current:

$$I^* = \frac{I \cdot R_r}{R_1 + R_{pipe} + R_2} \tag{21.15}$$

where overvoltage is discarded and R_r and R_{pipe} are respectively rail and pipe resistance, R_1 and R_2 are ground resistance. Corrosion damage, quantified through I^*, decrease as the rail resistance, R_r, decreases and the parallel soil path resistance increases.

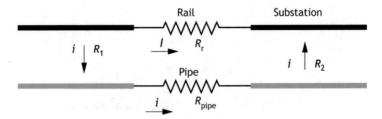

Fig. 21.17 Electrical scheme of non-stationary interference

21.4.3   Interference assessment

Stray current corrosion is assessed through potential measurements; indeed, the interference current cannot be measured. Different criteria apply for *stationary* and *non-stationary interference*, as follows:

- For *stationary interference*, the so-called *ON* and *OFF* potentials on the interfered structure are checked by switching the interference source on and off. The potential shift, $E_{on} - E_{off}$, quantifies the interference: positive at anodic zones and negative at cathodic ones. As reported in Fig. 21.18, on interfered structure, in the absence of CP, the OFF potential matches the free corrosion potential and the ON potential gives the sign of the anodic or cathodic interference

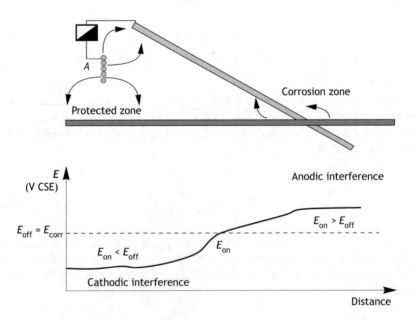

Fig. 21.18 Potential profile in case of interference

- *Non-stationary interference*, generated by DC traction systems, is checked and monitored through a 24-h potential recording. Interference is present if potential changes with time in either anodic (positive) or cathodic (negative) direction, as depicted in Fig. 21.14. However, it is often difficult to ascertain corrosion conditions from the potential recording, because there is a high *IR* drop contribution caused by stray current circulation in soil. For evaluating the true potential in the presence of stray currents, according to standards the use of coupons or potential probes is recommended.

21.4.4 Criteria for Interference Acceptance

The effect of stray current corrosion depends on interference current, which, as said, is not measurable. To assess interference conditions, the definition of an acceptable corrosion rate, generally agreed on the threshold rate of 10 µm/year, is based on the measurement of potential.

20 mV anodic potential shift criterion. Let's consider an interfered structure. On anodic zones the potential ennobles (more positive) and conversely on cathodic zones it becomes more negative than the free corrosion potential, E_{corr}, as shown in Figs. 21.19 and 21.20 for an active metal and a passive one in an ideal case study for which anodic and cathodic surface areas, S_a S_c, are of comparable size. Figure 21.21 illustrates the most common and dangerous case, when the cathodic surface area is much greater than the anodic one ($S_a \ll S_c$) and cathodic polarization becomes practically negligible.

For active metals, the increase in current density (and therefore in corrosion rate) can be estimated by measuring Ψ_a (Eq. 21.14). Assuming an anodic Tafel slope, *b*, of 50–100 mV/decade, an increase in potential of 50–100 mV corresponds to an increase in corrosion rate by one order of magnitude (for example, from 100 to 1000 µm/year). Accordingly, since the potential-current dependence is exponential, it is reasonable to consider 20 mV as maximum allowed potential increase. This is

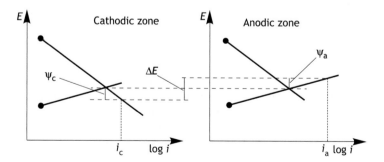

Fig. 21.19 Potential variation on anodic and cathodic zones for active material

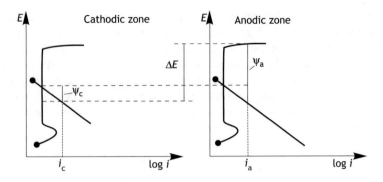

Fig. 21.20 Potential variation on anodic and cathodic zones for passive material

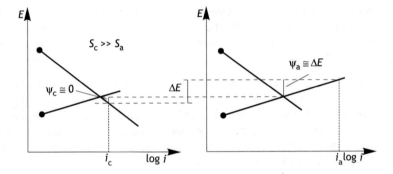

Fig. 21.21 Potential variation on anodic and cathodic zones for active materials when $S_c \gg S_a$

Table 21.7 Values of Ψ_a and Ψ_c in some practical cases (Lazzari and Pedeferri 2006)

Type of structure	Ψ_a (mV)	Ψ_c (mV)	$\Psi_a + \Psi_c$ (mV)
Bare steel	20	0	20
Coated steel	20	300	320
Steel in concrete (passive)	500–800	0	500–800
Steel in concrete (active)	20	0	20

the meaning of the 20 mV anodic potential shift criterion. Table 21.7 reports values of Ψ_a and Ψ_c for different practical cases.

Driving voltage criterion for concrete structures. Steel reinforcement of concrete structures resists corrosion because in passive condition due to the alkalinity (pH > 13) of the cement paste. Corrosion occurs when passivity is destroyed by acid attack, for example due to carbonation, or when chloride content exceeds a critical concentration (refer to Chap. 23) or to stray currents.

In the latter event, on passive steel, interference current can flow only if a driving voltage (ΔE) higher than 0.8 V for $S_a \cong S_c$ is available (refer to Fig. 21.20) which reduces to 0.5 V if cathodic surface area is large ($S_c \gg S_a$). So a high driving voltage is required because the anodic reaction is oxygen evolution, which occurs at a noble potential of +0.5 V CSE.

When interference is stationary, the continuous oxygen evolution breakdowns passivity due to the strong local acidification, then steel starts corroding as active metal. Non-stationary interference is generally not harmful because there is a sufficient time to neutralise the produced acidity during the short interference period. Once steel is active, the 20 mV anodic potential shift criterion applies. Table 21.7 summarizes the practical conditions.

21.4.5   Prevention and Control of Stray Current Corrosion

Depending on the type, different approaches are used to prevent and control interference. The main strategies are the following.

Stationary interference. Prevention methods of stationary interference follow two basic principles: the elimination of driving voltage and the increase in current path resistance. Driving voltage, ΔE, typically zeros by inserting a drainage, which is the most common, effective and economic method used for crossing pipelines. It consists of an electrical connection between the pipelines at the crossing point so that both pipelines are cathodically protected. A calibrated resistance (Fig. 21.22) is used when pipelines have different coatings, hence requiring different protection current density. Driving voltage can be reduced by installing galvanic anodes at the crossing, so most of the interference current leaves anodes and not steel, reducing corrosion rate. The resistance of the current interference path is increased by the use of insulating joints and coatings (Fig. 21.23). Insulating joints show best performance where the potential gradient in soil is low or minimum that is, generally far

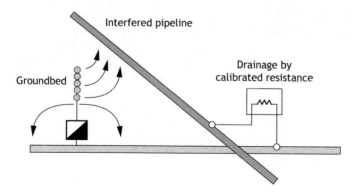

Fig. 21.22 Drainage at pipeline crossing

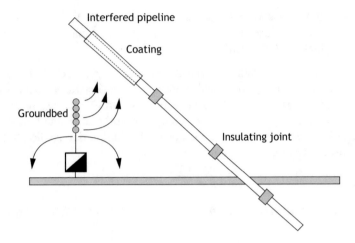

Fig. 21.23 Example of the use of insulating joint and coatings

away from both the crossing point and the interfering groundbed. Coatings are best effective when applied to cathodic zones.

Non-stationary interference. It is more difficult to prevent or reduce because changing in time. A general prevention measure is the adoption for rails of welded joints to reduce the ohmic drop within the metallic rails within below 1 mV/m. On the interfered structure, the reduction of interference current is achieved by increasing the pipeline resistance by means of insulating coatings on cathodic zones and by inserting insulating joints. Cathodic protection is an effective control method if operated by means of the so-called constant potential setting of the feeding system, with the aim to annihilate automatically interference effects; a reference electrode placed at the anodic zone is used to drive potential control.

Insulating Joint Interference
When protected pipelines convey electrolytes, particular attention should be paid to internal interference occurring at insulating joints. Figure 21.4 shows a typical case study, where cathodic protection current flows through the electrolyte, thus bypassing the joint and provoking an internal corrosion attack. Interference current (path 2) is the competing path of protection current (path 1).

Side effects of this inevitable interference are reduced by increasing resistance on the interference current path; this is achieved by installing an insulating spool, or by applying an internal coating for a suitable length on the protected side.

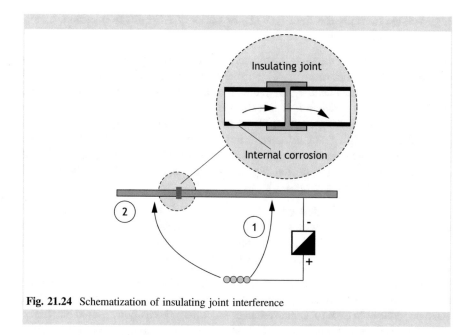

Fig. 21.24 Schematization of insulating joint interference

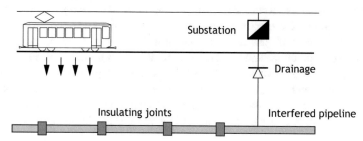

Fig. 21.25 Example of drainage and insulating joint insertion to mitigate non-stationary interference

An elegant method, alternative to CP, is the drainage, shown in Fig. 21.25, which consists of an electrical connection between the structure and the substation rail to drive the interference current through a metallic path. In order to avoid dangerous current inversion, as in the case of a temporary substation shutdown, the drainage must include a diode. Most often, drainage is provided by an ICCP system connected to the rail as groundbed and operated at constant potential.

21.4.6 Alternating Current Interference

Alternating current (AC) can cause corrosion attacks by an interference mechanism on buried metallic pipelines that run parallel to an AC interference source, as high-voltage transmission lines (HVTL) or tracks of AC railway traction systems. Interference mechanisms are as follows:

- *Interference by conduction*. It happens if AC current is spread in soil as typically from grounding networks of AC transmission lines and AC traction systems. AC currents may affect nearby buried metallic structures such as pipelines and tanks;
- *Interference by induction*. It takes place when a buried, well-coated pipeline with a high insulating coating, like extruded polyethylene or polypropylene, is parallel to a high voltage transmission line, typically 130 kV or higher. The alternating unbalanced magnetic flux established between cables and soil induces an AC in the coated pipeline (auto-transformer effect, which is not possible if HVTL is buried).

In both cases, the buried structure exchanges AC through coating defects or holidays. Nowadays, it is agreed that AC-induced corrosion happens in two specific conditions: freely corroding and overprotected conditions (i.e., pipeline potential below −1.2 V CSE) with AC density exceeding a threshold.

Influencing factors. The AC voltage, V_{AC}, on a pipeline is the driving force for the AC corrosion processes taking place on the steel surface at coating defects, so it must be reduced to avoid AC interference. AC voltage is easily measured between a metallic structure and a reference electrode placed in a remote position, i.e. where no further variations of the AC voltage are measured increasing the distance between the reference electrode and the structure. Moreover, corrosion damage depends on AC current density on a coating defect and on the level of DC polarization (IR-free potential and protection current density). However, in contrast to the AC voltage measurement, current density can be measured only through dedicated coupons simulating a coating defect with a known surface (1 cm^2) or it can be estimated based on measuring parameters, as remote voltage and soil resistivity. The spread resistance of a coating defect is given by the following equation:

$$R = \frac{\rho}{2\pi\phi} \qquad (21.16)$$

where R (Ω) is spread resistance, ρ (Ω m) is soil resistivity and ϕ (m) is coating defect size. The current density, i_{AC}, exchanged on the defect is given by:

$$i_{AC} = \frac{V_{AC}}{R \cdot A} = \frac{8\pi\phi V_{AC}}{\rho\pi\phi^2} = \frac{8V_{AC}}{\rho\phi} \qquad (21.17)$$

where i_{AC} (A/m^2) is current density, V_{AC} is remote voltage and other symbols are known.

According to international standard ISO 18086, the remote voltage on the pipeline, V_{AC}, and AC current density, i_{AC}, should be maintained lower than 15 V and 30 A/m^2, respectively, on a 1 cm^2 coating defect. By inputting the current density threshold of 30 A/m^2, the maximum allowable remote voltage of 15 V, and an average resistivity of 50 Ω m, the minimum coating defect diameter results about 10 cm. It follows that the main control method to decrease AC current density on smaller coating defects of a coated pipe is the grounding, which increases the exposed surface to soil.

Acceptable AC interference levels. In the last years, an extensive effort has been performed to provide acceptable criteria to evaluate AC corrosion likelihood. The recent international standard ISO 18086 reports that in the presence of AC inter-ference the criteria given by ISO 15589-1 (i.e. the −0.850 V CSE criterion) are not sufficient to demonstrate that steel is protected against corrosion. The standard provides the following acceptable interference levels, measured as an average over a representative period of time (e.g. 24 h):

- As a first step, the AC voltage on the pipeline should be decreased to 15 V R.M. S. or less;
- As a second step, effective AC corrosion mitigation can be achieved by meeting the cathodic protection potentials defined in ISO 15589-1, and:
  - Maintaining AC density, i_{AC}, lower than 30 A/m^2 on a 1 cm^2 coupon or probe, or
  - Maintaining the average cathodic current density lower than 1 A/m^2 on a 1 cm^2 coupon or probe if AC current density is more than 30 A/m^2, or
  - Maintaining the ratio between AC current density, i_{AC}, and DC current density, i_{DC}, less than 3.

In other words, while no AC density restrictions are defined for DC densities lower than 1 A/m^2, AC density is restricted to values lower than 30 A/m^2 with DC density in the range from 1 to 10 A/m^2. In a recent work by Brenna et al. (2015), a more conservative criterion has been proposed based on experimental tests on carbon steel under cathodic protection condition and in presence of AC stationary inter-ference. Laboratory tests showed that in overprotection condition, i.e., potential more negative than −1.2 V CSE, only a few A/m^2 of AC density could cause corrosion of overprotected carbon steel. Results showed that cathodic current densities lower than 1 A/m^2 in combination with AC density higher than 30 A/m^2 can lead to AC corrosion, as well as DC density higher than 1 A/m^2 in combination with AC density higher than 10 A/m^2.

AC Corrosion Mechanism
Various hypotheses about the mechanism by which AC produces and enhances corrosion of carbon steel (even in CP condition) have been pro-posed. Most interpretations are based on electrical equivalent circuits repre-senting the impedances existing between pipe and remote earth, on electrochemical and mathematical models considering the anodic semi-period effect of the AC signal, and on the effect of AC on the formation of passive layer on steel under cathodic protection condition.

Recent works by Goidanich et al. (2010a, b) and by Brenna et al. (2015, 2016) suggest a corrosion mechanism in which the effect of AC is twofold. A "mixed corrosion mechanism" was initially hypothesized, with a general decrease in both anodic and cathodic overvoltage and an increase in exchange current density of different metals (carbon steel, galvanized steel, zinc, and copper) in various environments (e.g., soil-simulating solution, artificial seawater) in the presence of AC. Generally, AC pushes the potential of carbon steel under cathodic protection toward more noble values, and reduces overvoltage contributions.

On the other hand, a two-step AC corrosion mechanism has been proposed (Brenna et al. 2015). In the first step, AC causes the weakening of the passive film formed under cathodic protection, due to electromechanical stresses. Electrostriction appears to be a convincing explanation of the passive film breakdown mechanism, because of the presence of high alternating electric field (of the order of 10^6V/cm) across the passive film. After film breakdown, high-pH *chemical corrosion* (i.e. potential independent) occurs in overprotection condition because of the high cathodic current density supplied to the metal.

21.4.7 Typical Cases of Improbable Interference

There is a tendency to attribute unexpected corrosion attacks to stray currents. Generally, this happens because of ignorance of corrosion principles or someone wants to avoid taking any responsibility and says: it is nobody's fault, it is just caused by stray currents! The following case studies are typical (for more details see Lazzari and Pedeferri 2006).

Corrosion in water piping and heaters. The perforation of water piping, heaters and heat exchangers in plants is sometimes attributed to stray currents and internal corrosion, for instance because of inadequate water treatment, is not taken into account. If stray currents were responsible for this damage, one should first identify their source. Often, stray currents are present in soil because of electric transit systems or ICCP systems, hence a question arises, that is, how they can affect the piping. Looking at this system, stray currents from rail or groundbed may be picked up by the grounding, then circulate in copper cables and eventually return to the soil, that is, to the transit substation or CP feeder. Such a current path (soil-grounding-copper-grounding-soil) can only produce corrosion of grounding rods where current leaves the metal to the soil. Other current paths are not possible. A hypothetical one, consisting of: soil-grounding-copper-pipe-water-pipe-copper grounding-soil, which potentially could produce corrosion, is not consistent since the internal path "copper-pipe-water-pipe-copper" cannot occur because the metal is equipotential. The conclusion is that the cause of corrosion is corrosion, and not stray current interference.

Corrosion of piping in buildings. Often in buildings, piping embedded in concrete suffers corrosion, causing trouble, high repair costs and litigation when the builder claims stray current effects, hence discarding responsibility. In this case, corrosion comes from the external surface so that stray current interference might seem to be responsible. Nevertheless, again, it is not. As in the previous case, stray currents may be present in the ground, and the grounding system may pick up some current. An alternative path is possible, such as: soil-grounding-pipe-concrete-soil, thus causing corrosion at the "pipe-concrete" interface, but this path is much more resistant by several orders of magnitude than metal paths, because bricks and concrete have a much higher resistivity. Therefore, some current approximately nano-amperes can leave bare pipe surfaces, but with no practical effects.

21.5 Applicable Standards

- AASHTO Soil Classification System, American Association of State Highway and Transportation Officials.
- ASTM G 51, Standard test method for measuring pH of soil for use in corrosion testing, American Society for Testing of Materials, West Conshohocken, PA.
- ASTM G 57, Standard test method for Field Measurement of Soil Resistivity Using the Wenner Four-Electrode Method, American Society for Testing of Materials, West Conshohocken, PA.
- ASTM G 187, Standard test method for Measurement of Soil Resistivity Using the Two-Electrode Soil Box Method, American Society for Testing of Materials, West Conshohocken, PA.
- EN 12501-1, Protection of metallic materials against corrosion—Corrosion likelihood in soil. Part 1: general, European Committee for Standardization, Brussels.
- EN 12501-2, Protection of metallic materials against corrosion—Corrosion likelihood in soil. Part 1: low alloyed and non-alloyed ferrous materials, European Committee for Standardization, Brussels.
- EN 50162, Protection against corrosion by stray current from DC systems, European Committee for Standardization, Brussels.
- ISO 11048, Soil quality. Determination of water-soluble and acid soluble sulphate, International Standard Organization, Geneva, Switzerland.
- ISO 15589-1, Petroleum, petrochemical and natural gas industries—Cathodic protection of pipeline systems—Part 1: On-land pipelines, International Standard Organization, Geneva, Switzerland.
- ISO 18086, Corrosion of metals and alloys—Determination of AC corrosion—Protection criteria, International Standard Organization, Geneva, Switzerland.

- ISO 21809, Petroleum and natural gas industries—External coatings for buried or submerged pipelines used in pipeline transportation systems, International Standard Organization, Geneva, Switzerland.
- NACE TM0106, Detection, testing, and evaluation of microbiologically influenced corrosion (MIC) on external surfaces of buried pipelines, NACE International, Houston, TX.

21.6 Questions and Exercises

21.1 Explain why in well-aerated soil corrosion rate of carbon steel decreases with time more rapidly than in case of poor aerated soil.

21.2 Consider Eq. 21.3 for the calculation of carbon steel corrosion rate in soil. Calculate corrosion rate after 1, 10 and 20 years in the following conditions: (a) chlorides 100 mg/L, sulphates 500 mg/L; (b) chlorides 1000 mg/L, sulphates 1000 mg/L.

21.3 Demonstrate by means of Evans diagram that the maximum driving voltage in differential aeration of carbon steel in soil is about 200 mV. [Hint: suppose that the Tafel slope of the anodic process is 100 mV/decade and 200 mV/decade in poor-aerated soil and well-aerated soil, respectively].

21.4 Consider the replacement of a pipe section in a sandy soil by a new spool. Make a corrosion assessment.

21.5 In free corrosion condition, carbon steel corrosion rate decreases with time. However, in the case of carbon steel corrosion due to galvanic coupling (for example with a grounding copper net), corrosion rate of carbon steel does not decrease with time. Which are the causes of this behaviour?

21.6 In a gas station, a double wall underground tank is connected to a copper grounding system. The tank external surface is coated with 1 mm thick polyester reinforced coating. Estimate the time of perforation of the outer wall (3.5 mm thick) in the presence of a small defect. Estimate oxygen content as a function of soil type and consider a surface area ratio of 30.

21.7 In a fuel station, three underground structures are electrically connected: a carbon steel tank coated with reinforced polyester (1 mm thick), a copper grounding network and a reinforced concrete foundation. Make a corrosion assessment. Explain why carbon steel reinforcements (that have the more negative free corrosion potential) do not undergo corrosion. Is the use of magnesium galvanic anodes effective?

21.8 How does an electrical drainage for interference prevention work?

21.9 A coated carbon steel pipe in soil (6 mg/L of oxygen) is interfered by a DC non-stationary source. What are the electrochemical reactions corresponding to the anodic and cathodic zones? Corresponding to the anodic zone, a positive potential shift of 50 mV with respect to the free corrosion potential has been measured. Calculate the corrosion rate in the anodic zone.

21.10 Consider the following interference conditions measured on a carbon steel corrosion coupon (1 cm^2), not in cathodic protection condition: (a) stationary interference (1.5 A/m^2 for 24 h); (b) non-stationary interference (10 A/m^2 for 5 min every hour). Which criterion would you use to compare the two cases? Calculate corrosion rate.

Bibliography

Brenna A, Lazzari L, Ormellese M (2015) AC corrosion of cathodically protected buried pipelines: critical interference values and protection criteria. In: Proceedings international conference corrosion/15, Paper N. 5753, ISSN 03614409, NACE International, Houston, TX

Brenna A, Ormellese M, Lazzari L (2016) Electromechanical breakdown mechanism of passive film in alternating current-related corrosion of carbon steel under cathodic protection condition. Corrosion 72(8):1055–1063

Elsener G, Jansch-Kaiser G, Sharp DH (1988) Soil (underground corrosion), vol 2. Dechema Corrosion Handbook. Frankfurt Am Main, FRG

Goidanich S, Lazzari L, Ormellese M (2010a) AC corrosion. Part 1: effects on overpotentials of anodic and cathodic processes. Corros Sci 52:491–497

Goidanich S, Lazzari L, Ormellese M (2010b) AC corrosion. Part 2: parameters influencing corrosion rate. Corros Sci 52:916–922

Harris JO, Eyre D (1994) Soil in the corrosion process. In: Shreir LL, Jarman RA, Burstein GT (eds) Corrosion Vol. I—metal/environment reactions, 3rd edn. Butterworth Edition, London, UK

Jailloux JM (1989) Durability of materials in soil reinforcement applications. In: Proceedings 8th Eurocorr conference, Vol. 1, Paper No. TR-086, Utrecht

Lazzari L, Pedeferri P (2006) Cathodic protection, ed. Polipress, Milan, Italy

Romanoff M (1986) Underground corrosion. Nat Bur Stand—Circular 579, U.S. Department of Commerce, Washington, 1957, NACE Int. Edition, Houston, TX

Trabanelli G, Gullini G, Lucci GC (1972) Sur la determination de l'agréssivité du sol. Ann Univ di Ferrara NS Sez V, III, (4), 43

Von Wolzogen Kuhr CAV, Van der Vlugt SS (1934) Graphitization of cast iron as an electrochemical process in anaerobic soil. Water (Den Haag) 18:147–165

Chapter 22
Atmospheric Corrosion

Keep up your bright swords, for the dew will rust them.
W. Shakespeare, Otello, I, 11

Abstract Metallic structures exposed to the atmosphere undergo corrosion when a thin liquid film forms at their surface. The extent of corrosion depends on chemical-physical properties of this film, hence on the parameters they are determined by, such as relative atmospheric humidity, temperature, composition as well as time of wetness. Many of these factors are difficult to quantify, and often have complex, contrasting effects on the corrosion process: this is the case of rain, wind and temperature. All of these factors are discussed in the chapter, together with the classification of atmospheric environments according to the ISO standards. The characteristics of most used metals are also reported with reference to their use in atmosphere.

Fig. 22.1 Case study at the PoliLaPP corrosion museum of Politecnico di Milano

© Springer Nature Switzerland AG 2018
P. Pedeferri, *Corrosion Science and Engineering*, Engineering Materials,
https://doi.org/10.1007/978-3-319-97625-9_22

22.1 Liquid Film

Electrochemical reactions only onset in the presence of a liquid film at the metal surface. The thickness of this film depends on the chemical and physical characteristics of the metal surface, including roughness and chemical contamination. On a clean surface, it would not exceed 1 μm when exposed to a relative humidity lower than 100%; when conditions are close to saturation it grows to thicknesses ranging from 1 to 10 μm, reaching some tens or hundreds μm in presence of condensation, and finally exceeds 500 μm in case of rain.

Considering thin films, oxygen diffusion in the electrolyte is not a rate-controlling step to determine corrosion rate. This is controlled by the scarce water presence when humidity drops below 80%, or by the diffusion of aggressive species (oxygen, water, chlorides) through corrosion products that cover the metal surface. Conversely, considering thick films, oxygen diffusion may control corrosion rate, as it happens in corrosion on immersed metals.

The worst conditions are found in presence of thin films, which do not obstacle oxygen supply, but thick enough to make anodic and cathodic processes easy (Fig. 22.1). Only for galvanic coupling, thick liquid films represent the worst conditions.

Film forms by condensation according to different chemical-physical phenomena, which can be summarized to four mechanisms:

- *Physical condensation*: it consists of water passing from vapour to liquid state at the metal surface, due to a decrease of the atmospheric temperature or because metal is colder than the surrounding atmosphere; film thickness is in the order of fractions of millimetre, and its composition is pure water
- *Adsorption condensation*: it is a purely physical phenomenon, caused by attraction forces between metal and water molecules. It produces pure water films with thickness ranging from few to hundreds of molecular layers at low relative humidity to 100% humidity, respectively
- *Chemical condensation*: it takes place in the presence of hygroscopic species at the metal surface, which dissolve in the water film creating highly conductive solutions. In many cases, for instance in presence of calcium or ammonium chloride—typical of marine environments—even very low values of atmospheric humidity can be sufficient to produce this type of condensation
- *Capillary condensation*: it is typical of rough surfaces, or coated by porous patinas.

22.2 Factors Affecting Corrosion

22.2.1 Relative Humidity

Corrosion rate rapidly increases with relative humidity when it exceeds a threshold defined as *critical relative humidity*, as shown in Fig. 22.2 for steel and copper.

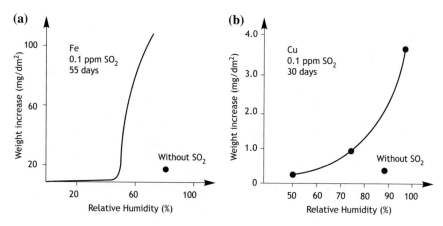

Fig. 22.2 Weight gain as a function of relative humidity in atmosphere containing 0.1 ppm of SO_2: **a** carbon steel after 55 days; **b** copper after 30 days

Table 22.1 Relative humidity producing condensation on salt contaminated surfaces

Salt	Relative humidity (%)
Na_2SO_4	93
$(NH_4)_2SO_4$	81
NaCl	78
$CaCl_2$	35
$FeCl_3 \cdot 12H_2O$	10

Critical relative humidity varies with the metal composition and surface finishing (for instance, shiny or opaque) and with the composition of corrosion products and contaminants present on the metal surface. In case of the presence of hygroscopic salts, as chlorides, critical relative humidity is very low[1] (Table 22.1); with very hygroscopic salts, surface wetting may result practically continuous.

22.2.2 Time of Wetness

Corrosion only happens if water is present at the metal surface,[2] therefore its rate depends on the time during which the surface remains wet, which is called time of wetness, τ. Time of wetness is correlated with the presence of high atmospheric

---

[1]It coincides with the value of relative humidity giving a vapour tension equal to that of a saturated solution of the same salts.

[2]Dry corrosion practically never happens at room temperature, excluding cases of slight oxidation or surface sulphuration, such as silver tarnishing (darkening) produced by traces of H_2S even in very low humidity conditions.

Table 22.2 Classification of time of wetness following ISO 9223

Category	Time of wetness		Examples
	(hours in a year)	(% in a year)	
τ_1	$\tau \leq 10$	$\tau \leq 0.1$	Internal microclimates with climate conditioning
τ_2	$10 < \tau \leq 250$	$0.1 < \tau \leq 3$	Internal microclimates without climate conditioning, excluding humid climates
τ_3	$250 < \tau \leq 2500$	$3 < \tau < 30$	External atmospheres in cold and dry climates and part of temperate climates: shielded and correctly aerated areas in temperate climates
τ_4	$2500 < \tau < 5500$	$30 < \tau \leq 60$	External atmospheres in all climates (excluding cold and dry areas); shielded and aerated humid areas; non-aerated temperate climates
τ_5	$5500 < \tau$	$60 < \tau$	Some zones of damp climates: shielded, non-aerated humid areas

relative humidity levels. In order to have a statistical evaluation on a sufficiently long period, typically over one year, it is used the time by which humidity exceeds a given value (80% according to ISO standard 9223). Table 22.2 reports the classification of time of wetness proposed by ISO 9223.

22.2.3   Temperature

Temperature plays a complex role on atmospheric corrosion. As it increases, the rate of electrochemical reactions increases; yet, at same water content relative humidity decreases, hence, jeopardizing the presence of the liquid film at the metal surface. Moreover, protective properties of corrosion products may change. If water freezes[3] corrosion stops because it loses its electrolytic properties.

Available data on tests performed at different European sites indicate an increase of the corrosion rate of carbon steel by approximately 1 μm/year per Celsius degree of mean annual temperature increase.

Table 22.3 reports estimated times of wetness of climatic zones characterised by different temperature and humidity.

---

[3]In contaminated atmospheres or inside pores, freezing is achieved some degrees below 0 °C.

Table 22.3 Estimated times of wetness of climatic zones characterised by different temperature and humidity [ISO 9223]

Climate	Min and max temperature (°C)	Max temperature with RH > 95%	Time of wetness (h/year)	Category
Very cold	−65/+32	+20	0–100	τ_1
Cold	−50/+32	+20	150–2.500	τ_2
Temperate Temperate, hot	−33/+34	+23 +25	2500–4200	τ_3
Hot, dry Very hot, dry	−20/+40 +3/+55	+27 +27	2500–5500	τ_4
Hot, humid	+5/+40	+31	4200–6000	τ_5

22.2.4 Atmosphere Composition

The composition of atmosphere is reported in Table 22.4. The concentration of main components (N_2, O_2) slightly varies from one region to another; conversely, that of minor components can vary consistently from one site to another, even daily or seasonally. For instance, the concentration of carbon dioxide is on average 380 ppm[4] but it may be higher inside highway tunnels, in poorly aerated parking or in particular environments, such as in crop silos, where it may reach 1%.

22.2.5 Contaminants

Table 22.5 reports annual releases of some contaminants in typical atmospheres. Pollutants have different origins as volcanic, from metabolism of vegetation and animals, from sea spray or dust carried by the wind, from exhausts of combustion of fossil fuels (carbon, oil, gas), from industrial emissions (for instance, chemical, metallurgical, cement industries). Finally, some substances are the result of reaction between pollutants and the atmosphere, triggered by ozone or ultraviolet radiation. Contaminants can accumulate on surfaces as dry deposits (which contribute to 70% of the total, approximately) or as liquid phase (small droplets of rain or fog, making up the remaining 30%).

Acid rains. This expression refers to all deposits (dry or humid) that cause an acidification of the metal surface. It is worth reminding that even in the absence of pollutants (SO_2, NO_X, HCl) rain is slightly acid due to the presence of carbon

---

[4]In the last thousands years and until two centuries ago, CO_2 concentration, estimated through the analysis of gas trapped in polar ice, was constant and equal to 270 ppm. Since the beginning of the industrial revolution it has been growing, first slowly, then faster, until reaching the current accumulation rate, equal to approximately 1 ppm per year.

Table 22.4 Average composition of natural atmospheres

Nitrogen (N$_2$)	78.1%	Hydrogen (H$_2$)	0.5 ppm
Oxygen (O$_2$)	20.9%	Nitrogen monoxide (NO)	0.3 ppm
Water vapour	0–5%	Ammonia (NH$_3$)	<0.1 ppm
Argon	0.93%	Sulphur dioxide (SO$_2$)	10^{-3} ppm
Carbon dioxide (CO$_2$)	380 ppm	Nitrogen oxides (NO$_x$)	10^{-3} ppm
Other rare gases	30 ppm	Hydrogen sulphide (H$_2$S)	10^{-3} ppm
Hydrocarbons (CH$_4$)	2 ppm		

Table 22.5 Indicative concentrations of some contaminants

Pollutant	Atmosphere		
	Industrial	Urban	Rural
SO$_2$ (μg/m^3)	100–200	30–80	5–20
NO$_x$ (μg/m^3)	100–200	20–40	<10
HCl (μg/m^3)	10–20	1–5	<1
Fumes and ashes (μg/m^3)	100–1000	20–50	<20
Continental Sea spray (mg/m^2 d)	30 Continental		
	1000 Coastal		
Other dusts (mg/m^2 d)	0.1–50		

dioxide in the atmosphere, with a pH of 5.5.[5] Taking into account the presence—even if minimal—of SO$_2$ and NO$_x$, even in non-polluted atmospheres, pH drops to five. Hence, the term acid rain refers to any deposit that causes a decrease in pH below five. In some periods of the year, the pH on metallic surfaces in industrial areas is between 4 and 4.5. It can be even lower in highly polluted areas.

Sulphur oxides. Among polluting substances, sulphur oxides (sulphur dioxide, SO$_2$, and sulphuric anhydride, SO$_3$) are the most dangerous. They can be present in large quantities in urban and industrial areas because of the use of sulphur containing fuels. When emitted, SO$_3$ content is generally 1–5% with respect to SO$_2$. The latter gradually oxidises to SO$_3$ with a permanence time in the atmosphere of 2–4 days. Sulphur oxides are the main responsible of acid rains. Just 100 μg/m^3 of SO$_2$ is enough to increase remarkably the corrosion rate of carbon steel. Such concentrations are found in industrial areas, especially in winter, and accelerate the deterioration of protective coatings such as painting and metallic coatings like galvanizing. Nowadays, SO$_2$ emissions in industrial countries are controlled and

---

[5]Carbon dioxide reacts with water in the following way: $CO_2 + H_2O \rightarrow H^+ + HCO_3^-$. The equilibrium constant of this reactions is: $K = [H^+][HCO_3^-]/[CO_2] = 10^{-7.6}$. Since $[H^+] = [HCO_3^-]$, then $[H^+]^2 = 10^{-7.6} \cdot [CO_2]$. For a CO_2 concentration of 380 ppm: $[H^+]^2 = 10^{-11}$, then $[H^+] = 10^{-5.5}$; pH is then 5.5.

there has been a sensible decrease in emissions in the last three decades. Anyway, they still represent the most important corrosion issue in urban and industrial areas.

Chlorides. Chlorides accelerate atmospheric corrosion processes both because they produce hygroscopic salts, which increase time of wetness, and because they make corrosion products less protective. Chlorides are present mostly in coastal areas. The mass of chlorides depositing on exposed surfaces decreases with distance from the coast. The parameters that play a role are not only the distance from the sea, but also wind speed and direction, surface orientation with respect to wind and to the ground, and height of surfaces. The Italian coastal situation is generally less aggressive than the average European one, due to the absence of strong permanent winds coming from the sea to the mainland, and consequently the influence of chlorides does not exceed a few hundred metres from the coast. Yet, in some areas affected by mistral winds—in particular west Sardinia and east Sicily—chlorides may be present in significant quantities also at higher distances. Chlorides are also present when de-icing salts, as sodium chloride, are used.

Solid particles. Air is also characterized by the presence of very small solid particles of different origin. Inorganic particles are lifted in air by the wind, for instance silica sand; vegetal organic particles, microorganisms and other organic substances; combustion by-products and ashes from industrial plants, domestic heating and combustion engines. The presence of these particles on the surface of metals can accelerate corrosion by favouring the time of wetness by capillary condensation or by chemical condensation in the presence of hygroscopic salts. They may also exert a depassivating action, as in the case of salts like $(NH_4)_2SO_4$ and $NaCl$, which dissolve in the liquid film and impede the formation of protective corrosion products. The effect of the presence of carbon powders is twofold: enhances galvanic corrosion (due to its electrical conductivity) and favours the absorption of corrosive gases as SO_2.

22.3 Classification of Environments

Factors that influence atmospheric corrosion of metals change consistently as a function of climatic zone, latitude, rainy or dry weather, rural or urban areas and proximity to the sea. To classify the corrosion behaviour in different geographical areas, the following four environments are generally considered:

- *Rural environment*: basically not polluted, far from industrial atmospheric exhaust releases and from coastal areas
- *Urban environment*: residential or commercial areas with light or moderate pollution, due for instance to car traffic or light industrial activities
- *Industrial environment*: characterized by relevant pollution, due to the presence of heavy industrial activities, mainly chemical and metallurgical ones
- *Marine environment*: areas close to the seaside.

Table 22.6 Classification of pollution levels based on SO_2 and chloride deposition as reported in ISO 9223 standard

Sulphur dioxide			Chlorides	
SO_2 class	Deposition rate (mg/m^2 d)	Concentration (μg/ m^3)	Chlorides class	Deposition rate (mg/m^2 d)
P_0	<4	<5	S_0	<3
P_1	4–24	5–30	S_1	3–60
P_2	24–80	30–90	S_2	60–300
P_3	80–200	90–250	S_3	300–1500

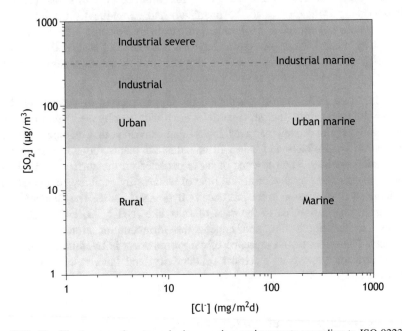

Fig. 22.3 Classification map for atmospheric corrosion environments according to ISO 9223

ISO 9223 standard classifies pollution levels as a function of SO_2 and chlorides deposition rate, following the scheme reported in Table 22.6. Figure 22.3 shows the map of corrosive atmosphere.

22.3.1   Microenvironments

The above reported definition of environments helps perform a first assessment of corrosion behaviour, but what really matters is the microenvironment. For instance, in areas close to chimneys or industrial exhausts, especially in conditions of thermal inversion, the concentration of pollutants and dusts is higher; conversely, if wind is present, leeward areas with respect to the emissions location—or coast position—are completely different from windward ones. Even between central and suburban areas of large cities, or between indoor and outdoor environments, or within one same building, the situation may change consistently.

22.3.2   Classification of Atmospheric Corrosiveness

ISO 9223, ISO 9224, ISO 9225 and ISO 9226 standards propose a classification of corrosiveness and a methodology to evaluate corrosion rate. Specifically, ISO 9223 defines six exposure classes: C1, C2, C3, C4, C5 and CX as a function of the increasing environment corrosiveness, and indicates for each class the intervals of corrosion rate values expected after one year of exposure of different metals. The methodology to evaluate corrosion rate in the first year of exposure is reported in ISO 9226, together with composition limits for considered metals (carbon steel, zinc, copper and aluminium).

Table 22.7 reports, for different classes of exposure, examples of corresponding outdoor environments (from ISO 9223 Annex) and corrosion rates in the first year of exposure in the specific environment for carbon steel and zinc.

When experimental data on corrosion rate are missing, a second approach proposed by standards allows estimating corrosion rate for the first year of exposure based on environment data. In this case, calculations are performed with dose-effect type equations and evaluations are based on the description of exposure conditions: relative humidity, SO_2 and chlorides deposition (P_d and S_d parameters, already described in Table 22.6 and evaluated based on ISO 9225). This method involves higher uncertainties in predictions.

Once the corrosion rate for the first year is known, it is then possible to estimate service life following ISO 9224 by applying the following equation:

$$s = C_{\text{rate},1} \, t^b \tag{22.1}$$

where s is the penetration of the corrosive attack, t the exposure time (years) and b a constant that depends on metal composition (on average, 0.5 for carbon steel and 0.8 for zinc, as reported in the standard) which underlines that these values were calculated on the basis of well-defined chemical compositions. In particular, for steel corrective factors are proposed as a function of alloying elements and their content. The standard suggests considering a validity of this equation up to a maximum of 20 years of service life, afterwards corrosion rate stabilizes and successive variations of corrosion penetration follow a linear trend.

Figure 22.4 shows corrosion penetration values calculated after 20 years of service life with both approaches suggested by the standard: values proposed for classes C1–CX are compared with values calculated with the dose-effect equation

Table 22.7 Classification of atmospheric environments based on first year corrosion rate in μm/year (according to ISO 9223)

Corrosiveness category		Outdoor environment	Carbon steel	Zinc
C1	Very low	Dry or cold zone, environment with very low pollution and time of wetness, for instance desert and Antarctic areas	1.3	0.1
C2	Low	Temperate zones, low polluted environments, e.g. rural areas, small towns Dry or cold zones atmospheric environments with short time of wetness, e.g., deserts, subarctic areas	1.3–25	0.1–0.7
C3	Medium	Temperate zones, environments with medium levels of pollution e.g. urban areas, coastal areas with low deposition of chlorides Subtropical and tropical zones, atmospheres with low pollution	25–50	0.7–2.1
C4	High	Temperate zones, environments with high pollution or substantial effects of chlorides, e.g. polluted urban areas, industrial areas, coastal areas without spray or salt water or exposure to strong effects of de-icing salts	50–80	2.1–4.2
C5	Very high	Temperate or subtropical zones, heavily polluted and/or with significant effects of chlorides, e.g. industrial areas, coastal areas, sheltered positions on coastline	80–200	4.2–8.4
CX	Extreme	Subtropical or tropical zones (very high time of wetness), environments with very high pollution and/or strong effects of chlorides, e.g. extreme industrial areas, coastal and offshore areas, occasional contact with salt spray	200–700	8.4–25

method. For P_0–P_2 and S_0–S_2 classes, the maximum values of pollutant concentration in each class were considered for P_3 and S_3 classes the maximum pollutant concentration used in the dose-effect correlation was considered.

22.3.3   Indoor Atmosphere

Most of the experimental data referred to atmospheric corrosion are related to outdoor exposure. The conditions of indoors exposure are different, because the metal is not exposed to rain and direct sunlight. Rusting depends on the condensation of the moisture, which may evaporate from the surface much more slowly in enclosed conditions; moreover, the corrosion products tend to remain on the surface and may build up in time to a thick layer.

According to the ISO 9223 standard, the indoor microclimates are included mainly in the category τ_1 and τ_2, as far as the time of wetness is concerned. In indoor atmospheres without conditioning, the categories τ_3 to τ_5 may apply if

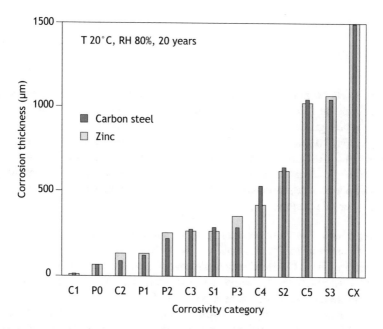

Fig. 22.4 Corrosion penetration values calculated after 20 years of service life: comparison between classes C1–CX and values calculated with the dose-effect equation method

sources of water vapor are present or in humid condition. Combining the values of time of wetness by that of pollution, the relevant corrosive classes can be defined. ISO 11844 classifies five categories of indoor atmospheres:

- IC 1—Very low indoor corrosivity
- IC 2—Low indoor corrosivity
- IC 3—Medium indoor corrosivity
- IC 4—High indoor corrosivity
- IC 5—Very high indoor corrosivity.

The upper limit of categories IC3 and IC5 correspond roughly to the upper limit of category C1 and C2, as per ISO 9223.

The corrosiveness of indoor atmospheres increases with the relative humidity, type and levels of pollution. Frequency of relative humidity and temperature variation, and frequency and time of condensation are also important. Accordingly, the time of wetness defined in ISO 9223 is not exactly applicable to indoor corrosiveness. Pollutant concentration are generally lower or similar in indoor environments, depending on the sheltering (i.e. filtration, conditioning) except for cases of internal sources of pollution.

22.4   Corrosion Behaviour of Most Used Metals

22.4.1   Carbon Steels

Carbon steels do not exhibit good atmospheric corrosion behaviour because rust, which forms on their surface, consists of a porous, non-protective layer.

The attack generally takes place as uniform corrosion, with a corrosion rate ranging from 20 μm/year in rural environments to 40–50 μm/year in urban environments, 60–100 μm/year in urban-industrial or marine ones and eventually 100–200 μm/year in industrial areas, and even higher in humid tropical marine climates.

Hence, carbon steel structures must be protected. Painting cycles or zinc coatings are used for this purpose.

Painting cycles. The use of paints is the most common practice to protect carbon steel structures exposed to the atmosphere. The general aspects of this technique are described in Chap. 17. Here only maintenance issues will be briefly considered. Except for extremely low aggressiveness conditions, a painting cycle is generally not able to ensure a duration sufficient to protect carbon steel for the whole service life (50 years for civil structures, 20–30 years for industrial plants). It is then necessary to proceed with periodical re-applications, which require the removal of the old paint and rust before applying the new coat, therefore these interventions are rather expensive. It is common practice to plan a preventative maintenance, which helps avoid expensive total painting replacements. A maintenance intervention consists of periodical repair of the damaged coating, whose frequency depends on environment aggressiveness, operation costs and inspections results (Fig. 22.5; Table 22.8).

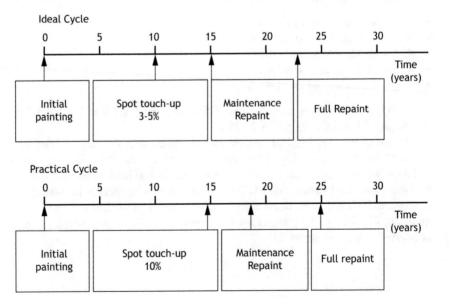

Fig. 22.5 Maintenance cycles for painted structures (from Brevoort and Roebuck 1993)

Table 22.8 Initial expected lifetime for some painting cycles in marine and urban environments

Cycle	Surface preparation	Minimum thickness (μm)	Marine environment (years)	Urban environment (years)
HB ST epoxy/ HB ST epoxy/ polyester urethane	SP6	300	12 (16)	14 (21)
Zn urethane/HB acrylic-urethane/ acrylic-urethane	SP10	250	12 (16)	14 (21)
Zn inorganic/HB epoxy/HB epoxy	SP10	280	12 (16)	15 (22.5)
Zn inorganic/HB acrylic urethane	SP10	280	12 (16)	15 (22.5)
Zn inorganic/HB epoxy/polyester urethane	SP10	225	12 (16)	15 (22.5)
Zn inorganic/HB epoxy/HB epoxy	SP10	225	12 (16)	14 (21)
Zn epoxy/HB epoxy/HB epoxy	SP10	280	12 (16)	14 (21)
Zn organic/HB acrylic urethane/ HB acrylic urethane	SP10	280	12 (16)	15 (22.5)
Zn epoxy/HB epoxy/polyester urethane	SP10	225	12 (16)	14 (21)
Zn inorganic/HB vinylic/HB vinylic	SP10	280	12 (16)	13 (19.5)
Zn inorganic	SP10	75	(15)	(17)
Zn organic	SP10	75	(6)	(5)

Values refer to an ideal maintenance cycle, or to a practical one (in brackets) (from Brevoort and Roebuck 1993)
HB high build (high thickness), *ST* surface tolerant, *SP6* commerical blast, *SP10* white metal sandblasting

These interventions are undoubtedly the cheapest way to extend the coating life. Yet, they are more delicate than the initial application, since a good adherence between old and new paint is often difficult to achieve. It is therefore recommended, when choosing the painting cycle in the design stage, to take into account the feasibility of maintenance operations.

Zinc coatings. Zinc coatings are the most widespread metallic coatings to protect carbon steel. Zinc corrosion rate, even in industrial and marine environments, is generally 10–30 times lower than that of carbon steel, since it tends to form a protective film, which is stable in the pH range of 6–12.5. Initially, zinc corrosion

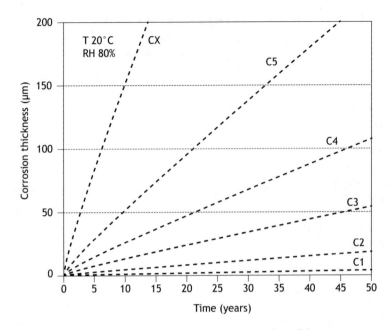

Fig. 22.6 Duration of the zinc coating in different environmental conditions

causes the formation of a hydroxide, which reacts with acid species present in the atmosphere (in particular, CO_2) and forms a protective film of basic zinc carbonate. In polluted areas, acid pollutants—first sulphur dioxide—produce a basic zinc sulphate, which is more soluble than carbonate, forming less protective corrosion products. An analogous situation can be found in marine areas, where a basic zinc chloride forms, which is more soluble as well. Consequently, zinc corrosion rate can vary significantly and increase even by one order of magnitude when passing from rural environments to those polluted by sulphur dioxide or to marine environments. Zinc corrosion rate depends on the environment, but it is almost constant in time (Fig. 22.6). Hence, the duration of the protection offered to steel depends on the environment and on coating thickness.

The protective characteristics of the coating are lost in case of scratches larger than some tenth of millimetre. A frequent damage is that caused by welding, as shown in Fig. 22.7.

The New Delhi Pillar

In New Delhi, India, there is a cast iron column, dating back to the IV century A.D. (the so-called New Delhi Pillar, Fig. 22.8), which has raised the curiosity of corrosionists. Its top part is perfectly conserved, showing no traces of rust, instead it appears coated with a bronze-bluish coloured oxide layer. Clearly, favourable environmental conditions (absence of chlorides and sulphur compounds) and the absence of sulphur and phosphorus in the metal

Fig. 22.7 Corrosion at a damaged zone of zinc coating caused by welding

allowed the formation and conservation of a compact, protective layer (maybe magnetite) instead of the typical porous, non-adherent rust. In its sixteen centuries of service life, the periodic exposure to high temperatures has favoured the growth of this film, which has reached a considerable thickness.

The good behaviour of the New Delhi Pillar cannot be used to support the widespread opinion that steels used in Medieval ages or in ancient times were more resistant than contemporary ones. The good conservation status of many iron inserts in medieval buildings should be ascribed mainly to the fact that they operated for centuries in presence of low amounts of water and no pollution, and only to a limited extent to the fact that the steel produced with wood carbon did not contain sulphur.

It is curious to notice that also in ancient times steel handcrafts were compared with older ones, saying that older steels were more resistant. As Pliny writes: "*It is said that in a city close to the Euphrate river, called*

Fig. 22.8 The New Delhi Pillar

> *Zeugma, there is a chain that Alexander the Great had used to build a bridge. Some rings of the chain were substituted. These ones corroded in short time, while the original ones remained intact."*

22.4.2 Weathering Steels (Cor-Ten)

Since the first decades of last century, it was known that by adding 0.2–0.3% of copper to a carbon steel, corrosion rate halves in atmospheric exposure conditions. In 1933 the U.S. Steel company launched a new low alloy steel containing 0.2–0.5% copper, 0.5–1.5% chromium and 0.1–0.2% phosphorus which doubled the atmospheric corrosion resistance compared with the known copper-alloyed steel; also the yield strength improved. In the following years, the composition was adjusted by introducing 0.4% nickel to minimise the damaging induced on the steel by hot working, and phosphorus was reduced to a content lower than 0.04% to avoid the formation of welding cracks. Small quantities of other elements (V, Zr, Mo) were also added to improve mechanical resistance, so that yield strength increased from 350 to 490 MPa.

These steels were called weathering steels or Cor-ten (corrosion tensile resistance steel). Initially, they were used with the addition of protective coatings. Applications like cargo carriages, trucks, farm machines, beams for bridges, showed service lives from 1.5 to 4 times higher than the same components made of carbon steel with the same protective coatings.

The first important application in the civil sector dates back 1964. It opened the way to several others, including the Chicago Civic Center, in the central square of Chicago, fronting the Picasso statue *La capra* (made of a mix of Cor-ten and stainless steel, Fig. 22.9), or the Italian Centro Sperimentale Metallurgico, close to Rome, which is also made of both Cor-ten and stainless steel. The most recent important applications are beams and scaffoldings of highway bridges (Fig. 22.10).

Weathering steels are self-protecting in many exposure conditions by the formation—in a period of 1 to 4 years—of a pleasant rusty patina, characterized by an outer porous layer and an inner thin film, amorphous and impermeable, rich in copper, chromium and phosphorus, which enhances corrosion resistance. During its formation, affecting parameters are wet-dry cycles, sunlight exposure (which probably exert a photocatalytic action), degree and type of pollution, therefore depending on geographical location, distance from the sea, and its shape. The importance of wet-dry cycles lies on the evidence that the patina does not form on permanently wet zones. Therefore, critical conditions are met at lower parts of structures in contact with soil and where stagnant water is formed.

In the presence of high chlorides concentration, as in marine environments or where de-icing salts are used, the patina can form in a non-correct way. Similarly,

Fig. 22.9 Picasso statue behind the Chicago Civic Center, made partially of Cor-ten, 1960

Fig. 22.10 Viaducts beams made of Cor-ten

high concentrations of SO_2 associated with high humidity conditions, as typical in industrial and marine-industrial areas in the years 1960s–1970s, lead to a non-protective patina as experienced in some unsuccessful applications observed in the 1970s.

Figure 22.11 compares the corrosion behaviour of carbon steel and weathering steel in industrial and marine environments. As the protective patina is formed,

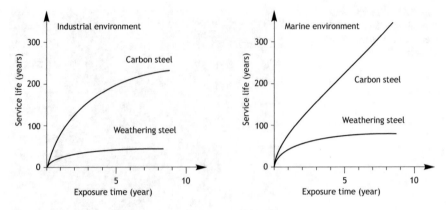

Fig. 22.11 Atmospheric corrosion trend of carbon steel and weathering steel

thickness loss is maximum after few years, then corrosion rate becomes negligible. In marine exposure, with continuous wetting by salty water, corrosion behaviour is similar to carbon steel.

22.4.3   Stainless Steels

Stainless steels offer an excellent behaviour in atmosphere if correctly selected. However, some concerns may arise, dealing with:

- *Surface staining* or *pinning*, in marine and industrial zones, especially in Mo-free stainless steels
- *Crevice corrosion* when gaps or dead spaces are present, favouring the entrapping of electrolyte
- *Galvanic coupling*, which affects less noble metals as carbon steel, aluminium and zinc coupled with stainless steels. In humid atmosphere and in the presence of condensation, insulation is recommended. Figure 22.12 shows an example of a galvanic corrosion in an atmospherically exposed galvanic coupling.

Figure 22.13 reports a guidance map for the choice of stainless steel based on environmental conditions as pollutants concentration.

Ferritic stainless steels. The classical 12 and 17% chromium stainless steels (AISI 410 and 430) as well as martensitic grades are the worst corrosion resistant stainless steels in atmosphere. In particular 12%Cr grade may show pinning and staining after a few years of exposure especially when a rough surface finishing is present. A regular cleaning of the metal surface can help avoiding these alterations, which anyway do not compromise mechanical resistance. These steels are not recommended in urban-industrial or marine environments, but they can be used in less aggressive conditions.

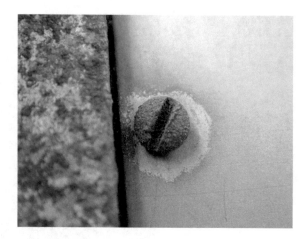

Fig. 22.12 Galvanic coupling: stainless steel screws on aluminium plate

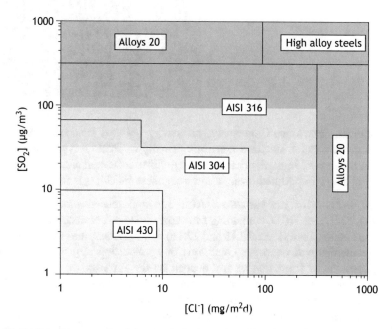

Fig. 22.13 Selection map of stainless steels in atmospheres based on pollutant concentration

Stainless steels without molybdenum. AISI 304 and related grades for welded structures, AISI 304L and AISI 321, are definitely the most used stainless steels. They do not undergo corrosion in rural or urban areas (Fig. 22.14). In the latter ones, after several years a slight opacification and some pinning can be noticed in

Fig. 22.14 Chrysler Building
(New York), made in AISI
304 stainless steel, 1930

areas shielded from rain. Conversely, in heavily polluted industrial and marine atmosphere they suffer localised corrosion, usually of shallow type. Hence, these steels can be adopted in rural, urban and low polluted industrial areas regardless the surface finishing. In polluted areas, a low roughness finishing is recommended.

Stainless steels with molybdenum. AISI 316 and related grades for welded structures, 316L and 316Ti, containing Mo show good performance also in polluted and marine atmosphere (Figs. 22.15 and 22.16). Nevertheless, it is still necessary to avoid geometries and exposure conditions or surface finishing that may favour water stagnation or formation of any deposit on the surfaces, or create gaps and crevices (as may happen, for instance, in the case of nets, badly performed welding or mechanical junctions). They suffer corrosion only in heavily polluted marine atmosphere due to the presence of chemical or metallurgical industries, or inside highway tunnels, where antifreeze salts carried by vehicles and high levels of SO_2 are simultaneously present. They also undergo corrosion if intermittent or continuous contact with brackish water or seawater takes place. In these cases, high-alloyed steels should be used. In the atmosphere that generates inside swimming pools, where water is maintained at temperatures higher than 30 °C and is treated with chlorine, also AISI 316 steels may present unacceptable corrosion attacks.

Fig. 22.15 Petronas Towers (Kuala Lumpur), made in AISI 316 stainless steel, 2000

Fig. 22.16 The "Bean" (Chicago), made in AISI 316 stainless steel, 2006

Austeno-ferritic (duplex) stainless steels. As far as pitting or crevice corrosion forms are concerned, the corrosion behaviour of these steels is similar to that of austenitic steels with same PREN index. On the other hand, they are much more resistant to SCC.

Corrosion of Stainless Steels in Swimming Pools

Stainless steels are largely employed in swimming pools: stepladders, handrails, aeration and ventilation pipes, trampolines, tie beams. In the last 20 years an increase in corrosion phenomena on these elements was registered: surface attacks with formation of rust (*staining*), localised spots (*pinning*), perforating corrosion (*pitting*) and even SCC with catastrophic consequences. In these acid conditions, SCC may take place on stainless steels at temperatures lower than 50 °C and close to room temperatures.

For instance, in 1985 the reinforced concrete false ceiling of a swimming pool in Zurich (built in 1972) collapsed, killing 12 people and injuring many others. The collapse was caused by the SCC-related failure of AISI 304 tie beams supporting the 200-ton false ceiling to the swimming pool roof. More recently, for the same reason, several other cases occurred, fortunately without tragic consequences; in fact, in 2001 and 2003 the suspended ceilings of two swimming pools collapsed, in Netherlands and Finland, respectively.

SCC occurs on aerial surfaces exposed to swimming pool atmospheres, but does not on surfaces that are immersed or continuously rinsed by the pool water. This occurs because of the formation of chloride containing acidic condensates.

When chlorine, which is normally used as sanitizing agent in swimming pools, is added to water, it dismutes and produces hypo-chloric acid and hydrochloric acid. The residual content of gaseous chlorine in water decreases to 1–2 ppm. At a temperature of 28–29 °C, this concentration usually generates a very low vapour pressure, hence a low chlorine content in the atmosphere, in the range 3–6 mg/m^3. This does not justify the high chloride contents observed on metallic surfaces where corrosion onsets: the mechanism is therefore more complex. Chlorine reacts both with water and with nitrogen and ammonia compounds present in water, released for instance by human perspiration or urine. The reaction causes the formation of chloramines (*mono-chloramine* NH_2Cl; *di-chloramine* $NHCl_2$; *tri-chloramine* NCl_3). These substances are highly volatile. Their evaporation and condensation on cold metallic surfaces explain the chloride-enrichment mechanism on aerial surfaces.

In the last 15–20 years a worse trend of failures has been recorded: the main reason is recognized on the increase of water temperature, which raised from 24 to 26–30 °C. Truly, the assessment of the incident occurred at the abovementioned Zurich pool revealed that the water temperature was raised to 37 °C three days a week as required and reserved for disabled people.

To prevent the SCC occurrence here described it is necessary to: limit water temperature; maintain relative humidity in the range 50–70%; clean stainless steel surfaces; avoid the use of AISI 304 or 316; choose SCC-resistant stainless steels as duplex (25–07) (with 3% Mo, PREN 35), AISI 904L (4.5% Mo, PREN 36), or alloys with 6% Mo (254 SMO, PREN 41).

22.4.4 Copper and Copper-Alloys

Copper and, in general, its alloys (bronze, brass) present a good atmospheric corrosion resistance. After a time that varies between few years and several years they become covered by a green patina (noble patina) mainly consisting of a mixture of $Cu_2CO_3(OH)_2$, $Cu_3(CO_3)_2(OH)_2$ and $Cu_4SO_4(OH)_6$, which strongly reduces corrosion rates (Fig. 22.17). In the presence of a protective patina, in rural areas corrosion rate is lower than 1 µm/year, in industrial and marine environments corrosion rate is a few µm/year. Corrosion attacks are possible only in presence of high pollutants concentration or stagnant water.

In civil building, copper is largely used for roofs and for rainwater drainage systems. After 1–2 years exposure to atmosphere, copper assumes a uniform dark colour, indicating the formation of a stable protective patina. In urban and industrial atmospheres, having a moderate content of sulphur dioxide, and in marine atmospheres, a green patina (based on copper hydroxyl-sulphate or copper hydroxy-chloride) forms in about 5–7 years. The green patina forms more slowly on vertical surfaces than horizontal or leaning surfaces, because the latter remain wet longer. In the presence of acidic water, as in industrial atmospheres near chimneys, the patina of hydroxyl-sulphate does not form; in these cases, the water drained from copper surfaces has a high copper ions content and gives rise to the blue colour of the wet surfaces. The same water, due to the reduction of the copper ions, can give rise to corrosion of steel and aluminium.

22.4.5 Aluminium

In a pH interval between 4.5 and 8.5, in environments with low pollution, a natural protective oxide film covers aluminium, thus preventing excellently corrosion with a corrosion rate lower than 1 µm/year. The protection properties of the passive film

Fig. 22.17 Parliament buildings (Ottawa), mansard roofs made in copper

Fig. 22.18 Warsaw
residential building in painted
anodized aluminium (Poland)

lower in the presence of SO_2, carbon particles and chlorides (marine environments) then triggering localised corrosion. To strengthen corrosion resistance, anodizing is used, which produces a thicker and robust film. Figures 22.18 and 22.19 show some use of anodized aluminium in polluted environments.

Aluminium is widely used for window frames and roofs. For these applications, anodised aluminium is normally installed. The use of pre-painted semi-finished products is also common. The most used paints are alkyd-melamine, polyester and polyvinyl-fluoride. Concrete, due to the high pH of water present in its pores, is highly corrosive for aluminium: when preparing mortars and concrete it is advisable to avoid splashes of wet cement on aluminium surfaces, since they cause stains. A good practice is to protect aluminium surface with removable plastic sheets.

22.4.6   Other Metallic Materials

Lead. The protection mechanism is similar to that of zinc, leading to the formation of a basic carbonate more resistant than the one that forms on zinc (Fig. 22.20).
Titanium. The appealing aspect of titanium and its resistance to atmospheric agents opened the way to applications of this metal in architecture for envelopes, external and internal coatings of relevant buildings. Titanium is used both in its *natural* grey

Fig. 22.19 Pirelli tower
(Milan), building envelope in
anodized aluminium, 1959

Fig. 22.20 Auditorium
(Rome), Renzo Piano, lead,
1996

colour and with anodic colours. To overcome cost related issues, it is used in very
thin sheets, often applied on polymeric supports.

A Catchy "Mono-Chromatism" Effect
Large grey titanium surfaces, at least in some finishing conditions, give
coloured glares whose hues change with observation angle and type of

Fig. 22.21 Guggenheim museum, Bilbao, F.O. Gehry, titanium, 1998

illuminant. This behaviour is well embodied by the Guggenheim Museum of Bilbao, opened in 1998. Architect Frank O. Gehry has realized an *amazing titanium envelope*: "*a sort of big metallic stripe*, as François Burkhardt writes, *which wraps the different 'boxes' the museum is made of and ensure its unity forming an incredible 'landscape' that—given the multitude of inclined or rounded planes—transports light, arising a different color in each single point, until it forms a catchy monochromatism effect that goes from gold to blue, and from pink to white*".

To be true, the aspect of 'natural' titanium can vary with surface finishing —drawing, sandblasting, etching or passivation. As Gehry writes: "*The rolling of the material is very delicate. It can lead to a dead surface or a wonderful light-receptive one.*" Whatever the treatment, the resistance to atmospheric agents remains very high. It is again Gehry who writes: "*The titanium is thinner than stainless steel would have been; it is a third of a millimetre thick. [...] It's ironic that the stability given by stone is false, because stone deteriorates in the pollution of our cities whereas a third of a millimetre of titanium is a hundred-year guarantee against city pollution. We have to rethink what represents stability (Fig. 22.21)."*

22.5   Corrosion and Protection of Metallic Cultural Heritage

The atmospheric corrosion of cultural heritage surfaces is highly influenced by the history of the metal and in particular by the presence of corrosion layers. Many statues and architectural elements, such as claddings and roofing, are originally designed to be exposed outdoor and are typically characterised by the presence of natural and/or artificial patina. Both natural and artificial patinas have a relevant historical and artistically value (see box "The noble patina"). The protection of outdoor objects is frequently obtained by the use of protective treatments. It has to be considered that there are specific requirements to be fulfilled by a product to be applied in the field of Cultural Heritage such as good protective efficacy, good optical properties, that is negligible modification of the appearance of the surface in terms of colour and gloss, chemical and photochemical stability and inertness towards the substrate. Moreover, retreatability of a material is an important requirement since protective treatments need to be maintained and, if necessary, to be removed and reapplied without altering the surface of the material.

In addition, there is a huge variety of objects that have remained underground or under the sea for centuries; a great variety of salts and corrosion products over the surface may be present. Unfortunately, it is not always possible to completely remove salts and destabilising species without irremediably damaging the object and therefore specific conservation strategies have to be considered since the presence of highly hygroscopic salts may favour condensation of water over the surface. It may happen, then, that atmospheric corrosion is active even at relative humilities that are normally considered safe. Environmental control and the used of closed showcases are frequent approaches for the preservation of archaeological objects. The closed environment of the showcases may present some issues, since different materials may constitute the objects (metal and wood for instance). They may require different environmental condition to be preserved. In the presence of hygroscopic salts, the metal may require a quite low relative humidity, which may result detrimental for the wood. In addition, wood may release substances, such as acetic or formic acid, which would cumulate into the showcase causing an acceleration of corrosion processes of bronze artefacts.

The Noble Patina
The term noble patina is reserved to the green patina that forms as corrosion product of copper or bronze. Some of the most ancient observations and

considerations on corrosion phenomena actually deal with this patina. In *De Pythiae Oraculis* Plutarch describes a statue at the entrance of the Apollo sanctuary, in Delphi. Although some centuries old, it appears covered with a brilliant patina of uniform colour; and he wonders whether it has been produced by ancient sculptor masters with special manufacturing, or by the action of atmosphere, which penetrate bronze and push corrosion out; or, eventually, if it is bronze itself that produce it by ageing.

In the Greek and Roman world, bronze works were preferred to show a finishing that helped the natural metal colour stand out, hence the problem was to avoid the formation of rust (*aerugo*, bronze rust) with periodical maintenance involving cleaning and oiling. Yet, sometimes artists broke with tradition and searched for the corrosion attack, instead of avoiding it. For instance, in San Marco horses in Venice, Italy, the mane gilding is engraved to weaken the brightness of gold with the colour of bronze appearing underneath, or better, of its corrosion products. In some cases, the corrosion product is not used to give a finishing to a complete work, but even to express feelings and moods. Pliny tells that the sculptor Aristonides, when representing Athamas shocked after having thrown his son Learchus from a cliff, used a copper-iron alloy, so that iron would rust inside the shiny copper surface and express the shame and dismay of the man.

In the Renaissance, the art of bronze melting finds new life and artificial patina is born. As Vasari writes in his treatise on sculpture: "*In time, bronze takes by itself a colour that turns to black, not to red as it happens when it is worked. Some make it black with oil, some make it green with vinegar, some give the black colour with paints, so that everyone changes it the way it pleases him most*". The diffusion of artificial patina on large scale dates to the beginning of the 20th century. Even if treatments and recipes increase, and with them the colours obtained, patina-making keeps being an art that relies on the skills and creativity of the artists that perform it, rather than on technical knowledge. Indeed, the artists themselves often realize it.

Even now, this treatment does not provide predictable and repeatable results: indeed, even a sculptor like Henry Moore defined it in 1967 as a very exciting intervention, but with very uncertain results. In fact, he writes: "*Bronze naturally in the open air (particularly near the sea) will turn with time and the action of the* atmosphere *to a beautiful green. But sometimes one can't wait for nature to have its go at the bronze, and you can speed it up by treating the bronze with different acids which will produce different effects. Some will turn the bronze black, others will turn it green, others will turn it red. Usually, when I prepare the cast I have an idea of the bronze finishing I want, dark or clear, and of the colour I want to obtain. When the cast comes back from the foundry, I give the patina, and sometimes it comes out good, but sometimes you cannot repeat what you have already done in other occasions. It is very exciting but poorly reproducible, this operation of bronze patination.*" (H. Moore, *Henry Moore on Sculpture*, 140, Philip James, New York, 1967)

22.6 Applicable Standards

- ASTM G 50, Standard practice for conducting atmospheric corrosion tests on metals, American Society for Testing of Materials, West Conshohocken, PA.
- ISO 8565, Metals and alloys—Atmospheric corrosion testing—General requirements, International Standard Organization, Geneva, Switzerland.
- ISO 9223, Corrosion of metals and alloys—Corrosivity of atmospheres—Classification, determination and estimation, International Standard Organization, Geneva, Switzerland.
- ISO 9224, Corrosion of metals and alloys—Corrosivity of atmospheres—Guiding values for the corrosivity categories, International Standard Organization, Geneva, Switzerland.
- ISO 9225, Corrosion of metals and alloys—Corrosivity of atmospheres—Measurement of environmental parameters affecting corrosivity of atmospheres, International Standard Organization, Geneva, Switzerland.
- ISO 9226, Corrosion of metals and alloys—Corrosivity of atmospheres—Determination of corrosion rate of standard specimens for the evaluation of corrosivity, International Standard Organization, Geneva, Switzerland.
- ISO 11844, Corrosion of metals and alloys—Classification of low corrosivity of indoor atmospheres, International Standard Organization, Geneva, Switzerland.

22.7 Questions and Exercises

22.1. Compare the atmospheric corrosion behaviour of stainless steel type AISI 316 (18% Cr, 8–10% Ni, 2–3% Mo) and titanium in the following conditions: rural environment, industrial environment (very polluted).
22.2. Compare the atmospheric corrosion behaviour of stainless steel type AISI 304 (18% Cr, 8–10% Ni) and galvanised steel (Zn thickness 80 μm) in the following conditions: urban environment, industrial environment (medium pollution).
22.3. Compare the atmospheric corrosion behaviour of "weathering" steel and Al–Si–Mg alloy (series 6xxx) in the following conditions: urban environment, marine environment.
22.4. Evaluate atmospheric corrosion behaviour for carbon steel in urban areas (temperate zones); select a suitable value of the corrosion rate in the first year ($C_{rate,1}$) and exponent b (Eq. 22.1) and evaluate the thickness loss after 20

and 50 years. To guarantee the minimum service life 50 years, with corrosion allowance 0.4 mm, is necessary to protect the steel? If yes, which protection method is suggested?

22.5. Repeat the Exercise 22.4 considering coastal areas in subtropical zones and rural areas in temperate zones.

22.6. Evaluate atmospheric corrosion behaviour for galvanised steel in industrial areas, medium pollution (temperate zones); select a suitable value of the corrosion rate in the first year ($C_{rate,1}$) and exponent b (Eq. 22.1) and evaluate the thickness loss after 20 and 50 years; which is the minimum zinc thickness able to guarantee the service life (50 years) without corrosion?

22.7. Repeat the Exercise 22.6 considering environment with low pollution (rural) and coastal areas in temperate zones.

22.8. A zinc layer is applied on carbon steel to protect it from atmospheric corrosion. Which is the influence of a defect in the zinc layer on steel corrosion behaviour? The effect would be the same if copper was used instead of zinc?

22.9. Why the performance of weathering steels is not satisfactory in case of immersion or continuous exposure to very high humidity (near to saturation)?

22.10. Explain the rationale behind the Fig. 22.13 (selection of stainless steels in atmospheric exposure).

22.11. Which is the main difference between the corrosion product formed in atmosphere by copper (and their alloys) and stainless steels? Is there any effect of this difference on the corrosion of the two metallic materials?

Bibliography

Brevoort GH, Roebuck AH (1993) A review and update of the paint and coatings cost and selection guide. Mater Perform 4:31–35

De la Fuente D, Castano JG, Morcillo M (2007) Long-term atmospheric corrosion of zinc. Corros Sci 49:1420–1436

De la Fuente D, Diaz I, Simancas J, Chico B, Morcillo M (2011) Long term atmospheric corrosion of mild steel. Corros Sci 53:604–617

Dillmann P (2007) Corrosion of metallic heritage artefacts. Investigation, conservation and prediction for long-term behaviour. In: European federation of corrosion (EFC) series, vol 48. Woodhead Publishing, London, UK

Dillmann P (2013) Corrosion and conservation of cultural heritage metallic artefacts. In: European federation of corrosion (EFC) series, vol 65. Woodhead Publishing, London, UK

Knotkova D, Kreislova K, Dean SW (2010) ISO CORRAG international atmospheric exposure program: summary of results. In: ASTM data series, vol 71. ASTM International, West Conshohocken PA, USA

LeyGraf C, Wallinder IO, Tidblad J, Graedel T (2016) Atmospheric corrosion. Wiley, New York, NY

Munger CG (1984) Corrosion prevention by protecting coating. NACE International, Houston, TX

Panchenko YM, Marshakov AI (2016) Long-term prediction of metal corrosion losses in atmosphere using a power-linear function. Corros Sci 109:217–229

Chapter 23
Corrosion in Concrete

> *You could not say that a piecework is perfect if it were useful for
> a short time.*
> Andrea Palladio, Italian architect (1508–1580)

Abstract This chapter deals with corrosion of rebars in concrete, specifying the types of corrosion forms with particular attention to the two most common ones, carbonation induced corrosion and chloride induced corrosion, due to the penetration in concrete of CO_2 and chloride ions, respectively. Hydrogen embrittlement of high strength steels and stray current corrosion are also introduced. The preventative and protection methods that can be adopted are discussed, starting with concrete quality and dealing with additional protection methods, namely cathodic prevention, stainless steel and galvanised steel reinforcements, concrete coatings and corrosion inhibitors. Finally, an outline on inspection and diagnosis is presented.

Fig. 23.1 Case study at the PoliLaPP corrosion museum of Politecnico di Milano

© Springer Nature Switzerland AG 2018
P. Pedeferri, *Corrosion Science and Engineering*, Engineering Materials,
https://doi.org/10.1007/978-3-319-97625-9_23

Fig. 23.2 Example of carbonation process (Policlinico Hospital, Naples, Italy, 1972)

23.1   Initiation, Propagation and Morphology of Corrosion

Romans built the Pantheon in Rome, Italy, about 2000 years ago and it is still perfectly conserved, just like structures which remained immersed in seawater for almost the same period of time. This demonstrates that concrete shows the durability of natural stones, provided that some specific causes of degradation, as acid rains, sulphate-enriched waters, freeze-thaw cycling and reactive aggregates are absent. The case of reinforced concrete structures is much different, as experience shows daily, and far from being "eternal", their durability is limited sometimes to tens of years, because of corrosion of steel reinforcements. In aerated, chloride-free alkaline solutions, pH > 11.5, iron passivates because a protective oxide film, a few nanometres thick, forms. In these conditions, corrosion rate is practically nil. This also happens to steel reinforcements embedded in sound, pristine concrete, which hosts in its pores a solution of sodium, potassium and calcium hydroxides with a pH between 13 and 13.5. Unfortunately, over time, concrete can lose its passivating properties for two main reasons[1]:

---

[1]Passive reinforcement can also become active if interfered by stray current, as discussed forward.

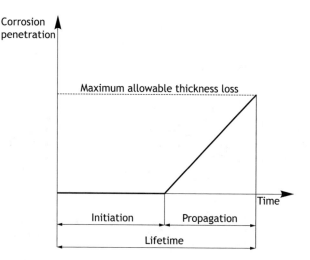

Fig. 23.3 Initiation and propagation stages for steel reinforcement corrosion in concrete (Tuutti's model 1982)

- *Chloride contamination* when concrete is in contact with chloride-containing environments. When chlorides penetrating the concrete cover exceed a critical concentration on the reinforcement surface, there is a local passive film breakdown (Fig. 23.1)
- *Carbonation process* of concrete due to a reaction with the carbon dioxide contained in the atmosphere, which brings the pH of pore solution from pH > 13 to pH < 9. This process starts from the concrete surface, then proceeds through the concrete cover until it reaches steel reinforcements, then causing passivity breakdown (Fig. 23.2).

Once the passive film breaks, corrosion occurs if water and oxygen are present on the reinforcement surface.

The service life of a reinforced concrete structure can be divided in two phases (Fig. 23.3): corrosion *initiation*, which is the time required for carbonation or chlorides penetration to reach the reinforcements, then producing passivity breakdown, and corrosion *propagation*, once corrosion attack has started.

Corrosion by carbonation appears uniformly distributed on the reinforcement surface; conversely, chloride-induced corrosion is mostly localized, with crater-like penetrating attacks, surrounded by passive, non-corroded areas, as in pitting corrosion mechanism; the attack spreads in case of large chloride contamination, especially when pH decreases. High-strength steel reinforcements in pre-stressed concrete structures can suffer from *hydrogen embrittlement* when peculiar environmental, electrochemical, mechanical and metallurgical conditions concur.

Reinforcement corrosion leads to a progressive reduction in safety margins established in design (Fig. 23.4), for instance:

- Reduces the cross section of reinforcements, then jeopardizing static and dynamic structural stability; in particular, corrosion of external supports, which first begins suffering depassivation by carbonation or chloride contamination,

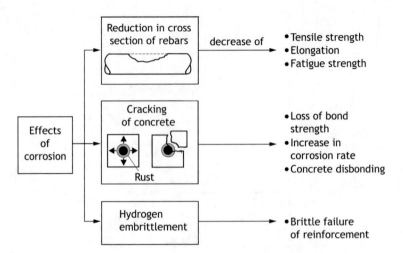

Fig. 23.4 Consequences of structural reinforcement corrosion in reinforced concrete structures

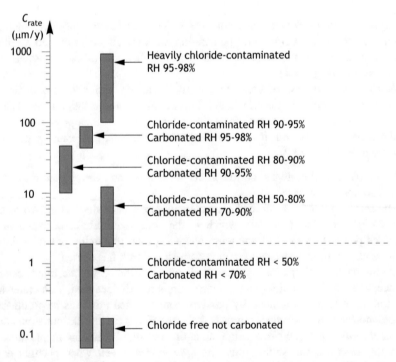

Fig. 23.5 Corrosion rate of steel reinforcements in various conditions adapted from Andrade (1988)

may compromise the confinement action, therefore weakening the main rein-
forcements stability
- Reduces adhesion between reinforcement and concrete, or even halts it if thickness loss is of the order of 50 μm
- Produces oxides (rust) with a volume 4–5 times higher than that of corroded iron, which cause cracks in concrete cover (above 50–200 μm corrosion thickness) and even its spalling or delamination
- Finally, on some types of high-strength steel, it can give brittle failure.

Corrosion rate is generally measured in μm/year, and varies with environmental and concrete conditions (Fig. 23.5). As long as corrosion rate is below 1.5–2 μm/year, there is no consequence on reinforcement integrity.

23.2   Corrosion by Carbonation

Atmospheric carbon dioxide reacts with alkaline components of the pore solution contained in concrete, as shown schematically by the following reaction:

$$CO_2 + Ca(OH)_2 \xrightarrow{H_2O, NaOH} CaCO_3 + H_2O \tag{23.1}$$

In reality, also intermediate reactions take place, involving sodium and potassium hydroxides.

Carbonation does not cause reduction of mechanical strength of concrete; however, there are important consequences on reinforcements, because the pH of pore solution decreases from initial values, typically between 13 and 14, to values close to neutrality, then causing passivity breakdown.

23.2.1   Carbonation Depth

Carbonation reaction starts on the concrete surface and proceeds through the cover, following a penetration rate derived from the equation below:

$$s = K \cdot t^{1/2} \tag{23.2}$$

where s is thickness of carbonated layer (in mm), and t is time (in years). The coefficient K equals mathematically the thickness of carbonated cover in the first year, and depends on environmental factors (temperature, humidity, carbon dioxide content) as well as on concrete-related factors, alkalinity and porosity: the most influencing parameters are humidity and porosity, as described in the following paragraphs.

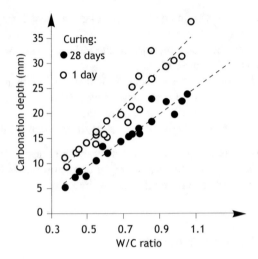

Fig. 23.6 Influence of W/C ratio and curing time on carbonated thickness (from Page 1992)

Relative humidity. Carbon dioxide diffusion within the concrete cover occurs through the pores, easily if filled with air, i.e., in gaseous phase, while very slowly if filled with water, about 10^4 times slower. Accordingly, diffusion rate decreases as relative humidity increases (more markedly above 80%) then it zeros in water-saturated concrete. On the other hand, carbonation takes place if water is present, and it zeros for relative humidity lower than 40%. For these two opposite limits, carbonation rate maximises in the relative humidity range 50–80%. It follows that microclimate, i.e., local temperature and humidity in differently exposed parts of a structure, outdoor or indoor, exposed or screened from rain, is of utmost importance.

Porosity. All porosity affecting factors (water/cement ratio, installation, compaction, curing) are crucial for determining carbonation rate. For example, Fig. 23.6 reports the effect of the water/cement ratio and curing on carbonated thickness, which decreases as water/cement ratio decreases and curing time increases.

23.2.2   Corrosion Rate

Once carbonation reaches reinforcements, the passive film breaks (i.e., initiation time ends) and corrosion starts if water and oxygen are present. Except in conditions of permanently water saturated concrete, a sufficient oxygen is always available for the corrosion process, therefore corrosion rate is governed by water content in concrete; the highest corrosion rate is of the order of several tens of µm/year at nearly water saturation conditions. The most common values for atmospherically exposed structures are between 5 and 10 µm/year, and drop to 1–2 µm/year for indoor concrete. Corrosion by carbonation becomes a concern only in conditions of high relative humidity or in the presence of alternating wetting, which affects moisture content on reinforcements.

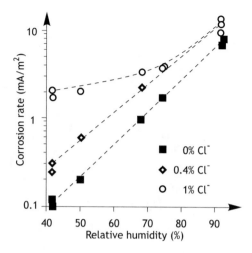

Fig. 23.7 Effect of low chloride content on corrosion rate by carbonation (from Glass 1991)

Fig. 23.8 Corrosion of reinforcements by carbonation in Milan, Italy, on a 1969 building (picture is dated 2006)

Fig. 23.9 Corrosion of
reinforcement by carbonation
in Milan, Italy (Marchiondi
Institute, 1957)

Corrosion harshens when chlorides are present in concrete even at content below
the threshold of depassivation, as shown in Fig. 23.7. Also the safe range of relative
humidity, at which corrosion rate is negligible, reduces. Figures 23.8 and 23.9 show
two examples of corrosion by carbonation.

23.3   Chloride-Induced Corrosion

Chlorides can penetrate concrete when chloride-containing solutions are in contact
with the concrete cover as for example in marine structures and in bridge slabs
when de-icing salts are spread. Corrosion starts when chloride concentration on
reinforcement exceeds a critical value, which ranges, for atmospherically-exposed
structures with carbon steel reinforcements, from 0.4 to 1% by cement weight. The
time required to reach such critical value is called *corrosion initiation time* and

depends on concentration of chlorides on external surface, concrete cover thickness and characteristics of cement paste that determines the chloride diffusion coefficient.

Experience shows that, in spite of the presence of several mechanisms, chloride concentration profile follows with good approximation an equation formally identical to the second Fick law by introducing the *apparent* (or *effective*) *diffusion coefficient*, D_{ce}, obtained experimentally:

$$\frac{\partial C}{\partial t} = D_{ce} \frac{\partial^2 C}{\partial x^2} \tag{23.3}$$

where C is chloride content (% by cement or concrete mass), time, t, is expressed in s, distance, x, from concrete surface is in cm, and apparent chloride diffusion coefficient, D_{ce}, is in cm²/s.

Assuming constant in time and space both chloride content on external surface, C_s, and diffusion coefficient (i.e., concrete is considered homogeneous), the solution of Fick law is:

$$C_x = C_s \left(1 - erf \frac{x}{2\sqrt{D_{ce}t}}\right) \tag{23.4}$$

which gives the chloride content C_x at depth x, at time t. Therefore, known D_{ce} and C_s, it is possible to figure out the time-based evolution of chloride concentration profiles in concrete cover and predict the time for corrosion initiation.

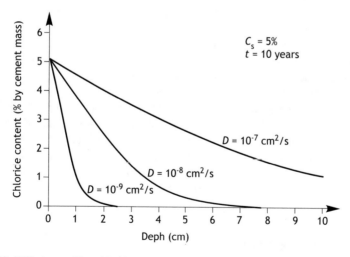

Fig. 23.10 Diffusion profiles, chloride concentration versus thickness, after 10 years of exposure, with different apparent chloride diffusion coefficients (chloride concentration: 5% on surface)

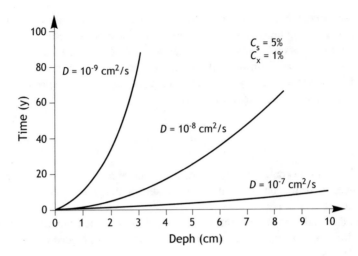

Fig. 23.11 Diffusion profiles, time versus penetration depth, with different apparent chloride diffusion coefficients (chloride concentration: 5% on surface, 1% in concrete) at the rebar surface

Figure 23.10 reports chloride concentration profiles after 10 years of exposure for different diffusion coefficients, while Fig. 23.11 shows time dependence of chloride penetration depth considering a chloride content of 1% by cement mass, both obtained by applying the second Fick law. Indeed, a pure diffusion mechanism applies on completely saturated concrete, only; in most practical situations, in addition to diffusion other transport mechanisms operate, for example, capillary absorption on dry concrete when in contact with chlorides containing solutions, often followed by evaporation which increases chloride concentration.

Because the penetration rate of chlorides depends on their concentration on the surface, C_s, it is important to input reliable values. For marine structures, after an initial transitory of a few months, chloride concentration on the surface is constant and reaches values higher than 3–4% in the splash zone. Often, prudentially, 5% is the value considered for assessment of expected life.

The diffusion coefficient, D_{ce}, typically varies from 10^{-9} to 10^{-7} cm^2/s depending on concrete properties, specifically on permeability and related factors as: water/cement ratio, vibration after casting, curing, presence of cracks and type of cement. The latter has a decisive role: passing from Portland cement to fly-ash mixed cements, the diffusion coefficient drops consistently, especially when ground granulated blast slag (GGBS) is used, and the effect lasts even tens of years.

23.3.1   Corrosion Rate

In atmospherically exposed structures, once the attack has started, corrosion rate can range from a few tens of μm/year to 1 mm/year, and increases as humidity increases from 70 to 95% and chloride content from 1 to 3% (by cement mass). Corrosion rate also increases with temperature, as proved by making a comparison between temperate and tropical weather. In practice, corrosion attacks in chloride-contaminated structures, once started, can result in unacceptable reductions of cross section of reinforcements in a very short time.

The lower limit of humidity, below which corrosion rate becomes negligible, depends on concrete characteristics, chloride content on surface and type of salt. In conclusion, it is much lower than that for corrosion by carbonation.

23.3.2 Structures at a Risk

Reinforced concrete structures at greater risk of chloride-induced corrosion are marine structures and highway bridges when exposed to deicing salts.

In marine structures, risky zones are: submerged, tidal and splash, and atmospheric zones. The most critical is doubtlessly the second one because both oxygen and chlorides are easily available, unlike the other two, where either oxygen or chlorides are not, as shown in Fig. 23.12. Corrosion attack is often localized, with a maximum penetration rate of some tenths of mm/year: accordingly, corrosion propagation time is discarded in corrosion assessment for the evaluation of service life. For structures near the sea, corrosiveness decreases as the distance from shoreline increases.

The use of de-icing salts on bridge decks leads to localized corrosion particularly on floor slabs, pier caps, columns and non-sealed joints; the lack or improper drainage systems, which cause ponding, increases the risk of reinforcement corrosion (Figs. 23.13 and 23.14). In critical areas, as viaducts joints, corrosion rate is as high as in the splash zone of marine structures, the producing a severe concrete spalling (Fig. 23.15).

Fig. 23.12 Chloride content profile in the most critical area of a marine structure, adapted from Bertolini (2013)

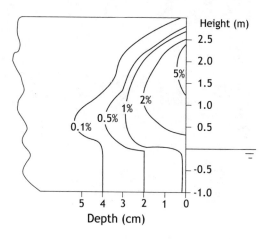

Fig. 23.13 Detail of localized corrosion caused by chlorides contamination (1980s Bologna—Florence highway, Italy)

Fig. 23.14 Localized corrosion caused by chlorides contamination (1980s, viaduct of Brennero highway, Italy)

Fig. 23.15 Spalling of concrete cover on bridge deck (1980s Milan-Genoa highway, Italy)

23.4 Hydrogen Embrittlement

Hydrogen embrittlement (HE, Figs. 23.16, 23.17 and 23.18) may occur on high strength steels with yield strength exceeding 700–900 MPa in the presence of atomic hydrogen. This one does not form on passive reinforcement in sound chloride-free concrete, instead it forms on de-passivated sites due to carbonation or chloride contamination or where passivation is absent because of local lack of concrete. Reinforced concrete structures susceptible of HE are pre-stressed and post-tensioned structures, where high strength steels are used (Figs. 23.19, 23.20 and 23.21). HE susceptibility is highly dependent on steel microstructure obtained by thermo-mechanical treatments to achieve the required mechanical strength. In practice, two processes are used: cold drawing followed by stress relieving or quenched and tempered steels. The latter, producing a more susceptible microstructure, is virtually out of production. Also perlitic steel is used for large diameter bars, obtained by hot working followed by cold deformation and stress relieving. In any case, HE susceptibility increases as mechanical strength increases.

HE-related failures of pre-stressed structures represent a small percentage with a decreasing trend because the most susceptible high-strength steels are no longer in production. Nevertheless, even in the 1990s, several structural collapses occurred, especially in Germany where the use of quenched and tempered steels was formerly widespread. A famous example is the failure of the Congress Hall in Berlin in 1980.

Fig. 23.16 Hydrogen embrittlement of high strength steel pre-stressing strands

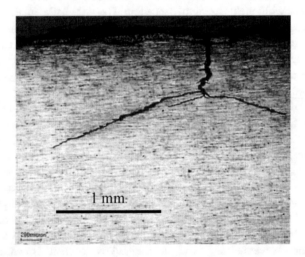

Fig. 23.17 Micrograph of cracks caused by hydrogen embrittlement of high strength steel pre-stressing strands

Fig. 23.18 Hydrogen embrittlement of a pre-stressed concrete cylinder pipe

Fig. 23.19 Hydrogen embrittlement of pre-stressed bars (Manfredonia Harbour, Italy, built on 1979)

Fig. 23.20 Post tensioned concrete structure collapsed in 1999 due to HE on carbon steel strands (S. Stefano Bridge, Messina, Italy—courtesy of Prof. E. Proverbio)

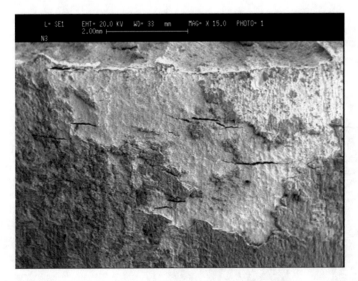

Fig. 23.21 Zone of fracture initiation caused by hydrogen embrittlement on a high strenght steel

23.5 Corrosion by Stray Currents

Very rarely, reinforcement of foundations or underground structures, such as railway tunnels, may suffer stray current corrosion, where the interference current leaves the reinforcement. Initially, on passive reinforcements, the anodic reaction where current leaves steel is oxygen evolution, but after some time as the result of the local acidification, which destroys the passive film, corrosion takes place. Accordingly, two conditions are necessary for stray current corrosion in sound alkaline and chloride-free concrete:

- *First condition.* Interfering electric field must be sufficiently strong to cause current circulation in concrete, overcoming ohmic and overvoltage barriers
- *Second condition.* Interference current must be stationary, i.e., flow "continuously" for a sufficient time to lead to passivity breakdown. Intermittent flows are less harmful because the acidity produced is neutralized by the surrounding alkalinity during any inactive period.

In practice, only the electric field generated in long railway tunnels, with reinforcements in electrical continuity, fits the first requirement; about the second one, as considered above, any interruption of the stray current flow delays or even halts the passive film breakdown. It can be concluded, in general, that *non-stationary interference*, as it occurs in electric transit railway systems, does not represent a real threat.

23.6 Prevention of Reinforcement Corrosion

Corrosion prevention must begin in design phase, involving conception, structural details, and materials; it then follows during construction through preparation, implementation, compaction and concrete curing. Finally, it continues throughout the operating life by planned maintenance and inspections.

This approach to durability of concrete structures has been called *holistic* in order to emphasize that prevention should not be considered as a mere set of possible prevention measures, but rather as articulated interventions from design to dismissing.

Preventative measures taken in design, construction and maintenance phase deal with the following rules of thumb:

- Easy access must be granted for inspection and maintenance activities
- Structure geometry has to minimize cracking
- Design of construction details, also with reference to the build-up phase, must avoid complex geometries, sharp edges, rebar congestion, water ponding as promoted by expansion joints and supports
- During transport and execution, necessary precautions must be taken to avoid segregation of mix. Vibration helps obtain maximum compaction possible

- Concrete cover thickness must comply with design data
- Temperature and humidity must be optimal for a sufficiently long period to allow and facilitate cement hydration.

In each of these phases all necessary controls need to be implemented to achieve a product that meets quality requirements. Since different subjects operate (designer, concrete producer, construction and maintenance companies) it is important to define battery limits for any controls and responsibility. Some of above issues are discussed in the following.

23.6.1 Quality of Concrete

The concrete cover provides protection of reinforcement against aggressive agents coming from outside, therefore it is sufficient that the surrounding concrete remains alkaline and chloride-free to ensure passive conditions. Accordingly, to prevent the corrosion of reinforcements, a low permeability concrete is required and achieved by following the updated knowledge of concrete technology.

The following parameters are of particular importance: water-to-cement ratio, content and type of cement, compaction, curing and presence of cracks. The European Standard EN 206 provides a guide for composition and properties of concrete to ensure 50-year lifetime. This standard deals with concrete prepared with Portland cements in accordance with EN 13670-1 standard, with concrete cover thicknesses complying with EN 1992-1-1 standard.

Water/cement ratio. Water-to-cement ratio is a key factor to determine the capillary porosity of the cement paste and thus its resistance to aggressive species penetration. Therefore, a low ratio is the primary choice to obtain an impervious concrete to either carbonation or chloride penetration.

Cement content. The increase in cement content, for a given water-to-cement ratio, on one hand leads to a higher amount of water, which increases workability of the mixture, but on the other hand also increases risk of cracking due to hydration heat or drying shrinkage or alkali-aggregate reaction.

Type of cement. The type of cement is particularly important for structures, which are exposed to chlorides. At same conditions, blended (except for limestone-based) and blast furnace cements are the most resistant to chloride penetration. Unfortunately, some blast furnace cements already contain 0.1–0.2% of chlorides, then weakening this advantage. Blended cements show also beneficial effects on the resistance to sulphate attack and alkali-aggregate reaction; a low hydration heat also characterizes them.

Installation and compaction. Installation and compaction of concrete strongly influence durability. The pouring of fresh concrete should ensure optimal formwork filling, avoiding air trapping: this is achieved by vibration. Voids due to segregation or insufficient vibration increase concrete permeability, jeopardize passivation of steel reinforcement and bring critical chloride content toward values lower than the typical range, 0.4–1% by cement mass. Excessive vibration can lead to segregation for high slump concrete. The addition of plasticizers becomes mandatory to achieve high workability with low w/c ratio. For structures with a complex geometry, it is possible to use self-compacting concrete, which enables to fill the formwork without requiring vibration thanks to a special composition.

Curing. Curing of concrete helps promoting cement hydration by a control of either moisture content or temperature a few days after installation. Adequate curing is often obtained by simply wetting the surface or reducing water removal and water evaporation or delaying the removal of formworks. Curing affects primarily durability and mechanical strength. The recommended minimum time of curing is available in ENV 13670 standard.

Cracking. When tensile stresses onset inside concrete, due to either mechanical loads or constraints, cracks may form. For example, internal or external constraints can prevent shrinkage due to a humidity change (plastic or hygrometric shrinkage). In the case of thick concrete casting, cracking can take place for excessive temperature changes between surface and core. Cracking can also occur when expansive compounds form inside a concrete slab, as in the case of steel reinforcement corrosion, alkali-aggregate reaction, sulphate attack or freeze-thaw cycling. Cracks can reduce corrosion initiation time, as they are preferential pathways for carbonation and chloride contamination. Within certain limits, cracks are not dangerous, provided that crack opening does not exceed 0.1 mm, because corrosion products tend to seal the cracks and favour local conditions for repassivation in the area close to reinforcements.

23.6.2 Cover Thickness

As cover thickness increases, corrosion initiation delays because time for carbonation or chloride contamination increases. Accordingly, based on concrete properties, carbonation and chloride contamination rates can be evaluated and, once the minimum time to trigger corrosion attack is fixed, the minimum cover thickness can be determined. Then, theoretically, it would be possible to ensure the durability of a structure by fixing a proper concrete cover thickness; in practice, however, cover thickness cannot exceed some limits for economic and technical reasons, for example, as concrete cover thickness increases, the risk of cracking strongly increases due to drying shrinkage.

As far as carbonation is concerned, the specified cover thickness should be ensured on the whole surface of the structure: in fact, since penetration of carbonation follows a parabolic law (Eq. 23.2) the relationship between thickness reduction and corrosion initiation time reduction is not linear; similar considerations apply to chloride induced corrosion.

23.6.3 Common Mistakes

Premature degradation of reinforced concrete structures deals with design and construction mistakes or lack of maintenance. Typical design casualties include:

- Construction procedure favouring local aggressive conditions (for example, an excessive number of joints or an unnecessarily complex geometry)
- Complex structural frame producing uneven stress distribution, then leading to cracking
- Not adequate details for construction, such as concrete cover thickness not suitable for aggressive environmental conditions, or geometric conditions favouring water ponding; rebar congestion impeding concrete compaction; sharp edges, which halve design life (Figs. 23.22 and 23.23).

Fig. 23.22 Example of insufficient concrete cover (Belvedere ceiling, 31st floor of Pirelli skyscraper, Milan, Italy, 1959—picture is dated 2005)

Fig. 23.23 Insufficient concrete cover due to segregation and rebar congestion (building La Nave, Politecnico di Milano, Milan, Italy, 1967—picture is dated 2008)

The most frequent mistakes in mixture proportion are: too high water/cement ratio, the use of Portland or sulphate resistant cements in chloride-containing environments. Typical mistakes in construction phase are addition of water, incorrect compaction, wrong or absent spacers for concrete cover thickness, bad or inaccurate curing.

23.7 Additional Protections

When design procedure, concrete cover thickness, choice of materials, mix composition, pouring, compaction and curing comply with regulations and avoiding abovementioned errors, durability has proved, in the majority of environmental conditions, to meet 50 years, as typically required for the design life of concrete structure.

However, for important structures exposed to chlorides, the compliance to regulations is not sufficient to avoid corrosion in shorter time, as it may happen for example on bridge decks where de-icing salts are spread, on splash zone of marine structure or on surfaces exposed to seawater.

On the other hand, a service life of 50 years may be too short for many important buildings, such as churches, monuments and public buildings that are expected to last, with minimum maintenance, 100 years or even more. In these cases, including conditions associated to severe environments, practical difficulty to build specified cover thickness or poor accessibility for maintenance or when there is no alternative for the use of low quality concrete, it becomes necessary to adopt specific additional preventative measures to increase durability as summarized in Fig. 23.24.

Figure 23.25 illustrates how additional preventative measures work by reducing aggressive species diffusion or by controlling the anodic corrosion process, since the cathodic process, i.e., oxygen reduction reaction, cannot be limited (in other words, today there are no available techniques to do it, unless for permanently water saturated structures).

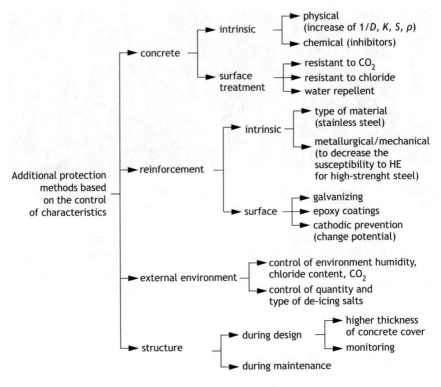

Fig. 23.24 Classification of additional preventative methods

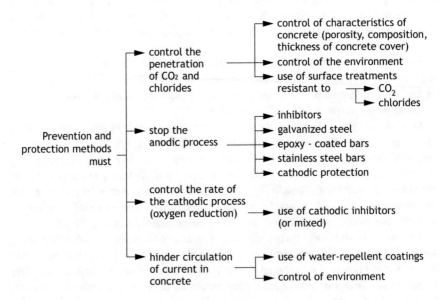

Fig. 23.25 Mechanisms of additional preventative methods

23.7.1 Concrete Surface Treatments

The main action of surface treatments is to slow down the penetration of aggressive species to increase initiation time, especially for chloride-induced corrosion. Once corrosion has started, effective treatments to reduce corrosion rate are those that enable to stop water penetration; this works also in the case of carbonation. It is possible to identify four main classes of surface treatments of concrete structures:

- Organic coatings
- Hydrophobic treatments
- Pores obstruction treatments
- High thickness cementitious coatings.

Organic coatings. These coatings must be compatible with concrete alkalinity; they are based on acrylic, polyurethane and epoxy binder types, containing suitable pigments, additives, solvents and diluents to ease their application. By forming a continuous film, typically 100–300 μm thick, on the surface, their action depends on porosity, i.e., on permeability to CO_2 and water vapour.[2]

Hydrophobic treatments. To reduce capillary absorption of water and dissolved aggressive substances, in particular chlorides, water repellent agents are used. They have no effect on gases such as carbon dioxide and oxygen, then useless to reduce carbonation. Hydrophobic surfaces can be obtained by impregnating concrete with compounds belonging to the family of silicones, silanes and siloxanes. These products are also absorbed by capillary action into pores for a few millimetres of thickness, therefore water remains on the surface of concrete as droplets. The deeper the capillary absorption of these hydrophobic substances, the longer their effective life. Ultraviolet radiation promotes degradation of these products. Application should be carried out when the concrete surface is not too damp.

Pore sealing treatments. To seal pores, silicates or silico-fluorides are used, which react inside pores with concrete to give clogging reaction products. Also some organic coatings based on epoxy or acrylic binders can be used, which penetrate into pores (sometimes impregnation is favoured by applying vacuum) and harden and seal pores.

Cementitious coatings. Cementitious coatings are typically made with a mortar or concrete containing polymeric materials to give high thickness and low permeability. These coatings exhibit: good adhesion to concrete, high flexibility, high permeability to water vapour, low permeability to water, high resistance to chlorides, sulphates and carbon dioxide diffusion and good resistance to aging due to

---

[2]There is a parameter, the equivalent air layer thickness, which measures permeability by indicating the equivalent meters of air that oppose the same resistance to gas diffusion as the coating. The minimum coating thickness should give a value of this parameter of at least 50 m versus CO_2 and lower than 4 m versus water vapour, to allow concrete to exchange vapour with the atmosphere.

sunlight. Because of their high deformability, these cementitious coatings are recommended for cracked concrete, since they seal cracks and withstand deformation, as well as, when compared to other coatings, high mechanical stresses, especially if variable in time.

23.7.2 Corrosion Inhibitors

Corrosion inhibitors are used primarily to prevent chloride-induced corrosion. A state of art report has been published by Elsener in 2001. Their efficiency depends on dosage, concrete permeability, pH of pore solution, temperature, chlorides and oxygen concentration. Two categories are used: inorganic (i.e., calcium nitrite) and organic.

Calcium nitrite. It is the oldest corrosion inhibitor used in concrete. It is added to the mixture as a solution that also contains suitable products for limiting side effects on curing and workability. Dosage depends on maximum expected chloride content, and is given by molar ratio $[NO_2^-]/[Cl^-]$ of about 1–1.25. For instance, to withstand a maximum chloride content of about 3% by cement weight, much above the critical interval 0.4–1%, a typically recommended dosage is 30 L/m^3. Calcium nitrite works in highest quality concrete only, as porous or cracked ones cannot prevent inhibitor leaching.

Organic inhibitors. Organic inhibitors include a broad class of products, which have been studied since 1980s. Mixtures of amines, alkanol-amines and their salts with organic acids, for instance benzoic acid, have proved to be effective. These inhibitors can be added to the mix or to mortars used for restoration, or sprayed on the surface to migrate to the reinforcement. Since they have been introduced recently, few reliable data on their long-term effectiveness are available; it is opinion that critical chloride content is increased to about 1.2%, therefore the life extension is limited (see Ormellese 2006; Bolzoni 2006). With reference to the effectiveness of migrating inhibitors applied on the concrete surface, opinions are still differing (Ormellese 2008).

23.7.3 Stainless Steel Reinforcements

To increase the corrosion resistance of reinforcements, stainless steels such as AISI 304L (EN 1.4307), AISI 316L (EN 1.4404) austenitic grades and AISI 318 (EN 1.4462) duplex grade have been used since a few decades. Table 23.1 reports their chemical composition. Nitrogen containing stainless steels are also used. As far as yield strength, elastic modulus and ductility properties of stainless steels are concerned, same values as carbon steel reinforcements must be ensured as well as weldability.

Table 23.1 Composition of stainless steels most commonly used in concrete, following EN, UNS and AISI designations (% by mass)

Types of steel			C max	S max	Cr	Ni	Mo
EN	UNS	AISI					
1.4301	S30403	304L	0.3	0.003	17.0–19.5	8.0–10.5	–
1.4436	S31603	316L	0.3	0.015	16.5–18.5	10.5–13	2.5–3.5
1.4462	S31803	318LN	0.3	0.0015	21–23	4.5–6.5	2.5–3.5

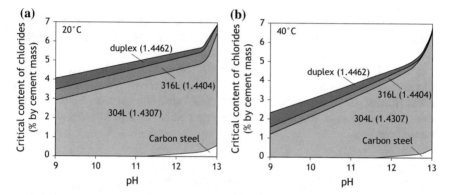

Fig. 23.26 Critical chloride content as function of pH and temperature for pickled stainless steels: **a** 20 °C; **b** 40 °C (Bertolini 2002)

Figure 23.26 shows the critical chloride content for mentioned stainless steels as function of pH: in concrete (pH $\approx$ 13) critical level is 5% by cement weight for molybdenum-free stainless steels and increases to 7–8% when molybdenum is present. These values drop down to about 3.5% when passive film is tinted (i.e., coloured) because of welding.

In carbonated concrete, with pH $\approx$ 9, the critical content significantly lowers especially if temperature reaches or exceeds 40 °C. Experience showed that in heavily cracked structures stainless steel reinforcements behave as in carbonated ones although concrete is still alkaline. Molybdenum containing stainless steels are recommended for hot climates, as well as for very cracked or carbonated chloride-contaminated concrete.

Sometimes stainless steel is used for external reinforcements (skin reinforcements) or critical zones only, while internal and remaining zones can host ordinary carbon steel. Stainless steel can be conveniently placed in high chloride content areas and carbon steel only where chloride content is expected not to exceed the critical threshold of 0.4%. As an important result, all reinforcements operate in passive conditions, so that no galvanic coupling onsets (in sound concrete the potential of passive carbon steel and passive stainless steel is the same).

23.7.4 Galvanized Steel Reinforcements

Galvanized reinforcements are obtained by hot dipping in molten zinc at approximately 450 °C. The higher the zinc layer thickness the longer the durability, but also the higher the coating tendency to crack during bending at yard; best compromise is a total zinc layer thickness of 80–120 μm with 10 μm minimum as pure zinc outer layer.

In alkaline concrete, zinc shows a non-acceptable corrosion rate if chloride content exceeds 1.2% by cement mass. In carbonated chloride-free concrete, corrosion rate of zinc layer is very low since zinc passivates.

Because of the excellent behaviour in carbonated concrete, galvanized reinforcements are recommended for precast elements, in lightweight or thin concrete cover structures not expected to be contaminated by chlorides.

Provided the zinc layer thickness is of 80–100 μm, carbonated structures with chloride contamination level below 1% by cement weight are expected to have service life of 100 years or more. Figure 23.27 shows a church in Rome, Italy, where galvanised steel was used to increase durability and also to avoid staining of exposed surfaces.

Fig. 23.27 Dives in Misericordia church, built with galvanized steel reinforcements (Rome, Italy, 2003)

23.7.5   *Cathodic Prevention (CPrev)*

The cathodic prevention, CPrev, introduced by Pietro Pedeferri in early 1990s, is the cathodic protection applied since the beginning throughout the entire service life of the structure. It is based on the dependence of pitting and repassivation potential of steel reinforcements on chloride content in concrete, as shown in the potential versus chloride content plot of Fig. 23.28, also called Pedeferri's diagram, which defines corrosion conditions of steel reinforcements. CPrev current density is in the order of 1–3 mA/m^2, which produces a potential decrease of 100–200 mV, then enabling an increase in critical chloride content of one order of magnitude, i.e., more than 4% by cement weight.

23.7.6   *Comparison of Additional Protections*

To choose an additional corrosion prevention method it is important to know thoroughly the exposure environment, design life, possible constrains for applicability (for instance power source for CP or CPrev) and associated costs (investment and operating costs).

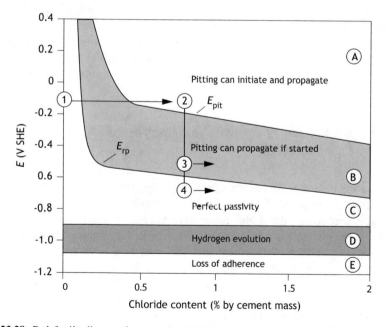

Fig. 23.28 Pedeferri's diagram for corrosion initiation and propagation conditions in concrete contaminated by chlorides, highlighting evolution paths of cathodic protection (see Sect. 23.9.2)

An example of limits of applicability in environments containing chlorides is the critical chlorides content, which is 0.4 to 1% by cement weight for carbon steel reinforcements and becomes:

- 1–1.2% for galvanized steel reinforcements and in case of use of organic inhibitors
- 1–3% range for calcium nitrite based inhibitors
- 3.5–8% range for stainless steels
- Even higher than 5–8% range for CPrev.

A guideline for expected operating life is as follows:

- 10–15 years for water-repellent treatments or coating barriers against carbonation or chlorides
- 100 years for galvanized steel reinforcements with zinc coatings (thickness 80–100 μm) and chloride content which does not exceed 1%
- No limits for stainless steel in alkaline concrete and less than 5% chlorides after removal of welding oxides
- CPrev with activated titanium anodes are designed to last up to 30 years and there are no limits for the concentration of chlorides.

Investment cost varies with the method selected; some general indications are available for comparison, for example, taking carbon steel reinforcement as reference, galvanized steel cost approximately doubles, while stainless steel cost is about 8–10 times greater for total replacement or 4–5 times more for partial replacement. To determine a correct and definitive comparison, LCC (Life Cycle Cost) of each technical solution has to be considered, because either investment or operating costs vary for each solution. For example, with stainless steel reinforcements concrete cover thickness can be much lower, then reducing the initial cost of the concrete cast, and there is no need for maintenance, then almost zeroing operating costs.

23.7.7   Evaluation of service life by performance based methods

As discussed before, at the beginning of this paragraph the recommendations given by EN 206 and Eurocode 2 (EN 1992-1-1) standards about concrete composition, placing, compaction, curing and cover have proven, in the majority of environmental conditions, to meet 50 years service life. In very aggressive environment, especially in presence of chlorides and longer service lives performance based approach should be used to estimate the service life by modeling deterioration mechanisms; moreover, these methods can provide much more information about the effect of different options, like the use of preventative techniques. Among performance based approaches, probabilistic ones are preferred because they are able to take into account the intrinsic variability of the influencing factors. This approach has been implemented in different models, among these the most relevant one is the Model code for service life design, issued by the International Federation for Concrete (fib) in 2006. The code models the effect of environment on the structure by using an approach derived from the equations used to describe transport of carbonation and chlorides into concrete, namely (23.2) and (23.3) and evaluate the probability that a defined limit state (initiation of corrosion, concrete cracking or spalling) is reached. Nowadays, the use of this model is limited and there are open questions to solve before more extended applications. First of all, comparison between the predictions and the real performance of structures is not easy because few number of structures were designed in agreement with the model. Moreover, the model does not provide sufficient indications for the determination of some input parameters, in particular the surface chloride concentration and the critical chloride content for rebars different from carbon steel.

23.8 Diagnosis

The management of a reinforced concrete structure needs a periodical checking of structure integrity, through diagnostic testing on both concrete and reinforcement degradation. Figure 23.29 reports a general procedure for inspection and control, which involves: gathering of information about location, exposure, design data (mix design, cover thickness, structural solutions), construction (origin of materials, construction techniques, qualification of contract company), history (accidental events, any evidence of oncoming degradation/decay). This information can help formulate and identify hypotheses on possible causes of degradation.

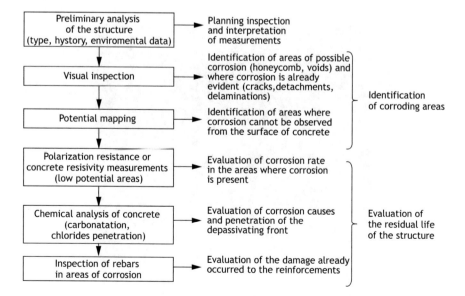

Fig. 23.29 Flow chart for concrete structure inspections

An inspection begins by a visual observation, aiming to estimate type and extent of damages such as: cracking along the reinforcements, concrete cover spalling, presence of corrosion products. Construction errors favouring the onset of corrosion are identified, as: cracking, non-uniform concrete cover thickness, lack of reinforcement coverage due to insufficient workability of fresh concrete, joints, macro-cavities produced by the concrete segregation.

Visual observation is often followed by simple checks to assess concrete conditions: hammer test to identify the presence of hidden delaminated zones based on the sound emitted, concrete hardness measurements and ultrasonic techniques.

23.8.1   Concrete Cover Thickness Measurements and Rebar Identification

There are instruments, called pachometers, based on magnetic principles, which help identify reinforcements position and cover thickness. By matching this information with the measurement of carbonation depth and chloride concentration profile, corrosion initiation time can be evaluated.

23.8.2   Analysis of Concrete

Analysis of concrete provides fundamental information on: cement content, water-to-cement ratio, additives, incorporated air.

Fig. 23.30 Phenolphthalein tests (sound concrete is the pink zone)

Phenolphthalein test. It is used to determine carbonation depth: the pH of a specimen is checked by spraying its surface with an alcoholic solution of phenolphthalein; carbonated concrete does not change colour, while sound alkaline concrete, not yet carbonated, shows a typical pink colour, as shown in Fig. 23.30. This test shall be performed on fresh surfaces on a perpendicular core specimen immediately after its extraction. By comparing carbonated thickness and concrete cover thickness, it is therefore possible to identify areas where carbonation induced corrosion is a risk.

Chloride content measurement. For structures exposed to chloride containing environments, chloride profile helps determine the risk of corrosion onset. Core specimens, cut into slices, or debris gathered at different depths are used for analysis. Samples are dissolved in nitric acid and chloride concentration is measured by chemical analyses. Chloride profile and chloride content on the surface, C_S, allow the calculation of the apparent diffusion coefficient, D_{ce}, by applying Fick law, which in turn allows to predict the evolution of chloride penetration into concrete (see Sect. 23.3).

23.8.3   Electrochemical Techniques

Some electrochemical techniques are used to determine the corrosion condition of steel reinforcements, whether passive, i.e., initiation period is still ongoing, or corroding as well as the corrosion rate.

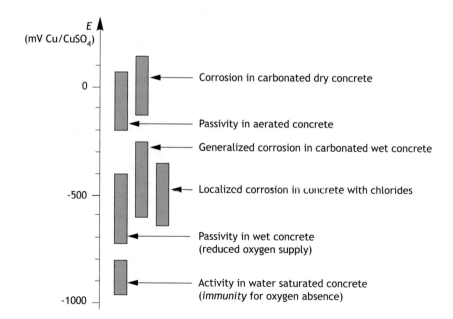

Fig. 23.31 Typical potential ranges of carbon steel reinforcements in concrete at different conditions

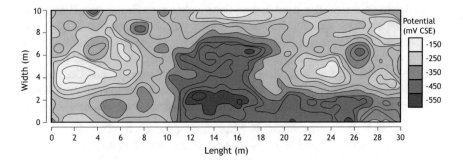

Fig. 23.32 Potential mapping of a bridge deck. Corrosion risk is in dark grey

Potential mapping. It consists of the measurement of the potential by locating the reference electrode on a prefixed grid on the external surface in order to determine anodic and cathodic zones. Figure 23.31 shows typical potentials measured in presence of carbonation, chlorides, and in water saturation, that is, for passive or active steel. To assure the electrical contact between reinforcements and the reference electrode, two checks are performed: the electrical continuity of reinforcements and the electrolytic contact by means of a wet sponge. Figure 23.32 shows a typical potential mapping of a bridge slab where de-icing salts are spread in winter. The interpretation of a potential mapping is based on ASTM Standard C876: potential values more noble than -0.2 V CSE have 90% probability of no corrosion, conversely, a value below -0.35 V CSE has 90% probability of corrosion; probability becomes 50% when potential falls within the interval. As a complementary evaluation criterion, since corrosion propagation occurs through a macrocell mechanism, the potential map has to show a clear potential difference of at least 150 mV to indicate that corrosion is occurring.

Linear polarization measurement. Where potential mapping shows an anodic site (i.e., a potential lower than -0.35 V CSE) corrosion rate can be estimated by the measurement of polarization resistance by the technique called Linear Polarization Resistance (LPR) described in Chap. 29.

Concrete resistivity. Once steel reinforcement has become active, corrosion rate, either uniform as in the case of carbonation or localized due to chloride contamination, depends on concrete resistivity. This is quite evident when a macrocell mechanism sets up as in case of chloride contamination, instead is more controversial for carbonation. Accordingly, empirical criteria are used, although they do not have general validity; for instance, they are not applicable when reinforcements are still passive and in water-saturated conditions. The most used guideline for corrosion rate prediction is reported in Table 23.2. Resistivity measurements are also used to identify changes in concrete, for example, carbonated or chloride contaminated zones.

Table 23.2 Empirical relationship between resistivity and corrosion rate

Resistivity (Ωm)	Corrosion rate
<100	High
100–500	Moderate
500–1000	Low
>1000	Negligible

23.9 Repair

To stop corrosion attacks, three approaches are used: restore passivity; apply cathodic protection; reduce moisture of concrete through a surface treatment. Different techniques are applicable for each approach.

23.9.1 Traditional Repair

Carbonated structures. Passivity destroyed by carbonation is generally restored by replacing the carbonated concrete in contact with reinforcements with fresh alkaline mortar. This intervention is cost-effective for limited zones, only. Suitable polymer modified mortars are nowadays available to ensure a good grip with concrete as well as a good compromise for drying shrinkage and risk of delamination or cracking. Typical properties are: low elastic modulus, low permeability and expansive action. Coatings slowing down carbonation rate are also often applied. This intervention becomes problematic when small levels of chlorides are present because, in spite of the mortar replacement, corrosion can start nearby the restored zones. When spalling and delamination have not yet occurred, it can be convenient to apply a water repellent treatment to reduce moisture content in concrete cover, which reduces considerably the corrosion rate. This treatment is practically useless if chlorides are present. These treatments have to be repeated periodically, for example every 10–15 years.

Chloride-contaminated structures. Interventions to stop corrosion induced by chlorides are more complex than those for carbonation, because repair has to be extended beyond areas where corrosion took place. Indeed, without doing this, corrosion starts at nearby repaired areas, where chloride content exceeds the critical value: it has to be noted that before repair corrosion did not occur there because of a beneficial galvanic effect exerted by corroding pit areas, as depicted in Fig. 23.33. After repair, a reverse macrocell starts operating, then simply moving the corrosion affected zones. A common mistake is the application of a concrete coating on old sound chloride-contaminated concrete. Another typical mistake is the use of so-called passivating primer without removing the adjacent chloride-contaminated concrete.

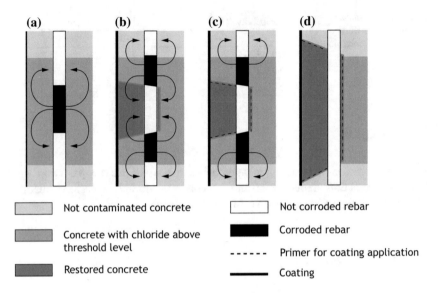

(a)   **(b)**   **(c)**   **(d)**

- Not contaminated concrete
- Concrete with chloride above threshold level
- Restored concrete
- Not corroded rebar
- Corroded rebar
- - - - - - Primer for coating application
- ———— Coating

Fig. 23.33 Evolution of corrosion zones after patch repair

For a long lasting repair, it is necessary to identify all areas where concrete, in contact with reinforcements, is already or will be in a near future chloride contaminated above the critical content; then, that concrete, even if mechanically sound, must be removed completely also in these areas.

Figure 23.34 shows the repair performed on a joint of a viaduct with an organic coating without removing all the concrete polluted by chlorides. Under the coating, corrosion continued to advance as before and, after a few years, the situation confirmed what explained in Fig. 23.33.

Repairing materials. Portland cement-based mortars can be used for conventional repair with and without addition of pozzolana, fly ash or silica fume to improve resistance to carbonation and chloride penetration. Also superplasticizers, fibres, polymers, corrosion inhibitors are added to increase, respectively, workability, resistance to cracking, adhesion and steel passivation. In any case, for best results a very low water/cement ratio (less than or equal to 0.4) and an appropriate curing are mandatory. To control plastic shrinkage, cracking and loss of adhesion, which are favoured by a high cement dosage, two approaches are used: shrinkage-compensating mortars, obtained by the addition of expansive chemical agents, or mortars with a low modulus of elasticity. The use in recent years of self-compacting concrete seems more advantageous from an economic viewpoint. To strengthen reinforced concrete structures, regardless the corrosion conditions of reinforcements, polymer matrix composites with carbon, glass or aramid fibres have become very popular. This intervention does not remove the cause of corrosion, which requires appropriate techniques as described above.

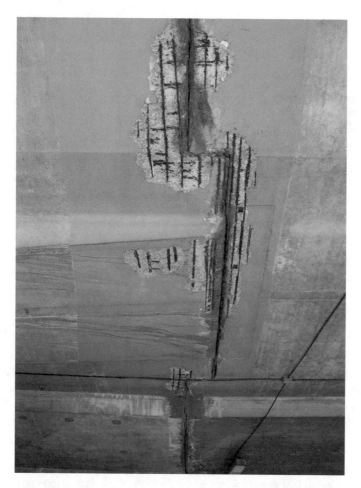

Fig. 23.34 Corrosion occurred after a few years from patch repair without removing chloride-contaminated concrete (E52 highway, Milan, Italy, now repaired)

23.9.2 *Electrochemical Repair Techniques*

Besides traditional methods, electrochemical techniques are applied to restore passivity of reinforcements in chloride contaminated or carbonated concrete, without removal of concrete. These techniques, which include cathodic protection, electrochemical re-alkalisation and electrochemical chloride removal, utilize the same electrochemical scheme, shown in Fig. 23.35, composed of a feeder and an anode, the latter consisting of an activated titanium mesh fixed on concrete surface, and reinforcements as cathode.

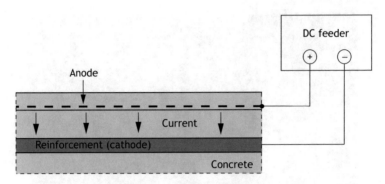

Fig. 23.35 Schematic of system employed to apply electrochemical repair techniques

Current circulation produces on steel reinforcement three effects:

- A decrease in potential
- A reduction or a halt of corrosion rate
- An increase in pH and a removal of chlorides, both facilitating repassivation of steel.

Potential lowering stops as circulation of current is interrupted, while chemical modifications are long lasting.

Cathodic protection of chloride-contaminated structures. As previously reported, Fig. 23.28 shows within Pedeferri's diagram the path of potential during chloride contamination and after the application of cathodic protection. Point ① represents initial passive condition when chloride content is zero; as chloride content increases corrosion sets up to point ② (region of corrosion). By applying CP, operating conditions can lead to point ④, where passivity is achieved, or point ③, where passivity is not fully restored but corrosion rate is strongly reduced, since the flowing cathodic current inhibits macrocell formation. Protection current density for atmospherically exposed structures is of the order of 2–20 mA/m² , as indicated in EN 12696 standard.

Electrochemical re-alkalisation. Electrochemical re-alkalisation is a temporary treatment for the restoration of concrete pH after carbonation, performed by applying a current density of 0.7–1 A/m² for a period ranging from 3 to 30 days by means of a temporary anode placed on the surface of the structure. As anode, a mesh made of activated titanium or steel, embedded in paper pulp soaked with sodium carbonate is used. On the surface of reinforcements, cathodic reactions produce hydroxyl ions, which restore alkalinity by bringing pH close to 14; meanwhile, due to electro-osmosis, diffusion and capillary absorption, alkalinity spreads also to concrete surface, then affecting the entire cover thickness.

Electrochemical removal of chlorides. By the same apparatus used for re-alkalisation, but with slightly higher current densities (1–2 A/m²) and longer times (from one to several months), electrochemical chloride removal can be carried out. In this case, besides the effect of concrete re-alkalisation, there is an ionic migration of chloride ions, from reinforcements towards the temporary anode, produced by the applied current.

23.10 Applicable Standards

- ASTM C876, Standard Test Method for Corrosion Potentials of Uncoated Reinforcing Steel in Concrete, American Society for Testing of Materials, West Conshohocken, PA.
- ASTM G109, Standard Test Method for Determining Effects of Chemical Admixtures on Corrosion of Embedded Steel Reinforcement in Concrete Exposed to Chloride Environments, American Society for Testing of Materials, West Conshohocken, PA.
- ASTM C1582/C1582 M, Standard Specification for Admixtures to Inhibit Chloride-Induced Corrosion of Reinforcing Steel in Concrete, American Society for Testing of Materials, West Conshohocken, PA.
- CEN/TS 14038-1, Electrochemical realkalization and chloride extraction treatments for reinforced concrete—Part 1: Realkalization, European Committee for Standardization, Brussels, Belgium.
- CEN/TS 14038-2, Electrochemical realkalization and chloride extraction treatments for reinforced concrete—Part 2: Chloride extraction, European Committee for Standardization, Brussels, Belgium.
- EN 206, Concrete—Specification, Performance, Production and Conformity, European Committee for Standardization, Brussels, Belgium.
- EN 1504, Products and systems for the protection and repair of concrete structures. Definitions, requirements, quality control and evaluation of conformity, European Committee for Standardization, Brussels, Belgium.
- EN 1992-1-1, Eurocode 2: Design of Concrete Structures. General Rules and Rules for Buildings, European Committee for Standardization, Brussels, Belgium.
- EN 12350, Testing fresh concrete, European Committee for Standardization, Brussel, Belgium.
- EN 12696, Cathodic Protection of Steel in Concrete, European Committee for Standardization, Brussels, Belgium.
- EN 13670, Execution of Concrete Structures, European Committee for Standardization, Brussels, Belgium.
- ISO 1920-11, Testing of concrete—Part 11: Determination of the chloride resistance of concrete, unidirectional diffusion, International Standard Organization, Geneva, Switzerland.
- ISO 1920-12, Testing of concrete—Part 12: Determination of the carbonation resistance of concrete—Accelerated carbonation method, International Standard Organization, Geneva, Switzerland.

23.11　Questions and Exercises

23.1　A reinforced concrete structure has to be built in Milan. The requested service life is 50 years. Determine the minimum concrete cover in order to assure the requested service life, being carbonation coefficient 4 mm/√year, corrosion rate equal to 5 μm/year and maximum corrosion penetration equal to 100 μm.

23.2　Due to error during casting of a reinforced concrete structure, in some areas the real cover is 10 mm, in some others is 40 mm. After how many years a repair is mandatory? Where? Consider the same data of exercise 23.1.

23.3　A reinforced concrete pillar is cast with a W/C ratio 0.5 and a concrete cover equal to 30 mm. The estimated carbonation coefficient is 4.5 mm/√year. Estimate corrosion initiation time. How does it change if, during casting, the concrete at the bottom of the pillar is not well compacted? Consider a W/C ratio equal to 0.7 and use Fig. 23.6, 28 days ageing, to estimate the new carbonation depth (data reported in the figure have been collected after 4 years of environmental exposure).

23.4　A jetty is cast with the following concrete mixture proportion: W/C ratio 0.45, cement 360 kg/m^3 and cover 35 mm (as suggested by standards). Galvanized steel rebars are used. With reference to Fig. 23.11 estimate the service life, being the chloride surface content equal to 5% versus cement weight, diffusion coefficient 10^{-8} cm^2/s, critical chloride content 1% versus cement weight.

23.5　With reference to Exercise 23.4, considering the same concrete cover, which is the maximum value of diffusion coefficient which allows to match the 50 year service life requirement? In which way the mixture proportions can be modified to reach this value?

23.6　A pillar of a reinforced concrete structure exposed to the atmosphere was subjected to measurements of carbonation penetration after 5 and 20 years from its construction. Measurements are reported in Table 23.3. The pillar is exposed to rain. Cover thickness is 30 mm. Determine corrosion initiation time in the weakest point of the structure. How does it change if the mean value of carbonation depth is considered?

23.7　The carbonation depth of a reinforced structure built in an urban environment equals its concrete cover after 30 years from its construction.

Table 23.3　Carbonation depth measured after 5 and 20 years of service life

Measurement	Depth after 5 years (mm)	Depth after 20 years (mm)
1	10	20
2	11	23
3	9	18
4	10	19

What should be the residual service life of the structure, considering a design service life of 80 y? Evaluate propagation time by choosing a suitable value for rebars corrosion rate, considering a relative humidity of 90%. Which is the propagation time of corrosion? Is a repair needed and, if so, when?

23.8 The concrete coating on an offshore platform riser was damaged so that a part of the steel surface in the splash zone was exposed. Explain which factors may cause extraordinarily high corrosion rates in this case.

23.9 Compare and discuss how the additional protection methods affect the critical chloride threshold. Consider also the duration of each method.

23.10 With reference the corroded viaduct reported in Fig. 23.15, describe in detail how a tradition repair system has to be applied.

23.11 The Tunnel under the English Channel is basically a reinforced concrete structure. Let's consider the reinforced concrete wall, 70 cm thick, separating seawater and the internal side of the tunnel. Cover is 40 mm on both the sides.

- Assuming rebars are in carbon steel, discuss the possible form of corrosion.
- Which additional protection method you suggest to guarantee 100 years service life? Describe how it should be installed.
- Considering the use of stainless steel rebars, where they have to be properly placed?

23.12. Is there any risk of galvanic corrosion in coupling carbon steel rebars with stainless steel rebar in concrete? Why? Describe the electrochemical behaviour using Evans diagram.

23.13. To face possible rebar corrosion due to DC interference, reinforced concrete tunnel of underground are typically electrically sectioned every 500 m. Why? (Refer also to Chap. 21).

Bibliography

Andrade C, Alonso MC, Gonzales JA (1988) An initial effort to use the corrosion rate measurements for estimating rebar durability, vol 29, ASTM, Baltimore, US

Bertolini L, Pedeferri P (2002) Laboratory and field experience on the use of stainless steel to improve durability of reinforced concrete. Corros Rev 20(1–2):129–152

Bertolini L, Bolzoni F, Pastore T, Pedeferri P (1996) Behaviour of stainless steel in simulated concrete pore solution. Br Corros J 31(3):218–222

Bertolini L, Gastaldi M, Pedeferri M, Pedeferri P (2000) Stainless steel in concrete. Annual progress report, European Community, COST 521 Workshop, References 269 Queens University, Belfast, pp 27–32

Bertolini L, Elsener B, Pedeferri P, Redaelli E, Polder RB (2013) Corrosion of steel in concrete: prevention, diagnosis, repair, 2nd edn, Wiley-VCH, Verlag GmbH & Co. KGaA, Weinheim

Bolzoni F, Goidanich S, Lazzari L, Ormellese M (2006) Corrosion inhibitors in reinforced concrete structures. Part 2: repair system. Corros Eng Sci Technol 41(3):212–220

Brenna A, Bolzoni F, Beretta S, Ormellese M (2013) Long-term chloride-induced corrosion monitoring of reinforced concrete coated with commercial polymer-modified mortar and polymeric coatings. Constr Build Mater 48:734–744

CEB (1992) Durable concrete structures. Bull Inf 183

Collepardi M (2006) The new concrete, Grafiche Tintoretto, Castrette di Villorba, Treviso, Italy

Elsener B (2001) Corrosion inhibitors for steel in concrete. State of art, EFC Publication no. 35. The institute of Materials, Maney Publishing, London, UK

Gastaldi M, Bertolini L (2014) Effect of temperature on the corrosion behaviour of low-nickel duplex stainless steel bars in concrete. Cem Concr Res 56:52–60

Glass GK, Page CL, Short NR (1991) Factors affecting the corrosion of steel in carbonated mortars. Corros Sci 32:1283

Ormellese M, Berra M, Bolzoni F, Pastore T (2006) Corrosion inhibitors for chlorides induced corrosion in reinforced concrete structures. Cem Concr Res 36:536–547

Ormellese M, Bolzoni F, Lazzari L, Pedeferri P (2008) Effect of corrosion inhibitors on the initiation of chloride-induced corrosion on reinforced concrete structures. Mater Corros 59 (2):98–106

Page CL (1992) Nature and properties of concrete in relation to reinforcement corrosion. In: International conference on corrosion of steel in concrete, Aachen, Germany

Pedeferri P (1996) Cathodic protection and cathodic prevention. Constr Build Mater 10:391–402

Pedeferri P (2004) Progresses in prevention of corrosion in concrete, Istituto Lombardo (Rend. Sc.) B 138, 129–143 (in Italian)

Tuutti K (1982) Corrosion of steel in concrete. Swedish Foundation For Concrete Research, Stockholm, Sweden

Chapter 24
Corrosion in Petrochemical Plant

My god, it takes an ocean of trust
In the kingdom of rust.

Doves

Abstract Production fluids extracted and transported in the oil and gas industry are multi-phase systems (oil, water and gas) with a variety of compositions. A necessary condition for corrosion is the metal wetting by an aqueous phase, which in turn depends on the fluid composition (water content) and on flow regime. If water wetting is effective on the metal surface, corrosiveness of the environment depends on its specific composition, for instance, on the presence or absence of CO_2 and H_2S in the gas phase, or oxygen in the liquid water phase that can accompany oil. The chapter deals with on corrosion problems concerning CO_2 and H_2S for both upstream, midstream and downstream, with a brief mention to some peculiar forms of corrosion

Fig. 24.1 Case study at the PoliLaPP corrosion museum of Politecnico di Milano

© Springer Nature Switzerland AG 2018
P. Pedeferri, *Corrosion Science and Engineering*, Engineering Materials,
https://doi.org/10.1007/978-3-319-97625-9_24

occurring in the refinery plant. The behaviour of most relevant metals in the different environments is also summarized.

24.1 Petrochemical Plants

In petrochemical plants, three sections can be distinguished: upstream production, where fluids are extracted from the reservoirs and transported to the wellhead; midstream, where fluids are transported from the wellhead to the plant for transformation; and downstream, where fluids are treated in the refinery plant.

Accordingly, the main components of a production plant can be divided in:

- Well area (production & injection wells) which comprises tubing, production casing, liner, downhole equipment, well-head
- Gathering system, including production flow-lines and production manifolds
- Downstream process area: production line, separators, crude oil treatment, gas treatment, production water treatment, water injection system, gas injection system, gas/oil storage or export systems.

Metallic structures suffer all the forms of corrosion already seen in the general chapters and some specific corrosion agents, which are the subject of this chapter (Fig. 24.1). Therefore, galvanic corrosion, pitting, crevice, corrosion-erosion and hot corrosion occurring in natural environments (seawater, industrial water and soil) will not be here resumed.

Production fluids have a variety of compositions. In general, they are a multi-phase system containing an oil phase, hydrocarbons, a gas phase and a water phase. Hydrocarbons range from the lighter (methane, CH_4) to the heavier (decanes, $C_{10}H_{20}$, tridecanes), to heavy compounds as asphaltenes and paraffins. Water can be present as formation water (brine) or condensed water. In vapour phase, corrosive agents can be present, such as carbon dioxide (CO_2) and hydrogen sulphide (H_2S). Production reservoir fluids are typically oxygen-free; however, oxygen can be present in some specific items, such as in water injection systems.

24.2 The Corroding Waters

Corrosion can occur if water is present, because an aqueous electrolyte is necessary for the corrosion process. Therefore, water cut, W_{CUT}, i.e. the percentage of water phase in production fluids, is the first factor to take into account for a corrosion assessment.

The water phase associated to hydrocarbons may have two origins:

- *Formation water*, associated with oil or multiphase hydrocarbons dragged from the reservoir
- *Condensation water*, originating from wet gas due to pressure/temperature variations.

Table 24.1 Comparison of composition of formation water of different reservoirs and seawater (according to ASTM D1141)

	Density (kg/L)	pH	TDS (g/L)	Na^+ (g/L)	K^+ (g/L)	Ca^{2+} (g/L)	Mg^{2+} (g/L)	Cl^- (g/L)	SO_4^{2-} (g/L)
Seawater	1.02	8.3	35	10.77	0.40	0.41	1.30	19.35	2.70
Brine 1	1.04	6.9	71	24.90	0.80	2.00	0.40	43.20	0.30
Brine 2	1.08	6.5	130	43.24	1.24	3.36	1.58	75.63	2.84
Brine 3	1.13	6.2	213	56.70	0.19	18.00	2.50	122.00	3.70

Formation waters from different reservoirs have similar composition, all characterized by high salinity, or total dissolved solids, TDS, and slightly acidic (pH 3–6). Regardless the specific salinity, typically the ratio among the content of the main ions, such as chloride, sodium, potassium and bicarbonate in formation water with respect to seawater is almost constant, about 4. Only Ca, Mg and sulphate ions do not follow this rule, since the relative sulphates precipitate. In other words, formation waters are a concentrated seawater.

Table 24.1 compares the standard composition of seawater, according to ASTM D1141, and the composition of formation water of some of thousands of reservoirs known in the world.

Wet hydrocarbon gas contains water in vapour phase: thermodynamic conditions, i.e., temperature and pressure, define its content. These two parameters are also linked by gas law and may vary during transport and treatment. Water content increases as temperature increases and pressure decreases. As a rule of thumb, at a fixed temperature, water content is inversely proportional to pressure. Corrosion occurs if water condensates, i.e. if working conditions, temperature and pressure, are below the dew point. Condensed water has a very low ionic content and a lower pH with respect to formation water, typically 3–4.

24.3 Water Wetting

Metal surfaces in contact with hydrocarbons may undergo corrosion only if they are wetted by the water phase in a permanent or intermittent way. The conditions giving rise to water wetting vary depending on the composition of the multi-phase fluid (hydrocarbon, water and gas) and the operating conditions, mainly temperature, pressure and flow regime.

A water wetting factor, F_{WW}, is considered to define wettability:

- $F_{WW} = 0$ when no water is wetting the metal surface (consequently, no corrosion is expected)
- $0 < F_{WW} < 1$ when water wetting is uncertain
- $F_{WW} = 1$ when there is a permanent condition of water wetting (maximum corrosiveness).

According to an empirical rule, water wetting ($0 < F_{ww} < 1$) occurs:

- In vertical flows, for example inside the tubing of wells, when the percentage of the aqueous phase exceeds 20% in volume
- In horizontal flows, for example inside transportation pipelines, when the percentage of water is higher than 1%
- In oil systems, if water percentage is higher than 40%. Water wetting is uncertain between 5 and 40% of water and is also a function of crude oil type and flow regime
- In wet gas, water wetting occurs only during water condensation.

For multiphase systems the water wetting factor, F_{ww}, is a function of water cut, flow velocity, flow pattern and oil/water emulsions. Figure 24.2 reports the water wetting factor as a function of water cut for a vertical flow, although often adopted also for horizontal one.

Water wetting is strongly influenced by flow pattern. Gas-liquid mixtures flowing in a pipe produce a variety of flow patterns. The particular "pattern" depends mainly on phase flow rates, fluid properties and tube sizes and orientation (vertical or horizontal).

The apparent velocities (in m/s), defined as the velocities a phase would exhibit if it was the only phase flowing alone through the whole pipe cross section, are used to estimate the flow pattern:

$$V_{SG} = \frac{Q_G/86{,}400}{\pi \cdot D^2/4} \tag{24.1}$$

$$V_{SL} = \frac{(Q_W + Q_O)/86{,}400}{\pi \cdot D^2/4} \tag{24.2}$$

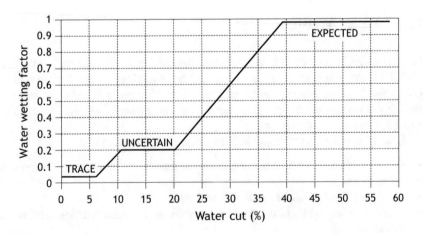

Fig. 24.2 Water wetting factor as a function of water cut

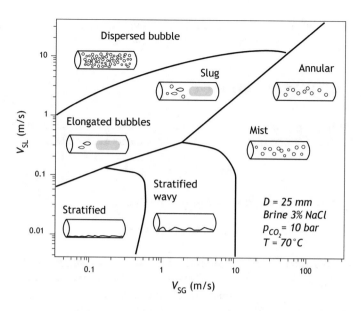

Fig. 24.3 Flow regime determination for an horizontal pipe

where V_{SG} is the gas apparent velocity, V_{SL} is the liquid (oil + water) apparent velocity, Q is the flow rate (in m^3/d) and D is the internal diameter of the pipe (in m).

Figure 24.3 reports a graph—only valid for horizontal pipes—where the flow regime is determined as a function of the apparent fluid velocities.

24.4 Corrosion Assessment

Several factors strongly influence corrosion occurrence and corrosion rate inside oil & gas facilities. In addition to water wetting factor and flow pattern, the following parameters must be considered: pressure, temperature, gas-oil ratio, CO_2 and H_2S content. The latter plays an important role; in fact, since production fluids are typically oxygen-free, CO_2 and H_2S are the two gases, which dissolve into water forming weak acids, which in turn are responsible for metal corrosion.

A production fluid is classified *sour* if H_2S is present, and *sweet* if it is not. Furthermore, it is called *CO_2—dominated* if the CO_2/H_2S ratio is greater than 200, otherwise it is considered *H_2S dominated*.

Corrosion rate is given by a simple relationship:

$$C_{rate} = F_{ww} \cdot C_{th} \tag{24.3}$$

where Cth is the theoretical corrosion rate when water wetting is 1, depending on the specific cathodic reactant (CO_2 or H_2S). Both water wetting factor and theoretical corrosion rate change during time, since water cut, pressure, temperature, fluid composition are changing during production.

24.5 CO_2-Related Corrosion

CO_2 is the most important cause of corrosion for carbon steel structures in oil and gas production. Carbon dioxide dissolves in water forming carbonic acid

$$H_2O + CO_2 \rightarrow H_2CO_3 \tag{24.4}$$

which is a weak acid, and dissociates to bicarbonates:

$$H_2CO_3 \rightarrow H^+ + HCO_3^- \tag{24.5}$$

At pH > 6, bicarbonates dissociate to form H^+ and carbonate ions, CO_3^{2-}. In production fluid, pH is lower than 6, so the latter dissociation occurs to a very small extent.

Frequently, fluid pH is unknown because it is difficult to measure. In presence of only CO_2, pH can be estimated based on CO_2 partial pressure, p_{CO_2}:

$$pH = 4 - 0.5 \log p_{CO_2} \tag{24.6}$$

A more correct pH estimation takes into account composition of the aqueous phase, partial pressure of CO_2 and H_2S and concentration of bicarbonates. At a fixed CO_2 partial pressure, Henry's law determines the amount of CO_2 dissolved into water (see box).

---

pH Estimation

Frequently, pH is unknown because it is difficult to measure: recourse is made to calculations based on composition of the aqueous phase, partial pressure of CO_2 and H_2S and concentration of bicarbonates (Figs. 24.4 and 24.5). The following empirical formulae are often used:

- In the presence of bicarbonates

$$pH = 4.4 - 0.475 \times \ln(p_{CO_2} + p_{H_2S}) + 0.5 \times \ln(HCO_3^-) + 0.00375 \times T$$

- In the absence of bicarbonates

$$pH = 3.8 - 0.195 \times \ln(p_{CO_2} + p_{H_2S}) + 0.00375 \times T$$

where p_{CO_2} and p_{H_2S} are the partial pressures of CO_2 and H_2S in the gas, T is the temperature (°C) and HCO_3^- is the concentration of bicarbonates in solution in meq/L.

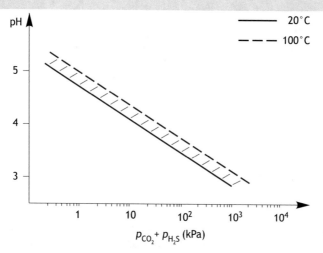

Fig. 24.4 pH of distillate (condensate) water as function of CO$_2$ and H$_2$S in the absence of bicarbonates

If the concentration of bicarbonates is unknown, it is assumed zero in the case of condensates alone (ionic strength $\mu < 0.5$) and equal to 10 meq/L in the presence of formation water (ionic strength $\mu > 0.5$). The ionic strength of a solution is a measure of the total concentration of ions present. It is a chemical-physical quantity that expresses the intensity of the electric field generated by the charges.

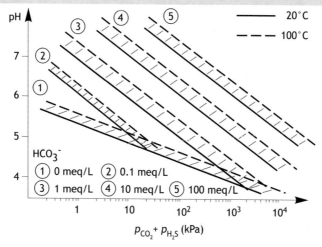

Fig. 24.5 pH of formation waters (brines) as function of bicarbonate, CO$_2$ and H$_2$S

To calculate the pH, reference can also be made to NORSOK M-506 Model (1998).

A later model has been proposed (De Waard et al. 7) which also takes into consideration the diffusion of CO_2. However, in order to apply this model, it is necessary to know the fluid dynamic conditions of the aqueous phase.

24.5.1 Corrosion Mechanism

Carbonic acid, though it is a weak acid, is extremely corrosive towards carbon steel. For example, comparing the corrosion rate of iron in carbonic acid and that in a strong acid such as hydrochloric acid, both with a pH of 4, the corrosion rate in carbonic acid is about ten times higher. The reason is the different kinetics involved in the hydrogen ion reduction reaction.

From an electrochemical point of view, at pH 6, the cathodic reaction is assumed to follow primarily the equation:

$$H_2CO_3 + e^- \rightarrow H_{ads} + HCO_3^- \qquad (24.7)$$

The hydrogen atoms adsorbed on the surface are there combined to form hydrogen gas:

$$2H_{ads} \rightarrow H_2 \qquad (24.8)$$

When carbon steel corrodes, the anodic reaction is:

$$Fe \rightarrow Fe^{2+} + 2e^- \qquad (24.9)$$

Then, the total corrosion reaction is:

$$Fe + H_2CO_3 \rightarrow FeCO_3 + H_2 \qquad (24.10)$$

Since the solubility of $FeCO_3$ is low and decreases with increasing temperature, $FeCO_3$ is deposited when temperature exceeds a limit that depends on CO_2 partial pressure. Accordingly, corrosion of carbon steel is:

- Generalized at temperatures below 80 °C, with a characteristic morphology known as mesa corrosion (Fig. 24.6).
- Generalized with significant localization in the temperature range 80–120 °C due to the formation of partially protective iron and calcium carbonates
- Negligible at $T > 120$ °C, since a compact protecting, passivating film forms.

Fig. 24.6 Example of carbon steel pipe suffering mesa corrosion

24.5.2   *Corrosion Rate Evaluation*

In Sect. 8.2.2, a critical discussion of corrosion in carbonic acid is reported, which helps understand the corrosion mechanism. For applications, the most used industry approach is based on laboratory experiments and field data, the corrosion rate of carbon steel mainly depends on three parameters:

- Partial pressure of CO_2, p_{CO_2}, or fugacity at high pressure
- Temperature
- pH of the aqueous phase, also influenced by bicarbonates.

The usual approach in industry to estimate CO_2 corrosion rate is mostly derived from field data and experimental testing. In the 1975, C. De Waard and D.E. Milliams proposed an equation to calculate corrosion rate in the presence of CO_2, subsequently corrected with U. Lotz (1991, 1995):

$$\log \left(C_{\text{rate},CO_2,\text{dWM}} \right) = 5.8 - \frac{1710}{T+273} + 0.67 \cdot \log \left(p_{CO_2} \right) \tag{24.11}$$

where $C_{\text{rate},CO_2,\text{dWM}}$ is the corrosion rate (mm/y), T is the temperature (°C), and p_{CO_2} is the CO_2 partial pressure (bar) in the gas phase.

Starting from this basic equation, correction factors have been introduced. The most important take into account gas fugacity at high pressures (system factor), formation of deposits (scale factor), pH (pH factor) and chromium content in the steel (Cr factor). Two minor factors are the glycol factor and the condensation factor.

The final corrosion rate is obtained multiplying the corrosion rate obtained by the basic De Waard and Milliams equation ($C_{\text{rate},CO_2,\text{dWM}}$) by the corrective factors:

$$C_{\text{rate,dWM,C}} = C_{\text{rate,dWM}} \times \left(F_{\text{system}} \times F_{\text{scale}} \times F_{\text{pH}} \times F_{\text{gly}} \times F_{\text{cond}} \times F_{\text{Cr}} \right) \quad (24.12)$$

Corrosion rate can be also estimated by applying a pure electrochemical approach based on the theory of corrosion in acid electrolyte (the so called *Tafel-Piontelli model*—see Sect. 8.1 and Lazzari 2017).

24.5.3   Metals for Sweet Condition

Carbon steel and low carbon steel alloys are the first material selection for pipeline transport of sweet production fluid. Typically, an extra thickness, the so-called , *CA*, beyond the value requested by mechanical requirement of the item, is used to take into account expected thickness loss due to CO_2 corrosion.

As a rule of thumb:

- *CA* = 1.5 mm is usually taken with non-corrosive fluids
- *CA* = 3 mm is used for weakly corrosive fluids or inhibited fluids
- *CA* = 6 mm is used for corrosive fluids (or non-inhibited fluids).

Corrosion allowance higher than 6 mm is seldom used.

If corrosion allowance higher than 6 mm is required, even in presence of inhibitors, corrosion resistant alloys (CRA) are used, with an increased cost, but a practical immunity to CO_2 corrosion; even Cr 13% is able to withstand sweet condition. Selection of proper CRA shall consider occurrence of other corrosion forms (i.e., SSC, SCC, pitting).

Correcting factors

At high operating pressures, higher than 100 bar, CO_2 partial pressure is substituted by its fugacity, f_{CO_2}, as follow

$$f_{CO_2} = a \cdot p_{CO_2}$$

where a is the fugacity coefficient, obtained as follows (T is temperature in °C, P is pressure in bar)

$$\log(a) = \left(0.0031 - \frac{1.4}{T+273} \right) \cdot P$$

The **system factor** is introduced to correct the non-ideality of the natural gas at operating pressures higher than 100 bar. It is estimated as follow:

$$\log\left(F_{\text{system}}\right) = 0.67 \cdot \left(0.0031 - \frac{1.4}{T+273} \right) \cdot P$$

The **scale factor** takes into account the scale formation (iron carbonate and magnetite) if operating temperature exceeds a scaling temperature, T_{scal}, calculated as a function of pCO_2. The presence of a scale reduces the rate of corrosion, since it reduces the capability of CO_2 to diffuse on the metal surface. Mitigating effects by this factor are not applicable when H_2S is present at high content.

The scaling temperature, T_{scale}, is determined as follow

$$T_{scale} = \frac{2400}{6.7 + 0.6 \cdot \log(f_{CO_2})} - 273$$

If actual T of the fluid is greater than T_{scale}, a scale factor has to be taken into account (in this case, pH factor, hereafter described, is taken equal to 1):

$$\log(F_{scale}) = \frac{2400}{T + 273} - 0.6 \cdot \log(f_{CO_2}) - 6.7$$

Otherwise, if real temperature is lower than the scaling temperature, scale factor is 1.

The **pH factor** compares the actual pH of the fluid with the saturation pH of iron carbonate or magnetite: if actual pH is more acidic a factor greater than 1 is applied, on the contrary if it is more alkaline the factor mitigates corrosion rate. pH factor applies only if fluid real temperature is lower than scaling temperature, i.e. scale factor is 1.

Saturation pH, pH_{sat}, is calculated for the precipitation of scales ($FeCO_3$ and Fe_3O_4), taking into account also scaling temperature.

$$pH_{sat, Fe_3O_4} = 1.36 + \frac{1307}{T + 273} - 0.17 \cdot \log(f_{CO_2})$$

$$pH_{sat, FeCO_3} = 5.4 - 0.66 \cdot \log(f_{CO_2})$$

The minimum value is considered to estimate F_{pH}.

If $pH_{sat} > pH_{act}$ $\log(F_{pH}) = 0.32 \cdot (pHsat - pHact)$

If $pH_{sat} < pH_{act}$ $\log(F_{pH}) = -0.13 \cdot (pH_{act} - pH_{sat})^{1.6}$

The **Cr factor** takes into account the fact that a small amount of Cr in carbon steel improves CO_2 corrosion resistance, since Cr makes iron carbonate film more stable. The addition of 0.5% Cr especially increases resistance to mesa corrosion, at temperatures below 90 °C, and it is very beneficial in combination with film forming corrosion inhibitors.

The Cr factor is estimated as follow

$$F_{Cr} = \frac{1}{1 + 1.4 \cdot \%Cr}$$

Considering the typical Cr content in low alloyed carbon steel (0.5%), the factor is close to 0.6.

The **glycol factor** takes into account effects of glycol on CO_2 corrosion rate. Glycol is introduced to prevent hydrates formation. The factor is estimated as follows:

$$F_{GLY} = A \times \log(W_{CUT}\%) - 2A$$

where A is a constant (1.6), weakly dependent on the alcohol type.

The **condensation factor** applies only to wet gas systems when cooling is expected. Water condensing under a high partial pressure of CO_2 is acid and quite corrosive, but iron bicarbonate $FeCO_3$ produced as corrosion product is highly soluble in the thin water film and behaves as an excellent pH controller. For the calculation of the condensation factor, two cases may be in general expected: slow cooling systems ($F_{cond} = 0.1$) and rapid cooling systems ($F_{cond} = 1$), as heat exchangers or coolers. Corrosion rate of steel exposed to water phase condensing from wet gas quickly decreases with time only in slowly cooling systems.

Henry's Law

Henry's law, formulated by the English chemist William Henry in 1803, regulates the solubility of gases in a liquid. It states that, at a constant temperature, the amount of dissolved gas is directly proportional to its partial pressure in the gas phase. The proportionality factor is called Henry's law constant.

Once the equilibrium is reached, the liquid is defined as saturated at that temperature. This state of equilibrium persists until the external pressure of the gas remains unaltered, otherwise, if it increases, other gas will enter into solution, if it decreases, the liquid will be in a state of over-saturation and gas will be released until pressures are balanced again.

There are several ways of describing the solubility of a gas in water. Usually Henry's law is written as follow

$$[GAS]_{water} = H \cdot p_{gas}$$

where $[GAS]_{water}$ is the concentration of a species in the aqueous phase (in mol/L), H is Henry's constant (in mol/atm L) and p_{gas} is the partial pressure of that species in the gas phase (in atm).

When the temperature of a system changes, the Henry's constant also changes:

$$H_T = H_{T^0} \cdot \exp\left[\frac{-\Delta H_{sol}}{R} \cdot \left(\frac{1}{T} - \frac{1}{T^0}\right)\right]$$

where H_T is the Henry's constant at a temperature T, H_T^0 is the Henry's constant at room temperature, ΔH_{sol} is the solution enthalpy, R is the gas constant, T is temperature (in K) and T^0 is room temperature (298 K).

Next table reports some values of Henry's constant and temperature dependencies.

Based on Henry's law, at room temperature, 1500 ppm$_{wt}$ of CO$_2$ are dissolved in a solution in equilibrium with a CO$_2$ partial pressure of 1 bar.

According to NACE MR0175-ISO 15156 standard, *sour condition* occurs when H$_2$S partial pressure is higher than 0.0035 bar, which corresponds to about 12 ppm$_{wt}$ of H$_2$S in solution at room temperature and 4 ppm$_{wt}$ at 75–80 °C.

If H$_2$S partial pressure is 1 bar, at room temperature, 3400 ppm of H$_2$S are dissolved into the solution.

	Unit	H$_2$S	CO$_2$	O$_2$	H$_2$
H at 25 °C	(mol/atm L)	10^{-1}	3.4×10^{-2}	1.5×10^{-3}	7.8×10^{-4}
$-\Delta H/R$	(K)	2100	2400	1800	500
H at 75 °C	(mol/atm L)	3.6×10^{-2}	10^{-2}	5.7×10^{-4}	6×10^{-4}

24.6   H$_2$S-Related Corrosion

According to NACE MR0175-ISO 15156 standard, *sour condition* occurs if the H$_2$S partial pressure is higher than 0.0035 bar. Three levels of corrosion severity are identified as a function of the in situ pH and H$_2$S partial pressure (Fig. 24.7). The most influencing parameter is the concentration of H$_2$S. The minimum effective H$_2$S concentration in the electrolyte is as little as 10 ppm, which corresponds, according to Henry's law, to a critical H$_2$S partial pressure of 0.0035 bar or 0.05 psi (see Henry's law box).

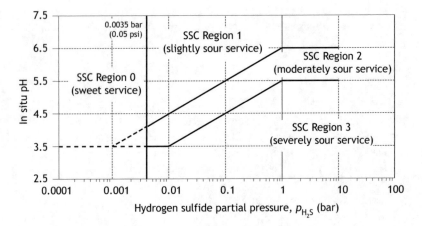

Fig. 24.7 Sour service condition according to NACE MR0175—ISO 15156

24.6.1 Corrosion Mechanism

Like CO_2, H_2S dissolves in water, forming a weak acid solution, with pH around 6:

$$H_2S \rightarrow H^+ + HS^-$$ (24.14)

The hydrogen ions are responsible for the corrosive nature of H_2S.

The most important cathodic reaction in a sulphide-containing environment can be expressed by:

$$2H_2S + 2e^- \rightarrow 2H_{ads} + 2HS^-$$ (24.15)

Considering carbon steel dissolution, the overall corrosion reaction is:

$$Fe + H_2S \rightarrow FeS + 2H_{ads}$$ (24.16)

The adsorption of HS^- ions on the steel surface forms iron sulphide films, FeS, which may inhibit the hydrogen atoms recombination (the so-called "poisoning effect"), allowing the diffusion of atomic hydrogen into the steel, and then promoting hydrogen damages. The latter is the only form of corrosion occurring in the bulk of the metal. All of the other corrosion processes happen at the metal surface.

Different forms of corrosion are possible in a sour fluid:

- Generalized corrosion of carbon steel
- Hydrogen Induced Cracking (HIC) with blister formation due to hydrogen atoms permeating into the steel in the absence of any internal or external stress
- Sulphide Stress Cracking (SSC) occurring on metals susceptible of hydrogen embrittlement in the presence of internal or external stresses.

24.6.2 Generalized Corrosion

H$_2$S is a weak acid. At pH 6, the corrosion rate of carbon steel should be low or even negligible; conversely, experience showed that carbon steel corrosion rate in a H$_2$S containing environment is remarkable (close to 0.5–1 mm/year, see also Sect. 8.2.3), because insoluble FeS forms, with an extremely low solubility product, in the order of 10^{-24}. This leads to a decrease in the anodic potential of iron dissolution and the consequent availability of a greater electrochemical driving force even in neutral solutions.[1]

The iron sulphide layer is not always compact and protective, and may lead to localized corrosion with a galvanic coupling mechanism due to the fact that FeS has an electronic conductivity and a practical nobility higher than iron.

The prediction of the corrosion rate is based on empirical rules, which can be summarized as follows:

- At $T < 60$ °C

  - Corrosion rate is negligible for H$_2$S concentrations in solution below 40 ppm
  - Significant corrosion (0.5 mm/year) occurs at H$_2$S concentrations above 3500 ppm (corresponding to a H$_2$S partial pressure of 1 bar)

- At $T > 60$°C

  - Corrosion rate is 1 mm/year at H$_2$S concentrations above 40 ppm. Higher corrosion rates are not possible because of the formation of a protective iron sulphide.

24.6.3 Hydrogen Induced Cracking (HIC)

The effect of H$_2$S as a cathodic poison for hydrogen recombination (see Sect. 14.1.2) allows the ingress of atomic hydrogen into the metal. Hydrogen atoms diffuse into the metal since they reach specific traps, such as inclusions, in particular elongated MnS$_2$, or micro-voids in the metal matrix. Once accumulated inside the traps, as no H$_2$S is present, hydrogen atoms recombine to form hydrogen gas, H$_2$. The hydrogen gas, too large to diffuse through the metal lattice, accumulates and generates extremely high internal pressures, sufficient to cause local plastic deformation of the metal and blister formation, as shown in Fig. 24.8.

---

[1]The driving force, ΔE, is given by the difference between equilibrium potential of the cathodic reaction, E_C, and equilibrium potential of the anodic reaction, E_A. In neutral conditions, $E_C = -0.059 \cdot$ pH $= -0.42$ V SHE. The equilibrium potential of iron dissolution is $E_A = -0.44 + 0.059/2 \cdot$ (log [Fe^{2+}]). In sour conditions, the formation of insoluble FeS reduces iron ion concentration at 10^{-12} mol/L. Then the driving force is about 0.4 V. At the same pH, in de-aerated water, the driving force is 0.2 V.

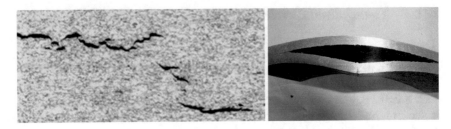

Fig. 24.8 Examples of HIC and blistering on a carbon steel pipe

The inclusions with the greatest impact on this phenomenon are those of manganese sulphide (MnS_2 type II), which during the hot rolling of the carbon steels used for pipelines and the sheets for pressurized containers are squeezed and arranged parallel to the direction of rolling, thus forming an easy trap for hydrogen atoms.

HIC is linked to the quantity of hydrogen atoms diffusing into the metal matrix and to time:

- For H_2S partial pressure above 0.1 bar (corresponding to 400 ppm in the aqueous phase), HIC occurs in a period of time comparable to the mean service life of an oil facility, 15–20 years
- For H_2S partial pressure below 0.1 bar, HIC occurs in a longer period of time.

From a practical point of view, when H_2S partial pressure is higher than 0.03 bar, two strategies can be adopted to limit HIC: (1) the use of corrosion inhibitors, which reduces the quantity of hydrogen produced; (2) the use of non-susceptible steels, for example steels treated with rare earth metals and steels with very low sulphur content (<30 ppm).

24.6.4 Sulphide Stress Cracking (SSC)

SSC occurs when atomic hydrogen diffuses into the metal remaining in solid solution in the crystal lattice. This reduces the ductility and deformability of the metal, promoting the hydrogen embrittlement of the metal (Fig. 24.9).

Several factors influence SSC: steel composition, thermal treatments, microstructure, mechanical resistance, pH, H_2S partial pressure, load applied, temperature and time. The onset of SSC is firstly evaluated based on the partial pressure of H_2S in the gas phase, which is proportional to its concentration in the aqueous phase. High yield strength steels (ultimate tensile strength above 700 MPa) are the more susceptible to SSC, since their microstructure is particularly sensitive to the effects of hydrogen. Ruptures are characterized by transgranular cracking. The cracking process is very rapid and has been known to take as little as few hours for a crack to form and cause catastrophic failure.

Fig. 24.9 Examples of SCC on a carbon steel tubing

The tendency for SSC to occur is increased by the presence of hard microstructures such as un-tempered or partly tempered low temperature transformation products (martensite, bainite). These structures may be inherently present in high strength low alloy steels, may result from inadequate or incorrect heat treatment, may arise in welds and particularly in low heat input welds in the heat affected zones.

Finally, SSC is a highly temperature dependent phenomenon: the maximum susceptibility occurs at room temperature.

24.6.5 Metals for Sour Service Condition

Carbon steel for sour service must have some very general mechanical and metallurgical characteristics to reduce susceptibility to hydrogen attack: cleanliness to avoid inclusions and segregation, homogeneity of composition, fine grain size to improve toughness and stability, i.e. complete release of internal stresses. Mn and S content should be very low to reduce the formation of a central segregation band. Steel can be treated with calcium or with rare-earth metals (an alloy is often used of Ce, La, Nd, Pr, so called Mischmetal alloy) to allow the formation of finely dispersed and spheroidal inclusions, much less harmful to steel with respect to the elongated ones. As regard mechanical properties, the maximum acceptable hardness for carbon steel working in severe sour service condition is 22 HRC (Rockwell C). Finally, to guarantee weldability in field, carbon equivalent should be less than 0.38, and post weld heat treatments are recommended both to reduce the hardness of weld deposit and heat affected zone and to release the residual stresses in the weldment.

Corrosion resistant alloys. Stainless steels are used in the presence of corrosive fluids, when carbon and low alloy steels are not recommended. As mentioned, all stainless steels resist to CO_2 related corrosion. Furthermore, sour service grades are available for any content of H_2S. Martensitic (13–15% Cr) stainless steels are the lowest grade. The interest in this stainless steel is twofold: high mechanical resistance, being martensitic, and low cost, because of low alloy content. However, they have high SSC susceptibility even with modest amounts of H_2S, and pitting and crevice corrosion may occur in chloride-containing environment at high temperature even with a very low amount of oxygen (Oldfield and Sutton 1978). Austenitic stainless steels are used because of many good properties: good weldability, resistance to SSC at low temperature; however, the presence of chlorides limits their resistance. Super austenitic stainless steels (6Mo and 904L grades) display higher resistance to pitting and stress corrosion cracking and higher mechanical properties. Duplex and superduplex stainless steels resist to Cl-induced SCC thanks to the ferritic microstructure and resist to hydrogen embrittlement thanks to the austenitic part. They show a good resistance to pitting at high T and Cl-concentrations. Ni-based alloys have an austenitic microstructure, are resistant to general corrosion and highly resistant to SSC. Highest grades, with high nickel and molybdenum content, are resistant to the most severe conditions.

Clad components are products comprising a carbon or low alloy steel base pipe with an internal layer (approximately 3 mm) of corrosion resistant alloy. The cheap carbon steel base gives the mechanical strength, while the internal CRA layer guarantees high corrosion resistance (Fig. 24.10). The internal layer may be metallurgically bonded (clad pipe) or mechanically bonded (lined pipe). Amongst the available options listed by API 5LD for the CRA layer, the solutions that found practical application and are available on the market are AISI 316, nickel alloy 825 and alloy 625.

24.7 Downstream Corrosion

24.7.1 Corrosion by S/H₂S Atmosphere

Steels used in oil refining processes undergo corrosion attacks at temperatures above 260 °C when the sulphur content (such as elemental sulphur, H_2S and

Fig. 24.10 Examples of carbon steel pipes with an internal clad in 625 Nickel alloy 3 mm thick

combined in mercaptans) is high. The extent of the attack depends on sulphur content, temperature and presence of gaseous hydrogen. At temperatures above 150 °C, mixtures of hydrogen and sulphur compounds lead to the formation of H_2S. The H_2S amount becomes significant at T higher than 260 °C, with a maximum at 370 °C. Carbon steel has a very low corrosion resistance. Small addition of chromium improves the resistance to attacks: at 370 °C, the addition of 5% Cr reduces corrosion rate from 3 to 1 mm/year. At higher chromium content, corrosion rate is further reduced to 0.5 mm/year if 9Cr–0.5Mo steel is used, to 0.1 mm/year when 12Cr steel or 18Cr steel are used. The corrosion resistance depends on the formation of a protective sulphide layer, which is more robust when the chromium content is increased.

24.7.2 Corrosion by Sulphur

Sulphur (melting point 113 °C and boiling point 445 °C) is present in some petroleum reservoirs in association with H_2S. The formation of sulphur may occur due to the reduction of sulphates to sulphur by methane

$$SO_4^{2-} + CH_4 \rightarrow S + 2H_2O + CO_2 \qquad (24.17)$$

or by catalytic decomposition of hydrogen sulphide or the oxidation of hydrogen sulphide by oxygen (due to contaminations).

Sulphur reacts with hydrogen sulphide to form polysulphides which in turn release sulphur to form deposits (slugs) and entrainment in the tubing of oil wells.

Sulphur is an effective cathodic reactant, acting as an electron acceptor through the catalytic action of the sulphide film which has electronic conductivity, and gives rise to high corrosion rates when it comes into direct contact with steels; if it is present in a hydrocarbon solution, corrosion rate is controlled by diffusion processes. At high temperatures (above 120 °C), sulphur dismutes to H_2S and sulphuric acid, causing the rupture of the passivation films and the onset of localized corrosion (crevice and SSC) on stainless steel alloys and corrosion resistant alloys. The sulphur corrosion rate on carbon and low alloy steels depends on temperature:

- At $T < 120$ °C and in direct contact with sulphur, corrosion rate is in the order of several mm/year and is only slightly influenced by temperature. The sulphides present on the surface catalyse the reaction and lead to macrocell conditions similar to corrosion by differential aeration
- At $T > 120$ °C corrosion rate increases due to the formation of sulphuric acid, but only up to about 150 °C
- At $T > 150$ °C corrosion rate begins to decrease due to the protective action of corrosion products
- At 180 °C corrosion rate is nonetheless so high (above 10 mm/year) that these materials cannot be used.

The presence of elemental sulphur generally leads to the localized corrosion of stainless steels: duplex stainless steels undergo generalized corrosion even at ambient temperature like carbon steels, whereas austenitic stainless steels present low resistance only above 120 °C. Martensitic stainless steels generally have low resistance. The presence of sulphur increases vulnerability to SCC: when T is higher than 120 °C, resistance to corrosion increases with increasing Ni, Cr and Mo content. Nickel alloys offering good resistance in severely aggressive conditions (S, H_2S, CO_2, Cl^-) at high temperatures must generally have a basic composition of the type Ni > 55%, Mo > 12%, Cr > 15% (typically Alloy G-50 22Cr–52Ni–11Mo– 0.7 W–0.8Cu). Grade 2 titanium gives rise to crevices when T > 130 °C; beta-C titanium alloys present greater resistance.

24.7.3   Corrosion by H_2/H_2S

In the presence of H_2/H_2S atmospheres, Cr–Mo steels do not resist at temperatures above 315 °C. It is necessary to use proper austenitic stainless steels, with an aluminium-enriched surface. In this case, corrosion rate is lower than 0.25 mm/year even at 500 °C.

24.7.4   Corrosion by Naphthenic Acid

In the presence of organic acids in crude oil, particularly those with naphthenic structure, carbon and low alloy steels undergo corrosion at temperatures between 200 and 400 °C with a maximum corrosion rate at about 270–280 °C. Aggressiveness is measured based on the neutralization number, or total acid number (TAN), which measures the acidity of the organic content. Once the acidity of the crude oil has been established, a crude oil is considered aggressive if the TAN is greater than 0.5 mg KOH/g. On a rough estimate, corrosion rate increases by three times every 100 °C increase in temperature above 230 °C, up to about 400 °C.

Corrosion is often associated with corrosion-erosion phenomena due to high turbulence of the fluids, for example in centrifugal pumps, in heaters (especially in the bends of coils), in the connecting line between heater and fractionating column and in the inlet section where the partially vaporized crude oil enters the column.

The corrosion mechanism involves the formation of iron complexes with organic acids.

Carbon steel has a good behaviour at temperatures below 220 °C. For higher temperatures cast iron and Cr–Mo steels with increasing chromium content up to 12% must be used. In very severe condition, stainless steels, AISI 316, AISI 309 and AISI 310, should be used. Monel, Inconel and Alloys B are suitable, but attention must be paid to the presence of sulphur and organic sulphur compounds.

24.7.5 Hydrogen Attack

Hydrogen atmospheres above 200 °C and pressures above 7 bar cause hydrogen damage with blister formation and decarburization of steel and also a reduction of mechanical properties. Cr–Mo steels are resistant to hydrogen damage (due to the stability of carbides). The resistance is verified using the Nelson curves; for more details please refer to Chap. 14.

The formation of blisters and cracks is due to the formation of methane by the reaction of hydrogen with free carbon.

In the presence of acid attack and hydrogen sulphide (H_2S), the atomic hydrogen penetrates into the crystal lattice of the iron causing:

- Cracking (Step Wise Cracking) or swelling (blisters). This phenomenon is often referred to as HIC (Hydrogen Induced Cracking) for C-Mn steels even without tensile stresses
- At low temperature, hydrogen embrittlement on susceptible materials (high strength steels) in the presence of tensile stresses above a critical threshold.

24.7.6 Organic Acid Corrosion

Organic compounds present in some crude oils decompose in the crude furnace to form low molecular weight organic acids which condense in distillation tower overhead systems. The low molecular weight organic acids that are formed include formic acid, acetic acid, propionic acid, and butyric acid. They may also result from additives used in upstream operations or desalting. Corrosion is a function of the type and quantity of organic acids, metal temperature, fluid velocity, system pH, and presence of other acids. Formic acid and acetic acid are the most corrosive. They are soluble in naphtha and are extracted into the water phase, once water condensates, and contribute to a reduction of pH. The presence of organic acids will contribute to the overall demand for neutralizing chemicals but their effects may be completely masked by the presence of other acids such as HCl, H_2S, carbonic acid and others. Corrosion is most likely to be a problem where relatively non-corrosive conditions exist in an overhead system and if there is a sudden increase in low molecular weight organic acids. The latter reduces the pH of water in the overhead system, requiring a potentially unexpected increase in neutralizer demand. The corrosion mechanism of organic acids is reported in Chap. 8.

All carbon steel piping and process equipment in crude tower, vacuum tower and coker fractionator overhead systems including heat exchangers, towers and drums are susceptible to damage where acidic conditions occur. Corrosion tends to occur where water accumulates or where hydrocarbon flow directs water droplets against metal surfaces. Corrosion is also sensitive to flow rate and tends to be more severe in turbulent areas in piping systems.

24.7.7 Polythionic Acid Stress Corrosion Cracking

It is a form of stress corrosion cracking normally occurring during shutdowns, start-ups or during operation when air and moisture are present. This type of corrosion occurs on austenitic stainless steels, alloy 600/600H and alloy 800/800H. A combination of environment, material, and stress are required. Cracking is due to sulphur acids forming from sulphide scale, air and moisture acting on sensitized austenitic stainless steels. It is usually adjacent to welds or high stress areas. Cracking may rapidly propagate through the wall thickness of piping and components in a matter of minutes or hours.

24.7.8 High Temperature Sulphidation

Elevated temperature sulphidation in crude oil processing units is one of the most relevant materials problems encountered in the refining industry. It generally occurs in the temperature range 250–550 °C. This phenomenon is encountered in distillation units processing crude oils, which contain a significant concentration of sulphur compounds (such as mercaptans, sulphides, disulphides, etc.) with total sulphur higher than 0.6%, by weight. Most severe sulphidation attack in crude distillation units occurs in flash zones of towers, furnace tubes, and transfer lines.

The relative corrosiveness of different sulphur compounds for carbon steel increases with temperature, chemistry of the sulphur functional group and type of the organic group. Thiophene was found to be the least aggressive compound. Above 450 °C the aggressiveness of this attack starts to decrease, owing to the decomposition of reactive sulphur organic compounds and to the formation of a protective coke layer on steel.

Iron sulphide scales are only partially protective and do not eliminate further attack. At long exposure times the iron sulphide scales increase in thickness and eventually spall, resulting in fresh metal surface exposed to the sulphidizing environment. The cycle of growth and spalling of sulphide scales is periodically repeated. Velocity also plays a role, particularly in turbulent vapour/liquid mixtures, which are most prone to continuously erode the sulphide scale and significantly accelerate the rate of attack.

Steel alloyed with Cr exhibits a two-layer scale: a mixed inner scale composed by iron sulphide and a sulpho-spinel $FeCr_2S_4$, and outer scale by $Fe_{1-x}S$. When Cr content increases, the inner layer tends towards single-phase sulpho-spinel $FeCr_2S_4$. It is generally thought that this scale is more protective than iron sulphide.

Poisoning Effect of H₂S on Human Beings

H_2S is considered a broad-spectrum poison, which means it can damage various body organs and apparatuses. At high concentrations, it paralyzes the olfactory system making it impossible to perceive its unpleasant odour and can cause unconsciousness within a few minutes. Typical thresholds are as follows:

- 0.0047 ppm is the recognition threshold, the concentration at which 50% of human beings can perceive the characteristic odour described as "rotten eggs"
- <10 ppm is the exposure limit 8 h a day, without permanent damages
- 10–20 ppm is the limit beyond which eyes are irritated
- 50–100 ppm cause permanent ocular damages
- 100–150 ppm paralyze the olfactory nerve after few inhalations, preventing the smell from being detected and therefore danger is not recognized
- 320–530 ppm cause pulmonary edema with high risk of death
- 530–1000 ppm strongly stimulate the central nervous system and accelerate breathing, causing even more gas to be inhaled and causing hyperventilation
- 800 ppm is the deadly concentration for 50% of human beings for 5 min of exposure (LD_{50})
- >1000 ppm cause the immediate collapse with suffocation, even after a single breath ("blow of lead of the barrel workers", so called because the victims were the workers using barrels in the tanning of skins).

24.8 International Standards

- API RP 14-E, Design and installation of offshore production platform piping system, American Petroleum Institute, Dallas, TX.
- API RP 571, Damage Mechanisms Affecting Fixed Equipment in the Refining Industry, American Petroleum Institute, Washington, DC.
- API RP 941, Steels for Hydrogen Service at Elevated Temperatures and Pressures, American Petroleum Institute, Dallas, TX.
- EFC 16, European federation of corrosion publications (1996) Number 16, Guidelines on materials requirements for carbon and low alloy steels for H₂S—Containing environments in oil and gas production. The Institute of Materials.
- EN ISO 3183, Petroleum and natural gas industries—Steel pipe for pipeline transportation systems

- ISO 15156, Petroleum, petrochemical and natural gas industries—Materials for use in H_2S-containing environments in oil and gas production. Part 1: General principles for selection of cracking-resistant materials; Part 2: Cracking-resistant carbon and low alloy steels, and the use of cast irons; Part 3: Cracking-resistant CRAs (corrosion-resistant alloys) and other alloys.
- ISO 21457—Materials selection and corrosion control for O&G production systems
- NACE MR0103, Materials Resistant to Sulfide Stress Cracking in Corrosive Petroleum Refining Environments", NACE international, Houston, TX.
- NACE MR0175, Sulphide stress cracking metallic material for oil field equipment, NACE international, Houston, TX.
- NACE TM0177, Laboratory Testing of Metals for Resistance to Sulfide Stress Cracking and Stress Corrosion Cracking in H_2S Environments" NACE international, Houston, TX.
- NACE TM0284, Evaluation of Pipeline and Pressure Vessel Steels for Resistance to Hydrogen-Induced Cracking" NACE international, Houston, TX.
- NORSOK M-001, Materials Selection, Oslo, Norway.
- NORSOK M-506, CO_2 corrosion rate calculation model, Lysaker, Norway.

24.9   Questions and Exercises

24.1   The working condition of a carbon steel pipeline (with a tensile strength of about 500 MPa, i.e. approx. 70 ksi) transporting a multiphase fluid are as follow: P 100 bar, T 80 °C, in situ pH 4.3 (no bicarbonates), water cut 25%, horizontal flux, fluid velocity 2 m/s, CO_2 molar fraction 1%, H_2S molar fraction 0.01%. Design life is 20 years.

- Which corrosion do you expect?
- Estimate the corrosion rate using De Waard Milliams approach
- Is the corrosion allowance a possible solution?
- Calculate the corrosion rate and the corrosion allowance in the presence of a corrosion inhibitor with a 95% efficiency.

24.2   Referring to the same working condition reported in Ex 24.1, how corrosion rate and corrosion allowance change if a carbon steel alloy with 0.6 Cr is used?

24.3   The working conditions of a carbon steel pipeline transporting a multiphase fluid are as follow: P 80 bar, T 75 °C, in situ pH 4.0 (no bicarbonates), fluid velocity 1.5 m/s, CO_2 molar fraction 1.3%. H_2S is absent. Water cut is less than 5% for the first 5 years, 20% from year 6 to year 11, 25% from year 12 to year 17, 35% till 20 years (design life). Estimate the thickness loss corrosion rate using De Waard Milliams approach.

24.4 What is the difference between *sweet service* and *sour service*?

24.5 Describe in a qualitative way the mathematical approach used to estimate CO_2-corrosion rate. Referring to Chap. 8, make some numerical example of corrosion rate evaluation comparing the De Waard Milliams approach with the Tafel-Piontelli model.

24.6 A multiphase fluid is transported through a carbon steel pipeline. Working conditions are as follow: T 70°C—P 80 bar—CO_2 3.5%—H_2S 0.6%—pH 3.9. According to NACE MR0175—ISO 15156 (Fig. 24.7) which is the corrosion condition?

24.7 Which are the additional requirements for carbon steel (according to ISO, NACE and EFC standards) in order to guarantee a safe use of the pipeline in the declared working conditions. Justify each of them and give values of the relevant parameters.

24.8 A new pipeline has to transport a gas hydrocarbons stream under the following conditions: required capacity 10^8 Nm^3 per month (density 1.5 kg/Nm^3); condensed water (20% by weight); CO_2 content 5% by volume on the separated gas; chlorides in the formation water 30 g/L; temperature 40 °C; pressure 7 MPa. Make a corrosion assessment in order to perform a proper material selection. [Hint: fix a pipe diameter, estimate a wall thickness; suggest toughness, consider the use of inhibitor, if the case, compare carbon steel, copper alloys and stainless steels].

24.9 A cladded pipe (base metal carbon steel, clad in Alloy 625) is used to convey a sour fluid from a platform to the onshore plant. After two years a severe leakage is detected at a weld. Make a corrosion assessment.

24.10 Describe the main differences among the localized corrosion form in the presence of H_2S.

24.11 A carbon steel pipe (24″ in diameter, nominal OD 609.6 mm, thickness wall 12.70 mm) suffered internal localised corrosion attacks located at 3 and 9 o'clock position. Perforation occurred after 10 years. Working condition were as follow: P 50 bar, T 30 °C, CO_2 1.3%. The pipeline was carrying formation water and methane. Corrosion was observed at the water line, in a portion of the line where the pipe was half filled with formation water. Which is the cause of corrosion?

Bibliography

Corrosion Data Survey (1985) Metal section, 6th edn. NACE International, Houston, TX

Corrosion in the Oil Refining Industry Conference (1998) NACE Group Committee T-8. NACE International, Houston, TX, 17–18 Sept 1998

De Waard C, Milliams DE (1975) Carbonic acid corrosion of steel. Corrosion 31:5

De Waard C, Lotz U, Milliams DE (1991) Predictive model for CO_2 corrosion engineering in wet natural gas pipelines. Corrosion 47(12):976

De Waard C, Lotz U, Dugstad A (1995) Influence of liquid flow velocity on CO_2 corrosion: a semi-empirical model. Corrosion, 95, paper n. 128, NACE International, Houston, TX

Kane RD (2006) Corrosion in petroleum refining and petrochemical operations. In: ASM handbook. ASM International, vol 13C, pp 967–1014

Lazzari L (2017) Engineering tools for corrosion. Design and diagnosis. In: European federation of corrosion (EFC) series, vol 68. Woodhead Publishing, London, UK

Oldfield JW, Sutton MH (1978) Crevice corrosion of stainless steels. Br Corros J 13:104

Chapter 25
Corrosion in the Human Body

Human blood seeks revenge upon iron, In fact once
encountered it, it tends to get rusty, faster and faster.

Plinio, Nat., 34, 146

Abstract Metallic materials can find many kinds of applications in the human body: for example, in orthopaedics, for hip and knee prostheses, for osteosynthesis devices, in the cardiovascular sector, for endovascular prostheses, cardiac valves, pacemakers; in stomatological areas and for osteointegrated dental implants. Herein, some corrosion problems linked to the metallic materials used in the human body are examined, focusing in particular on orthopaedic materials. Failure mechanisms of these materials is briefly revised, dealing with fatigue, general and localised corrosion, fretting. Finally, a brief outline of the surface finishing treatments is presented.

Fig. 25.1 Case study at the PoliLaPP Corrosion Museum of Politecnico di Milano

© Springer Nature Switzerland AG 2018
P. Pedeferri, *Corrosion Science and Engineering*, Engineering Materials,
https://doi.org/10.1007/978-3-319-97625-9_25

25.1   Characteristics of Metals for Orthopaedic Purpose

All materials applied in the human body need to be resistant to degradation and biocompatible (Fig. 25.1); metals used for orthopaedic prostheses or for osteosynthesis, beyond to be corrosion resistant, need to have also high mechanical resistance and fatigue resistance and an adequate modulus of elasticity.

25.1.1   Mechanical Resistance

The cross section area of an osteosynthesis device that fits the femoral head can also be 10 times smaller than the one of the osseous structure; since the bone mechanical resistance is of 90–120 MPa, the metal with which the nail is made needs to have a resistance higher than 1000 MPa. Likewise, to realize a hip prosthesis stem, since its resistant section is considerably inferior to the bone's one, it is necessary to use a material with tensile resistance of at least 600–800 MPa. The request of such high mechanical characteristics limits the choice of employable metals and excludes the use of ceramic or polymeric materials. Only some composite materials may be used, but their clinical employment is still far.

25.1.2   Fatigue Resistance[1]

The bone capacity of regenerating itself ensures that it is not subject to fatigue phenomena, even if it is submitted to frequent cyclical loadings. This does not happen with synthetic materials (metals, polymers). Also people that conduct a sedentary lifestyle load their weight on each leg from 10^5 to more than 10^6 times per year. Taking also into account the reduced cross sections of the implants, these conditions can cause fatigue problems. Besides the application of cyclic loads, the onset of these phenomena depends on implant design and surface finishing conditions. It is particularly favoured by the fact that implants present discontinuities of various nature introduced for design requirements (abrupt section variations, holes, etc.), defects introduced in fabrication (inadequate surface finishing) or during application (defects due to surgical instruments, or to the need to modify the implant shape in order to adapt it to the patient) or finally arisen during service (localized forms of corrosion).

---

[1]The term *fatigue* is commonly used even though, since the phenomenon occurs in an aggressive environment, it would be more appropriate to say *corrosion-fatigue*.

25.1.3   Resistance to Generalized Corrosion

Human body fluids, characterized by the presence of chlorides and rich in oxygen, are very aggressive. Only some noble metals like gold and others belonging to the platinum family are immune from corrosion in human body; yet, in the orthopaedic field these metals cannot be employed due to their poor mechanical characteristics. For orthopaedic applications, active-passive metals are used, characterized by a low —but not nil—corrosion rate. A uniform corrosion rate can be hypothesized as high as 0.03 $\mu g/cm^2$ day, so the amount of metal ions released by a synthesis tool or by a prosthesis is less than 0.5 mg/year (Table 25.1), which does not cause significant problems to the patient, in absence of other types of corrosion.

25.1.4   Resistance to Crevice Corrosion

Crevice corrosion is the most widespread corrosion form for osteosynthesis tools devices. Activation sites mostly consist of the matching of the screws (that fix the implant) and their housings, i.e., they are localized where friction is present between metallic surfaces, which damages the passive film. The presence of crevice corrosion in these areas can increase by even 100 times the quantity of metal ions released in the tissues (Table 25.1). The analysis of corrosion cases encountered in removed implants shows that the susceptibility to crevice corrosion is high for stainless steel, low for cobalt alloys, nil for titanium and its alloys.

25.1.5   Resistance to Fretting Corrosion

The conditions existing in contact areas between two metal surfaces, in particular in the conical coupling between femoral head and stem, and under the heads of screws that fix ostheosythesis tools, are the cause of fretting corrosion. In these areas the contact surfaces are subjected to continuous relative movements of very small slip (even 10^{-2} μm).

Table 25.1 Metal ions release in the human body from orthopaedic implants

AISI 316L screw/plate		Ti6Al4V hip prosthesis	
Ion release after 1 year in passive conditions	500 μg	Ion release after 1 year in passive conditions	400 μg
Ion release in 1 year in presence of 5 triggers (2 mm^2 each) of crevice corrosion	50 mg	Ion release in conditions of fretting corrosion (1 cm^2 × 0.1 mm)	40 mg

In the case of stainless steel, particularly susceptible to crevice corrosion, friction intensifies this phenomenon, as it causes a continuous mechanical damage to the protective film. In cobalt alloys friction makes crevice corrosion possible. Considering titanium and its alloys, fretting corrosion is the main cause of degradation and the main limit of these materials, also because cracking due to fatigue or simply brittle cracking are often triggered in the corroded areas. Table 25.1 also shows how fretting corrosion can determine an increase by 100 times in the amount of metallic ions released from a hip prosthesis in Ti6Al4V.

25.1.6   Corrosion for Galvanic Coupling

Currently, it is rare to use implants of different nobility that may cause galvanic corrosion. Anyways, not all couplings turn out to be dangerous. For example, it is dangerous to couple titanium and cobalt alloys with stainless steel, conversely the combination of titanium and cobalt alloys is safe.

25.1.7   Biocompatibility

All of the metals employed in the field of orthopaedics undergo corrosion to some extent, therefore releasing ions in the tissues around the implant. Some of these ions are eliminated by the organism through physiological mechanisms (iron for example), others (above all chromium, nickel and cobalt) tend to concentrate in specific organs (liver, kidney, spleen). It is therefore mandatory that ions released are tolerated by the organism, without giving problems of local irritation, allergic reaction, carcinogenicity, mutagenicity. If this is verified, materials are called biocompatible. Biocompatibilty can be defined as the characteristic of a material to be well accepted in the human body; the degree of biocompatibility is measured by the entity of the provoked reactions.

For each metallic ion there exists a limit of tolerance. The presence of various elements can lead to a synergic action. For example, let's consider nickel, an element often present in alloys used in human body. The release of its ions in the tissues that surround the implant can determine local irritations and systemic effects. The most widespread reaction is allergy, which affects a relevant number of patients, especially women. In fact, about 30% of women report phenomena of cutaneous allergy just by having a simple contact with objects containing nickel (costume jewelry, watch cases, glasses) against the 3% of men. A study on cutaneous sensitivity to various metals has been carried out on patients that had reported nickel allergy: results are reported in Table 25.2. Data demonstrate that no patient shows allergy by touching stainless steel AISI 316, which shows a low corrosion rate in synthetic sweat (<1 $\mu A/cm^2$), while practically all of them do if in contact with materials coated with nickel, which presents a high corrosion rate (>1 mA/cm^2).[2]

25.2   Classes of Metals Employed in Orthopaedics

As a matter of fact, only few metals have adequate characteristics in order to be employed in orthopaedics, among them some kinds of austenitic stainless steels, some cobalt- chromium-molybdenum alloys and some titanium alloys.

25.2.1   Austenitic Stainless Steels

The majority of metallic components for temporary applications (osteosynthesis devices) are made of these alloys, as well as a little part of permanent implants (prostheses). The main advantages are: low cost, good mechanical properties if hardened, ease of production by plastic deformation, ease of shaping by machining. The main defects are the presence of nickel and the susceptibility to crevice corrosion, in particular at low percentages of molybdenum and nitrogen. In the past, various types of austenitic stainless steels were employed; currently, the ISO standard 5832-1 allows the use of three classes of steel (Table 25.2).

The most traditional steel is the ISO 5832-1D, which corresponds to AISI 316L steel with higher molybdenum, which contains chromium (17–19%), nickel (13–15%), molybdenum (2.25–3.5%) and nitrogen (<0.10%) and it is characterized by low levels of sulphur, phosphorus, and inclusions. It is the most economic material indicated by the standard and can be easily processed by plastic deformation and chip removal. Mechanical characteristics, not particularly high at the solubilized state, (tensile strength equal to 690 MPa, yield strength equal to 190 MPa) can be increased by means of cold working (reaching 1100 and 690 MPa, respectively). Unfortunately, this steel is sensitive to crevice corrosion. It is the most employed material for ostheosythesis devices, as they are generally removed once they have fulfilled the task.

ISO 5832-1E steel contains chromium (17–19%), nickel (14–16%), molybdenum (2.35–4.2%) and nitrogen (0.1–0.2%); compared to the traditional steel ISO

---

[2]The EU directive 94727, 1994 imposes that metal parts aimed to skin contact must not release more than 2 mg/cm^2 week of nickel. The fulfillment of this directive has brought to a sensitive reduction of dermatitis due to nickel contact. Unfortunately, European Bank technicians did not take this into account if what *Nature* (2002) writes is true, namely, that 1 and 2 euro coins have a nickel release rate 230–320 times higher than the limit level, for this reason they induce allergic reactions in individuals sensitive to this metal.

Table 25.2 Allergic reactions due to prolonged skin contact on patients that had previously witnessed nickel allergy

Steel	Passivation current	Ni released in artificial sweat (μg/ cm^2 week)	Patients with allergic reaction (%)
AISI 316L	<1 μA/cm^2	<0.05	0
AISI 303	>1 μA/cm^2	≈0.5	14
Ni coating	>1 mA/cm^2	>70	96

5832-1D it is much more resistant to crevice corrosion—even though it is not immune—and has higher mechanical characteristics.

ISO 5832-9 steel contains chromium (19.5–22%), nickel (9–11%), molybdenum (2–3%), manganese (2–4.25%) and nitrogen (0.25–0.5%). The high level of nitrogen ensures a better resistance to crevice corrosion, especially if molybdenum content is near 3%, and better mechanical characteristics both in the solubilized state (tensile strength 740 MPa, yield strength 430 MPa) and after cold working (for cold worked bars tensile strength can be as high as 1800 MPa). Such better characteristics come with the drawbacks of a higher cost and a more difficult processing.

During the past years, stainless steels with high nitrogen and manganese and practically lacking nickel (even lower than 0.1%) have been introduced, with mechanical resistance and resistance to localized corrosion comparable to the steel ISO 5832-9 ones.

25.2.2   Cobalt Alloys

Cobalt alloys used in orthopaedics may be of two types: cast alloys and wrought alloys. Cast alloys (ISO 5832-4) contain, apart from cobalt, chromium (26.5–30%) and molybdenum (4.5–7%). They exhibit a tensile strength of 655 MPa and yield strength equal to 450 MPa. The main advantages of cobalt cast alloys are: high mechanical properties, excellent resistance to corrosion, in particular to fretting corrosion. The main disadvantages are: high cost, low fatigue resistance, impossibility of both plastic deformation and machining.

Wrought alloys (ISO 5832-5, 6, 7, 8) are expensive, more than titanium and characterized by the presence of nickel. The advantages of these alloys are: excellent mechanical characteristics, good resistance to corrosion. The disadvantages are: high cost, complex production technology and presence of nickel.

Wrought cobalt alloy ISO 5832-12, which has been developed more recently, contains chromium (28%) and molybdenum (6%), and other elements (Ni, Mn, Si lower than 1%). After cold working it reaches high values of mechanical resistance (yield strength 830 MPa) with a good toughness (12% elongation). This alloy is currently the most employed for the production of orthopaedic tools by plastic deformation, in particular for stems and articular heads.

25.2.3 Titanium and Titanium Alloys

Titanium is considered the most biocompatible metal. Pure titanium has poor mechanical properties, therefore, for some applications it is replaced by titanium alloys. The most commonly used titanium alloy is Ti6Al4V (ISO 5832-3), with a tensile strength of 860 MPa in the annealed state, which can be increased with quenching and aging treatments. In order to replace vanadium, which has arisen many doubts on its biocompatibility, Ti7Al8Nb (ISO 5832-11) and beta alloy Ti15Mo5Zr3Al have been introduced. These alloys have mechanical characteristics and corrosion resistance similar to Ti6Al4V, but at present they are less widespread and more expensive.

The sensitivity to fretting corrosion is the main limit of titanium and its alloys, especially because corrosion sites often trigger fatigue failures or simply brittle fractures.

The main advantages of titanium are: good biocompatibility, especially since it does not obstacle osseointegration, good workability by machining, possibility of hot plastic deformation. Disadvantages are: low mechanical properties, sensitivity to fretting corrosion, difficulty of cold plastic deformation. Titanium alloys are more expensive, less biocompatible, but with better mechanical properties compared to pure titanium.

The near-equiatomic Ni–Ti alloy (55% by weight of nickel and 45% by weight of titanium) is particularly interesting for bioengineering applications: it is not commonly employed in orthopaedics yet, but it is currently being studied for its super-elasticity and shape memory properties. This alloy has been used in the orthodontic industry for a few decades, being responsible for the main improvements in the field, and is becoming an important material for endovascular stents (see Box). However, there is still concern (although not shared by everyone) about its biocompatibility, due to its high nickel content.

25.3 Surface Finishing Treatments

Before being implanted, certain materials may require surface finishing treatments such as barrel finishing, electro-polishing, passivation and anodising or other specific processes.

25.3.1 Barrel Finishing

Tumbling, or barrelling, is a mechanical finishing operation consisting in inserting the metal components in vibrant or rotating machines, known as barrels, along with inert materials of specific shape. It eliminates surface defects caused on metal components by previous operations (turning burrs, small pressing defects, sharp edges) and is used to obtain very fine surface finishing (for example, such as that necessary in polishing hip prostheses metal heads). This process can occur in different stages by using inert elements of progressively decreasing size and it can allow to obtain surface finishing comparable to those obtained by hand polishing with diamond polishing pads.

25.3.2   Electropolishing

It is an electrochemical process that allows to obtain a fine surface finishing of metal components (polishing). The component is immersed in an appropriate bath and acts as the anode. Process effectiveness and speed depend on type of bath, temperature, time, current density and metal alloy that has to be treated. Unlike the other mechanical finishing processes, this technique does not cause deformations, inclusions or contaminations on the treated surface. It is commonly used for stainless steel finishing, but it can be used for any metallic biomaterial. After the electropolishing process, the metallic components are generally passivated.

25.3.3   Passivation

It is possible to apply to metal components some treatments that increase the thickness of the protective passive film. These treatments can be carried out by immersion and permanence of the metal components in specific baths, for instance containing concentrated nitric acid.

25.3.4   Titanium Anodising

In order to improve corrosion resistance, titanium and titanium alloys can be oxidised through electrochemical methods. As the applied potential increases, thicker titanium dioxide films can be obtained.

Endovascular Prostheses
All metals which are to be implanted into the human body must be corrosion resistant and biocompatible. Depending on the type of application, other characteristics are required. For example, endovascular prostheses—stent—must be easy to extend, radiopaque, rigid, fatigue and compression resistant and possibly not too expensive, therefore sometimes corrosion resistance is

not considered as first requirement. To learn more, let's interview a technician working in this field: "Dear Paolo, can you give me a few brief information about stents used today, the materials employed and the specific problems that this application causes to them?"

Answer: *"Dear dad, this is not a simple matter. In coronary arteries (where 75% of stents are used) stents are implanted through angioplasty balloon inflation. This type of stent (Fig. 25.2) has a greater radial force, lower risks of misplacement and can be seen with fluoroscopy. Unfortunately, it is neither very elastic nor flexible (this also depends on stent design). It is good for vessels such as coronary or renal arteries, which are positioned in depth and do not risk external compression or deformation. The most used material for this type of stent is stainless steel, typically AISI 316L, even if recently, more corrosion-resistant—but especially more radiopaque and more rigid—cobalt alloys have been introduced. Cr–Co stents allow to maintain the same level of radial resistance, flexibility and radiopacity, with smaller mesh thickness, with subsequent smaller risks of restenosis (a partial or total narrowing of a blood vessel).*

Stents of larger diameter (5–10 mm vs. 2–4 mm of the coronary arteries), both self-expanding and balloon-expandable stents, are used in non-coronary arteries (e.g., iliac, femoral, renal, carotid arteries). Balloon-expandable stents are used to treat deep arteries (ex. renal, iliac arteries) and are similar to coronary stents, while self-expanding stents, which do not need the balloon technique to expand—even if in more calcified artheriosclerotic plaques they may require a successive post-dilation of this type—are very flexible and elastically deformable. The frequent artery movements, especially in legs, can cause fatigue cracks with hemodynamic and clinical consequences in some stent models available commercially.

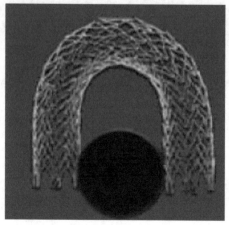

Fig. 25.2 Self-expanding stent (diameter 3 mm)

Self-expanding stents are made of a nickel-titanium alloy. Corrosion and fatigue resistance improves if the electro-polishing surface finishing is properly performed. Other alloys have been discarded because they make the stent difficult to expand, non-radiopaque, non-biocompatible, rigid or too expensive to produce.

Drug-eluting stents able to inhibit cell proliferation, which leads to restenosis, are now implanted. In the future, we aim to develop bio-absorbable stents: magnesium prototypes are currently undergoing a clinical evaluation (God bless corrosion in this case!), even though drug delivery systems based on polylactic acid are the most promising. The first experiments on men are now being carried out after a number of preclinical studies".

Two Memories

I have two precise memories from my job within the field of orthopaedics.

The first one is of the early 70s, when I observed stress corrosion cracking in prostheses and osteosynthesis plaques coming from a clinic and two hospitals in Lombardy, Italy. They were nominally AISI 316 stainless steel implants, about thirty of which were removed from patients for various reasons in the previous ten years after different in vivo periods, spanning from six months to three years. Some of them suffered very harsh corrosion (one example is reported in Fig. 25.3). Chemical analyses showed that different components, in particular some screws, contained insufficient molybdenum levels to be classified as AISI 316. Yet, at that time the essential role of this element in increasing corrosion resistance in chloride-containing environments, such as human body, was already known. Today, international standards provide the chemical composition, structure, surface and mechanical characteristics of implanted alloys, and most importantly require strict controls. Consequently, a patient who undergoes total hip replacement or has a plaque implanted can be sure (at least, almost sure) not to run the same risk as his forty-years-ago fellows, namely, the risk of corrosion.

The second memory dates back to the 1978 spring-summer season that I spent in Connecticut State University as a guest of Professor N.D. Green. During those years, among other things, the professor was responsible for corrosion problems within human body and he collaborated with his colleagues in biology who were investigating the potential side effects of corrosion of stainless steel implants. In fact, there was concern, which ultimately turned out to be unfounded, about a potential carcinogenicity of nickel released during corrosion along with iron, chromium and molybdenum. I spent some hours at the Department of Biology where some researchers

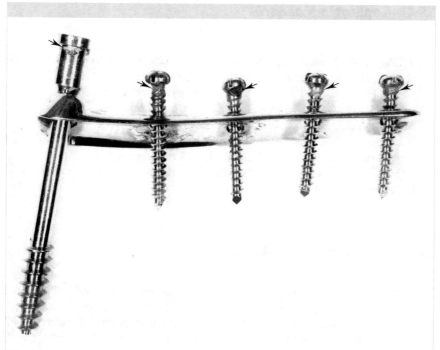

Fig. 25.3 Crevice cracking in osteosynthesis device for femoral fractures, occurring at the screw head-plaque coupling

were working on this topic. In the previous months, two plates of AISI 316 were implanted into the superior neck region of dozens of test animals; the plates were welded to create a crevice, together with an AISI 304 plate to accelerate the corrosion attack (no corrosion was starting on a single plate and, therefore, in the absence of crevice). By placing a small electrode near the plates, they measured its potential for weeks and, once the corrosion was triggered, they began to perform periodic in vivo measurements of corrosion rate using the linear polarisation technique. That was what they were doing in that moment. After a few months, they would have sacrificed the test animals to determine the concentration of chromium, nickel and molybdenum in their lung, pancreas and spleen and to provide cancer evidence, which actually was never found. I participated in some measurements of corrosion rate. They required time, because the animal was subjected to local anaesthesia, and they were boring and repetitive, so they used to work with the radio on. I still remember the newscast announcing the assassination of Aldo Moro, an eminent Italian politician, during one of those measurements.

Pietro Pedeferri's memories

25.4   Applicable Standards

- ASTM A967, Standard specification for chemical passivation treatments for stainless steel parts, ASTM International, West Conshohocken, PA.
- ASTM F-86, Standard practice for surface preparation and marking of metallic surgical implants, ASTM International, West Conshohocken, PA.
- ASTM F-2129, Standard test method for conducting cyclic potentiodynamic polarization measurements to determine the corrosion susceptibility of small implant devices, ASTM International, West Conshohocken, PA.
- ISO 5832, Implants for Surgery—Metallic Materials, International Standard Organization, Geneva, Switzerland.

25.5   Questions and Exercises

25.1   Discuss the galvanic coupling between stainless steels and cobalt-based alloys used for prostheses. Consider the influence on crevice corrosion.

25.2   In vivo studies on the use of NiTi orthodontic devices have reported several cases of severe inflammatory reactions resulting in contact dermatitis. Which degradation mechanism would you envision? How can you increase the service life of these materials?

25.3   Due to issues related to possible flammatory responses, the maximum ion release allowed from the acetabular cup of a hip prosthesis with surface area 80 cm^2 is 0.6 mg/year. Calculate the maximum corrosion rate allowed for a stainless steel AISI 316L, in the simplified hypothesis that all elements are released with the same rate (use an average molecular mass of 56 g/mol).

25.4   The same alloy of the previous exercise is used to manufacture bone plate and screws of an osteosynthesis device. Crevice corrosion onsets. To which extent does corrosion rate change? Which influence does it have on the implant service life? First try to hypothesize reasonable values based on the new corrosion mechanism. You can then refer to Table 25.1 to have reliable values of crevice corrosion rate.

25.5   The EU directive 94727, 1994 imposes that metal parts aimed to skin contact must not release more than 2 $\mu g/cm^2$ week of nickel. Calculate the related maximum corrosion rate of an AISI 303 stainless steel, with Ni content 9%, in the simplified hypothesis that all alloy elements are released with the same rate (i.e., Ni release accounts for 9% of the overall corrosion current).

25.6   A 65-year old man undergoes dental surgery. Which implant material would you select to meet an expected service life of 20 years?

25.7   Which is the role of surface treatments in extending a metal implant service life?

25.8 Which are the advantages of using cobalt-based alloys or titanium alloys with respect to austenitic stainless steels?

Bibliography

Cigada A, Chiesa R, Pinasco RM, Hisatsune K (2002) Metallic materials. In: Barbucci R (ed) Integrated biomaterials science. Kluwer Academic Press, Plenum Publisher, New York, USA

Hansen DC (2008) Metal corrosion in the human body: the ultimate bio-corrosion scenario. The Electrochem Soc Interface 17(2):31–34

Chapter 26
High Temperature Corrosion

Iron, made glowing by the action of fire, gets corrupted.
Pliny, Hist. Nat., 34

Abstract A metal in contact with a hot gas, typically at temperatures above 400 °C, in absence of liquid water phase, can suffer corrosion, also called hot corrosion. While aqueous (wet) corrosion processes are of electrochemical nature, hot corrosion is a chemical process, i.e., governed by chemical process kinetics in gas phase. Nevertheless, the oxide layer that forms at the metal surface is influenced by ionic diffusion and electronic conductivity within the oxide, as typical of an electrochemical mechanism. Corrosion attacks include: thinning due to the formation of non-protective scale, corrosion products and metal evaporation, metal degradation by molten salts, erosion-corrosion assisted by entrained solid particles, localized attack at grain boundaries, embrittlement. In this Chapter, the properties of oxides, as morphology, conductivity, protectiveness are described, together with the oxidation behaviour of metals and alloys; other processes (sulphidation, carburisation) and different environments, like steam and combustion gases, are briefly outlined.

Fig. 26.1 Case study at the PoliLaPP corrosion Museum of Politecnico di Milano

© Springer Nature Switzerland AG 2018
P. Pedeferri, *Corrosion Science and Engineering*, Engineering Materials,
https://doi.org/10.1007/978-3-319-97625-9_26

26.1 Corrosive Gases

Hot gas, typically from combustion and chemical processes, contains oxygen, nitrogen and water vapour, plus other gases, for example: H_2, CO, CO_2, H_2S, SO_2, SO_3, NH_3, S, HCl, in variable concentrations, giving rise to specific alteration processes called carburization, sulphidation, chlorination and nitriding (Fig. 26.1). Moreover, in particular when sulphur, sodium and vanadium are present, salts may form, which at the operating temperatures can melt. These conditions are encountered in many industrial applications, petrochemical, nuclear and metallurgical, as in examples shown in Table 26.1.

26.2 Thermodynamics and Kinetics

Metals, except gold, oxidize spontaneously when exposed to oxygen because the standard energy variation, ΔG^0, is negative. ΔG^0 varies with temperature as follows (activities of metal and relevant oxide are by definition unitary):

Table 26.1 Examples of hot corrosion in industrial plants

Temperatures	Application	Types of hot corrosion
1100 °C	Ethylene-pyrolysis furnaces	Carburization, oxidation
	Gas turbines	Oxidation, sulphidation, sulphate/chloride assisted corrosion, ash-related corrosion
1000 °C	Reformer catalyst tubes for production of ammonia, methanol, oxo-alcohol, etc.	Oxidation, carburization
	Reformer catalyst tubes for hydrogen production	Oxidation, combustion induced corrosion
950 °C	Reactor and catalyst support grid for nitric acid production	Nitriding/oxidation
	Superheater supports in oil fired refinery boilers	Fuel ash corrosion and oxidation
700–850 °C	Furnace tubes in carbon disulphide production	H_2S sulphidation, carburization
	Heater tubes in hydrodealkylation	H_2S/H_2 corrosion
550–650 °C	EDC cracker tubes in vinyl chloride monomer production	Halide gas corrosion
	Convertors, ammonia plant, reactors for hydrocracking	Nitriding, hydrogen attack H_2S/H_2 corrosion
450–550	Gas combustion, boilers	Oxidation

$$\Delta G^0 = -RT \ln\left(\frac{1}{p_{O_2}}\right) \qquad (26.1)$$

The plot of ΔG^0 as a function of T, which is a set of straight lines, is called Ellingham[1] diagram.

From a thermodynamic point of view, hot corrosion cannot be avoided, but rather slowed down by ensuring the formation of a *protective scale*, which should satisfy these following general requirements:

- High thermodynamic stability
- High melting temperature
- Low growth rate
- Good adhesion to the metal surface
- Good healing properties when damaged or cracked
- Thermal expansion coefficient close to the one of metal
- Good erosion resistance.

Among corrosion products that may form, only oxides satisfy these requirements, specifically: Al_2O_3, SiO_2 and Cr_2O_3. This suggests that the metal, at least at the substrate surface, must contain high enough quantities of one of these elements.

26.3 Scales

The oxidation resistance is therefore necessarily linked to the formation of an oxide scale, its covering power and its adherence to the substrate, therefore preventing or slowing down further oxidation. To predict the protection properties of the oxide, the *Pilling–Bedworth Ratio* (PBR) is used. This index is the ratio between volumes of oxide and oxidized metal, respectively. An index lower than one indicates that the oxide is not protective (i.e., oxide volume is not sufficient to cover the metal surface); conversely, if the index is much greater than one (i.e., higher than 1.8) and the oxide reaches high thickness, its volume is too big, then giving rise to mechanical stresses within its layer, which lead to the oxide rupture. In practice, compact and adherent oxides can form only if their PBR is about 1.5. Table 26.2 shows the PBR for most common metals.

---

[1]Harold Johann Thomas Ellingham (1897–1975) was a British physical chemist and is best known for the diagrams named after him that plot the change in standard free energy with respect to temperature for reactions like the formation of oxides, sulphides and chlorides of various elements.

Table 26.2 Pilling–Bedworth ratio for some metal oxides

Oxide	PBR	Oxide	PBR
CaO	0.6	MnO	1.8
BaO	0.7	TiO$_2$	1.9
MgO	0.8	CoO; Cr$_2$O$_3$	2.0[a]
CdO; CeO$_2$	1.2	Fe$_3$O$_4$; Fe$_2$O$_3$; SiO$_2$	2.1[a]
Al$_2$O$_3$	1.3	Ta$_2$O$_5$	2.3
Pb$_3$O$_4$	1.4	Nb$_2$O$_5$	2.7
NiO	1.5	V$_2$O$_5$	3.2
ZnO; BeO; PdO; ZrO$_2$	1.6	MoO$_3$	3.4
FeO; CuO	1.7	WO$_3$	3.4

[a]The PBRs of Cr$_2$O$_3$ and SiO$_2$ are higher than 2, which would indicate a non-protective scale; yet, these oxides are so protective that the actual quantity of oxide forming is very limited, which hinders oxide cracking and spalling due to the accumulation of large volume oxide

26.3.1   Non Protective Oxides

When a metal oxidizes and forms an oxide with Pilling–Bedworth Ratio lower than one, the metal is continuously exposed to hot gas, then oxidation proceeds at constant rate; the metal thickness loss, x, is given by:

$$x = C_1 t \qquad\qquad (26.2)$$

where t is time, and C_1 is a constant of the metal. Oxide thickness to time relationship is linear (Fig. 26.2).

An example of this behaviour is Mg. Also Mo follows the same kinetics although for a different cause, which is because its oxide, MoO$_3$, is volatile (melting temperature 795 °C and boiling temperature 1100 °C).

Fig. 26.2 Kinetic trend of mass change of samples subject to hot corrosion

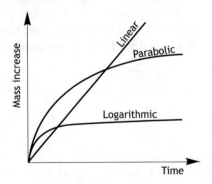

26.3.2 Protective Oxides

When the oxide is protective, i.e., Pilling–Bedworth Ratio exceeds one, the corrosion rate decreases with time by following two trends: parabolic and logarithmic (Fig. 26.2):

$$x = C_2\sqrt{t} \tag{26.3}$$

$$x = C_3 \ln(C_4 t + 1) \tag{26.4}$$

where t is time, and C_2, C_3, and C_4 are constants related to the metal.

Some common metals, such as aluminium, beryllium, zinc and chromium, show a logarithmic kinetics of growth. The reason for this deviation from the parabolic behaviour is complex: in the aluminium and beryllium is the low mobility of electrons participating in the oxidation process, while for zinc and chromium it is the slower diffusion rate of ions. Aluminium and chromium oxides have a crystalline structure compliant with that of the underlying metal and therefore are adherent and protective.

A parabolic dependence is achieved when diffusion of oxygen ions, O^{2-}, is slow and determines process rate. In these cases, the growth rate is proportional to the ion flux, J_{ion}, given by Fick's law, hence giving a parabolic dependence. Since the diffusion coefficient has an Arrhenius type dependence with temperature, oxide growth rate increases with temperature accordingly.

The parabolic oxidation rate constants, C_2, for some metals at a temperature of 1000 °C are reported in Table 26.3. As an example, the thickness loss after 10 year exposure time for Fe, Cr and Al is 18 mm, 180 μm and 72 μm, respectively. As temperature changes, oxidation rate changes through the variation of the constant, C_2, with temperature. As mentioned, the relationship is of Arrhenius type, that is:

$$C_2 = C_2^0\, e^{-\frac{Q}{RT}} \tag{26.5}$$

where C_2^0 is constant at a reference temperature. By plotting constant C_2 *versus* temperature, Fig. 26.3 is obtained.

Table 26.3 Parabolic oxidation rate constants, C_2, for some metals (1000 °C, $p_{O_2} = 1$ bar)

Metal	C_2 (g^2/m^{4}s)	C_2 (m/s$^{1/2}$)	Metal	C_2 (g^2/m^{4}s)	C_2 (m/s$^{1/2}$)
Fe	50	1	Cr	5×10^{-3}	0.01
Co	2	0.15	Si	10^{-3}	0.01
Ti	1	0.2	Al	10^{-4}	0.004
Ni	2×10^{-2}	0.02	Ta, Nb	2×10^{-4} (g/m^{2}s) (linear)	

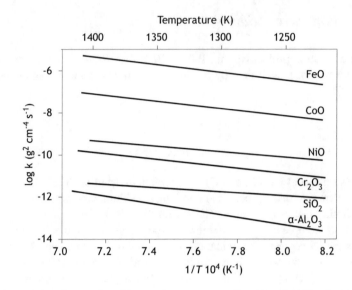

Fig. 26.3   Variation of parabolic oxide growth rate constant, C_2, with temperature for some oxides

26.4   Wagner Theory

In 1933 C. Wagner[2] proposed an electrochemical mechanism for oxide growth based on:

- An anodic reaction that takes place at the metal-oxide interface that produces metal ions and electrons
- A cathodic reaction of oxygen reduction into O^{2-}, by consuming electrons. This reaction can take place at the oxide-atmosphere interface or within the growing oxide
- A transport of ions inside the oxide film (positive ions or anionic vacancies from metal toward gas atmosphere and in opposite direction negative ions or cationic vacancies)
- A transport of electrons inside the oxide film.

Figure 26.4 compares the electrochemical mechanism for aqueous and hot corrosion assuming the transport of ions and electrons, only. Then, the metal works as anode, the oxide provides both electronic and ionic conductivity, the cathode is the oxide surface in contact with the atmosphere.

---

[2]Carl Wagner (1901–1977) is also remembered as the "father of solid-state chemistry" for his pioneering work in a variety of fields including tarnishing reactions, catalysis, photochemistry, fuel cells, semiconductors, and defect chemistry.

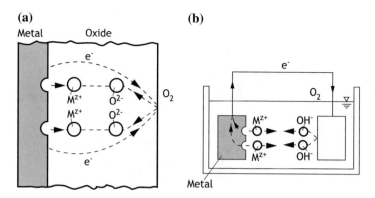

Fig. 26.4 Corrosion mechanism comparison: **a** hot corrosion; **b** aqueous corrosion (from Fontana)

As in aqueous corrosion, the slowest of the four processes determines the corrosion rate. Since in hot corrosion overvoltage of anodic and cathodic processes are negligible, because of the high temperature, the kinetically controlling process is electronic or ionic transport within the oxide.

A laboratory test that proves the electrochemical mechanism of hot corrosion is the growth of silver bromide on a silver strip exposed to a bromine atmosphere at high temperature: this case study involves a salt instead of an oxide, but the mechanism is the same. Silver bromide is a solid electrolyte with ionic conduction by Ag^+ ions but not an electronic conductor, therefore the bromide film does not grow. However, by placing a platinum grid on the film surface, electrically connected with the silver strip, electrons can circulate from silver to platinum and corrosion can proceed with formation of AgBr (Fig. 26.5).

This case study shows that both ionic and electronic conductivity is necessary for oxide film growth: one of the two is not sufficient condition for metal oxidation. Silver bromide is a good ionic but not electronic conductor; conversely, magnetite is a good electronic but not ionic conductor, therefore both form protective films.

An oxide film can be considered protective if, provided it is uniform and compact (PBR higher than one), it is either ionically or electronically non-conductive: this is a sufficient condition to ensure low oxidation rate. Therefore, oxide film resistivity is a further index for protection degree evaluation. The most protective oxides are those of aluminium, beryllium, zirconium and silicon, which are precisely the most resistive oxides. Also calcium and magnesium oxides have high resistivity.

26.4.1 Oxide Conductivity and Lattice Defects

To better understand how oxides grow, it is necessary to analyse the mechanism of ionic and electronic conduction.

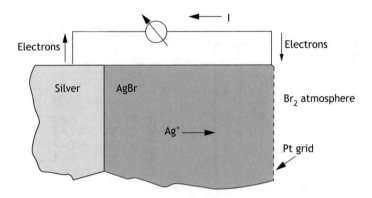

Fig. 26.5 Corrosive system consisting of silver foil-silver bromide-platinum in an atmosphere of high temperature bromine (from Bianchi-Mazza)

Table 26.4 Oxide types of several metals

Oxide type	Oxides
Metal excess (n-type)	BeO, MgO, CaO, SrO, BaO, CeO_2, ThO_2, UO_3, TiO_2, ZrO_2, V_2O_5, Ta_2O_5, MoO_3, WO_3, Fe_2O_3, ZnO, CdO, Al_2O_3, SiO_2, SnO_2, PbO_2
Metal deficient (p-type)	Cr_2O_3, UO_2, FeO, NiO, CoO, Co_3O_4, PdO, Cu_2O, Ag_2O

Perfect ionic crystals are not conductive. In the presence of lattice defects as interstitial ions and ionic vacancies, always present, ionic crystals become conductive through either movement of electrons and electron vacancies (electronic conductivity) or movement of ions and ionic vacancies (ionic conductivity).

Oxides are not fully stoichiometric, i.e. oxygen and metal in the lattice do not have exact ratio as in their chemical composition. For example, in copper oxide (Cu_2O) the number of Cu^+ ions is not exactly twice the number of O^{2-} ions, instead it is slightly lower. The opposite occurs for zinc oxide (ZnO) where the number of Zn^{2+} cations slightly exceeds that of O^{2-} ions.

For electroneutrality requisites, an excess of electrons (e^-O)[3] or electron vacancies ($e^- \square$) is present to counterbalance stoichiometry defects. When conductivity is due to electron vacancies (i.e., formally as positive charge movement) oxides are semiconductors of p-type (Cu_2O, NiO, FeO, CoO, Ag_2O, MnO and SnO are of this type); when electrons are involved, oxides are semiconductors of n-type (Table 26.4). The latter can be created also by oxygen ion vacancies as in zirconia, ZrO_2. Other n-type oxides are: CdO, Al_2O_3, V_2O_5, TiO_2.

Figure 26.6 shows the oxide growth mechanism for copper, zinc and zirconium, respectively of p-type (Cu_2O) and n-type (ZnO and ZrO_2). For copper oxide

---

[3]Symbols ($\square$) and (O) indicate vacancy and interstitial, respectively.

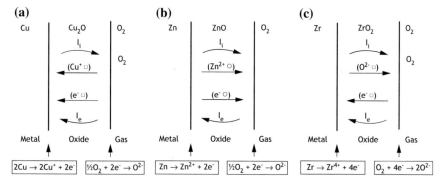

Fig. 26.6 Growth mechanism of oxide on copper (p-type oxide), zinc (n-type oxide) and zirconium (n-type oxide)

growth, charge carriers are electron vacancies, (e⁻ □), moving in the same direction of current and ionic vacancies, $(Cu^+□)$, which move in the opposite direction (Fig. 26.6a). In case of zinc oxidation, which gives an *n*-type oxide, carriers are electrons, (e⁻○), which move in the opposite direction to current, and interstitial ions, $(Zn^{2+}○)$, which move in the same direction of the current (Fig. 26.6b). Finally, for zirconium oxidation, which gives an *n*-type oxide for oxygen deficiency, carriers are electrons moving in the opposite direction of current and oxygen vacancies, $(O^{2-}□)$, which move in the same direction (Fig. 26.6c). In summary, the oxide grows by ion migration from the metal-oxide interface to the oxide surface and electron movement in opposite direction, or by ionic vacancies migration from outer to inner oxide surface and electronic vacancies in opposite direction.

26.5 Morphology of Oxide Films

Oxide and metal maintain a good adhesion if the respective crystal lattices fit one another. When the oxide film is very thin, adherence is good in spite of a high mismatch between the two lattices because the oxide assumes an amorphous structure. However, as soon as the oxide grows, the amorphous structure, which has a high internal energy, tends to crystallize and spalls off if no crystalline rearrangement between oxide and metal takes place. Oxidation process proceeds at high rate following an almost linear trend through parabolic sections, as Fig. 26.7 shows.

> **Hauffe's Rules**
> The oxidation rate is determined by the conduction mechanism which has the lowest rate within the oxide. The latter can be modified by adding some impurities in the oxide.

Fig. 26.7 Pseudo-linear
dependence of mass change
against oxidation time

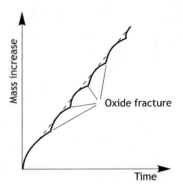

With n-type oxides, the presence of an impurity cation with a higher valence than the oxide reduces either the concentration of oxygen vacancies (for oxygen deficient, MO_{1-x}), or the concentration of interstitial metal cations (cation excess, $M_{1+x}O$), hence reducing the conductivity of the oxide and then oxidation rates. The opposite happens by adding lower valence impurity cations.

With p-type oxides, the presence of an impurity cation with a lower valence than the oxide reduces either the concentration of metal cation vacancies (for metal cation deficient, $M_{1-x}O$), or the concentration of interstitial oxide anions (oxygen excess, MO_{1+x}), hence reducing the conductivity of the oxide and then oxidation rates. Again, the opposite occurs by adding higher valence impurity cations.

Another cause of oxide spalling is internal stress in the film. Figure 26.8a shows the case of an oxide that grows by metal ions diffusion: the metal-oxide interface grows inward while the external oxide surface grows outward. If the involved surface is flat, no stresses arise, while the opposite occurs with a curved surface (Fig. 26.8b): as corrosion proceeds, a convex metal surface causes compression stresses inside the film, hence the film remains adherent, conversely, concave surface causes tensile stresses which spall off the film.

Figure 26.9a shows the opposite case, when the oxide grows by anions diffusion: the oxide grows at metal-oxide interface: if the PBR is higher than unity (i.e., oxide volume is greater than that of metal) and the surface is not flat, convex surfaces lead to a tensile stress inside the film (new oxide formed pushes the present layer) and concave surfaces exert a compressive one.

In general, metal oxides are brittle, then resisting compression better than tensile stress. Furthermore, under compression, internal micro-voids tend to form by coalescence of vacancies or blisters aroused at metal-oxide interface as shown in

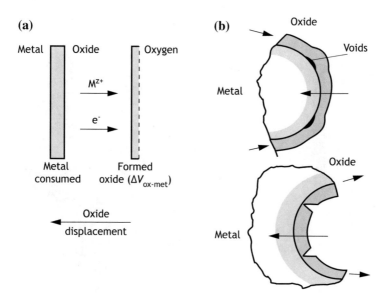

Fig. 26.8 a Ions moving in the oxide during its growth by cation transport; **b** stress induced by the oxide growing on curved surfaces (from Shreir)

Figs. 26.8b and 26.9b, and contribute to reduce stresses. When the metal works at constant load and at a continuous or cycling high temperature, creep induces a tensile stress to the oxide if thermal expansion coefficients of metal and oxide are different. Oxides can withstand these stresses, i.e., ability to deform by allowing dislocation movement; conversely, there is the need of a high hardness to resist abrasion due to solid particles in the gas (ash, dust, condensed water drops), low volatility and high melting temperature.

The corrosion behaviour is influenced by working temperature conditions, i.e., constant or cycling. In the former case (i.e., equipment operating at a constant temperature) an oxide layer easily forms, stable and adherent, assisted by internal stress rearrangement. In case of equipment subject to frequent thermal cycling, allotropic transformation of oxides can occur, causing a variation of structure and volume of oxide, which may be a further cause of the layer spalling, together with the mismatch between the metal thermal expansion coefficient and that of the related oxide.

Alloys containing molybdenum, tungsten and vanadium suffer *catastrophic oxidation* because oxides that form on them have a low melting temperature, therefore they can be easily removed due to either gas turbulence or evaporation. This catastrophic attack tends to localize because where it starts there is a local increase in temperature, hence enhancing the oxidation process.

(a)   **(b)**

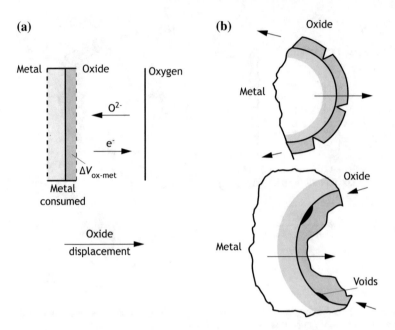

Fig. 26.9 **a** Ions moving in the oxide during its growth by anion transport; **b** stress induced by the oxide growing on curved surfaces (from Shreir)

26.6 Oxidation of Metals

The factors influencing the oxidation rate for a metal or alloy are different. Concerning the metal: chemical composition, impurities, crystal lattice orientation, surface finishing, geometry and thickness; from the hot gas standpoint: composition, impurities, pressure, flow rate, temperature and its variations. In the following, the behaviour of some relevant metals is highlighted.

Nickel. It forms a stable p-type oxide, $Ni_{1-x}O$, where x is 10^{-4} at 900 °C at an oxygen partial pressure of 1 bar. The oxide grows as columnar grains at the oxide-gas interface by migration of metal ions. The presence of impurities influences oxide structure, favouring the formation of porous, fine grains at the metal-oxide interface.

Iron. Iron forms a multi-layered scale consisting of the following three stable oxides: hematite, Fe_2O_3, magnetite, Fe_3O_4 and wüstite, FeO, as shown in Fig. 26.10. In practice, the following scales form:

- below 570 °C, the sequence of oxides is: $Fe/Fe_3O_4/Fe_2O_3/O_2$,
- above 570 °C the sequence becomes: $Fe/FeO/Fe_3O_4/Fe_2O_3/O_2$.

Wüstite, FeO, is stable at temperatures above 570 °C only, and because the mobility of Fe^{2+} within FeO is very high, oxidation rate is very high, too. Moreover,

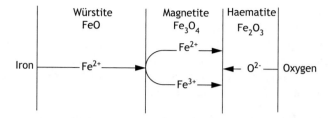

Fig. 26.10 Schematic illustration of iron oxide formation

Fig. 26.11 Phase diagram of
the Fe–O system

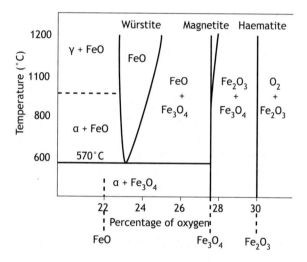

FeO has poor protective properties, so based on that, iron and low alloy steels can be used at temperatures below 570 °C, only; furthermore, below this temperature the diffusion of ions, Fe^{2+}, into magnetite is slow, hence resistance to oxidation is further increased (Fig. 26.11).

Chromium. Chromium forms Cr_2O_3 oxide (corundum with spinel structure) of p-type (although at low oxygen pressures it seems that it becomes of n-type). $Cr_{2-x}O_3$ has a value of $x = 9 \times 10^{-5}$ at 1100 °C and oxygen partial pressure of 1 bar. Since the oxide is relatively stoichiometric (low concentration of defects) its transport mechanism is affected by diffusion at grain boundaries. Above 900 °C in an atmosphere rich in oxygen, it is oxidized to hexavalent oxide, CrO_3, volatile, with loss of protection properties.

Aluminium. It forms the oxide Al_2O_3 which is very stable and very protective, since it is stoichiometric. Some alloys are designed to form a film of Al_2O_3 that offers protection up to 1300 °C.

Silicon. As aluminium, it forms an oxide, SiO_2, very stable and very protective because stoichiometric. New alloys are designed to form a layer of SiO_2 that offers protection up to 1200 °C.

Titanium. Titanium oxidation appears complex due to the formation of many stable oxides (Ti_2O, TiO, Ti_2O_3, Ti_3O_5, TiO_2). At temperatures below 1000 °C and oxygen partial pressure of 1 bar only TiO_2 is formed. At temperatures above 600 °C the growth kinetics is parabolic and can become pseudo-linear after long exposures. At high temperature oxygen dissolves in the metal in significant quantities causing the formation of cracks and exfoliation of the metal.

Molybdenum. The oxidation of molybdenum leads to the formation of volatile oxides (MoO_3 melts at 795 °C). These oxides are not protective and oxidation has a catastrophic trend.

26.7 Oxidation of Alloys

In case of alloys, the different affinity with oxygen of components determines the formation of oxides of more reactive metals, as for silicon, aluminium and beryllium, which also produce high resistance oxides; their presence, even in low content, allows the formation of a protective oxide film.

To highlight the complexity of high temperature corrosion of alloys, let's consider the simplest case of oxidation of a binary alloy AB, in which A is the solvent metal and B the solute one.

26.7.1 Oxidation of Only One of Two Metals in Alloy

We take into account two cases: one in which the alloy element (solute) oxidizes and one in which the solvent oxidizes.

In case the alloy element oxidizes, this can occur internally to the matrix, which consists of a solid solution of a solute B in a solvent A (it is the case of silicon–silver alloys, where silica globules, SiO_2, are formed, dispersed in non-oxidized silver, Fig. 26.12a); or it may undergo external oxidation (in case of iron–chromium alloys in which, when the partial pressure of oxygen is lower than that of iron oxide dissociation, it forms a film entirely made of Cr_2O_3, Fig. 26.12b).

In case the solvent metal oxidizes, the film consists of solvent metal oxide, in which particles of solute element B are dispersed. This is the case of copper–gold, Au–Cu, or copper–silver, Cu–Ag, alloys (Fig. 26.12c). The metal B may sometimes form a layer at the alloy surface. This is for example the case of the Ni–Pt alloys, in which oxidation forms nickel oxide supported by a layer of practically pure platinum (Fig. 26.12d); it is also the case of steels where impurities such as copper, tin or silver are concentrated on the alloy surface in a layer in direct contact with the oxide.

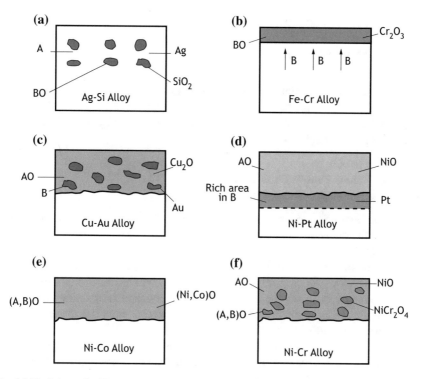

Fig. 26.12 Schematic illustration of the various possibilities of oxidation of the alloy AB described in the text: **a** and **b** oxidation of the metal solute; **c** and **d** oxidation of the metal solvent; **e** and **f** oxidation of the alloy of the two metals

26.7.2   Oxidation of Both Metals in Alloy

It is possible to observe two more cases, depending on whether the oxides that are formed are insoluble or soluble in one another. The first is that of copper–nickel (Cu–Ni), copper–zinc (Cu–Zn), copper–aluminium (Cu–Al) and copper–beryllium (Cu–Be) alloys. The second is the case of nickel–cobalt (Ni–Co) alloys which form a film consisting of solid solutions between NiO and CoO (Fig. 26.12e).

If the two metals form a double oxide AO-BO, the film is generally biphasic in that the second oxide is found dispersed in the oxide of the base metal. This is the case of the oxidation of Ni–Cr alloys (Fig. 26.12f).

26.8 Other Processes

26.8.1 Sulphidation

Similarly to oxygen, sulphur reacts with metals to form a sulphide scale (i.e., $M + \frac{1}{2} S_2 = MS$), following the same steps consisting of nucleation and growth of sulphide through an internal sulphidation reaction. The sulphidation rate is the result of the following processes:

- Sulphur supply
- Metal cations transport inside the scale
- Electrons transport inside the scale.

Sulphur ions, S^{2-}, do not migrate because of their big size. The controlling process is the transport of metal cations, since electronic conductance of sulphides is generally high.

Three important issues differentiate sulphides from oxides: sulphides of the main elements in alloys for high temperature applications are more stable than corresponding oxides, their volume is greater than the corresponding oxide volume, and the melting point is lower. Accordingly, sulphidation is more dangerous and more severe than oxidation.

Table 26.5 reports melting temperatures of eutectics between a metal and its sulphides for some metals of interest for high temperature applications. If the scale melts, there is no protection effect.

As shown in Table 26.6, sulphidation rates are five orders of magnitude higher than oxidation rates. In practice, metals and alloys used for hot corrosion applications to resist oxidation are not suitable for sulphidation. A possible temperature range for applications is below 500 °C.

26.8.2 Carburization

Metals exposed to gas mixtures containing CH_4 (and other hydrocarbons), CO_2, CO, H_2 and H_2O at temperatures above 800 °C may suffer *carburization* and *metal dusting* due to the deposition of elemental carbon, which forms by decomposition reactions, easily catalysed by the metal itself.

Table 26.5 Melting temperature of eutectic between metals and their sulphides

Eutectic	Temperature (°C)	Eutectic	Temperature (°C)
$Ni-Ni_3S_2$	645	Fe–FeS	985
$Co-Co_4S_3$	880	Cr–CrS	1350

Table 26.6 Constant rate comparison between oxidation and sulphidation (800 °C, $pO_2 = 1$ bar or $pS_2 = 1$ bar)

Metal	Oxidation rate $(g^2/m^4\ s)$	Sulphidation rate $(g^2/m^4\ s)$	Metal	Oxidation rate $(g^2/m^4\ s)$	Sulphidation rate $(g^2/m^4\ s)$
Ni	6×10^{-3}	200			
Co	5×10^{-2}	90	Co–20% Cr	10^{-5}	1
Cr	10^{-5}	3	Fe–20% Cr	10^{-5}	1
Ni-20% Cr	10^{-5}	1.5	Fe–20% Cr–5% Al	0.5×10^{-5}	0.5

Carburization is the result of diffusion of carbon into the alloy, taking place quickly at temperatures exceeding 900 °C. It is common in ethylene furnaces after ethylene pyrolysis. As carbon diffuses into the alloy, it reacts with alloy elements to form isolated carbides on the metal surface and internal precipitates. Typically, in iron–chromium alloys, carbides are $(Fe/Cr)_7C_3$ and $(Fe/Cr)_{23}C_6$ and others when Nb and W are present. Internal carbides decrease the mechanical properties of the alloy.

As often happens, the presence of small quantities of oxygen inhibits carbon deposition by forming an oxide scale. Because in many process environments, oxygen quantity is too low to form protective metal oxides (NiO, CoO, Fe_2O_3 and also Cr_2O_3), elements similar to oxygen as Si and Al are added. For instance, the presence of minimum 2% of silicon helps prevent carburization of AISI 314 (25% Cr, 20% Ni, 2% Si). Grinding the metal surface, which increases dislocation density, helps the development of a protective oxide layer of SiO_2, and facilitating the nucleation spreading of the protective SiO_2 layer; this treatment has become common practice for ethylene pyrolysis tubes.

Metal dusting is a severe damage of all iron and nickel-based alloys at temperatures in the range 450–800 °C in atmospheres containing CO_2, CO, H_2 and H_2O, where carbon disintegrates the metal surface into dust consisting of carbides, oxides and metal particles. As for carburization, the presence of an oxide scale helps inhibiting metal dusting; however, oxides form hardly at these relatively low temperatures. It is suggested that damage starts from local flaws in the oxide scale and proceeds locally as pit-type appearance, leaving undamaged the surrounding surface. The mechanism of metal dusting is not fully understood, yet.

26.8.3   Halogenation

In the combustion of waste, hot gases may contain halogens, typically HCl and Cl_2. Reactions between metals and halogens are similar to oxidation and sulphidation, but products, i.e., metal halides, are volatile, hence cannot develop any type of protection and kinetics approaches the linear dependence.

26.9   Environments

26.9.1   Oxygen and Air

Oxygen is the most common oxidant, which is present in air and process gas, often entraining ash. Typical metals operating in air at high temperature are resistors of electric heaters and machinery for heat treatments. In these conditions, the oxidation rate is influenced by the presence of pollutants and velocity of gas. About the latter, high flow rates cause a corrosion rate decrease by locally reducing the temperature of the metal surface. In oxygen-rich atmospheres a local overheating can lead to metal self-ignition.

26.9.2   Steam

Corrosion by steam occurs in power plants, either conventional, thermal or nuclear. The corrosion product is a metal oxide which protects the metal. In the case of low alloy steels containing small amounts of Cr–Mo–V, the oxide also resists decarburization caused by hydrogen produced in the oxidation reaction, $Fe + H_2O \rightarrow FeO + H_2$. A negative effect is caused by drops of condensation on turbine blades due to mechanical shock on the oxide layer.

The most severely stressed point in the steam circuit of a thermal power plant is in superheater tubes, where, however, most critical problems are on the external surface exposed to hot combustion gases, which may contain sulphur compounds, ash and others elements.

26.9.3   Sulphur Compounds

In reducing environments, i.e., without oxygen, sulphur-containing compounds form sulphides that alter the protection properties of surface layers of exposed metals. The greatest danger comes from the formation of liquid phases, which— similarly to the molybdenum oxide—do not allow the formation of a protective scale. The case of nickel is remarkable, as it forms a eutectic metallic nickel–nickel sulphide, Ni_3S_2, with melting temperature 645 °C, which excludes the use of nickel and nickel-based alloys in sulphur containing environments. In this case, iron–nickel alloys with high chromium content (about 20%) or cobalt alloys are used when high mechanical properties are not required.

Elemental sulphur and hydrogen sulphide are the most dangerous substances with regard to this type of attack. Sulphur oxides are less dangerous since they form protective oxide layers.

Sulphur and sodium chloride containing fuels lead to another severe form of attack due to the formation of sodium sulphate. Sodium chloride may be present in fuels for several reasons: for example, for the presence of brackish water coming directly from production wells, or seawater used to wash crude oil tanks, or introduced with the combustion air in aircraft engines. Sodium sulphate has a melting temperature in the range between 700 and 850 °C, hence it forms a liquid phase causing oxide film rupture and rapid metal wastage.

26.9.4 Combustion Gas

Combustion gases of clean fuels are mixtures of CO, CO_2, H_2O, N_2, NO_x and O_2; in case of partial combustion, hydrocarbons, carbon and hydrogen may also be present. The composition, or better the content ratio, determines the aggressiveness of the hot gas, whether there is an excess of C, CO and hydrocarbons (carburizing condition), or conversely a high content of H_2O, H_2 or CO_2 (decarburizing condition). In the first case, carbon can form carbides within the metal matrix with carbon-affinity elements as chromium, titanium, niobium, causing two detrimental effects: a mechanical one, i.e., embrittlement due to the formation of precipitates at grain boundaries, and one on corrosion resistance, because the precipitation of chromium carbides depletes chromium content, hence reducing oxidation resistance.

There is a particular corrosion form called *green rot*, which occurs on iron–chromium–nickel alloys in an environment that is oxidizing for chromium, but not for iron, when carburizing characteristics change over time; during carburizing periods, carbon enrichment takes place and successively, during the oxidation phase (i.e., decarburizing condition), an internal oxidation happens on the chromium depleted matrix. It follows that internal oxidation spreads through grain boundaries, then allowing the formation of bulky nickel oxides, which give a typical green colour to the fracture surface.

Conversely, in decarburizing environments, enriched in CO_2, H_2O or H_2, decarburization can occur through the diffusion of carbon to the metal surface followed by its reaction; carbon depletion in the matrix leads to the loss of mechanical properties.

When the decarburizing species is atomic hydrogen, this diffuses easily within the metal and reacts with carbon or with carbides according to the reaction: $C + 4H \rightarrow CH_4$. The methane formed is not soluble in the metal, so it accumulates in micro cavities then increasing internal pressures which deform the metal. On carbon steel, methane begins to form at temperatures around 300 °C, while for low alloy steels at 400–500 °C.

The combustion of coal and fuel produces substances that are solid at room temperature, called ashes; these could be metal oxides formed in the combustion process or also already present in the fuel. Oxides of aluminium and silicon mixed with coal and vanadium oxides are examples of ashes. The action of solid ashes is the abrasion of the oxide layer or, conversely, a screening effect; melted ashes are more dangerous.

26.9.5 Nitridation

It occurs in ammonia plants when internal nitrides form as a result of nitrogen diffusion, similarly to carburization. Among those most commonly used in practice, the most susceptible element to form nitrides is chromium.

26.10   Materials for Use at High Temperatures

To improve resistance to high temperature corrosion, two ways can be envisaged: alloying and cladding. When using alloying elements, neither mechanical nor structural characteristics should be jeopardized. Effective alloying elements are chromium, aluminium and silicon, which form stable and adherent oxides on the metal surface. When increasing chromium content in iron and iron–nickel alloys to improve the resistance to oxidation and sulphidation, there is the risk of formation of a brittle phase. Accordingly, the composition of alloys for high temperature is always a compromise between mechanical and corrosion resistance requirements, where one of the two will prevail according to the need of application.

In the following, iron alloys are considered, only. Carbon steel resists oxidation up to 570 °C regardless the addition of small amounts of alloy elements (Cr, Mo, V) for improving mechanical properties through heat treatments. To increase the temperature limit, higher contents of chromium and silicon must be added, as summarized in Table 26.7, with focus on alloys used in refineries.

Among stainless steels, the best behaviour is given by cast austenitic ones. Corrosion resistance and structural stability increase with both chromium and nickel content; AISI 309 and 310 steels show the best performance among semi-finished wrought and can withstand significant stresses at temperatures up to 950 °C, for example in hydrocarbons pyrolysis plants. The current trend is the use of cast alloys

Table 26.7 Scaling temperature of steels used for high temperature applications in refineries

Metals	Temperature (°C)
Carbon steel (0.1% C)	570
5 Cr–0.5 Mo	620
7 Cr–0.5 Mo	650
9 Cr–1.0 Mo	680
AISI 410 (12 Cr)	760
AISI 304, 321, 347 (18 Cr–8 Ni)	900
AISI 316 (18 Cr–10 Ni–3 Mo)	900
AISI 446 (27 Cr)	1030
AISI 309 (22 Cr–20 Ni)	1100
AISI 310 (25 Cr–20 Ni)	1150
Alloy 218 (17 Cr–8 Mn–8.5 Ni–4 Si)	980

with higher carbon content for better mechanical characteristics and high silicon content (up to 2.5%) to improve carburization resistance.

The application limit of these alloys is the mechanical resistance, which becomes unacceptable at approximately 950 °C, while hot corrosion resistance is extended to temperatures as high as 1100–1150 °C due to high content of alloying elements. Above 900 °C, also nickel–cobalt-iron superalloys are used, often with addition of aluminium (about 1.5%) or chromium to improve the resistance to cyclic oxidation or carbides formation.

26.11   Questions and Exercises

26.1   Discuss the different oxidation kinetics of metals in hot environment: which characteristics of the metal oxide determine oxidation kinetics? Make examples of linear oxidation rate and parabolic oxidation rate.

26.2   Which properties of the metal oxide determine whether the oxide is protective or not, and why? Based on these considerations, explain why silicon dioxide is more protective than hematite.

26.3   With the help of Table 26.3, calculate the thickness loss of iron, nickel and chromium after exposure to an oxidizing environment at 1000 °C for 1 month in the hypothesis of a parabolic oxidation rate. Use these data to draw considerations on the composition of high temperature alloys.

26.4   If the oxidation of a steel alloy produces mostly wüstite, FeO, with an oxidation rate constant, C_2, equal to 50 g^2/m^4 s and PBR of 1.7, calculate the weight gain that the metal experiences in 1 year. Compare it with the thickness loss provided in the text (Sect. 26.3.2).

26.5   An alloy is exposed to an oxidizing environment at 1000 °C. The weight gain due to oxidation is 0.0015 mg/cm^2 after 1 h, 0.005 mg/cm^2 after 12 h, 0.007 mg/cm^2 after 24 h, and 0.01 mg/cm^2 after 48 h. To which growth mechanism does this alloy adhere?

26.6   Consider data from the previous exercise. In the hypothesis of a parabolic oxidation, estimate the oxidation rate constant C_2 (in $\mu m/s^{1/2}$), knowing that oxide density is 6.67 g/cm^3 and PBR is 1.5.

26.7   Which metal or metal alloy would you select for a gas turbine? Explain why.

26.8   Why is sulphidation more dangerous than oxidation?

26.9   Steels suffer sulphidation, but the addition of chromium and aluminium decreases the sulphidation rate, as proved by sulphidation rate constants reported in Table 26.6. Why?

Bibliography

Bianchi G, Mazza F (1989) Corrosione e protezione dei metalli, 3rd edn. Masson Italia Editori, Milano (in Italian)

Birks N, Meier GH, Petit FS (2006) Introduction to the high-temperature oxidation of metals, 2nd edn. Cambridge University Press

Fontana M (1986) Corrosion engineering, 3rd edn. McGraw-Hill, New York, NY

Rapp RA (ed) (1983) NACE-6, High temperature corrosion. NACE International, Houston, TX

Pilling NB, Bedworth RE (1923) The oxidation of metals at high temperatures. J Inst Met 29:529–591

Shreir LL, Jarman RA, Burstein GT (1994) Corrosion. Butterworth-Heinemann, London, UK

Chapter 27
Prevention of Corrosion in Design

[for a civil structure] one euro spent in design with the aim to prevent corrosion produces same benefits as in construction phase by paying five euros, or after construction by spending twenty-five euros, one hundred twenty-five euros just before corrosion initiation and eventually six hundred twenty-five euros after corrosion had occurred.

W. R. De Sitter (Law of five, 1984).

Abstract This chapter deals with the principles that should guide the design of a structure based on corrosion prevention. This includes a series of preventative measures that can be chosen once the environment and its criticalities are identified, as well as a careful metal selection. Information on the main classes of metallic alloys and their corrosion resistance are provided, as a help to guide the selection.

Fig. 27.1 Case study at the PoliLaPP Corrosion Museum of Politecnico di Milano

© Springer Nature Switzerland AG 2018
P. Pedeferri, *Corrosion Science and Engineering*, Engineering Materials,
https://doi.org/10.1007/978-3-319-97625-9_27

The main preventative methods that should be adopted in the construction, storage and operating phases are also examined.

27.1 Design Life and Reliability

Corrosion prevention aims to preserve stability, functionality, characteristic of a structure exposed to an aggressive environment (for instance, a bridge, an equipment, a car body, any plant item) with the scope of maintaining reliability and quality as required.

The design life of a structure is the expected operating time that would require no extraordinary maintenance. It can be a few years for goods, 10 years for a car, 20 for a chemical plant, 30 or more for thermal power plant, an offshore platform or industrial building, 50 years for a civil building and more than 100 years for a public infrastructure such as a bridge, a church or a public building.

As far as corrosion is concerned, the reliability of a structure or a plant is the probability that corrosion does not happen. The more severe the consequences of corrosion, the higher must be that probability.

Reliability depends on materials adopted, corrosion monitoring and corrosion control on the design, construction and operational phase (Fig. 27.1) together with maintenance strategy: this is an economic matter. Ensuring an absolute reliability is nonsense; conversely, it is worth fixing a required reliability for the specific application, which implies to search the so-called optimum as the minimum cost given by the sum of the cost of material and corrosion prevention measures (design expenses) and the cost of remedial actions (corrosion related costs), necessary when corrosion takes place. A high reliability is chosen when consequences are severe, implying either high prevention costs or design costs (i.e., highly corrosion resistant metals) and the opposite applies when corrosion-related costs are modest.

27.1.1 How to Choose Reliability and Related Solutions

Let's consider, as often happens, that many different technical options can ensure a chosen reliability; a question arises: which solution fits better? For example, one can argue: is it wiser to select cheap metals and plan a frequent substitution (for instance, yearly) or to choose an expensive corrosion resistant metal or even a combination of corroding metal and an additional corrosion prevention method (CP or a corrosion inhibitor)? Obviously, some options are not applicable either technically (for instance, a corrosion allowance cannot be used for tubes of heat exchangers) or operationally (for instance, a shutdown may be required for material substitution).

There are constrains due to the process. For instance, reactors for chlorination of toluene could be made of carbon steel, from a corrosion viewpoint, but steel catalyses secondary reactions, therefore it cannot be employed; lead cladding must be used. Similarly, copper alloys cannot be used in plants for soap production because copper ions catalyse the oxidation reaction; furthermore, corrosion products of nickel interact with polyethylene reaction, impeding the correct polymerization.

Among various technical solutions, characterized by the same and necessarily agreed reliability, the final choice is based on the economic appraisal, through the *Life Cycle Cost* approach, which returns the cheapest, although reliable, option. A more sophisticated analysis is a probabilistic Life Cycle Cost approach by introducing the expected probability of failure; in this case, rather than a minimum cost option, a distribution of LCC is obtained, which shows the expected probability of expenditure that is, by this approach, equivalent to the reliability.

27.2 Prevention in Design Phase

As reference, let us consider a chemical plant, although these considerations are general. First of all, corrosion prevention starts from design, for which the following must be done:

- Definition of the environment in all parts of the plant from the corrosion viewpoint. Evaluation of need and possibility to change operating conditions
- Selection of materials on the basis of agreed reliability through LCC analysis. Further considerations have to take into account the market purchasing and mechanical properties
- Evaluation of construction constrains and grounding requirements to assess any possible galvanic coupling.

As learned from experience, about 90% of juvenile failures related to corrosion are because of lack of one of above issues.

27.2.1 Evaluation of Aggressiveness

The aggressiveness of an environment depends on many parameters (for instance, for waters: temperature, oxygen content, flow regime and others) which vary in time and space. For example, crude oil can be aggressive on the basis of water cut, carbon dioxide and hydrogen sulphide content, naphthenic acids, flow regime, gas-oil-ratio, pressure and temperature, vertical or horizontal flux and others; all these factors may change well by well and as a function of the exploitation grade.

27.2.2 Reduction of Aggressiveness

In industrial plants, operating conditions are often modified to allow the use of less expensive metals, ensuring the agreed reliability. For example, for water circuits, by removing oxygen and keeping water near neutral or slightly alkaline, carbon steel can be successfully used.

In plants for production of diluted nitric acid by using sulphuric acid, acid ratio is fixed in order to reduce corrosion rather than on the basis of a process optimization. It follows that sulphuric acid concentration is maintained above 70%, then allowing the use of cast iron for piping and steel for tanks after cooling down.

Similarly, to avoid chloride-induced SCC on austenitic stainless steels, temperature is kept below 50 °C then avoiding the use of nickel alloys. More generally, compatibly with the process, when an inorganic acid is required, sulphuric or nitric acids are preferred instead of hydrochloric acid, because less expensive metals can be adopted.

27.2.3   Local Conditions

In a plant, different corrosion conditions can establish locally for a couple of reasons: different hydrodynamics and water phase separation, either for unmixing or for condensation. The former occurs in production and refinery of hydrocarbons and the latter typically from combustion gas and exhausts. Condensed water is acidic when CO_2, SO_2, HCl and others are present in the vapour. These species are not aggressive in gas phase: for example, carbon steel is used in presence of steam containing carbon dioxide at 300 °C, provided no condensation takes place. Accordingly, shutdowns are dangerous because condensation happens forming carbonic acid; corrosion is controlled by injecting ammonia or amines in steam to neutralize possible condensates. In critical zones, more resistant metals can be employed.

27.2.4   Homogeneity Is Preferred

Often, aggressiveness depends on non-uniform conditions of oxygen content, pH, chemical concentration, temperature, flowing or stagnant conditions; or presence of not uniform scales. An example is the following. A vertical heat exchanger serves for cooling a process fluid by means of water flowing at the shell side. On top of the heat exchanger, gas and steam separate causing an increase in temperature because cooling water cannot be effective. This leads to concentrate salts giving rise to chloride-induced SCC of austenitic stainless steel, when industrial water is used (Fig. 27.2). With a horizontal heat exchanger this occurrence does not happen because cooling water wets continuously all tubes.

Inside fissures and interstices, aggressiveness often increases; typical conditions are: threaded joints, flanges, deposits and scales. Heat transfer further increases aggressiveness, as it occurs between tube sheet and tube where tightness is obtained by mechanical expansion. On the other side, at tube inlet, local turbulence

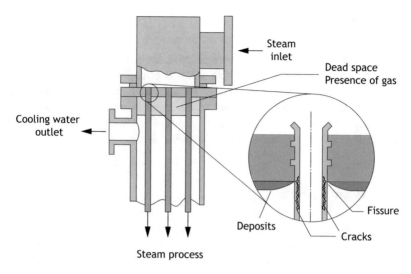

Fig. 27.2 SCC on top of tubes in a vertical heat exchanger (from Fontana)

conditions increase aggressiveness, jeopardizing the resistance of copper alloys. Pumps, cross section area variations, agitators, weld beads and the presence of a heater may generate local turbulence conditions.

When aggressiveness changes downstream a chemical process and different metals are used, galvanic corrosion conditions can arise; on the other hand, one-way valves can help avoid fluid reversal during shutdowns.

27.2.5 Change of Aggressiveness in Space and with Time

Aggressiveness can change with time. There are typically three moments characterizing a structure lifetime: testing and starting, normal operating and shutdown: materials and prevention measures have to cope successfully with all these situations.

Concerning location, inlet and outlet zones can show different turbulence conditions as well as screened zones and interstices may face increased concentration of some species.

Unfortunately, during operation, exposure conditions can change, then jeopardizing the resistance of materials selected; this is inevitably a risk, hard to foresee. Sometimes, impurities may rise corrosion concern. For example, an AISI 316 grade stainless steel reactor, designed for neutralisation of hot solutions (up to 90 °C) of sulphuric acid ($\approx$3%) and oxygenated water ($\approx$1–2%) with cobalt carbonate, worked properly for almost two years; however, as the carbonate supplier changed for a cheaper, chloride-polluted product, pitting occurred with perforation of the reactor bottom, 15 mm thick, in a few months.

27.3   Metal Selection

In design phase, corrosion engineers select resisting metals at specified operating
conditions, complying with design life at fixed reliability. Obviously, for those
many applications for which corrosion is not a concern, materials are selected on
mere mechanical requisites, while corrosion engineering should assist material
selection when corrosion is an important issue, starting from so-called *basic criteria*
and taking into account so-called *technological criteria*.

27.3.1   Basic Criteria

They deal with basic knowledge of corrosion principles, starting from thermodynamics, i.e., Pourbaix diagrams, to establish corrosion or immunity or passive
condition, then following with kinetics to determine corrosion rate. Evaluation of
applicable and helpful corrosion prevention methods, as injection of inhibitors and
cathodic or anodic protection, should be included in the corrosion assessment study.

27.3.2   Technological Criteria

Criteria are based on standards (NACE, ASTM, ISO, CEN, UNI, DIN, BS) or by
companies as proprietary standards, which cover almost all applications. Rather
than a theoretical approach, standards highlight guidelines based on knowhow and
experience. The inherent limit of these documents is that instructions are synthetic
and without justification; for this, it is important that users patronize the basic
criteria to avoid unwise mistakes.

Besides corrosion requirements, material selection takes into account other
requirements, as for example mechanical, thermal and electrical ones, as well as
weldability. An economic appraisal is generally the last step for final choice.

27.4   Some General Features of Used Metals

27.4.1   Carbon and Low Alloy Steels

These metals are the most used because they meet cost and mechanical requirements. About cost, they are the cheapest ones; parametrically by mass, if carbon
steel cost is one, for zinc is two; for aluminium and lead is four; for copper and
stainless steel is ten; for chromium and tin is twenty; for nickel alloys and titanium
is thirty to fifty. Mechanical properties are good and can be changed by alloying,
heat and mechanical treatments.

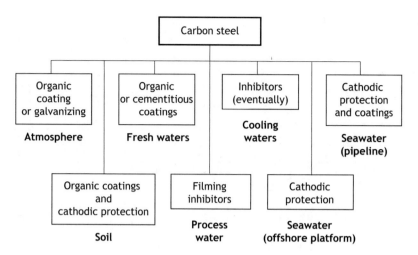

Fig. 27.3 Guideline for the use of carbon steel in neutral and low aggressiveness environments

Figure 27.3 summarises the use of carbon and low alloy steels for industrial processes or fluids, provided the adoption of a suitable prevention method, as coatings, cathodic protection and corrosion inhibitors. Typical applications are for atmospheric exposure structures (by using COR-TEN® or with the addition of painting), buried structures (with coatings and cathodic protection), water plants and water carrying piping (coatings, oxygen removal and cathodic protection).

In alkaline solutions, steel is passive, so it can be used bare, as in sound, pristine concrete, i.e., neither carbonated nor chloride-contaminated; in boilers after deoxygenation and slight alkalization. In heating circuits, no treatment is required, provided there is no refilling after start-up (water is spontaneously deoxygenated, being oxygen depleted by corrosion in early stages). Potable fresh wasters are generally non-corrosive provided they are scaling (i.e., positive saturation index). There are specific environments, as sulphuric and nitric acids, generally considered highly corrosive, to which steel fairly resists because it passivates.

Carbon and low alloy steels suffer SCC in specific environments and operating conditions: for instance, when temperature exceeds 70 °C in solutions containing nitrates, hydroxides, carbonates/bicarbonates, or at room temperature in the presence of sulphides by hydrogen embrittlement mechanism for high strength steels (tensile strength excceeding 750 MPa).

27.4.2 Stainless Steels

These alloys fit the requirements relating to reliability, safety and durability in almost all industries: manufacturing, chemical, energy, transportation, constructions. As far as civil industry, furniture goods, health and food industry are

Table 27.1 Chemical composition (% by mass) of most used stainless steels

Type	C (max)	Cr	Ni	Mo	Others	PREN
Austenitic AISI 304	0.08	18–20	8–11			18
Austenitic AISI 304 L	0.03	18–20	8–12			18
Austenitic AISI 321	0.08	17–19	9–12		Ti = 5 x %C	18
Austenitic AISI 347	0.08	17–19	9–13		Nb + Ta = 10x%C	18
Austenitic AISI 316	0.08	16–18	10–14	2–3		24–28
Austenitic AISI 316Ti	0.08	16–18	10–14	2–3	Ti = 5 x %C	24–28
Austenitic AISI 317L	0.03	18–20	11–15	3–4		28–32
Austenitic AISI 309	0.20	22–24	12–15	–	Low C as L and EL grades	22–24
Austenitic AISI 310	0.25	24–26	19–22	–	Low C as L and EL grades	24–26
Super-austenitic	0.02	19.5–21.5	17.5–18.5	6–6.5	N = 0.5–1	42–43
	0.02	19–21	24–26	6–8	N = 0.2–0.3	44–48
Duplex 2304	0.03	22–24	5.5–7.5	0.1–0.6	N = 0.1–0.5	24–26
Duplex 2205	0.03	24–26	6–8	2.7–4.5	N = 0.1	36–38
SuperDuplex 2507	0.003	25	7	4	N = 0.25	36
Ferritic AISI 405	0.15	11.5–13.5	–	–	–	12
Ferritic AISI 430	0.15	15–17	–	–	–	16
17-4 PH	0.07	15–17	3–5		Cu = 3.0; Nb = 5 x % C – 0.45	17

concerned, operating, aesthetic and economic targets are achieved. Table 27.1 reports main commercial stainless steels, while Fig. 27.4 summarizes the development philosophy of most common compositions.

Their corrosion resistance relies on the passive film, a few nanometres thick, made of chromium oxide mainly. Provided the passive film is flawless and self-healing if locally destroyed for either mechanical or chemical reasons, stainless steel behaves as a noble, corrosion-resistant metal; conversely, when the passive film is permanently destroyed locally, localized corrosion happens.

Chromium is fundamental for stainless steels to build a corrosion resistant passive film. The minimum Cr content is 11%: this stainless steel resists oxidation in non-polluted atmospheres. To enhance corrosion resistance, a minimum of 18% is needed, as for ordinary and most common stainless steels. To further increase resistance, a higher Cr content in combination with Mo (generally 2–4%) and N is specified. Further resistance is achieved by anodic or cathodic protection.

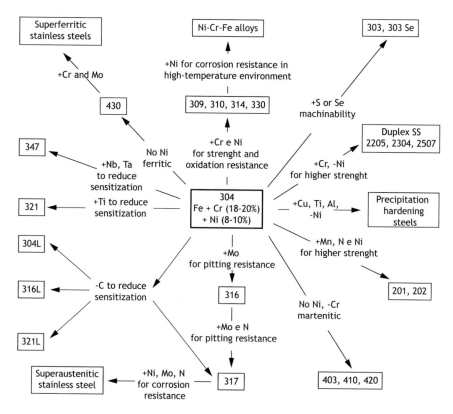

Fig. 27.4 Logical graph for stainless steel composition, AISI designation (from Sedriks)

The addition of nickel (8–10%) stabilizes the austenitic (FCC) structure, enhancing low temperature toughness, ductility and high temperature corrosion resistance. By adjusting the Ni concentration to about a half, i.e., 4–5%, both austenitic and ferritic phases stabilize to about 50% each, obtaining the so-called duplex stainless steels. These grades show enhanced mechanical resistance and an improved resistance to SCC, either chloride-induced, through an anodic mechanism, or hydrogen embrittlement, for instance in the presence of hydrogen sulphide; the addition of Mo and N increases the resistance to pitting.

Ferritic stainless steels, although cheaper, are less used because they are more brittle on welds. As general feature, they offer the advantage to be chloride-induced SCC resistant but a weakness associated to the high susceptibility to hydrogen embrittlement. Other stainless steel types are the martensitic ones and those obtained by precipitation hardening.

To predict the structure of a stainless steel on the basis of composition, the Schaffler diagram can be used (Fig. 27.5). It is obtained by plotting two parameters,

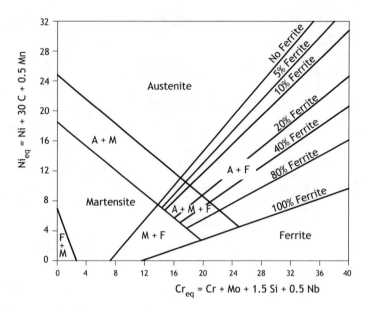

Fig. 27.5 Schaffler diagram

Ni-equivalent and Cr-equivalent, where Ni-equivalent reflects the austenizing elements and similarly Cr-equivalent for ferritic ones. Unlike carbon, other elements tend to stabilize their own structure (for instance, Cr, as BCC, stabilizes the ferritic phase and Ni, as FCC, the austenitic one, and so on for other elements).

27.4.3   Nickel Alloys

The FCC structure of nickel alloys, which gives high ductility at low temperature, in combination with a high resistance to corrosion (chloride-containing, alkaline and reducing solutions), leads to a huge use in chemical and petrochemical industry. In addition, pure Ni is also used for sodium and potassium hydroxide solutions, provided they are ammonia-free.

Table 27.2 reports the most used grades of Ni-based alloys. Alloy-20 derives its name from the composition—approximately 20% Ni–20% Cr (plus Mo)—which is one step forward from ordinary stainless steel (for instance, 316 grade); alloy B and alloy C, known with the commercial name Hastelloy, contain high Mo concentration and resist very aggressive environments, from strong acids to high chloride content solutions. Alloy 600 resists oxidizing solutions and high temperature gas. Alloy 400 and k-500 (also known as Monel) are Ni-Cu alloys, which implies both a noble behaviour (i.e., potential above hydrogen equilibrium potential) and a passive state; they are used for deaerated hydrofluoric acid.

Table 27.2 Chemical composition (% by mass) of nickel alloys

Type	Fe	C (max)	Cr	Ni	Mo	Others	Notes
Alloy 20	Balance	0.07	19–21	32–38	2–3	Cu, Nb, Ta, Mn, P, S, Si	
Alloy 400 (Monel)	2.5	0.3		Balance		Cu = 28–34; Mn; S; Si	
Monel K-500	2	0.25		Balance		Cu = 27–33; Al = 2.3–3.15; Mn; Ti; S; Si	Precipitation hardened
Alloy 600	6–10	0.15	14–17	Balance		Cu, Mn, Si, S	
Alloy B	≤2	0.01	1	65–69	26–30	Co, Mn, Si, P, S	
Alloy C	1.5–6	0.01	16–23	56–65	13–18.5	Al, Cu, Mn, Si, Ti, V	

27.4.4 Aluminium Alloys

Aluminium and aluminium alloys resist corrosion in solutions with pH in the interval 4–9 because they passivate, forming a strong, protective passive film, which—unlike that of stainless steels—is an insulator, then reducing electron transport. The passive film can be strengthened by anodizing. Strong acids and strong alkaline solutions destroy the passive film, which causes then a fast uniform attack. Impurities are always detrimental.

Corrosion resistance depends on composition, as Table 27.3 reports. The Cu containing 2xxx series, widely used in aerospace industry, suffers corrosion because heat treatments separate a noble phase, triggering localized attack. Mg-containing series, as 3xxx and 5xxx, show a better corrosion resistance because the separated phase is less noble. To improve corrosion resistance, plating (cladding) with pure Al and anodizing are widely used.

Table 27.3 Corrosion behaviour of aluminium alloys (wrought)

Series	Alloy elements	Uniform Corrosion	Pitting	Exfoliation	SCC
1xxx		O	O	O	I
2xxx	Cu	M	P	P–M[a]	VS–R[a]
3xxx	Mn, Mn + Mg	O	O	O	R
4xxx	Si	M	G	G	G
5xxx	Mn, Mg, Cr	O	G	G	I–R
6xxx	Mg, Si	O	G	O	I
7xxx	Zn, Mg, Mn, Cu	M	M	M–P	VS–R[a]

O optimal; *G* good; *M* mediocre; *P* poor; *I* immune; *R* resistant; *S* susceptible; *VS* very susceptible
[a]Depending on thermal treatment

27.4.5 Copper Alloys

Copper and copper alloys show a good corrosion resistance due to a passivation process: in urban atmosphere, a basic sulphate (brocantite) forms, in marine atmosphere, basic chloride (atacamite) forms. Also in waters, copper alloys passivate.

Brass is largely used for seawater applications although suffering from corrosion-erosion, SCC if in presence of ammonia, and selective corrosion (i.e., dezincification).

Brass and bronze are employed in all environments (acidic, neutral and alkaline) provided oxidants, like nitrates and chromates, or complexants, like ammonia, or chemical species reacting with copper, like sulphur, mercury and hydrogen sulphide, are not present.

Table 27.4 reports composition and properties of the main copper alloys.

27.4.6 Titanium and Its Alloys

Titanium passivates by means of its oxide that is highly resistant to oxidizing electrolytes and high chloride concentration. It suffers from corrosion in the presence of fluoride ions and some organic acids because of complexing effects, crevice corrosion in acids at temperature above 70 °C and hydrogen embrittlement for formation of titanium hydrides.

Commercial purity titanium is used in 4 grades (namely, 1, 2, 3 and 4, with different oxygen content, in the range 0.18–0.45%) with tensile strength from 240 MPa for grade 1–550 MPa for solubilized grade 4, which is less ductile and less resistant to corrosion.

Titanium alloys classify in three types: α (compact hexagonal), β (BCC) and biphasic, α + β. The most used alloy is grade 5, biphasic type, Ti6Al4 V (6% Al, 4% V) which finds applications in aerospace industry thanks to its mechanical resistance. Pd containing alloys, typically 0.2% Pd, resist both oxidant and reducing environments due to so-called cathodic alloying (see Sect. 7.1.4).

27.5 General Philosophy for Metal Selection in Industry[1]

A rational approach to material selection for most of industrial process fluids is discussed in the following.

---

[1]This paragraph is based on Giuseppe Faita training Courses.

Table 27.4 Copper alloys for pipes and tube-sheet

Type	Cu	Zn	Ni	Sn	Al	As	Fe	Mn	v_{max}, seawater (m/s)	Notes
Alloys for pipes										
Brass 70/30	70	30				0.04			1	
Admiralty brass	70	29	1			0.04			1.2	Highly resistant H$_2$S
Al brass	76	22			2	0.04			2.5	Good resistance to impingement
CuNi 70/30	68.5		30				0.7	0.8	3	Better resistance to impingement than Al brass Resistant to ammonia and to SCC
CuNi 70/30 modified	66		30				2.0	2	4	Resistant to ammonia and to SCC
CuNi 90/10 modified	87		30				2.0	1	3	Resistant to ammonia and to SCC
Al brass mono-phase	95			5					3	
Alloys for tube-sheet										
Munts metal	58–61	Bal.		0.25				0.15		Suffer dezincification Do not couple with pipes in Al brass
Naval brass	59–62	Bal.		0.5–1	6–8			0.10		
Al brass (type D)	88–92						1.5–3.5			
Type E	78–85		4–7		8–11			0.5–2		Promotes galvanic corrosion on pipes in admiralty brass and Al brass

27.5.1 Alkaline Solutions

In alkaline solutions, most metals and alloys are passive or passivated because of the formation of oxide and hydroxide scales, therefore resisting corrosion. A guideline for material selection is reported in Fig. 27.6.

Carbon and low alloy steels are used in diluted alkalis, with pH in the range 9.5–10.5 as typical in boilers, provided a complete oxygen depletion, at almost all operating temperatures.

In concentrated alkalis, carbon steel is employed at temperatures below 100 °C to avoid amphoteric dissolution; ordinary stainless steels, as 18Cr-10Ni, only resist up to slightly higher temperatures. For high temperatures, up to 350 °C as for melted soda, Ni has to be used.

Attention must be paid to the risk of SCC on carbon and low alloy steels, as well as on austenitic stainless steels, when temperature exceeds 50 °C and tensile stress, even locally, is close to the yield strength, as shown in Fig. 27.7. Stress relieving carried out at about 600 °C, especially after welding, reduces local tensile stress fields, then reducing the risk of SCC occurrence. For austenitic stainless steels, since stress relieving can cause carbide precipitation, the treatment is questionable. Ni and Ni-based alloys are not susceptible of SCC in alkalis at any temperatures.

27.5.2 Chloride-Free Acidic Solutions

Carbon and low alloy steels cannot be used in acidic solutions (pH < 5) unless corrosion inhibitors are added. To select resisting metals, a first check is required: presence or not of chlorides, followed by a second check: presence of nitric acid or other strong acids. Organic acids are considered separately.

Almost all stainless steel grades resist chloride-free acids at pH exceeding 2. At lower pH, i.e., highly concentrated acids, in the presence of oxidizing species, primarily oxygen, high PREN stainless steels are required (i.e., high Cr and Mo content) or low iron content alloys (alloy 20 and nickel-based alloys). As rule of thumb, PREN > 25 and Fe% < 50% should be selected. When oxidizing species are absent, although alloy 20 and nickel-based alloys can resist corrosion attacks, the best choices are alloy 400 and alloy B. However, these two latter alloys fail in the presence of oxidizing species, even if present accidentally for short exposure time.

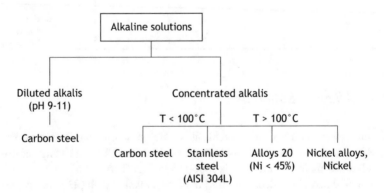

Fig. 27.6 Guidelines for selection of metals in alkalis (from Faita)

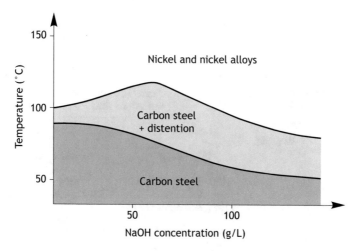

Fig. 27.7 Application map for selection of steels and Ni alloys in alkalis (Graver, 1985)

To resist nitric acid, Mo-free stainless steels have to be used (AISI 304, AISI 347, duplex 2304 and AISI 310 for high temperatures, as low carbon grade (L) and extra low carbon grade (EL)).

Figure 27.8 summarizes the indications to be followed in chloride-free acidic environments.

27.5.3 Chloride-Containing Environments

Figure 27.9 shows the guideline for metal selection in chloride-containing environments. Two conditions have to be considered: alkaline solutions (pH $\geq$ 7) and acidic solutions (pH < 7).

pH $\geq$ 7. For chloride content above 100 ppm and in the presence of oxidants, low PREN stainless steels (below 26) suffer localized corrosion and even SCC (for austenitic grade) for temperature exceeding 60 °C when under tensile stresses. To resist pitting, a stainless steel grade with proper PREN must be used, such as alloy 20 and alloy C. To avoid SCC occurrence, duplex stainless steels, 2205 and 2507 grades, can be used. For high chloride contents, copper alloys and titanium are best choices. Without the presence of oxidants, even with high chloride concentrations, depending on pH, carbon steel, copper alloy and low PREN stainless steels can be conveniently considered.

pH < 7. In practice, since acidic conditions increase aggressiveness, even for relatively low chloride content exceeding 100 ppm the use of highly resisting metals becomes mandatory, as alloy C, B and titanium.

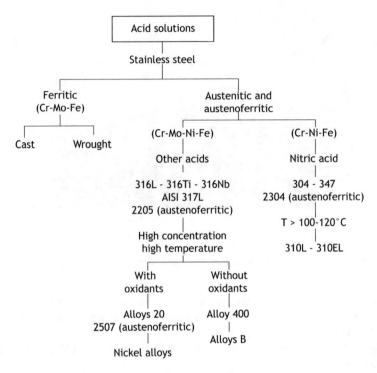

Fig. 27.8 Guidelines for selection of metals in chloride-free acids (from Faita)

27.6 Prevention by Design

Non-homogeneous conditions, either metallurgical or environmental, are the most frequent causes for corrosion attacks, accordingly they must be carefully avoided in design. For example, galvanic couplings, local mechanical stresses, turbulence conditions, local condensations, should be avoided. Some hints are reported in Figs. 27.10, 27.11 and 27.12. Figure 27.13 deals with heat exchangers conditions and provides a practical guide to avoid most common corrosion attacks.

27.7 Prevention in Construction

The construction phase may require mechanical work (for instance plastic deformation), thermal treatments, welding or mechanical assembling, which can cause damages on coatings (if present) or induce residual stress or phase precipitation. Non-destructive tests can be used to check the metal status during construction and before commissioning.

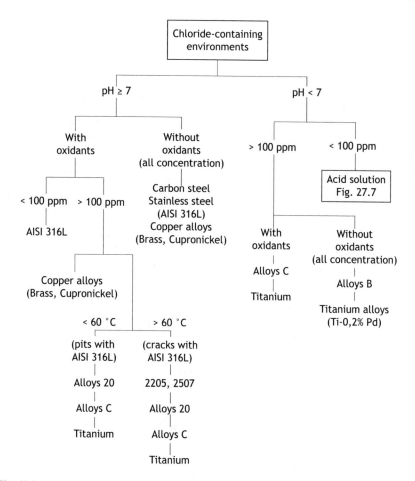

Fig. 27.9 Guidelines for selection of metals in chloride-containing environments (from Faita)

To facilitate final assembling, each component should be properly identified especially when various metals are used.

27.8 Prevention in Storage

Before final assembling, components often require to be stored. Accordingly, attention should be paid to avoid chloride contamination in chloride-containing atmospheres, such as in marine locations.

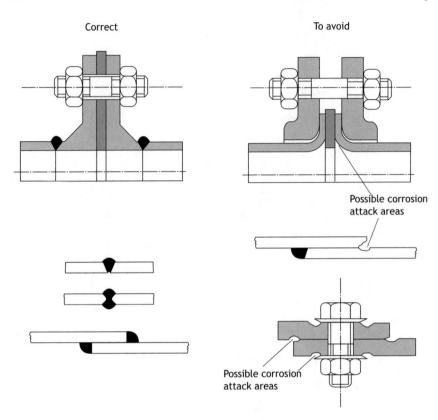

Fig. 27.10 Schematic representation of correct joints and joints to be avoided

27.9   Commissioning and Start-up

Hydrotesting before and during commissioning is twofold important. Firstly, corrosion must be avoided during testing, by employing treated non-corrosive fluids, as inhibited or deaerated and low chloride-containing water. Secondly, it is important to avoid conditions that can trigger corrosion once production has started. This is a typical situation occurring on stainless steel piping in food industry plants where, after hydrotesting, water is not totally drained, so that in some zones entrained water concentrates by evaporation and critical chloride concentration for pitting initiation is reached. At production start-up pitting can eventually proceed, because already initiated in this former phase.

Also in boilers, localized corrosion can happen for similar reasons once operating starts. As it appears evident, attention is crucial during hydrotesting and following drainage operation, which must be thoroughly done by drying with dry nitrogen flow.

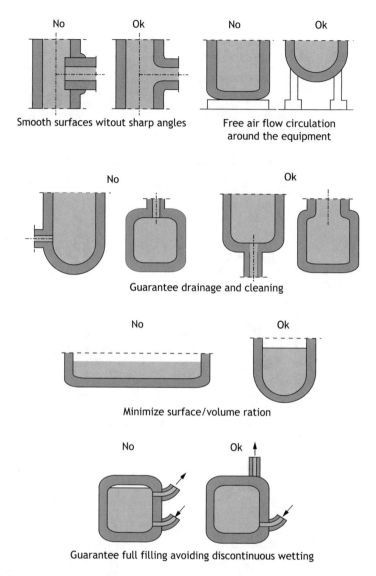

No   Ok   No   Ok

Smooth surfaces witout sharp angles   Free air flow circulation
                                              around the equipment

No   Ok

Guarantee drainage and cleaning

No   Ok

Minimize surface/volume ration

No   Ok

Guarantee full filling avoiding discontinuous wetting

Fig. 27.11 Schematic representation of correct equipment details and elements to be avoided

Start-up is a critical phase when passive conditions are involved, too. Passive films must form uniformly with suitable structure, composition and protective properties. Often, before starting production, a passivating treatment is required; for stainless steels, nitric acid is used, followed by a neutralizing treatment with sodium hydroxide. Another example is the pre-passivation of copper alloys before starting seawater circulation, obtained by a treatment with ferrous sulphate solution, which helps the passivation of brass and nickel-copper alloys.

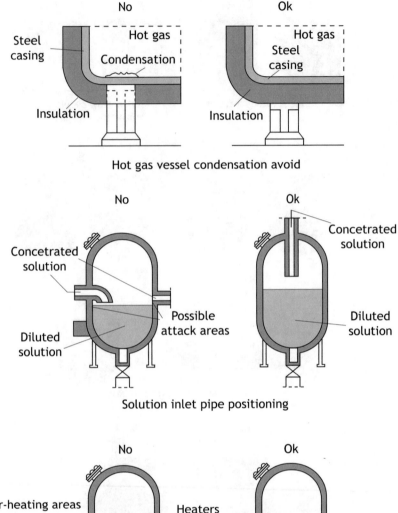

Fig. 27.12 Schematic representation of correct equipment details and elements to be avoided

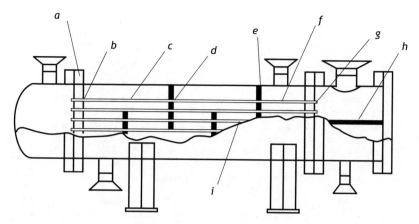

Fig. 27.13 Corrosion forms typical of heat exchangers and their position: a) crevice corrosion under joints; b) crevice corrosion between tube and tube sheet in the back of the tube sheet; c) erosion-corrosion grooves on the outside of tubes due to excessive fluid velocity; d) crevice corrosion or fretting corrosion of tubes in the diaphragms contact area; e) corrosion by differential aeration due to preferential deposition of sludge; f) erosion-corrosion grooves outside tubes; g) corrosion caused by turbulence at the tube inlet; h) corrosion by differential aeration under deposit in the distributor; i) SCC in all positions where tensile stresses originate due to non-allowed thermal expansions or other reasons

27.10 Prevention During Operation

During operating or production, corrosion control is achieved by monitoring and planned inspections. As far as monitoring is concerned, corrosion rate is measured in combination with environment-related parameters, such as temperature, fluid velocity, pH, cathodic reactant concentration, as oxygen, chlorine and others.

27.11 Planned Maintenance

Almost every year, or in specified periods, a corrosion-oriented inspection is carried out on plants. This requires a preparation of the plant and facilities, which normally consists of removal of internal fluids, often followed by a purge with inert atmosphere. When etching is required to prepare internal surfaces for visual inspection, conditioned solutions should be used to avoid damages; generally, an inhibited, acidic, chloride-free solution is recommended, also considering any potential side effect at start-up conditions.

When possible, emptying the plant is discouraged because process fluid is often less corrosive than the uncontrolled transitory conditions set up during fluid discharge. A typical situation is when fluid in entrapped zones concentrates, then leading to local aggressive conditions.

Final Remark

Professor Roberto Piontelli in his pioneering book, Elementi di teoria della corrosione a umido, Ed. Longanesi, 1961, never translated, stated:

L'arte di scegliere correttamente i materiali e i metodi di prevenzione della corrosione in relazione alle condizioni in cui andranno a operare; di evitare accuratamente le cause di eterogeneità o di procedere a neutralizzarne le conseguenze; di evitare le condizioni di spazio morto, le azioni abrasive, di prestare attenzione alla presenza in ambiente anche apparentemente poco aggressivi di sollecitazioni di trazione o a fatica; di controllare le condizioni ambientali chimiche o di moto relativo, e tutte le altre forme di insidia generali o specifiche che una rassegna incompleta come quella sviluppata in questo volume può servire ad additare, a spiegare, a riconoscere; ecco il primo requisito di un corrosionista.

The art of selecting correct materials and corrosion prevention methods on the basis of operating conditions; of avoiding heterogeneity or select prevention measures; of avoiding recesses, abrasion and turbulence conditions and paying attention to mechanical stresses and fatigue also for fairly aggressive fluids; of specifying chemical parameters and flow regime and foreseeing general and specific insidious events, some of them this book illustrate; all this is the primary role of a corrosion engineer.

27.12 Questions and Exercises

27.1 As reported in Sect. 27.2.5, the bottom of a chemical reactor was perforated in a few months because of an unexpected chloride contamination. Explain the pitting mechanism occurred and the prediction of time to perforation. Highlight a strategy to avoid such inconvenient.

27.2 Draw Evans diagrams for alloy B and C in the presence and absence of oxidants.

27.3 Try to demonstrate why Cu containing series 2xxx suffers corrosion, and Mg-containing series, 3xxx and 5xxx, resist corrosion basis on separated phases, which are more noble in series 2xxx and less noble in the others.

27.4 Which are the candidate materials for a heat exchanger? The cooling fluid is seawater. Distinguish among water box, tube sheet and tubes. Justify your selection. Which corrosion phenomena do you want to avoid?

27.5 A water injection plant is used to inject at high-pressure huge quantities of water into the reservoir in order to increase oil recovery from an oil and gas formation. The following five (5) components may be identified: water supply pump (lifting pump); flow-line from supply well area to injection area (with a length from 100 m up to few km); booster pump to increase pressure; distribution system (manifold); injection wells. Considering the use of carbon steel, please for each plant unit indicate and justify the corrosion-related problems considering the three types of waters:

- Low salinity water from supply wells (for instance, fresh water, underground water table or river water)
- High salinity water, such as formation water from hydrocarbons (Total Dissolved Salt higher than 200 g/L, as NaCl)
- Seawater

27.6 For the same plant of Ex. 27.5, make a justified material selection in order to avoid any corrosion problem, considering also the possibility to add treatments such as oxygen removal (physical and/or chemical treatment), anti-bacteria treatment, erosion control system.

27.7 A plant is composed of two distinct sections. Section one for the treatment of an acidic aqueous solution; section two for the treatment of a basic aqueous solution. Each section is constituted by a tank, a pump, a heat exchanger and a mixer for the dilution with water. The two streams are finally mixed in a reactor to obtain a neutral salt soluble in water. For at least one of the two sections indicate: materials for the main components (tanks, pumps, transfer pipes, heat exchangers) justifying their choice (taking into account chlorides, salinity, pH, oxygen, chlorine, etc.); any corrosion prevention treatments. Operating conditions: atmospheric pressure, tanks at room temperature, acidic solution at 40 °C, basic solvent heated at 100 °C, demi water used for preparation and dilution; final product temperature 80 °C.

27.8 Market lead companies to replace in heat exchangers copper with aluminium for both cost and material issues. When it possible to do this?

27.9 What kind of insulated panels (base metal) would you use for zootechnical applications?

27.10 After production of a carbon steel fire protection system, a hydrotest has been performed to verify the quality of the welds. Which water you would suggest to use? After the test, what would you suggest to do for the internal water: complete or partial emptying, constant pressurized filling? Justify your answer. If the fire system piping is in stainless steel, do you suggest the same water and the same after-test treatment?

Bibliography

Fontana M (1986) Corrosion engineering, 3rd edn. McGraw-Hill, New York, NY
Graver D (1985) Corrosion Data Survey, NACE, Houston
Piontelli R (1961) Elementi di teoria della corrosione a umido dei materiali metallici. Longanesi, Milano (in Italian)

Chapter 28
Monitoring and Inspections

In the Venetians' arsenal as boils
Through wintry months tenacious pitch, to smear
Their unsound vessels; for the inclement time
Seafaring men restrains, and in that while
His bark one builds anew, another stops
The ribs of his that hath made many a voyage,
One hammers at the prow, one at the poop,
This shapeth oars, that other cables twirls,
The mizzen one repairs, and main-sail rent;

Dante, *Inferno*, XXI, 7–15

Abstract Corrosion of industrial equipment can lead to risk conditions for economic losses and safety of personnel. To minimize that risky, besides the correct selection of material and the use of proper preventative methods, the design of a corrosion monitoring system plays an important role, followed by programmed inspections during operation. The analysis of monitoring and inspection results is also important to plan maintenance activities. This Chapter deals with the selection of the correct monitoring strategy and its application to operating systems: the most common monitoring techniques, as the use of corrosion coupon, electrical probes, linear polarisation resistance, galvanic probes, corrosion potential are presented. A brief outline of the most important non-destructive techniques is also reported.

Fig. 28.1 Case study at the PoliLaPP Corrosion Museum of Politecnico di Milano

© Springer Nature Switzerland AG 2018
P. Pedeferri, *Corrosion Science and Engineering*, Engineering Materials,
https://doi.org/10.1007/978-3-319-97625-9_28

28.1   Corrosion Monitoring

To operate safely in the presence of corrosive fluids, it is necessary to control the corrosion process by following how it proceeds over time, based on the expected damage as, for instance, uniform thinning or localized attack (Fig. 28.1). This activity is known as corrosion monitoring.

Corrosion rate is the main and typical parameter measured, continuously or periodically; it is also used to optimize the dosage of corrosion inhibitors. Type of measurement, frequency and devices depend on the expected corrosion attack. The following classification can be adopted:

- *Thinning*: measurement of corrosion rate by means of coupons or electro-chemical techniques
- *Localized corrosion*, as galvanic corrosion or pitting and crevice of stainless steels: measurement of electrical parameters, i.e., potential and current
- *Cracking due to SCC or corrosion-fatigue*: unlike other corrosion forms, no monitoring techniques are used, although sophisticated methods such as electrochemical noise and acoustic emission have been proposed.

Corrosion monitoring greatly contributes to increase the safety level of a plant by gathering important information on the corrosion process that takes place in the period between inspections, which are usually planned during shutdowns; moreover, corrosion monitoring is a very cheap activity compared to an inspection.

Nowadays, available monitoring systems deal with general corrosion in easily accessible locations, only; therefore a lot must be done for either localized corrosion or harsh operating conditions, for example at high temperature or in deep oil and gas wells. Most recent progress is mainly related to hardware or software for electronic devices to gather and elaborate data, then improving on-line remote control.

Corrosion monitoring is essentially based on intrusive systems that require the insertion of sensors or devices inside equipment, vessels, reactors or piping, to be exposed to the process fluid. A non-intrusive method is based on the measurement of residual thickness from the accessible external surface by ultrasound technique. As a general philosophy, corrosion monitoring should be based on at least two methods or measuring techniques to increase the reliability of gathered data; at least the corrosion coupon technique should be used.

Corrosion rate depends on many factors, as environment and metal chemical composition, operating conditions (temperature, pressure and flow rates), presence of galvanic couplings. Accordingly, a monitoring system can be based on the measurement of one of these factors, as temperature to check conditions favourable to SCC, or concentration of corroding metals as iron in the fluid, or of species that provide the cathodic process (pH, oxygen content, chlorine).

28.1.1 Selection of Monitoring Locations

Only a few points of a plant are monitored for either technical availability or economics. Selected locations have to be representative of the plant, at least according to a criterion of conservativeness (most severe corrosion condition). For instance, the most adopted criteria to select monitoring posts are the following:

- Zones where turbulence is higher (for instance, downstream elbows or sharp cross section variations) in case of risk of erosion-corrosion or in general where the corrosion rate is under of oxygen diffusion control
- Stagnant zones, in case of pitting of stainless steel, since initiation is favoured
- Presence of galvanic couplings (hazardous contact with different materials)
- High stress zones or zones with different microstructure (as in case of welds) in case of risk of SCC
- Areas where corrosion inhibitors, or chemicals in general, are added
- Anodic zones in case of buried structures, which can be affected by stray currents.

Once a monitoring point has been selected, the exposure to the fluid must also be considered, such as centre or bottom or upper position inside a line, taking into account any possible stratification of phases.

28.2 Common Monitoring Methods

To monitor a corrosion process, two distinguished approaches are followed:

- *Direct methods* that provide direct measurement of corrosion rate. This is achieved through the use of corrosion coupons and spools
- *Indirect methods*, which allow the calculation of the corrosion rate through the measurement of parameters useful to assess the corrosion process, such as electrochemical techniques (potential and linear polarization resistance measurements) when fluid is an electrolyte, electrical resistance probe and hydrogen probe.

28.2.1 Corrosion Coupon

A corrosion coupon is a specimen made of the same construction metal with suitable shape, such as a strip or a disk. The disk shape coupon better reproduces the real exposure conditions, since it replaces a portion of the equipment wall. Coupons are inserted inside the process lines for a fixed period, then recovered for visual examination and weighing, for calculation of average corrosion rate.

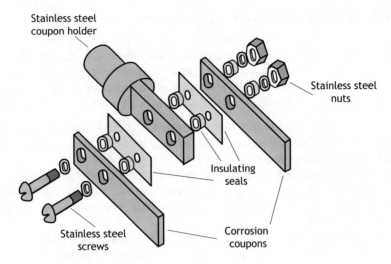

Fig. 28.2 Assembly of strip specimen type

Insertion and recovery of corrosion coupons are carried out either during shutdowns or during operation by suitable pressurized retrievable devices. Attention must be paid to two aspects:

- The risk of galvanic contact with the coupon holder, typically made of stainless steel, if not electrically insulated (Fig. 28.2)
- The metal composition, which has to be the same of the exposed one; for carbon steel structures, mild steel or pure iron is usually used for a more conservative approach.

The main drawback of this method is that the information on corrosion rate is delayed and available once corrosion has already occurred, because necessarily exposure time must exceed one month or so to allow mass loss measurement. Figure 28.3 shows schematically two types of corrosion coupons: two strips immersed in the process fluid and a disk flash mounted at the bottom of the pipe.

28.2.2   Corrosion Spool

A spool is a span of line usually 0.5–1 m long, inserted by flanges, which can be retrieved periodically during a shutdown or by means of a dedicated by-pass for visual examination. Compared to the use of corrosion coupons, it shows the advantage to reproduce the true exposure condition, while showing the same disadvantage of providing only delayed information.

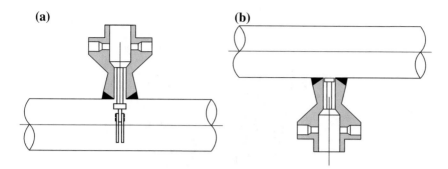

Fig. 28.3 Types of corrosion coupon: **a** strip; **b** flash-mounted type (used when water separates at the bottom of a line)

28.2.3 Electrical Resistance Probe

The electrical resistance probe consists of a metal coupon of the same material of the pipe, which allows the calculation of corrosion rate through the measurement of its electrical resistance (Fig. 28.4). As corrosion proceeds, the electrical resistance of the coupon increases because the corrosion attack reduces its cross section area, according the second Ohm's law. By plotting the electrical resistance against time, a straight line is obtained with a slope proportional to corrosion rate. Advantages of the method are many: no retrieve operation is required because the measurement is easily carried out externally; corrosion rate (i.e., the slope) can be calculated after very short exposure time (a few days is enough) and it works in any H_2S-free environment. In the presence of hydrogen sulphide, the electrical resistance is influenced by the formation of iron sulphide, which has an electronic conductivity, therefore readings are erratic. Since resistance is temperature-dependent, the device is provided with a calibration system to take into account the effect of temperature. The main drawback of this method is that in case of very low expected corrosion rate, the thinning of the coupon and consequently the electrical resistance variation are small, so that a great accuracy of the electrical instrumentation is required.

28.2.4 Linear Polarisation Resistance

LPR gives the instant corrosion rate, as will be discussed in Chap. 29 for laboratory testing techniques. The method can be applied only in the presence of an electrolyte, for instance water-handling plants. Two devices are available (Fig. 28.5): a three-electrode device and a two-electrode one. The latter system is often used because simpler: counter and reference electrodes are made of same material.

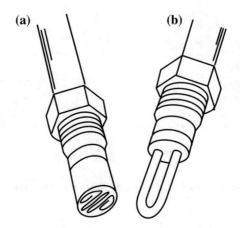

Fig. 28.4 Electrical resistance probe: **a** flash-mounted type; **b** coupon type

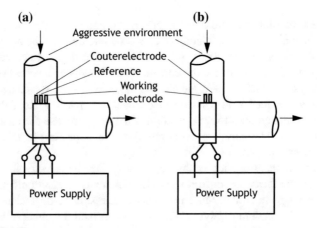

Fig. 28.5 LPR measurement on field: **a** three-electrode device; **b** two-electrode device

Sometimes as counter-electrode the base metal of the structure is used. Working electrode is made of same metal of the line. Figure 28.6 shows some types of electrodes: flash-mounted (Fig. 28.6a) and strip type (Fig. 28.6b), where the former is advantageous because reproducing the real exposure conditions and avoiding risk of obstruction.

28.2.5   Galvanic Probe

This simple device is used when the cathodic process is oxygen and chlorine reduction in waters. The cathode is gold plated to limit either overvoltage or

Fig. 28.6 LPR device: **a** flash-mounted type; **b** strip electrode type

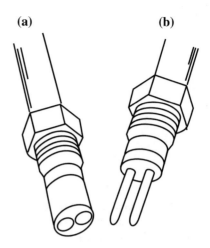

(a)   **(b)**

scaling. The parameter measured is the current flowing between line and golden specimen, by means of a zero-ammeter. The sensitivity of the method is 0.05 ppm of oxygen content (i.e., about 0.6 μm/year corrosion rate of mild steel) corresponding to a current of 1 μA on a cathode with 20 cm^2 as surface.

28.2.6   Potential Measurement

The potential measurement of a metallic structure is simple and provides information about the degree of corrosion or protection of the metallic component. For instance, the potential measurement is used to verify the cathodic protection of metallic structures, as buried pipelines or tanks, to control the passive condition of concrete reinforcements, or to measure the polarization of structures in zones affected by stray currents. The potential measurement is closely related to the corrosion rate if correctly interpreted. For instance, considering mild steel in neutral-pH condition where it shows an active behaviour, a potential variation of 100 mV corresponds to a variation of corrosion rate of about one order of magnitude, being the Tafel slope about 100 mV/decade.

The potential measurement is carried out by means of a high impedance voltmeter and a reference electrode. An ideal voltmeter has infinite internal impedance, meaning that it would draw zero current within the circuit, hence not affecting the measurement. Truly, the current flowing in the measurement circuit is not zero, instead is settled by the potential to the circuit resistance ratio, where the resistance practically coincides with the voltmeter impedance. An impedance higher than 10 MΩ is recommended in order to avoid the polarization of the reference electrode, as discussed in Chap. 19, Sect. 19.2.3.

Fig. 28.7 Scheme of
measurement of potential

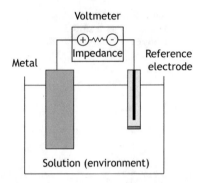

The reference electrode, connected to the negative pole of the voltmeter, shall maintain its potential constant and it is placed in contact with the environment of the structure, while the latter is connected to the positive pole (Fig. 28.7).

The reference electrode is composed of a metal strip immersed in an electrolyte. On its surface there is an electrochemical equilibrium that determines a potential, which remains constant provided there is no current circulation on its surface during measurement. Table 28.1 lists the reference electrodes usually used to measure potential in the laboratory and in the field:

- Ag/AgCl/seawater (SSC) and Zn/seawater (ZN) for seawater applications
- Cu/CuSO$_4$ (CSE) for soil applications
- Manganese dioxide MnO$_2$ (MN) and mixed-metal-oxide activated titanium (MMO) for reinforced concrete structures.

Table 28.2 reports reference electrode equivalencies.

Potential measured through a reference electrode depends on electrode position with respect to the structure and on the presence of circulating current. The presence of ohmic drop into the electrolyte may affect the reading and its correct interpretation, as in the case of CP monitoring: for details refer to Chap. 19.

28.2.7 Bio-probe

This device was introduced in early 1990s when it was realized that biofilm in seawater, where pitting corrosion on common stainless steels is a threat, produces on stainless steel an ennoblement of oxygen reduction cathodic process up to 300 mV (i.e. a the reduction of the activation overvoltage) then exceeding pitting potential and triggering pitting initiation (see Chap. 11, Sect. 11.1). The device consists of an electrode made of stainless steel and a reference electrode (typically, in seawater iron is used) for continuous measurement of potential: an increase in the reading indicates the setup of biofilm and therefore the high probability of pitting initiation on susceptible stainless steels or a signal that biocide water treatment failed.

Table 28.1 Reference electrodes used for potential measurement

	Type of electrode (electrode reaction)	Potential at 25 °C (V vs SHE)	Use and notes
SCE	**Calomel** $(Hg/Hg_2Cl_2; Cl^-)$ $Hg_2Cl_2 + 2e^- \rightarrow 2Hg + 2Cl^-$	$E = 0.268 - 0.0591 \cdot \log [Cl^-]$ KCl_{sat} $E = 0.244$ Temp. Coeff. $= -0.65$ mV/°C	Used in laboratory as reference electrode to calibrate reference electrodes used in field
SSE	**Mercury/sulphate of mercury** $(Hg/Hg_2SO_4; SO_4^{2-})$ $Hg_2SO_4 + 2e^- \rightarrow 2Hg + SO_4^{2-}$	$E = 0.615 - 0.0295 \cdot \log [SO_4^{2-}]$ K_2SO_4 sat. $E = 0.710$	Used in laboratory
SSC	**Silver/silver chloride** $(Ag/AgCl; Cl^-)$ $AgCl + e^- \rightarrow Ag + Cl^-$ Obtained by anodic behaviour in a NaCl solution	$E = 0.222 - 0.059 \cdot \log [Cl^-]$ 0.1 M KCl $E = 0.288$ 1 M KCl $E = 0.222$ Seawater $E = 0.250$ Temp. Coeff. -0.6 mV/°C	Used in seawater also as fixed reference electrode, although sensitive to pollution and unstable. It requires periodical refreshment (anodic behaviour). In brackish waters reference value must be calculated (Cl^- concentration dependence)
CSE	**Copper/copper sulphate** $(Cu/CuSO_4; Cu^{2+})$ $Cu^{2+} + 2e^- \rightarrow Cu$ Consisting of a copper rod immersed in a saturated $CuSO_4$ solution. The ionic contact is achieved by saline bridge made of wood or porous ceramic	$E = 0.34 - 0.0295 \cdot \log [Cu^{2+}]$ Saturated $CuSO_4$ $E = 0.318$ For practical use $E = 0.3$	Used in soils. Easy to prepare. Very stable. Chlorides may contaminate the electrode, hence its use in seawater is not recommended. Used in soil as fixed reference electrode

(continued)

Table 28.1   (continued)

	Type of electrode (electrode reaction)	Potential at 25 °C (V vs SHE)	Use and notes
ZN	**Zinc/seawater** Free corrosion reaction	E = −0.80 (Zn free corrosion potential)	Used in seawater as fixed reference electrode. Also used in soil as fixed reference electrode
MN	**Manganese dioxide** (MnO_2; KOH electrolyte, pH 13)	E = 0.35	Used in concrete as fixed reference electrode
MMO	**Activated titanium** Titanium coated with mixed metal oxides of noble metals (Ir, Rh). Alkaline electrolyte at constant pH and oxygen	E = 0.20 (embedded in concrete, pH $\cong$ 13)	Used in concrete and soils as fixed reference electrode provided it is encapsulated in constant electrolyte. At constant pH it may be used as oxygen probe

Table 28.2   Equivalencies of common reference electrodes (mV)

	SHE	SSC	SCE	CSE	ZN	MN	MMO
SHE	0	−250	−240	−300	+800	−350	−200
SSC	+250	0	+10	−50	+1050	−100	+50
SCE	+240	−10	0	−60	+1040	−110	+40
CSE	+300	+50	+60	0	+1100	−50	+100
ZN	−800	−1050	−1040	−1100	0	−1150	−1000
MN	+350	+100	+110	+50	+1150	0	+150
MMO	+200	−50	−40	−100	+1000	−150	0

Example Value vs CSE = value measured vs RE used + value of column CSE

28.2.8   Hydrogen Probe

It works in acidic solutions where cathodic process is hydrogen evolution if a poison of atomic hydrogen recombination to form hydrogen molecule is present. The principle is as follow: a high percentage of atomic hydrogen produced by the cathodic process (i.e., by corrosion) enters steel because of the poison, typically hydrogen sulphide, H_2S. Atomic hydrogen diffuses into steel and is captured in a

Fig. 28.8 Hydrogen probe (principle)

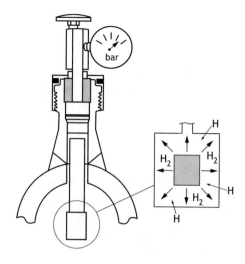

suitable trap. Within the trap, atomic hydrogen combines to form hydrogen molecules, then producing an increase in pressure, which can be measured by a pressure gauge; the rate of increase in pressure is proportional to the corrosion rate. Figure 28.8 illustrates schematically what a hydrogen probe consists of: in practice, it is made of a hollow cylinder made of mild steel, in which hydrogen accumulates.

There is another device, the *patch pressure probe*, which measures diffusing hydrogen directly. It consists of an electrochemical cell, directly fixed to the external surface of a line, where the line surface works as anode, the electrolyte is an alkaline solution and the cathode is a metal with low overvoltage for hydrogen evolution. As soon as diffusing atomic hydrogen reaches the cell, it is oxidized to H^+ by imposing a voltage between anode, i.e., external surface, and cathode, where H^+ is reduced to hydrogen: the current flowing in the cell measures the diffusing hydrogen, or in other terms, corrosion rate.

Hydrogen probe is typically used in oil and gas industry to monitor corrosion in H_2S-containing process fluids, where electrical resistance probe does not work. Rather than to measure corrosion rate, it is more often used to drive the dosage of corrosion inhibitor.

28.3 Other Methods

Methods used in laboratory testing, such as electrochemical noise, electrochemical impedance spectroscopy and acoustic emission, were proposed for field applications. Except for acoustic emission, the other two methods are still in a pioneering stage.

28.3.1 Electrochemical Noise

The method consists of the continuous measurement of potential and current exchanged between two identical coupons. It works better for passive metals, because current fluctuation increases when passive film breakdowns as localized corrosion starts. It can indicate when operating conditions change to trigger localized corrosion. It has been used in water cooling circuits.

28.3.2 EIS (Electrochemical Impedance Spectroscopy)

This method was tested on plants with deluding results; therefore, it is still at an experimental stage. It seems to be promising to monitor coating efficiency and check the passive film breakdown and to measure corrosion rate and inhibitor efficiency.

28.3.3 Acoustic Emission

It is used in hydro-testing of equipment operating at high pressure. The goal of the test is to localize the presence of defects within the metal, which may grow if they exceed the critical size. This is important because a defect can grow during testing without leading to unstable rupture, which can happen later during operation. To localize the growing defect, at least three sensors have to be used to allow triangulation of signals. The size of the defect is successively obtained through ultrasonic testing to define if defect size is critical, then deciding its removal.

28.4 Plant Inspection

Inspection of plants includes:

- *Visual inspection* of exposed surfaces before and after a careful surface cleaning (sandblasting would be necessary) to check the presence of any corrosion attack, either generalized or localized, and cracking. Visual inspection is a preliminary activity which is carried out before proceeding with others, with the aim to optimize time and efforts
- Use of *Non-Destructive Techniques* (NDT) for residual thickness measurements or detection of cracks by means of various methods as magnetic particles, dye penetrant, X-ray or γ-ray, ultrasonic testing, eddy currents.

The planning of inspections follows the *Risk Based Inspection* approach, which consists of the evaluation of risk (technical and economical) associated to the occurrence of a corrosion event. A so-called risk matrix is used to help prioritizing the inspection types; the risk matrix consists of two axes, one for the probability of the event (i.e., corrosion) occurrence and the second one to set the entity of consequences if the considered event takes place.

28.4.1 Liquid Penetrant

The use of liquid penetrant is a possible way to improve visual inspection. After a surface cleaning to remove oxides and solid contaminants, a solution with surface-active and coloured liquid is sprayed to form a continuous film. The sprayed solution penetrates inside all discontinuities open to the surface, as porosity, cracks, shrinkage areas and laminations. Excess penetrant is removed from the surface and the liquid, which has entered discontinuities, is made visible by a developer. Different penetrants and different developers are used, in particular there are visible penetrants which produce red contrast at visible light or fluorescent penetrants that are normally green and that can be seen using ultraviolet light. The liquid penetrant testing has a very high sensibility, showing cracks with width even lower than 0.1 mm.

28.4.2 Magnetic Particles

Another method used to improve visual inspection is the Magnetic Particle Inspection (MPI). It can only be applied to ferromagnetic materials and allows detecting surface or sub-surface discontinuities. The tested component is magnetised to produce magnetic field on the surface. A surface or sub-surface discontinuity, which lies in a direction transverse to the direction of the magnetic field, creates a distortion of the magnetic field lines. The distortion can be made visible by means of small magnetic particles sprayed over the material surface. A defect, which lies transversally the magnetic field lines, causes a large distortion of the lines and gives a strong signal.

28.4.3 Radiographic Testing

Radiographic testing is based on the property of high-energy electromagnetic waves to penetrate thick materials. A source of X-rays or γ-rays is placed on one side of the material to be tested, a photographic film is placed on the other side. The radiation is partially absorbed by the tested material and more radiation will pass through a region where material is thinner or a cavity is present, less radiation where material is thicker or a denser phase is present. The transmitted radiation is recorded on the film that is developed as negative: then a darker area indicates more incident radiation, i.e., thinner or less dense material, a lighter one indicates less radiation, i.e., thicker or denser material.

X-ray and γ-ray are used to detect both internal and surface defects, as blisters, voids and welding defects.

28.4.4   Ultrasonic Testing

The most used NDT is the ultrasonic testing (UT) because it is easy, fast and precise. Figure 28.9 illustrates the principle of the method and a practical example of application. A beam of high frequency sound waves is introduced by an UT transducer (the Transmitter) into the test material for the detection of surface or internal flaws. The sound waves travel through the material with some loss of energy (attenuation) and they are reflected at the interfaces. The reflected beam is received by another UT transducer (the Receiver), displayed on a screen and then analysed to define the presence and location of flaws or discontinuities.

The UT probe must be coupled to the test object surface with a coupling medium, normally a liquid or a gel, which allows transmission of elastic waves from the probe to the piece and vice versa. UT results can be visualized in different ways: the simplest way, called A-scan, is the representation on a time axis of the amplitude of reflected waves. The B-scan represents defects on a plane thickness-probe travelling direction and it gives information on the position of a defect in the specimen thickness.

High-speed automatic ultrasonic inspection systems have been developed to improve UT performance in particular in welding control during plant construction when the weld geometry is regular and repetitive, e.g., pipeline girth welds or plate welding of storage tanks. Automated UT is faster and more reproducible and reliable than manual UT and data can be recorded as well as processed in more complex ways, like 3-D tomography.

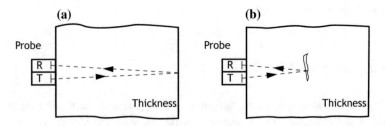

Fig. 28.9 a Principle of UT for thickness measurement; **b** example of thickness measurement by UT

The inspection of pipelines is performed by the use of tailored tools, called *intelligent PIG (Pipeline Inspection Gauge)*, based on UT method or electro-magnetic principle to measure continuously the thickness and check the presence of cracks or corrosion attacks.

28.4.5 Eddy Current Method

Eddy current inspection is used in a variety of industries to detect surface and near surface defects. In fact, on thin materials, as tubing and sheet stock, eddy currents can be used to measure the thickness of the material. This makes eddy current a useful tool for detecting corrosion damage and other damages that cause a thinning of the material. The technique is typically used to make corrosion thinning measurements in the tubing walls in heat exchangers. Eddy current testing is also used to measure the thickness of paints and other coatings.

In eddy current testing, a circular coil carrying an AC current is placed in close proximity to an electrically conductive specimen. The alternating current in the coil generates a changing magnetic field, which induces eddy currents within the wall. Variations in the phase and magnitude of these eddy currents can be monitored using a second "search" coil, or by measuring changes to the current flowing in the primary "excitation" coil. Variations in the electrical conductivity or magnetic permeability, or the presence of any flaws, will cause a change in eddy current flow and a corresponding change in the phase and amplitude of the measured current. This is the basis of standard (flat coil) eddy current inspection, the most widely used eddy current technique.

28.5 Applicable Standards

- API 1104, Welding of pipelines and related facilities, American Petroleum Institute, Northwest Washington, DC.
- API RP 579, Fitness-for-Service, American Petroleum Institute, Northwest Washington, DC.
- BS 7910, Guide on methods for assessing the acceptability of flaws in metallic structures, British Standard.
- EN 473 Non-destructive testing. Qualification and certification of NDT personnel, European Committee for Standardization, Brussels.
- EN 13455, Unfired pressure vessels. Part 5 - Inspection and testing, European Committee for Standardization, Brussels.
- ISO 9712, Non-destructive testing. Qualification and certification of personnel, International Standard Organization, Geneva, Switzerland.
- SNT-TC-1A, Recommended practice, personnel qualification and certification in non-destructive testing.

28.6 Questions and Exercises

28.1 Lists pros and cons of direct and indirect monitoring methods.

28.2 Which forms of corrosion do you monitor by corrosion coupons?

28.3 Describe the principle of the potential measurement. What devices do you need? With what features?

28.4 The free corrosion potential profile of a buried pipeline, 5 km long, has been measured, a reading every 200 m. Based on the profile, it is possible to estimate the corrosion rate? Why? Make some numerical examples.

28.5 Imagine to calibrate an SCE electrode with an SSC (1 M KCl) electrode. Which is the potential difference in absolute value? And between a CSE and ZN electrode?

28.6 A corrosion inhibitor has been added in a pipeline. How LPR can be used to check its efficiency and to optimise its dosage?

28.7 Both corrosion coupon and LPR are used to estimate corrosion rate. Which are the main differences between the two methods? When LPR should be preferred to coupon?

28.8 Lists the benefits of non-destructive testing.

28.9 Choose a non-destructive technique to verify internal defects in the welds. Describe how it works and which information are detectable. Can penetrant liquid be used? Why?

28.10 Why intelligent PIG are periodically used on pipelines transporting oil?

28.11 A 50 km long cladded pipe (base metal carbon steel, clad in alloy 625) is used to convey a sour fluid from a platform to the onshore plant. All circumferential welds has to be checked. Which NDT would you suggest? Why? Which information each technique is able to give?

28.12 The internal bottom of a carbon steel tank (40 m in diameter) suffered generalized corrosion due to the presence of a corrosive liquid. How to monitor the residual thickness? Has the measurement to be performed on the entire internal surface? How to select the area to be analysed?

Bibliography

1. API RP 580, Risk-based inspection methodology. American Petroleum Institute, Washington, DC
2. API RP 581, Risk-based inspection methodology. American Petroleum Institute, Washington, DC
3. ASM Metals Handbook (2001) Non-destructive evaluation and quality control, vol 17. ASM International, Northern New England
4. Hellier C (2003) Handbook of nondestructive evaluation, 2nd edn. McGraw-Hill Education, New York
5. Lazzari L, Pedeferri P (2006) Cathodic protection. Polipress, Milan, Italy
6. Shull PJ (2001) Nondestructive evaluation. Marcel Dekker, New York, NY

Chapter 29
Testing

There's something with these tears
Turning me to rust.

Echo & the Bunnymen

Abstract Corrosion tests are an important instrument used to clarify the mechanisms of corrosion process, to develop new materials and new methods of protection, to carry out quality control tests, to follow the behaviour of materials in operation and, finally, when corrosion has occurred, to study the causes and the remedies. The classification of corrosion tests adopted in this chapter provides the division in two macro categories: exposure tests and electrochemical tests. This chapter wants to give some examples of the many possibilities of existing corrosion tests.

Fig. 29.1 Case study at the PoliLaPP Corrosion Museum of Politecnico di Milano

© Springer Nature Switzerland AG 2018
P. Pedeferri, *Corrosion Science and Engineering*, Engineering Materials,
https://doi.org/10.1007/978-3-319-97625-9_29

29.1   Test Classification

Corrosion tests can be classified in different ways as witnessed in the specific bibliography which does not acclaim any univocal classification. The most general classification distinguishes the tests according to form of corrosion, environments (natural and industrial) or materials, or again in short, medium and long-term tests (accelerated or in service tests). To validate the metal selection carried out in design, exposure testing is often required. It consists of testing candidate metals in process fluid, following two conditions: same process condition (in service) or so-called accelerated conditions (Fig. 29.1).

The accelerated tests clearly contain the quality control tests and pass-fail tests. The first type of test consists of standard tests required and carried out to check compliance with acceptance criteria. Test conditions can be in accordance with international standards, as CEN, ASTM, ISO, NACE, national regulation, as UNI in Italy, or according to company standards. For example, the most common and known are Huey Test, Strauss Test and Streicher Test for intergranular corrosion susceptibility of stainless steels and nickel alloys, or $FeCl_3$ tests for pitting corrosion. Many localized corrosion attacks, follows a behaviour described by a *pdf* of Log N type or very close to it. This implies that the occurrence of the attack happens during the first period of exposure (also said infant mortality) rather than after long exposure time. In other words, should the attack take place, this does occur very soon or never. Based on this behaviour, so-called pass-fail tests are performed for a pre-determined fixed exposure time: test is passed if within the exposure time no failures, or damage below a threshold, occurred.

The classification adopted in this chapter provides the division in two macro categories: exposure tests and electrochemical tests. The exposure tests consist of direct exposure of the metal to an environment that simulates real aggression conditions, or makes it more severe. Exposure can take place in natural environments as well as artificial environments (for example climatic chamber, autoclave...), in immersion (continuous or intermittent) in natural or synthetic (aggressive) solutions. Electrochemical test are the basic laboratory tests that provide the determination of basic corrosion-influencing parameters that are published in journals and books: thermodynamic and kinetic parameters, as anodic and cathodic characteristics, corrosion potential in referred electrolytes, tendency to passivation, oxidant power. Since test conditions are rigorously controlled, results show a high reproducibility.

This chapter far from being exhaustive describes the most important tests, their purposes and the possible applications.

29.2 Accelerated Tests and Statistics

Aim of corrosion test is the achievement of a result in short time, if possible, and the extrapolation to a longer time as design life or even greater. Figure 29.2 illustrates the simplest meaning of an extrapolation from laboratory testing results to the shorter time.

To accelerate the test, a corrosion-related parameter or factor is conveniently increased. Extrapolation is possible, hence acceptable, if the following conditions fit:

- Corrosion mechanism is the same (for instance, same cathodic process)
- There is a relationship between rate (i.e., time) and varied parameter
- Probability density distribution of results is the same.

Last condition derives from reliability and is clearly represented in Fig. 29.3. In normal operating, no failure is expected, instead by increasing an influencing parameter failure occurs because resistance is overwhelmed. Arrhenius firstly recognized that in accelerated tests the relationship between time and the affecting parameter is logarithmic as follows:

$$MTTF = A \cdot e^{-B(\sigma - C)} \tag{29.1}$$

MTTF is generic time-to-failure (or testing time); A, B and C are experimental constants and σ is the affecting parameter. The plot in semi-logarithm scale, ln *MTTF* versus σ, is depicted in Fig. 29.2:

- Affecting parameter, σ, varies in the range $\sigma > C$
- When $\sigma = C$, MTTF > A
- B is slope of the straight line.

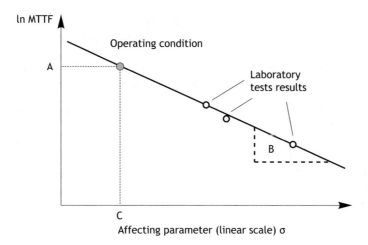

Fig. 29.2 Principle of extrapolation from testing results to operating conditions (from Lazzari 2017)

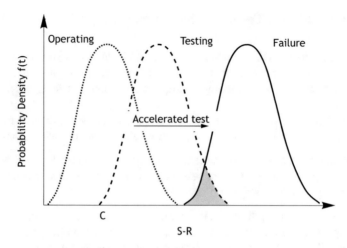

Fig. 29.3 Comparison of probability density distribution of failure-related parameters in testing and operating (from Lazzari 2017)

MTTF is a distribution. In Fig. 29.2 the relevant value is plotted, as mean in case of uniform distribution or most likely a minimum or maximum value given by the extreme value statistics, generally Weibull or Gumbel. Statistics is described in Chap. 30. For more details on the use and interpretation of accelerated tests, refer to Lazzari (2017).

29.3   Exposure Tests

Tests are carried out by direct exposure of metal coupons to the environment, which can be the same as in service or modified to increase aggressiveness (accelerated tests).

Typical parameter measured is mass loss for uniform corrosion. In localized corrosion related testing, other parameters are measured or controlled:

- In pitting corrosion tests: initiation time, critical chloride concentration, critical pitting temperature (CPT), pit depth, pit density
- In crevice tests: percentage of sites attacked, depth of attacks, critical crevice temperature (CCT)
- In intergranular tests: presence of attacks, checked by micrographic examination and quantified by mass loss
- In SCC tests: presence of cracks or time-to-failure.

To test coating and painting performance, the standardized salt spray test (also known as salt fog test) is often used for comparison or ranking, although its theoretical basis is doubtful.

29.3.1 Mass Loss

It is the simplest corrosion test. These tests is carried out by immersion in a solution, in natural or artificial atmosphere and inside equipment or plants. It consists in measuring the mass variation to determine the corrosion rate by the exposure of a metallic coupon to an environment for a defined time. The sample is weighed before and at the end of exposure after removing, usually by pickling, the corrosion products. The mass loss of the sample, Δm, is then evaluated and, given the exposure time (t) and the exposed surface of the sample (S), the mass loss rate per unit area, $C_{rate,m}$, and the thinning rate, C_{rate}, are calculated with the formulas seen in Chap. 1.

This test is standardized; in particular ASTM G1 is specific for preparing specimens, for removing corrosion products after the test and for evaluating the corrosion damage that has occurred. A complete set of tables shows the cleaning procedure (chemical or electrolytic) to remove corrosion products without significant wastage of base metal; for each class of metals is indicated the solution, the temperature and the duration. This allows an accurate determination of the mass loss of the metal or alloy that occurred during exposure to the corrosive environment. These procedures, in some cases, may apply to metal coatings. However, possible side effects from the substrate must be considered.

ASTM G31 describes in details apparatus, sampling, test specimens and test condition (composition, temperature, agitation aeration) and test duration to conduct immersion tests. As regards the latter parameter, as rule of thumb, the minimum exposure time (in hours) should comply with the ratio $50/C_{rate}$ (in mm/year) to obtain reliable and measurable data. To reduce the test times, the aggressiveness of the exposure environment can be increased: this is the case of the salt spray tests. These tests are used as quality control tests, to quickly compare different materials. Rarely, however, the results of these tests can be used to predict the actual behaviour of a metal (see Sect. 29.3.8).

29.3.2 Pitting Corrosion

In the presence of localized corrosion, mass loss is useless. Most commonly the number of localized attacks (pits) per unit area or the pit depth are detected, or the maximum penetration rate (obtained by dividing the maximum depth of the pit for the exposure time) is calculated.

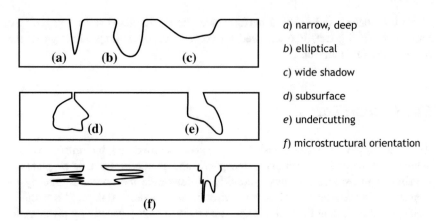

a) narrow, deep

b) elliptical

c) wide shadow

d) subsurface

e) undercutting

f) microstructural orientation

Fig. 29.4 Variations in the cross-sectional shape of pits (according to ASTM G46)

ASTM G46 covers the procedures for the identification and examination of pits and for the evaluation of pitting corrosion to determine the extent of its effect. The standard specifies the type of the inspection, visual or non-destructive (radiographic, electromagnetic, sonic and penetrants), how to determine the pith depth and how to describe the pitting in terms of density, size, and depth. Examples of analysis of pitting-related testing taken from ASTM G46 are reported in Figs. 29.4 and 29.5.

One of the well-known test of localized corrosion is represented by the ferric chloride ($FeCl_3$) test, ASTM G48, to determine the susceptibility to pitting (or crevice) corrosion of stainless steels and nickel alloys, and in particular to define the critical temperature of pitting and crevice.

The ASTM G48 standard provides the method to expose the material to a concentrated oxidizing chloride-containing environment. The solution simulates the chemical composition of the environment inside a pit in a stainless steel. The standard describes six methods depending on the form of localized corrosion (pitting or crevice) and on the metal (stainless steel or nickel-based and chromium-bearing alloys). The specimen is exposed to a $FeCl_3$ solution for a relatively short time (72 h) at a certain temperature (function of the metal composition). The test must be performed with 3–5 specimens at least and at the end of the test, both mass loss and localized attacks are checked. It is a comparative test: no extrapolation can be done under different environmental conditions.

29.3.3   Crevice Corrosion

Also for crevice corrosion, mass loss measurements only are useless. For this form of corrosion, it is necessary an apparatus that reproduces the operating conditions that favor corrosion. Typically the device is generally called "crevice former" or

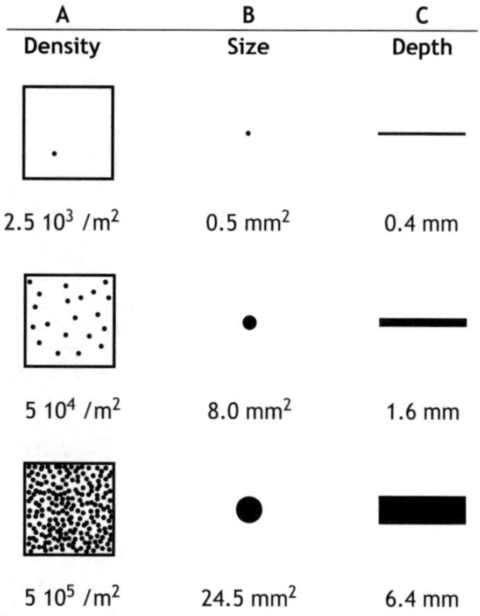

Fig. 29.5 Standard rating charts for pits (adapted from ASTM G46)

A Density	B Size	C Depth
2.5 10³ /m²	0.5 mm²	0.4 mm
5 10⁴ /m²	8.0 mm²	1.6 mm
5 10⁵ /m²	24.5 mm²	6.4 mm

"crevice assembly" (ASTM G48 reports some example). Examples of crevice formers are shown in Figs. 29.6, 29.7 and 29.8.

The apparatus was originally developed by Anderson: it consists of two serrated segmented washers, of inert material (PTFE), with 20 crevice sites (slot) created beneath each washer. Two of this washers when attached with a nut and bolt to a flat test specimen generate 40 individual crevice sites.

Since ASTM G48 has already been discussed, another example of crevice test standard is here discussed. ASTM G78 standard describes the crevice corrosion test for samples of different geometries (flat and cylindrical). Although the test is suitable for seawater applications, it can be used in other aqueous chloride-containing environments. This standard, in addition to different geometries of specimens, describes the possible apparatus to create the gap, for example coupons, strips, O-rings, blocks continuous and segmented washers. The severity of the test can vary according to the crevice former, which differ in size and degree of tightness. Metal samples are exposed to the environment for a standardized time, typically 30 days. Susceptibility to crevice corrosion is assessed through three parameters: mass loss, corroded area measurement and penetration depth. Figures 29.9 and 29.10 show the result of a crevice testing on stainless steel in seawater.

Fig. 29.6 Crevice former designs

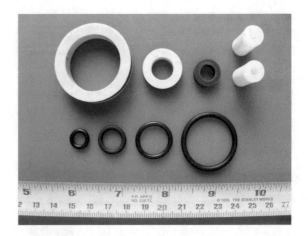

Fig. 29.7 Example of new crevice former

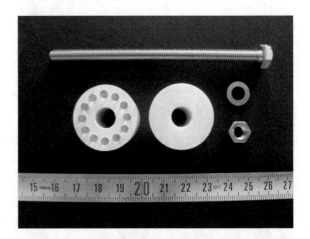

Fig. 29.8 Example of assembly of new crevice former

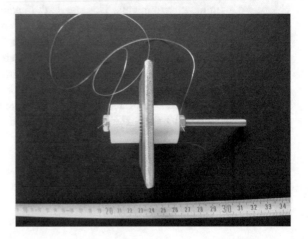

Fig. 29.9 Example of a
result of crevice testing

Fig. 29.10 Crevice corrosion
on laboratory samples of
stainless steel AISI 316

29.3.4   Galvanic Coupling

The test method for galvanic corrosion consists in immersing the metallic coupling
in the environment of interest. Obviously, the material surface condition, the
environment and the geometry, above all the ratio between the cathodic and the
anodic areas (see Chap. 10) should simulate the real application. Particular attention
to the electrical connection should be given.

ASTM G71 covers conducting and assessing galvanic corrosion tests to describe
the behaviour of two different metals in electrical contact, in an electrolyte under
low-flow conditions. The body of the standard fully describes material and specimens preparations as well as the environment (laboratory or field), the method of
exposure, the procedure and the method for evaluating the results. During the

exposure, galvanic current and potentials measurements have to be recorded. At the end of exposure, if possible, mass loss and visual inspection are recommended.

29.3.5 Integranular Corrosion

Mass loss test is used also for intergranular corrosion tests combined with visual observations. ASTM A262 standard practice for detecting susceptibility to intergranular attack in austenitic stainless steels is used to verify their susceptibility to intergranular corrosion. The standard provides five test methods, each with a different solution (Table 29.1) and time of test ranging from few minutes to 10 days depending on the composition of stainless steel examined. At the end of the test, the weight loss is determined, the presence of corrosion attacks along the weld are observed and the maximum depth of attack is measured.

ISO 3651 Part 1 and Part 2 concern the Huey test and the Strauss test, respectively.

ASTM G28 covers the intergranular corrosion test for wrought, nickel-rich, chromium-bearing alloys. Two practices are reported: method A, ferric sulphate-sulfuric acid test; method B, mixed acid-oxidizing salt test. The solutions are still boiling and the time of test is 24–120 h long, depending on the type of metal. At the end of the test, weight loss is determined, metallographic examination is observed if necessary, and the maximum depth of attack is measured.

Table 29.1 Description of the five practice, according to ASTM A262

Practice	Test name	Temperature	Time
A	Oxalic acid etch test for classification of etch structures of Austenitic SS: "*Screening Test*"	Ambient	1.5 min
B	Ferric sulfate–sulfuric acid test for detecting susceptibility to intergranular attack in austenitic SS: "*Streicher Test*"	Boiling	120 h
C	Nitric acid test for detecting susceptibility to intergranular attack in austenitic SS: "*Huey Test*"	Boiling	48 h (5 times)
E	Copper–copper sulfate–sulfuric acid test for detecting susceptibility to intergranular attack in austenitic SS: "*Strauss Test*"	Boiling	15 h
F	Copper–Copper sulfate–50% sulfuric acid test for detecting susceptibility to intergranular attack in molybdenum-bearing austenitic SS	Boiling	120 h

29.3.6 Stress Corrosion Cracking

To determine the susceptibility to SCC, different types of tests can be performed. Tests are classified as follows:

- Constant load
- Constant strain
- Slow strain rate
- Fracture mechanics test (pre-cracked specimens).

Laboratory tests were developed to accelerate the SCC response of specific metals and for developing solutions that simulate the exposure service conditions. ASTM G36 is the most known standard. It describes the procedure for conducting stress corrosion cracking tests in boiling magnesium chloride at about 155 °C. This specific testing environment offers an accelerated method to classify the degree of susceptibility to stress corrosion cracking in environments containing chlorides. This test is applicable to stainless steels and susceptible alloys as castings, welded and plastic worked products and is a method for detecting the effects of composition, heat treatment, surface finishing (Fig. 29.11).

Constant load test. A specimen, generally cylindrical in shape, is loaded by a tensile test machine while exposed to the environment, at a constant nominal stress level. Aim of the test is to determine the threshold stress level, below which SCC does not occur, and the time-to-failure. Figure 29.12 shows a typical σ-time plot.

Constant strain test. A pre-stressed specimen, to have a predetermined strain, is immersed in the solution test for an agreed exposure time, generally one month. Specimens are C-ring or U-bend or bend beam types. At the end, failure or presence of cracks is checked. By a series of specimens, stressed at different levels, threshold

Fig. 29.11 Strauss test

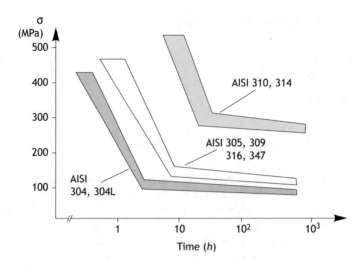

Fig. 29.12 Time-to-failure as a function of the applied nominal stress for different austenitic stainless steels in a boiling solution of magnesium chloride (42% by weight)

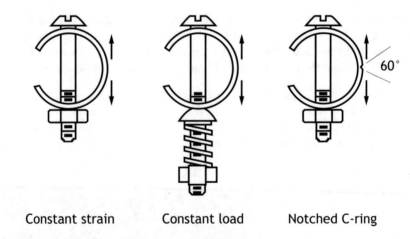

Fig. 29.13 Methods of loading C-rings (adapted from ASTM G38)

stress is determined. ASTM G38 standard deals with the design and processing features, and the procedures for stressing, exposing and inspecting the specimens (C-ring) to be subjected to stress corrosion. The standard includes the methods, formulas and tables used for the determination of stress according to the imposed deformation. Figure 29.13 shows examples of how to load the C-rings. ASTM G30

Fig. 29.14 Examples of U-bends (adapted from ASTM G30)

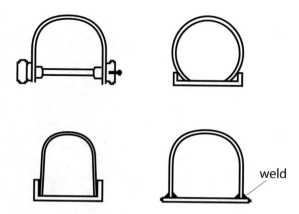

weld

covers techniques for making and using U-bend specimens. Usually, the U-bend specimen is a stripe which is bent 180° around a predetermined radius and maintained in this constant strain condition during the test. Examples of configurations of the U-bend samples are reported in Fig. 29.14. In H_2S-containing environment, pass-failed criterion is used. NACE TM0177 is an example of test method for testing metals immersed in low-pH aqueous environments containing hydrogen sulphide (H_2S) and subjected to tensile stresses. The test method is one of the most famous relatively to sulphide stress cracking (room temperature and pressure) and stress corrosion cracking (elevated temperatures and pressures). It describes specimens, test solutions and operating conditions, and test vessels and fixtures. Four practices are reported: Method A—Tensile Test, Method B—Bent-Beam Test, Method C—C-Ring Test and Method D—Double-Cantilever-Beam (DCB) Test.

Slow strain rate test. This test is similar to the one at constant load, while operated dynamically by applying a constant strain rate in the range 10^{-7} to 10^{-4} s^{-1}. Stress-strain curves obtained in inert environment, generally in air or oil, and in reference environment are compared through: maximum load, elongation and reduction of cross surface area. A visual inspection checks fracture surface feature and the presence of secondary cracks.

Fracture mechanics test. Previous tests utilize smooth specimens while this test is based on notched specimens. The notch is generally a fatigue crack. Test is carried out at constant load or constant strain; in the latter, specimen used is WOL or WOL-modified (WOL means Wedge Opening Loaded), as shown in Fig. 29.15. At the end of test, crack growth is measured, then allowing the calculation of $K_{I\text{-SCC}}$: in fact, crack growth stops as soon as K_I lowers to $K_{I\text{-SCC}}$.

Fig. 29.15 WOL specimen
for SCC test at constant strain

29.3.7   Erosion, Cavitation and Fretting

Many standards regarding erosion, cavitation and fretting wear exist, as ASTM
G32, ASTM G73, ASTM G76 and ASTM D4170. It is important to recognize the
mechanism and consequently the main factors that control the phenomenon in order
to select the proper test condition. Apparatus, test materials, sampling, procedure
and data analysis are carefully described in the standard. For example, in ASTM
G76 for conducting erosion tests, geometry and velocity of solid particles, specimen
orientation relative to the impinging stream, temperature of the specimen and
particles carrier gas and test duration are defined. The steady state erosion rate is
determined from the slope of the mass loss versus time plot. The average erosion
(mm^3/g) is calculated by dividing erosion rate (mg/s) by the abrasive flow rate (g/s)
and then dividing by the specimen density.

29.3.8   Artificial Atmosphere—Cabinet Test

Since 1900, cabinet tests have been used to evaluate coatings performance and to
carry out accelerated corrosion tests. Cabinet testing takes its name from the closed
chamber in which tests are conducted and where the conditions of exposure are
controlled. This type of test is generally used for corrosion performance of metals
used in natural atmospheres as pass-fail test.

The environment produced inside the chamber combines usually: salt fog,
humidity, hot and low temperature, corrosive gases and ultraviolet exposure.

In order to correlate test results with service performance, it is necessary to
establish acceleration factors and to verify that the corrosion mechanisms are indeed
following the same paths.

The salt spray (or salt fog) test is one of the most widespread, long established
and standardized test method. Usually, coated metals are tested to verify the degree

of corrosion protection of the coating to the underlying metal. The appearance of corrosion products (rust or other oxides) is evaluated after a pre-determined period of time. Test duration depends on the corrosion resistance of the sample.

The apparatus for testing consists of a closed testing cabinet or chamber, where a salt water (5% NaCl) solution is atomized by means of spray nozzles using pressurized air. A corrosive environment of dense salt water fog is produced in the chamber, so that test samples exposed to this environment are subjected to severely corrosive conditions. Chamber volumes vary from supplier to supplier. If there is a minimum volume required by a particular salt spray test standard, this will be clearly stated and should be complied with. There is a general historical consensus that larger chambers can provide a more homogeneous testing environment.

ASTM B117 was the first internationally recognized salt spray standard, originally published in 1939. Other important relevant standards are ISO 9227, JIS Z 2371 and ASTM G85.

ASTM B117 reports apparatus, procedure and conditions required to produce and maintain the Neutral Salt Spray test (often abbreviated to NSS); type of test specimen, exposure periods or the interpretation of the results are specified or mutually agreed between the purchaser and the seller. Results are represented generally as testing hours in NSS without appearance of corrosion products.

ASTM G85 describes five modified salt fog tests: acetic acid-salt spray test (ASS); cyclic acidified salt spray test; seawater acidified test, cyclic (SWAAT); SO_2 salt spray test (cyclic test), dilute electrolyte cyclic fog dry test. The more severe test is performed in acetic acid with copper chloride (CASS).

Humidity test and corrosive gas test are also conducted in cabinet according to specific standards.

29.4 Electrochemical Tests

Electrochemical test are typically laboratory tests used to define the basic corrosion-influencing factors, as thermodynamic and kinetic parameters, anodic and cathodic characteristics, corrosion potential in referred electrolytes, tendency to passivation, oxidant power. Since test conditions are rigorously controlled, results show a high reproducibility.

Instrument used to plot the anodic and cathodic curves measure a total current density, equal to the algebraic sum of the anodic and cathodic current densities. At the free corrosion potential, the measured external current is zero (anodic current equals cathodic current). At potential more noble than free corrosion potential, the anodic current density exceeds the cathodic one, the opposite occurs at less noble potential. When far from the free corrosion potential (at least 50 mV), measured current density coincides with anodic or cathodic one, then following Tafel law (Fig. 29.16).

Polarisation curves can be obtained by two methods: galvanostatic (i.e., by imposing the current) and potentiostatic (i.e., by imposing the potential). Figure 29.17 shows the principle of the two methods. The galvanostatic method is

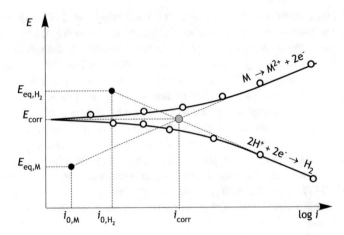

Fig. 29.16 *E*-Log *i* diagram of a corroding metal in acidic solutions (experimental curves and Tafel straight lines)

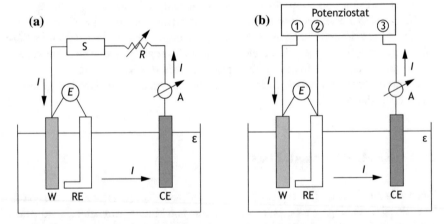

Fig. 29.17 Experimental set-up to obtain the characteristic curves: **a** galvanostatic mode; **b** potentiostatic mode (W is working electrode; RE is reference electrode, CE is counter-electrode)

easy and requires simply a generator, an ammeter and a voltmeter: once fixed the circulating current (i.e., the current density) the electrode potential is measured. The measurement is carried out by increasing steps, often manually. The potentiostatic method is more sophisticated since it requires a tailored instrument, which is the potentiostat, introduced by Edeleanu at University of Cambridge, UK, in 1954. An electronic circuit that allows to vary the potential independently from the circulating current characterizes the potentiostat. The instrument works as depicted in Fig. 29.18. The potential, E, of the so-called working electrode is continuously increased and simultaneously compared to a floating pre-fixed value, E_C, in order to

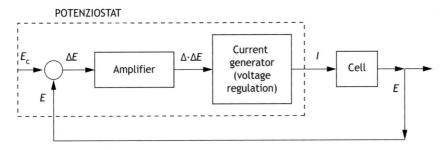

Fig. 29.18 Block scheme principle of a potentiostat

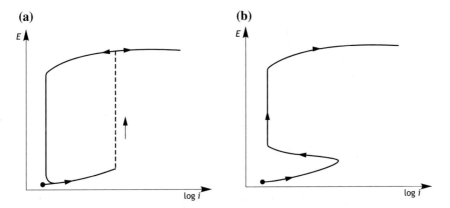

Fig. 29.19 Active-passive curve obtained by: **a** galvanostatic mode; **b** potentiostatic mode

measure the difference E-E_C; the potentiostat imposes a current which zeros continuously and dynamically the measured difference of the potentials. By this "trick", current can either increase or decrease.

The potential variation can be performed by steps (potentiostatic mode) or continuously (potentiodynamic mode). In conclusion, potential is the variable and current is the output.

The two methods, galvanostatic and potentiostatic, are not equivalent: to characterize an active-passive behaviour, potentiostat method is more suitable. It should be considered that by applying, step-by-step, galvanostatic method, the entire curve cannot be obtained (Fig. 29.19a) because once reached the critical passivation current, the potential jumps directly to transpassive zone, then losing the passive interval and the passivity current density value. In other words, Log i-E plot is less complete than E-Log i one; on the other hand, potentiostatic method acts as a corrosion process, where driving voltage determines the corrosion current and not the opposite (Fig. 29.19b).

Often, also in this textbook, Evans diagram is used with "log i" as abscissa, although the variable is the potential, E (Fig. 29.20).

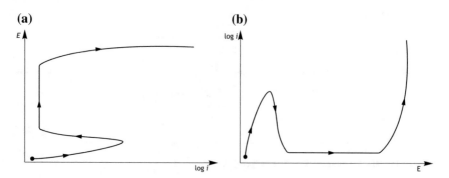

Fig. 29.20 Evans diagram representation: **a** E-log i plot; **b** log i-E plot

When running a test, either galvanostatic or potentiostatic, attention should be paid on ohmic drop contribution in high resistivity electrolytes. To minimize it, Luggin capillary or a Piontelli probe are conveniently used. Furtherly, it must be remembered that polarization curve depends on sweep/scanning rate; international standards recommend a rate not higher than 20 mV/min.

29.4.1   Uniform Corrosion

Tafel extrapolation method. As shown in Fig. 29.16, from the anodic or cathodic curves, often from one of them only, by extrapolating Tafel straight line to the corrosion potential, E_{corr}, corrosion rate, i_{corr}, is obtained. Exchanged current densities are easily obtained by extrapolation to the calculated equilibrium potential according to the metal under corrosion of the cathodic predominant process. Tafel slopes may be estimated directly from the graph. This method is simple, easy, quick and accurate when single anodic and cathodic processes occur. Accordingly, potential is scanned from −200 mV to +200 mV with respect the measured free corrosion potential. No standards currently exist for this method.

Linear polarization resistance (LPR method). Also called *Stern-Geary method*. The principle of the method is the following: a metal is polarized from its free-corrosion potential, E_{corr}, toward anodic or cathodic direction by imposing an external current density, i_e. The polarization ΔE (=$|E - E_{corr}|$) is in the range of maximum 20 mV. The $\Delta E/i_e$ ratio, called *polarization resistance*, R_p (Ω m^2), is inversely proportional to the corrosion rate through the Stern-Geary equation (see box):

$$i_{corr} = \frac{B}{R_p} \qquad (29.2)$$

where B is a constant, estimated as follow:

$$B = \frac{1}{2.3} \cdot \frac{b_a b_c}{b_a + b_c} \tag{29.3}$$

b_a and b_c are the Tafel slopes of the anodic and cathodic process, respectively. Constant B is typically 0.026 V/decade for active metals in acids ($b_a = b_c = 0.12$ V/decade) and 0.052 for active metals in aerated solutions under diffusion control ($b_a = 0.12$ V/decade; $b_c = $ infinite).

As said, the method is reliable for small polarizations, i.e., $\Delta E \leq 20$ mV, typically 10 mV. The method appears as more precise as corrosion rate is low, so it is more attractive and useful than mass loss coupons, when long exposure time would be required, or when on-time interventions is necessary as for inhibitor dosage adjustment. Furthermore, it allows measurements without coupon retrieval and the continuous monitoring on plants. It is applied in a variety of environments, as process fluids, waters and concrete. Some conditions jeopardize the accuracy of the measurements as corrosion potential too close to equilibrium potential, high resistivity, i.e., high ohmic drop; however, rather than absolute values, variations in time can be conveniently used. The most evident drawback of the method is that it works for uniform corrosion, only.

ASTM G59 covers an experimental procedure for polarization resistance measurements, which can be used for the calibration of instruments and verification of experimental technique. The test method can provide reproducible corrosion potentials and potentiodynamic polarization resistance measurements.

Stern-Geary Equation

To cause a small cathodic polarization, ΔE_c, an external current density, i_e, is applied, which is given by $i_e = i_c - i_a$, where i_c and i_a are cathodic and anodic current density, respectively.

From the figure, the following can be written:

$$CB = OC/b_c; \quad CB = \log i_c - \log i_{corr} = \log(i_c/i_{corr}) = \Delta E_c/b_c$$
$$AC = OC/b_a; \quad AC = \log i_{corr} - \log i_a = \log(i_{corr}/i_a) = \Delta E_c/b_a$$

$$i_c = i_{corr} \cdot 10^{\Delta E_c/b_c}; \quad i_a = i_{corr} \cdot 10^{-\Delta E_c/b_a}; \quad i = i_c - i_a = i_{corr} \cdot [10^{\Delta E_c/b_c} - 10^{-\Delta E_c/b_a}]$$

Because ΔE_c is small (about 10 mV) compare to b_c and b_a (higher than 100 mV) the following approximation is acceptable:

$$10^x = 1 + (\log 10)x + (\log 10)^2 x^2/2 + \cdots \cong 1 + (\log 10)\, x$$

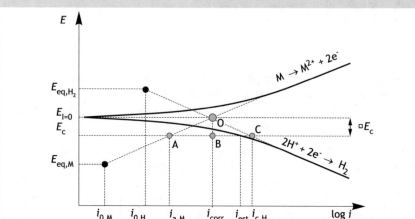

Fig. 29.21 Polarization curves to explain Stern-Geary extrapolation method of corrosion rate

therefore:

$$i = i_{\text{corr}} \cdot \left[1 + 2.3\frac{\Delta E_c}{b_c} - 1 + 2.3\frac{\Delta E_c}{b_a} \right]$$

$$i_{\text{corr}} = \frac{1}{2.3} \cdot \frac{b_a b_c}{b_a + b_c} \cdot \frac{I}{\Delta E} = \frac{1}{2.3} \cdot \frac{b_a b_c}{b_a + b_c} \cdot \frac{1}{R_p}$$

where ΔE is either cathodic or anodic and R_p is polarization resistance (Fig. 29.21).

Electrochemical Impedance Spectroscopy (EIS). In 1960s I. Epelboin (1916–1980) and his research group in Paris introduced the method with the aim to study the processes occurring at the electrode interface. It consists of injecting an alternating current between a working electrode and a counter electrode and plotting the potential and the current measured at different frequency (from this the name spectroscopy). Potential $E(t)$ and current density $i(t)$ are linked through impedance $Z(w)$, which are function of frequency, f ($\omega = 2\pi f$):

$$E(t) = Z(\omega) \cdot I(t) \tag{29.4}$$

where $E(t) = E_o \sin \omega t$; $I(t) = I_0 \sin (\omega t + \theta)$; $\omega = 2\pi f$; θ is angle phase and f is frequency (hertz).

Anodic and cathodic processes occurring on surface electrode influence the angle phase, θ, which is important for the interpretation of the process. Circuit impedance is a complex parameter, therefore composed of a real component $Z'(\omega)$ and an imaginary one $Z''(\omega)$, in phase opposition:

$$Z(\omega) = Z'(\omega) + Z''(\omega) = R - \frac{j}{\omega C} \qquad (29.5)$$

It appears that electronic circuit, composed of resistances (real component) and capacities (imaginary component), can simulate processes. The capacity can represent the delay, i.e. θ, of the signal.

To ease the interpretation, Nyquist diagrams, $Z''(\omega)$ against $Z'(\omega)$ are used, associated to an equivalent electronic/electrical circuit. Two simple typical case studies are considered.

The first one is uniform corrosion where only activation overvoltage applies. Equivalent electrical circuit is simply a resistor and a capacitor in parallel, representing the activation overvoltage, and a resistance in series, which represents electrolyte resistance (proportional to resistivity). The circuit is known as *equivalent Randles circuit*. Nyquist diagram is a semicircle (Fig. 29.22) where frequency increase is anticlockwise. Either at high or at low frequency, imaginary component $Z''(\omega)$ zeros, then allowing the measurement of electrolyte resistance R_Ω at high frequency and $R_\Omega + R_p$ at low frequency, where R_p is transfer charge resistance.

The second case study is a uniform corrosion process under diffusion control or in the presence of coatings or scales. The equivalent electrical circuit contains so-called Warburg impedance, W, revealed at low frequency by the presence of a $45°$ slope line (Fig. 29.23), which is a finger print of the type of process.

To carry out an EIS measurement a tailored instrumentation is required and consisting of a feeding system in a wide range of frequency, a precision potentiostat and a signal analyser. ASTM G106 standard covers an experimental procedure to check instrumentation and technique for collecting and presenting electrochemical

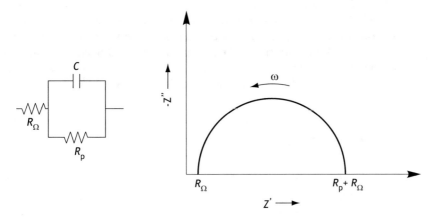

Fig. 29.22 Equivalent electrical circuit and Nyquist diagram for a corrosion process driven by activation overvoltage

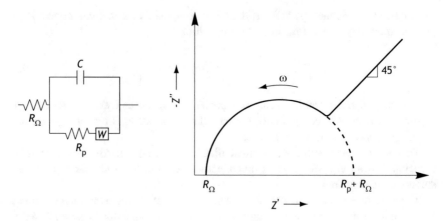

Fig. 29.23 Equivalent electrical circuit and Nyquist diagram for a corrosion process driven by diffusion overvoltage

impedance data. It provides standard material, electrolyte and procedure for performing EIS at the open circuit or free corrosion potential. This practice may not be appropriate for collecting impedance information for all materials or in all environments.

29.4.2 Pitting Potential and Repassivation Potential

To determine the susceptibility to pitting of stainless steels, a cyclic potentiodynamic polarisation test is performed by imposing a potential scan from the free corrosion potential (or a little more cathodic value) forward the anodic direction until current density reaches $10 \ A/m^2$ (or other agreed value). Potential is then reversed until passivity is again established. The two potentials are *pitting potential* and *repassivation potential*, respectively (Fig. 29.24). Both potential scan rate and current density threshold influence the result: pitting potential lowers as scan rate decreases and repassivation potential lowers as current density threshold increases. Accordingly, tests are performed following standards, as ASTM G61.

29.4.3 Galvanic Coupling

Two different metals are connected each other with a zero resistance ammeter and immersed in the same solution. The galvanic corrosion rate is directly estimated as a function of time. Calculated corrosion rate must be added to the corrosion rate in free corrosion conditions. Standards do not exist for direct measurements of galvanic current.

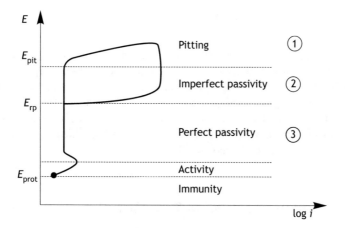

Fig. 29.24 Pitting and repassivation potential measurement

29.4.4 Intergranular Corrosion

An electrochemical test, called EPR, *Electrochemical Potentiokinetic Reactivation*, is used to check sensitization of stainless steel and nickel-based alloys. ASTM G108 standard is typically used. After surface preparation of specimen, a potential polarization +0.2 V SCE is imposed by a potentiostat in a 0.5 M H_2SO_4 + 0.01 M KSCN solution at 30 °C. After that, a potential scan toward cathodic direction is recorded as shown in Fig. 29.25. Test result is interpreted by comparison with non-sensitized specimen behaviour: in general, the higher the nose area the more sensitization degree.

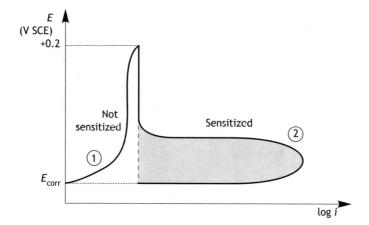

Fig. 29.25 EPR test interpretation (according to ASTM G108)

29.4.5   Stress Corrosion Cracking

For SCC by slip-dissolution mechanism, the critical intervals of potential can be obtained by comparison of anodic polarization curves, respectively worked out at low and high potential scan rate. The latter, i.e. high scan rate, shows the metal behaviour when not yet completely covered by a film, while the former shows the opposite. Critical intervals are easily found by comparison where anodic dissolution rate in scan rate test exceeds the other one. Figure 29.26 shows an example for carbon steel in carbonate/bicarbonate solution.

29.4.6   Other Electrochemical Techniques

Potentiostatic. Potentiostatic polarization can be used alternatively or in addition to potentiodynamic polarization techniques, in cases where conditions of stable state are desirable. It consists in applying through a potentiostat a constant potential at the metal-solution interface and measuring the resulting current as a function of time. Usually after obtaining a potentiodynamic polarization curve it may be convenient to conduct potentiostatic tests at potentials of particular interest. For example, where the value of the perfect passive potential is close to the free corrosion potential, it is convenient to obtain long-term data in the region between the perfect passivity potential and the pitting potential. In addition, to interpret potentiodynamic polarization curves with anomalous trend it may be useful to have data obtained with potentiostatic tests, which are certainly more reliable. The main

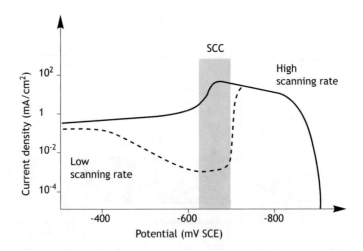

Fig. 29.26 Potentiodynamic anodic curve of C-Mn steel in 1 N carbonate/bicarbonate solution, at 90 °C. Dotted zones are critical potential intervals for SCC

disadvantage of this technique is given by the long times required for the measurement and by the number of tests to be carried out.

Cyclic voltammetry. In a cyclic voltammetry experiment, the working electrode potential is ramped linearly versus time with a high potential scan rate. Once the set potential is reached, the working electrode potential is reversed to return to the initial value. Cycles of ramps in potential may be repeated as many times as needed. The current at the working electrode is plotted versus the applied voltage to give the cyclic voltammogram trace. Cyclic voltammetry is generally used to study the electrochemical properties of electrolyte, to analyse the anodic and cathodic processes, to determine the stability of reaction products, or the presence of intermediates in redox reactions.

29.5   Applicable Standards

- ASTM A262, Standard practices for detecting susceptibility to intergranular attack in austenitic stainless steels, American Society for Testing of Materials, West Conshohocken, PA.
- ASTM B117, Standard practice for operating salt spray (fog) apparatus, American Society for Testing of Materials, West Conshohocken, PA.
- ASTM D4170, Standard test method for fretting wear protection by lubricating greases, American Society for Testing of Materials, West Conshohocken, PA.
- ASTM G1, Standard practice for preparing, cleaning, and evaluating corrosion test specimens, American Society for Testing of Materials, West Conshohocken, PA.
- ASTM G4, Standard guide for conducting corrosion tests in field applications, American Society for Testing of Materials, West Conshohocken, PA.
- ASTM G28, Standard test methods for detecting susceptibility to intergranular corrosion in wrought, nickel-rich, chromium-bearing alloys, American Society for Testing of Materials, West Conshohocken, PA.
- ASTM G30, Standard practice for making and using U-bend stress-corrosion test specimens, American Society for Testing of Materials, West Conshohocken, PA.
- ASTM G31, Standard practice for laboratory immersion corrosion testing of metal, 1American Society for Testing of Materials, West Conshohocken, PA.
- ASTM G32, Standard test method for cavitation erosion using vibratory apparatus, American Society for Testing of Materials, West Conshohocken, PA.
- ASTM G36, Standard practice for evaluating stress-corrosion-cracking resistance of metals and alloys in a boiling magnesium chloride solution, American Society for Testing of Materials, West Conshohocken, PA.
- ASTM G38, Standard practice for making and using C-ring stress-corrosion test specimens, American Society for Testing of Materials, West Conshohocken, PA.

- ASTM G46, Standard guide for examination and evaluation of pitting corrosion, American Society for Testing of Materials, West Conshohocken, PA.
- ASTM G48, Standard test methods for pitting and crevice corrosion resistance of stainless steels and related alloys by use of ferric chloride solution, American Society for Testing of Materials, West Conshohocken, PA.
- ASTM G61, Conducting cyclic potentiodynamic polarization measurements for localised corrosion, American Society for Testing of Materials, West Conshohocken, PA.
- ASTM G71, Standard guide for conducting and evaluating galvanic corrosion tests in electrolytes, American Society for Testing of Materials, West Conshohocken, PA.
- ASTM G73, Standard test method for liquid impingement erosion using rotating. Apparatus, American Society for Testing of Materials, West Conshohocken, PA.
- ASTM G76, Standard test method for conducting erosion tests by solid particle impingement using gas jets, American Society for Testing of Materials, West Conshohocken, PA.
- ASTM G78, Standard guidecrevice corrosion testing of iron-base and nickel-base stainless alloys in seawater and other chloride-containing aqueous environments, American Society for Testing of Materials, West Conshohocken, PA.
- ASTM G85, Standard practice for modified salt spray (fog) testing, American Society for Testing of Materials, West Conshohocken, PA.
- ASTM G59, Standard test method for conducting potentiodynamic polarization resistance measurements, American Society for Testing of Materials, West Conshohocken, PA.
- ASTM G78, Standard guide for crevice corrosion testing of iron-base and nickel-base stainless alloys in seawater and other chloride-containing aqueous environments.
- ASTM G102, Standard practice for calculation of corrosion rates and related information from electrochemical measurements, American Society for Testing of Materials, West Conshohocken, PA.
- ASTM G106, Standard practice for verification of algorithm and equipment for electrochemical impedance measurements, American Society for Testing of Materials, West Conshohocken, PA.
- ASTM G108, Standard test method for electrochemical reactivation (EPR) for detecting sensitization of AISI type 304 and 304L stainless steels, American Society for Testing of Materials, West Conshohocken, PA.
- ASTM G150, Standard test method for electrochemical critical pitting temperature testing of stainless steels, American Society for Testing of Materials, West Conshohocken, PA.
- ISO 3651-1, Determination of resistance to intergranular corrosion of stainless steels—part 1: austenitic and ferritic-austenitic (duplex) stainless steels—corrosion test in nitric acid medium by measurement of loss in mass (Huey test), International Standard Organization, Geneva, Switzerland.

- ISO 3651-2, Determination of resistance to intergranular corrosion of stainless steels—part 2: ferritic, austenitic and ferritic-austenitic (duplex) stainless steels —corrosion test in media containing sulfuric acid, International Standard Organization, Geneva, Switzerland.
- ISO 6509-1, Corrosion of metals and alloys—determination of dezincification resistance of copper alloys with zinc—part 1: test method, International Standard Organization, Geneva, Switzerland.
- ISO 6509-2, Corrosion of metals and alloys—determination of dezincification resistance of copper alloys with zinc—part 2: assessment criteria, International Standard Organization, Geneva, Switzerland.
- ISO 9227, Corrosion tests in artificial atmospheres—salt spray tests, International Standard Organization, Geneva, Switzerland.
- ISO 11845, Corrosion of metals and alloys. General principle for corrosion testing, International Standard Organization, Geneva, Switzerland.
- ISO 17475, Corrosion of metals and alloys—Electrochemical test methods— Guidelines for conducting potentiostatic and potentiodynamic polarization measurements, International Standard Organization, Geneva, Switzerland.
- NACE TM0177, Laboratory testing of metals for resistance to sulphide stress cracking and stress corrosion cracking in H_2S, NACE International, Houston, TX.
- NACE TM0284, Evaluation of pipeline and pressure vessel steels for resistance to hydrogen-induced cracking, NACE International, Houston, TX.

29.6 Questions and Exercises

29.1 Design an accelerating testing for carbon steel operating in fresh water at 40 °C, fluid velocity 1 m/s and oxygen content 0.1 ppm. Calculate the intensification index.

29.2 Explains the philosophy behind corrosion tests.

29.3 Describe the typical parameter measured in exposure test. Differentiates for the main forms of corrosion.

29.4 Which is the most famous test for pitting corrosion? Why that solution is used?

29.5 Salt spray fog is used to compare coatings. How test results can be used?

29.6 Explain the principle of Stern-Geary method. Is the method applicable to study the corrosion of passive metals?

29.7 What do you get from the Tafel extrapolations method?

29.8 Explain the difference between potentiodynamic, potenziostatic and cyclic voltammetry electrochemical techniques.

29.9 EPR test runs by the application of a potentiodynamic scan from passive interval downward less noble, i.e., cathodic, potentials. Does the application of cathodic protection, for example by means of a galvanic anode (for instance pure iron) reveal the same behaviour? Explain.

29.10   How do you study the susceptibility of a stainless steel to pitting? Explains both exposure and electrochemical tests and the variables that influence the results.

29.11   An AISI 304 stainless steel tank has to store a chloride-containing solution. Which test do you recommend to verify the metal-solution compatibility?

29.12   A pharmaceutical plant has to convert the production to a new medicine. Which test do you recommend to check the metal compatibility with the new reagents?

29.13   Which tests are mandatory to qualify a carbon steel pipe for sour service condition?

29.14   A new welding procedure for an austenitic AISI 316 has been put in place. Which test would you recommend to check is the procedure is correct form a corrosion point of view?

29.15   Suggest and develop a series of tests to verify the localised corrosion resistance of three stainless steels to a commercial syrup.

Bibliography

Baboian R (2004) Corrosion tests and standards: application and interpretation. ASTM International, West Conshohocken

Electrochemical Techniques for corrosion Engineering (1986) Baboir Editor. NACE International, Houston, TX

Lazzari L (2017) Engineering tools for corrosion. Design and diagnosis. European Federation of Corrosion (EFC) Series, vol 68. Woodhead Publishing, London

Stern M (1958) A method for determining corrosion rates from linear polarization data. Corrosion 14:60

Stern M, Geary AL (1957) Electrochemical polarization. A theorical analysis of the shape of polarization curve, J Electrochem Soc 104:56

Chapter 30
Statistical Analysis of Corrosion Data

The plates [of the ship] were pitted till the men that were paint,
paint, painting her, laughed at it.
R. Kipling, *The Day's Work, Bread upon Waters.*

Abstract Statistical analysis—from data sampling to interpretation of results—is fundamental to all branches of science and engineering, as well as in the field of corrosion. Once corrosion data are obtained from testing (i.e. laboratory and/or field investigation), monitoring and inspection activities, statistical analysis can be very helpful to interpret such results, providing a rational, engineering approach. Nowadays, the amount of corrosion data has continuously increased. In spite of this, the statistical approach is not widely used in corrosion science and engineering even if proper methodologies are available to organize corrosion information and to improve industrial plant design and maintenance. In this chapter, the basic concepts

Fig. 30.1 Case study at the PoliLaPP Corrosion Museum of Politecnico di Milano

© Springer Nature Switzerland AG 2018
P. Pedeferri, *Corrosion Science and Engineering*, Engineering Materials,
https://doi.org/10.1007/978-3-319-97625-9_30

of corrosion probability and statistical treatment of corrosion data are discussed. The chapter does not cover detailed description of statistical methods, rather considers a range of approaches with applications in corrosion testing.

30.1 Fundamentals of Statistics

30.1.1 Mean and Variability of Data Distribution

When measuring values associated with the corrosion of metals, as corrosion rate or time-to failure of a corroded component, several factors act to produce a scattering of data. The pattern in which data are scattered is called *distribution*. Usually, these factors, mainly related to environmental and metallurgical properties, act in a random way so that the average of several values approximates the expected value better than a single measurement. Generally, the need for a statistical treatment of data is felt especially for localized corrosion phenomena (Fig. 30.1), as pitting corrosion and stress corrosion cracking (SCC), which are characterized by an initiation (or incubation) period and by a propagation time. Statistical analysis can be applied to both of them, although the interest on the initiation time remains predominant from an engineering point of view.

When working with a large and discrete data set, it can be useful to represent it with a single value that describes the "middle" or "average" value of the entire set. In statistics, this single value is called the central tendency and mean, median and mode are the ways to describe it. In corrosion, average values are generally useful in characterizing corrosion rates. In cases of corrosion penetration due to pitting and cracking, failure is defined as the first through-penetration and average penetration rates or mean times have poor meaning. In these cases, extreme value analysis is used. The variability of a data distribution can be defined by some parameters; the most used are discussed in the following.

Mean. The mean, μ, or expected value, is calculated by summing all data points, x_i, and dividing by the total number of data points, N:

$$\mu = \frac{\sum_{i=1}^{N} x_i}{N} \tag{30.1}$$

Median. The median is the middle value in a set of data. It can be found by ordering all data points and selecting the one in the middle of the rank. If there are two middle numbers, the median is the mean of them.

Mode. The mode is the value that occurs the highest number of times.

Standard deviation. It is a measure used to quantify the variation or dispersion of a set of data. A low standard deviation indicates that the data points tend to be close to the mean of the set, while a high standard deviation indicates that the data points are spread out over a wider range of values. Standard deviation, σ, measures the variation of a set of data around the mean value, μ, and it is given by:

$$\sigma = \sqrt{\frac{\sum_{i=1}^{N}(x_i - \mu)^2}{N - 1}} \tag{30.2}$$

Variance. It is the expectation of the squared deviation of a random variable from its mean. It measures how far a set of data are spread out from their average value. The variance, σ^2, is the square of the standard deviation (Eq. 30.2).

Coefficient of variation. The coefficient of variation is defined as the standard deviation divided by the mean ($C.V. = \sigma/\mu$). This measure of variability is particularly useful in cases where the size of the errors is proportional to the magnitude of the measured value so that the coefficient of variation is approximately constant over a wide range of values.

Range. The range is defined as the difference between the maximum and minimum values in a set of replicate data values. The range provides an indication of statistical dispersion and it is measured in the same units as the data.

Precision, repeatability and reproducibility. *Precision* is closeness of agreement between randomly selected individual measurements or test results. The standard deviation of the error of measurement may be used as a measure of imprecision. *Repeatability* is the ability of one investigator or laboratory to reproduce a measurement previously made at the same location with the same method. Another aspect of precision concerns the ability of different investigators and laboratories to reproduce a measurement. This aspect is called *reproducibility*.

30.1.2 Statistical Distributions of Scatter Data

The scatter of a data distribution is often displayed through a histogram, by dividing the range of data obtained from testing into equal intervals. As reported on ISO 14802 *Corrosion of metals and alloys—Guidelines for applying statistics to analysis of corrosion data*, the number of intervals, k, is proportional to the number of data, N:

$$k = 1 + 3.32 \cdot \log N \tag{30.3}$$

where 3.32 is the logarithmic base 2 of 10. In a *frequency distribution* plot, each bin contains the number of values that lie within the range of values that define the bin.

In a *cumulative distribution*, each bin contains the number of values that fall within or below the bin.

Histograms provide a discrete analysis of data; the corresponding continuous function, which represents data distribution best-fit, is the *probability density function*, $f(x)$, which provides the probability of the random variable falling within a particular range of values. Accordingly, the cumulative distribution function, $F(x)$, is defined as:

$$F(x) = \int_{-\infty}^{x} f(u)du \qquad (30.4)$$

30.1.3   Reliability and Hazard Functions

Reliability, $R(x)$, is defined as the ability of a system to perform its required functions under stated conditions for a specified time. It is related to the cumulative density function, Eq. 30.4, as follows:

$$R(x) = 1 - F(x) = 1 - \int_{-\infty}^{x} f(u)du = \int_{x}^{\infty} f(u)du \qquad (30.5)$$

The hazard function, $h(x)$, represents the probability of a failure incident to take place corresponding to an instant x. The cumulative hazard function, $H(x)$, is related to reliability, $R(x)$, and $F(x)$ as follows:

$$h(x) = \frac{f(x)}{R(x)} \qquad (30.6)$$

$$R(x) = 1 - F(x) = e^{-H(x)} \qquad (30.7)$$

The hazard function ranges from 0 (when $R = 1$) to $+\infty$ (when $R = 0$).

30.2   Probability Distributions Observed in Corrosion

A variety of distributions are observed in corrosion (Shibata 2000), as the normal distribution (pitting potentials), lognormal (SCC failure time), exponential (induction time for pit generation), Poisson distribution (two-dimensional distribution of pit), and extreme-value statistics, Gumbel and Weibull distributions (maximum pit depth, SCC failure time, corrosion fatigue crack depth).

Some example of fundamental distribution for corrosion phenomena (obtained from empirical observations by several authors in literature) are reported in Table 30.1 (Kowaka 1994). Moreover, in the last years, an effort has been done by introducing statistical treatment of corrosion data in the international standard ISO 14802 *Corrosion of metals and alloys—Guidelines for applying statistics to analysis of corrosion data*, in order to share recent progresses and provide a guideline for operators in the field of corrosion. Nevertheless, this standard only deals with basic and rough aspects of statistical analysis of corrosion data.

30.2.1 Normal (Gaussian) Distribution

Normal distribution is bell-shaped and symmetrical with respect to the mean value, μ, which is at the same time mode and median of the distribution, as defined in Sect. 30.1.1. The normal distribution is non-zero over the entire real line. It follows that it may not be a suitable model for variables that are inherently positive or strongly skewed. Such variables may be better described by other distributions, such as the lognormal distribution. The probability density function of normal distribution is:

$$f(x) = \frac{1}{\sigma\sqrt{2\pi}} \cdot e^{-\frac{(x-\mu)^2}{2\sigma^2}} \tag{30.8}$$

Table 30.1 Examples of fundamental distribution for corrosion phenomena obtained from empirical observations by several authors in literature (Kowaka 1994)

Fundamental distribution	Example
Poisson distribution	– Number of pits
Exponential distribution	– Time to failure of H-charged 0.9% C steel under constant load – Incubation period for pitting and crevice initiation of stainless steels in NaCl solution
Normal distribution	– Pitting potential of AISI 304 stainless steel in 3.5% NaCl solution – Pit depth for carbon steel in fresh water supply pipe – Rate of activation for sensitized AISI 304 stainless steel in electrochemical potentiodynamic reactivation test – Intergranular SCC depth of sensitized AISI 304 stainless steel in high temperature pure water
Lognormal distribution	– Time to SCC failure of Al alloys in 3% NaCl solution – Time to SCC failure of sensitized stainless steels in high temperature pure water – Time to SCC failure of AISI 310 stainless steel in $MgCl_2$ solution – Time to SCC failure of carbon steel wire in $Ca(NO_3)_2 + NH_4NO_3$ solution

The area under the curve provides the probability of occurrence, calculated by the cumulative probability function, $F(x)$:

$$F(x) = \int_{-\infty}^{x} f(x)\mathrm{d}x = \frac{1}{\sigma\sqrt{2\pi}} \int_{-\infty}^{x} \mathrm{e}^{-\frac{(x-\mu)^2}{2\sigma^2}} \mathrm{d}x \tag{30.9}$$

Since cumulative probability cannot be calculated analytically for each normal distribution, it is common practice to convert a normal to a *standard normal distribution* and then use the standard normal table to find probabilities. A standard normal table, also called Z-table, is used to find the probability that a statistic is observed below, above, or between values of the standard normal distribution, with zero mean and a unitary standard deviation. If x is a random variable with mean μ and standard deviation σ, its Z variable is calculated as follows:

$$Z = \frac{x - \mu}{\sigma} \tag{30.10}$$

An example in corrosion of a normal distributed variable is pitting potential of stainless steel in chloride-containing solution.

Example. Consider a data set of pitting potentials of an austenitic stainless steel in seawater, normally distributed with mean value +0.460 V SCE and standard deviation 0.055 V. Data were obtained from laboratory tests. In seawater, due to the presence of microbiological activity, free corrosion potential is +0.350 V SCE. What is the probability for pitting corrosion to occur?

Pitting corrosion initiation occurs if $E_{pit} < E_{corr}$. The probability to have pitting corrosion corresponds to the probability P ($x < +0.350$ V SCE), where x is pitting potential, normally distributed. The standard variable Z is calculated according to Eq. 30.10: $Z = (x - 0.460)/0.055 = (0.350 - 0.460)/0.055 = -2.00$.

It follows that the probability P ($x < +0.350$ V SCE) can be converted to P ($Z < -2.00$). From the standard probabilities table, easily available in any statistical book, it follows that the pitting probability is lower than 2%.

Example. Consider the same data set of pitting potential reported in the previous example ($\mu = +0.460$ V SCE; $\sigma = 0.055$ V). Calculate the potential corresponding to a corrosion probability lower than 5%.

From the standard probabilities table, a cumulative standard probability of 5% corresponds to a Z value of -1.64. The corresponding normally distributed pitting potential is $x = \mu + \sigma \cdot Z = +0.460 + 0.055 \cdot (-1.640) = +0.370$ V SCE. In other words, corresponding to a free corrosion potential of +0.370 V SCE, the pitting corrosion probability is 5%.

30.2.2 Lognormal Distribution

A lognormal distribution is a continuous probability distribution of a random variable, x, whose logarithm is normally distributed. Considering the exponential function $x = \exp(W)$, if W is normally distributed, then $\ln(x)$ is normally distributed, as $\ln(x) = W$. A log normally distributed variable takes only positive real values. The probability density function, for $x > 0$, is:

$$f(x) = \frac{1}{x\omega\sqrt{2\pi}} \cdot e^{\left(-\frac{(\ln(x)-\theta)^2}{2\omega^2}\right)} \tag{30.11}$$

Mean value, μ, and standard deviation, σ, are:

$$\mu = e^{\theta + \frac{\omega^2}{2}} \tag{30.12}$$

$$\sigma = \sqrt{e^{2\theta + \omega^2}(e^{\omega^2} - 1)} \tag{30.13}$$

where θ and ω are the mean value and standard deviation of $W = \ln(x)$. The conversion with the standard normal distribution of the variable Z (Eq. 30.10) is possible for the random variable W normally distributed. An example in corrosion of a variable lognormal distributed is SCC failure time.

Example. Consider a data set of SCC failure time (x) for stainless steel in boiling MgCl$_2$ at 154 °C (Shibata 2000). Mean value is 114 min and standard deviation is 25 min. The mean value and standard deviation of $W = \ln(x)$ are 4.7 and 0.2, respectively. What is the probability that the failure time is higher than 100 min?

The probability to have failure in times higher than 100 min is $P(x > 100 \text{ min}) = P(\ln(x) > \ln(100)) = P(W > \ln(100)) = 1 - P(W \leq \ln(100)) = 1 - P(W \leq 4.6)$. Being W normally distributed, the standard variable Z is calculated according to Eq. 30.10: $Z = (W - 4.7)/0.2 = (4.6 - 4.7)/0.2 = -0.5$. It follows that the probability is $P(x > 100 \text{ min}) = 1 - P(W \leq \ln(100)) = 1 - 0.308 = 69\%$.

30.2.3 Poisson and Exponential Distributions

The Poisson distribution expresses the number of times an event occurs in an interval of time or space. The probability function of the Poisson distribution gives the probability that a discrete random variable is equal to a value x, where $x > 0$:

$$f(x) = \frac{e^{-\alpha} \cdot \alpha^x}{x!} \tag{30.14}$$

where α is mean value of events in the interval. The standard deviation of the distribution, σ, is $\alpha^{0.5}$. Poisson distribution is considered an appropriate model when the occurrence of one discrete event does not affect the probability that a second event will occur, that is, events occur independently, and when the rate at which events occur is constant.

Mears and Brown, and then Shibata (1996), reported that a Poisson distribution could describe the distribution of pits on the metal surface. Considering a mean rate of pit generation, $\lambda(s^{-1})$, the expected mean value of pits on the surface after a time t is $\alpha = \lambda t$:

$$f(x) = \frac{e^{-\lambda t} \cdot (\lambda t)^x}{x!} \tag{30.15}$$

where x is the number of pits generated after the time t. The survival probability represents the reliability of the system, $R(t)$, calculated by imposing $x = 0$, i.e. no corrosion attacks on the surface:

$$R(x = 0, t) = e^{-\lambda t} \tag{30.16}$$

Thus, the exponential distribution defined by Eq. 30.16 is the distribution of pit generation time, t, when random pit generation occurs. Equation 30.16 can be written as:

$$\ln R(t) = -\lambda t \tag{30.17}$$

where the meaning of symbols is known. The corresponding hazard cumulative function, $H(t)$, can be calculated according to Eqs. 30.7 and 30.17:

$$H(t) = -\ln R(t) = \lambda t \tag{30.18}$$

As expected, the reliability of the system increases as the pit generation rate decreases, which is considered constant and independent on the operating time of the system (the process is "without memory"). Nevertheless, in a macrocell corrosion phenomenon, as pitting corrosion, the occurrence of a localized corrosion attack on the metal surface affects the probability that a second event will occur, because of the presence of the macrocell current with creation of anodic and cathodic zones, so that λ cannot be strictly considered constant with time and independent on the position on the metal surface, unless in the initial period. The Mean Time To Failure (MTTF) is calculated as follows:

$$MTTF = \frac{1}{\lambda} \tag{30.19}$$

Example. Let's assume that the initiation time of pitting corrosion of carbon steel bars of a marine reinforced concrete structure follows an exponential distribution.

From previous experience, the mean initiation time is 10^5 h (about 11 years). What is the reliability after 7 years of service life? And what is the time before the first maintenance service considering a minimum allowed reliability of 90%?

The mean rate of pit generation is considered constant and equal to $\lambda = 10^{-5}$ h^{-1}. Reliability after 7 years ($\approx 0.6 \times 10^5$ h) can be calculated by Eq. 30.16:

$$R(7y) = e^{-\lambda t} = e^{-\left(0.6 \times 10^{-5} \times 10^5\right)} = 0.55 = 55\%$$

To assure the specified reliability (90%), the minimum time of monitoring and maintenance is:

$$\ln R = \ln(0.9) = -\lambda t = -0.1$$

$$t = \frac{0.1}{\lambda} = \frac{10^{-1}}{10^{-5}} = 10^4 \, h \cong 1 \, \text{year}$$

This approach considers a constant mean rate of pit generation and that reliability decreases as time increases. Conversely, pitting initiation can be better described by a distribution in which the failure rate decreases with time (infant mortality), as discussed in the following.

30.2.4 Generalized Extreme Value Statistics

Extreme value theory deals with the stochastic behaviour of the extreme values in a process. In corrosion engineering, extreme value statistics provides a powerful method for analyzing localized corrosion data, and especially for estimating pit depth or minimum time-to-failure. Indeed, from a practical point of view, the maximum pit depth is more important than the average pit depth because the deepest pit causes perforation.

The extreme value theory is defined by the so-called *generalized extreme value distribution*, GEV, which is a family of continuous probability distributions developed to combine the Gumbel, Fréchet and Weibull distributions also known as type I, II and III.

Introducing the location parameter, α, and the scale parameter, β, the cumulative distribution function of the GEV distribution is:

$$F(x) = \exp\left(-\left(1 + \gamma \cdot \left(\frac{x - \alpha}{\beta}\right)\right)^{-\frac{1}{\gamma}}\right); \gamma \neq 0 \qquad (30.20)$$

$$F(x) = \exp\left(-\exp\left(-\frac{x - \alpha}{\beta}\right)\right); \gamma = 0 \qquad (30.21)$$

where γ is the shape parameter, which governs the tail behavior of the distribution. The sub-families defined by $\gamma = 0$, $\gamma > 0$ and $\gamma < 0$ correspond, respectively, to the Gumbel, Fréchet and Weibull distributions. Type I distribution for the largest values and Type III distribution for the smallest values are the most observed in corrosion and are discussed in the following.

30.2.5 Gumbel Extreme Value Statistics

Extreme value statistics using Gumbel distribution (type I) is recommended for estimating the maximum corrosion depth, as described by Gumbel (1958). For pitting corrosion, a standardized procedure has been proposed in order to analyse the maximum pit depth distribution. Gumbel's cumulative function, Eq. 30.21, can be re-written as:

$$- \ln(- \ln(F(x))) = \frac{1}{\beta} x - \frac{\alpha}{\beta}\qquad(30.22)$$

The cumulative probability, $F(x)$, can be plotted as a straight line on Gumbel probability plot, which reports the values of cumulative probabilities on the vertical axis and the maximum penetration depths, x, on horizontal axis. The cumulative probability can be calculated as:

$$F_i = \frac{i}{1+N}\qquad(30.23)$$

where i is the ith position of the ordered value of x, in ascending order, and N is the total number of samples. The scale parameter, β, and the location parameter, α, can be calculated by the slope and the intercept of the straight line in the Gumbel probability plot, according to Eq. 30.22. The mean value and standard deviation of Gumbel distribution are calculated as follows:

$$\mu = \alpha + 0.58 \cdot \beta\qquad(30.24)$$

$$\sigma = 1.28 \cdot \beta\qquad(30.25)$$

where 0.58 is the Eulero-Mascheroni constant.

For the pit depth distribution, the *return period* provides an estimation of maximum pit depth over the entire surface from the information obtained for minute area of sampling, providing that the environmental and metallurgical conditions are the same over the surface (Kowaka 1994). The return period, T, is a measurement of the average recurrence interval over an extended period of time and it is defined as:

$$T = \frac{1}{1 - F(x)} \tag{30.26}$$

By introducing Eq. 30.23 and considering the average rank method, the return period can be written as:

$$T = \frac{1+N}{1+N-i} = \frac{1}{1-f} \tag{30.27}$$

where f is the confidence. For example, to achieve a confidence of 90%, at least 11 samples are required. The value of x (pit depth) at a given T is the maximum pit depth, x_{max}, for the T times larger surface area, S, compared with the small sample area, s.

Example. The piping system of a cruise ship is made of AISI 304 stainless steel. After few months from delivery, frequent pitting corrosion attacks occurred on pipe joints due to the application of a lubricant containing chlorides. In order to evaluate the corrosive state of the piping system, a corrosion investigation has been carried out on 37 pipe joints removed from the ship after 24 months. The joints were analyzed by OES (Optical Emission Spectroscopy) in order to measure the number of pits and the depth of corrosion attacks. Calculate the probability to find corrosion attacks at 24 months deeper than the maximum acceptable threshold (800 μm).

Table 30.2 reports the maximum pit depth of samples extracted from the field after 24 months. Data are ordered in ascending order and the Gumbel cumulative probability is calculated by means of Eq. 30.23. A Gumbel probability curve is plotted by reporting the cumulative probability and the maximum pit depth, x (Fig. 30.2).

The scale parameter, β, and the location parameter, α, are calculated by the slope and the intercept of the regression line in the Gumbel probability plot:

$$0.0039 = \frac{1}{\beta} \Rightarrow \beta = 256\,\mu m$$

$$0.847 = \frac{\alpha}{\beta} \Rightarrow \alpha = 217\,\mu m$$

The mean value (Eq. 30.24) and standard deviation (Eq. 30.25) of Gumbel distribution are:

$$\mu = \alpha + 0.58 \cdot \beta = 217 + 0.58 \cdot 256 = 365\,\mu m$$

$$\sigma = 1.28 \cdot \beta = 1.28 \cdot 256 = 328\,\mu m$$

Considering a maximum acceptable thickness reduction of 800 μm, defined by mechanical resistance requirements, the reliability function that a pit has a depth over this threshold is:

Table 30.2 Maximum pit depth of AISI 304 samples extracted after 24 months from field and calculation of Gumbel cumulative probability

Sample	Max. pit depth (µm)	$F_i = i/(1 + N)$	$-\ln(-\ln(F_i))$
1	0	0.03	−1.29
2	0	0.05	−1.08
3	0	0.08	−0.93
4	0	0.11	−0.81
5	0	0.13	−0.71
6	0	0.16	−0.61
7	100	0.18	−0.53
8	150	0.21	−0.44
9	150	0.24	−0.36
10	150	0.26	−0.29
11	150	0.29	−0.21
12	200	0.32	−0.14
13	200	0.34	−0.07
14	250	0.37	0.00
15	250	0.39	0.07
16	250	0.42	0.15
17	300	0.45	0.22
18	300	0.47	0.29
19	300	0.50	0.37
20	300	0.53	0.44
21	300	0.55	0.52
22	350	0.58	0.60
23	350	0.61	0.69
24	350	0.63	0.78
25	350	0.66	0.87
26	400	0.68	0.97
27	500	0.71	1.07
28	500	0.74	1.19
29	500	0.76	1.31
30	600	0.79	1.44
31	700	0.82	1.59
32	700	0.84	1.76
33	800	0.87	1.96
34	800	0.89	2.20
35	800	0.92	2.50
36	800	0.95	2.92
37	1210	0.97	3.62

Fig. 30.2 Regression line of Gumbel cumulative probability versus maximum pit depth of data reported in Table 30.2

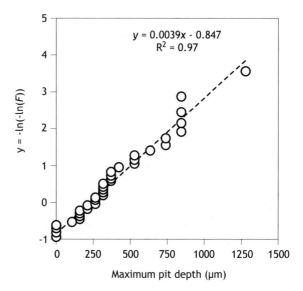

$$R(x = 800\,\mu\text{m}) = 1 - F(x = 800\,\mu\text{m})$$

$$R = 1 - \exp\left(-\exp\left(-\frac{800 - 217}{256}\right)\right) \cong 10\%$$

30.2.6 Weibull Extreme Value Statistics

The cumulative distribution function of the Weibull distribution (type III) for the smallest value is:

$$F(x) = 1 - \exp\left(-\left(\frac{x - \alpha}{\beta}\right)^{-\frac{1}{\gamma}}\right) = 1 - \exp\left(-\left(\frac{x - \alpha}{\beta}\right)^{\delta}\right) \qquad (30.28)$$

where α is the location parameter, β is the scale parameter and $\delta = -\gamma^{-1}$ is the shape parameter. Frequently, the location parameter, α, is not used, and the equation reduces to the two-parameter Weibull distribution. The Weibull distribution is widely used in reliability and life data analysis due to its versatility related in particular to the shape parameter, δ, also known as Weibull slope, which defines the failure rate, i.e. failure events for unit time, $\lambda(x)$:

$$\lambda(x) = \frac{\delta}{\beta}\left(\frac{x - \alpha}{\beta}\right)^{\delta - 1} \qquad (30.29)$$

where the variable x assumes the meaning of operating time and the other symbols are known. In reliability engineering, the cumulative distribution function corresponding to a *bathtub curve* may be analysed using a Weibull chart where the value of the parameter δ controls the failure rate with time (Fig. 30.3):

- $\delta < 1$ for the early failures, where the failure rate decreases with time (infant mortality)
- $\delta = 1$ for the random failures, where failure rate is constant, $\lambda = 1/\beta$, during "useful life"
- $\delta > 1$ for the wear-out failures, where the failure rate increases with time at the end of the design lifetime.

In other words, in the early life of a component adhering to the bathtub curve, the failure rate is high but rapidly decreases. In the mid-life, the failure rate is low and constant. In the late life of the component, the failure rate increases rapidly.

Estimation of the Weibull parameters. The estimation of the parameters of the Weibull distribution can be found graphically via probability plotting paper, or analytically. Only the graphical method is discussed because it provides a better understanding of how data are distributed. To better illustrate this procedure, consider the following example of initiation time of localized corrosion of ten reinforced concrete samples exposed to laboratory tests to reproduce the exposure condition of concrete structure in the splash zone of a marine environment. From previous experience, the accelerated testing conditions are characterized by an intensification index (Lazzari 2017) of 3, which means that the expected initiation times in field are three times those measured in laboratory tests. The free corrosion potential of each concrete bar has been measured periodically in time in order to monitor corrosion initiation.

As a first step, corrosion initiation times are listed in ascending order, as reported in Table 30.3. For each time-to-corrosion, the median rank is calculated using the following equation:

$$F(i) = \frac{i - 0.3}{N + 0.4} \tag{30.30}$$

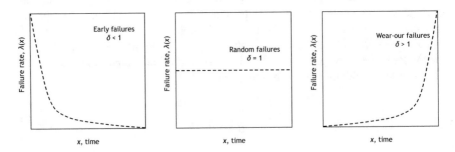

Fig. 30.3 Failure rate, $\lambda(x)$, with time as a function of the shape parameter, δ

Table 30.3 Initiation time of localized corrosion, x_i, of ten reinforced concrete samples exposed to chloride-containing solution to reproduce the splash zone of a concrete structure

i	x_i (days)	$\ln(x_i)$	$F_i = (i - 0.3)/(N + 0.4)$	$R_i = 1 - F_i$	$\ln(\ln(1/R_i))$
1	525	6.26	0.067	0.933	−2664
2	546	6.30	0.163	0.837	−1723
3	560	6.33	0.260	0.740	−1202
4	590	6.38	0.356	0.644	−0822
5	609	6.41	0.452	0.548	−0509
6	620	6.43	0.548	0.452	−0230
7	630	6.45	0.644	0.356	0033
8	640	6.46	0.740	0.260	0299
9	650	6.48	0.837	0.163	0594
10	670	6.51	0.933	0.067	0993

Equation 30.28 can be rewritten (considering $\alpha = 0$) by introducing the reliability, $R = 1 - F$, as follows:

$$R(x) = 1 - F(x) = \exp\left(-\left(\frac{x}{\beta}\right)^{\delta}\right)$$

$$\ln\left(\ln\left(\frac{1}{R(x)}\right)\right) = \delta \ln(x) - \delta \ln(\beta)$$

(30.31)

In a Weibull plot (Fig. 30.4), which reports $\ln(\ln(1/R_i))$ as a function of $\ln(x)$, Eq. 30.31 is represented by a straight line with slope δ and intercept $-\delta \ln(\beta)$, where the scale parameter, β, represents the initiation time corresponding to which R is equal to 37%.

The linear regression of the data set provides the values of the Weibull slope, δ, and of the scale parameter, β:

$$\delta = 13.8$$
$$\beta = 610 \text{ days}$$

According to Fig. 30.3, the failure rate increases with time for $\delta > 1$, and the corresponding failure mechanism is the wear-out failure at the end of the design lifetime. In this example, initiation time of localized corrosion depends strongly on the penetration rate of chlorides through the cover concrete by diffusion and other transport phenomena. Then, once chloride content at the rebar reaches the critical threshold, corrosion propagates by a macrocell mechanism. Considering a relia-bility (a survival probability) of 99%, the corrosion initiation time is 437 days, obtained by Eq. 30.31 by introducing the calculated parameters of Weibull distri-bution. Being the intensification index equals to 3, which means that the expected

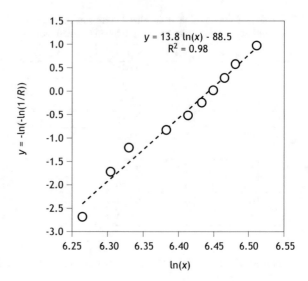

Fig. 30.4 Weibull probability paper of data reported in Table 30.3

initiation times in field are three times the times measured in laboratory tests, it is possible to conclude that in field reliability is maintained higher than 99% for times lower than about 4 years.

30.3   Sample Size and Curve Fitting

30.3.1   Sample Size

Sample size determination assumes a crucial role in a statistical analysis of data and represents the number of observations or replicates to include in a data set. The goal of selecting the proper sample size is to make inferences about a population from a sample. In corrosion testing, it is essential to select a, the area of each sample, and N, the number of samples, carefully. While small a and N make measurement easy, the error margin is increased. Otherwise, larger sample sizes generally lead to increased precision when estimating unknown parameters.

30.3.2   Curve Fitting

Once data are obtained from laboratory testing or field investigations, it is desirable to determine the best algebraic expression to fit the data set by minimizing the variance between the measured value and the calculated value. The regression equations, including linear, polynomial and multiple-variable regression equations, can be sometimes used, with some constrains, for extrapolations that refer to the use of a fitted curve beyond the range of the observed data.

Linear regression. Data are fitted to a linear relationship, $y = mx + b$. The best fit is given by:

$$m = \frac{N \sum xy - \sum x \sum y}{N \sum x^2 - \left(\sum x\right)^2} \tag{30.32}$$

$$b = \frac{1}{N}\left[\sum x - m \sum y\right] \tag{30.33}$$

where m and b are the slope and y-intercept of the estimated line and N is the number of observations on x and y.

Polynomial regression. This analysis is used to fit data to a polynomial equation of the following form:

$$y(x) = a + bx + cx^2 + \cdots \tag{30.34}$$

where a, b, c, etc. are the constants used to fit the data set. Polynomial regression fits a nonlinear relationship used to describe nonlinear phenomena.

Multiple regression. This analysis is used when data sets involve more than one independent variable. In case of a linear regression, the best fit is given by the general equation:

$$y = a + b_1 x_1 + b_2 x_2 + b_3 x_3 + \cdots \tag{30.35}$$

where a, b_1, b_2, b_3, etc. are the constants used to fit the data set; x_1, x_2, x_3, etc. are the observed independent variables.

For instance, the international standard ISO 9223 *Corrosion of metals and alloys. Corrosivity of atmospheres, Classification, determination and estimation* reports an estimation of the corrosivity of atmospheres based on corrosion rate calculated from environmental data or from information on environmental conditions and exposure situation.

In particular, the proposed functions for four standard metals (carbon steel, zinc, copper, aluminium) describe the corrosion rate (C_{rate}, μm/year), after the first year of exposure in open air as a function of SO_2 deposition (P_d, mg/(m^2 d)), chloride deposition (s_d, mg/(m^2 d)), temperature (T, °C) and relative humidity (*RH*, %). Regression equations are generally not linear; for carbon steel, the multiple regression equation is:

$$C_{rate} = 1.77\, P_d^{0.52} \cdot \exp(0.020\, RH + f_{St}) + 0.102\, S_d^{0.62} \cdot \exp(0.033 \cdot RH + 0.040\, T) \tag{30.36}$$

where f_{St} is a function of temperature.

30.4   International Standard

- ASTM G 16—Standard guide for applying statistics to analysis of corrosion data
- ISO 14802—Corrosion of metals and alloys—Guidelines for applying statistics to analysis of corrosion data.

Bibliography

Gumbel EJ (1958) Statistics of extremes. Columbia University Press, New York

Kowaka M (ed) (1994) Introduction to life prediction of industrial plant materials: application of the extreme value statistical method for corrosion analysis, Allerton Press, New York (originally published in Japanese by The Japan Society of Corrosion Engineers, 1984)

Lazzari L (2017) Engineering tools for corrosion. Design and diagnosis. European Federation of Corrosion (EFC) Series, vol 68. Woodhead Publishing, London, UK

Shibata T (1996) W.R. Whitney award lecture: statistical and stochastic approaches to localized Corrosion. Corrosion 52(11):813–830

Shibata T (2000) Corrosion probability and statistical evaluation of corrosion data. In: Revie RW (ed) Uhlig's corrosion handbook, 2nd edn. Wiley, Hoboken, NJ, pp 367–392

Glossary

Accelerated corrosion testing Corrosion test carried out by imposing more severe test parameters to reduce substantially its duration.

Acidic corrosion Corrosion attack in acidic solutions for which cathodic process is hydrogen evolution.

Active (1) The high tendency of a metal to react. (2) A state of a metal that is corroding.

Active metal A metal which is active and has a tendency to react (or corrode).

Aerobic Presence of air or oxygen as dissolved gas.

Aerobic environment An environment with oxygen.

Ammeter Instrument that measures the intensity of electric current, in A, characterized by very low internal impedance.

Ampere (A) Unit of measurement of electric current in the International System of Units (SI).

Amphoteric metal A metal that is susceptible to corrosion in both acid and alkaline environments.

Anaerobic Free of air or oxygen.

Anaerobic environment An environment free of oxygen or air.

Anion A negatively charged ion.

Anode The electrode on which oxidation takes place releasing electrons.

Anodic current Current flowing across the metal-electrolyte interface from the metal surface to the electrolyte, causing an anodic polarisation.

Anodic oxidation Formation of an oxide film at a metal surface (usually aluminium and titanium) by an electrochemical process.

© Springer Nature Switzerland AG 2018
P. Pedeferri, *Corrosion Science and Engineering*, Engineering Materials,
https://doi.org/10.1007/978-3-319-97625-9

Anodic polarisation Or anodic overvoltage. The change of the electrode potential in the noble (positive) direction caused by an anodic current.

Anodic protection Electrochemical technique to protect active-passive metals by supplying an anodic current.

Anodic reaction Oxidation reaction that produces cations, generally metallic ions, releasing electrons.

Atmosphere The mass of air that surrounds the Earth, composed by roughly 78% N_2, 21% O_2, 0.3% Ar, 0.33% H_2O and 0.04% CO_2.

Austenitic stainless steel A stainless steel with austenitic micro-structure at room temperature.

Auxiliary electrode See Counter Electrode.

Backfill Material used around galvanic anodes and impressed current anodes to reduce anode-soil resistance.

Barrier effect The property of coatings or layers with low permeability to liquids and gases.

Bimetallic corrosion See Galvanic Corrosion.

Bimetallic coupling See Galvanic Coupling.

Biocide Chemical substance that destroys or inhibits the growth or activity of living organisms, used to eliminate microbiological degradation processes.

Biofilm A formation of a bacteria-enriched layer on a metal surface immersed in seawater.

Blast-furnace cement Cement obtained by mixing ground granulated blast furnace slag and Portland cement.

Blended cement Cement obtained by adding substances with pozzolanic activity to Portland cement.

Blistering Swelling of paint, or formation of a blister inside steel due to hydrogen ions diffusion and recombination within the metal to form hydrogen molecules, creating high internal pressure.

Boundary element method (BEM) A numerical technique for solving field equations based on imposing the constrains on anodic and cathodic surfaces.

Brackish water Water with content of salts higher than 2 g/L.

Brittle cracking Cracking with limited or no plastic deformation.

Buffer A substance that prevents pH changes when present in the electrolyte.

Calamine Layer of mixture of iron oxides formed at the surface of steel at high temperature or in poorly aerated environments.

Calcareous deposit A calcium carbonate scale formed in hard freshwater or deposited on a metallic surface in seawater when cathodic protection is applied.

Carbon steel (CS) Steel containing basic elements only, as carbon and manganese.

Carbonation Loss of alkaline pH of concrete due to carbon dioxide reaction with calcium hydroxide in concrete and mortars.

Cation A positively charged ion.

Cathode The electrode on which a reduction reaction takes place by gaining electrons.

Cathodic current Current flowing across the metal-electrolyte interface from the electrolyte to the metal surface, causing a cathodic polarisation.

Cathodic disbonding Destruction of adhesion between coating (or paint) and metal surface due to products of cathodic reactions (hydrogen evolution).

Cathodic polarisation Or cathodic overvoltage. The change of the electrode potential downward caused by a cathodic current.

Cathodic prevention (CPrev) Electrochemical technique as Pietro Pedeferri (1938–2008) named to protect passive metals from pitting consisting on applying a cathodic current before passive film breakdown.

Cathodic protection (CP) Electrochemical technique to protect metals, consisting on lowering the metal potential by supplying a cathodic current.

Cathodic protection by immunity Cathodic protection condition which brings metals to immunity.

Cathodic protection by passivity Cathodic protection condition which brings and maintains metals to passive state.

Cathodic protection by quasi immunity Cathodic protection condition which brings metals to negligible corrosion rate (below 10 µm/yfor carbon steel and cast iron in waters and soil).

Cathodic reactant Chemical species whose reduction provides the complementary process to the metal anodic dissolution.

Cathodic reaction Reduction reaction that gains electrons released by the anodic reaction.

Caustic brittleness Stress corrosion cracking of steel in hot alkaline solutions.

Cavitation Damage to a metallic material under conditions of severe turbulent flow.

Cell See Electrochemical Cell and Macrocell.

Cement Hydraulic binder; finely ground inorganic material that forms with water a paste that hardens thanks to hydration processes.

Chemical equivalent Atomic or molecular mass of an element divided by its valence.

Clad, cladding material Layer of a corrosion resisting metal used as corrosion prevention for vessels and pipelines; backing metal is typically carbon and low alloy steel.

Clay Finely grained natural rock contained in soils with high capacity to retain water.

Coating Physical barrier to separate the metal surface from the environment.

Coating efficiency See Efficiency.

Complexing species Chemical species that forms stable complex (for example cyanides) or insoluble products with metal ions (for example oxides, hydroxides or sulphides) decreasing ions concentration in solution.

Concentration cell An electrochemical cell composed of two identical electrodes immersed in two electrolytes that differ by their concentration.

Concentration polarisation Overvoltage contribution produced by concentration changes in the electrolyte.

Concrete Construction material obtained by mixing in suitable proportions cement, water and stone aggregates.

Conductivity Quantitative expression of the attitude of a conductor to allow the flow of an electric current, measured in S/m. It is the inverse of resistivity.

Conductor Medium that makes ions or electrons available as charge carriers, which can migrate under the action of an electric field.

Corrosion Deterioration of a material, usually a metal, by a chemical or electrochemical reaction with its environment.

Corrosion allowance Extra wall thickness which can be consumed by corrosion without affecting the integrity and resistance to a tensile load or a pressure.

Corrosion coupon Piece of material of the same metal alloy of the plant, exposed to the same environment, in order to monitor corrosion and protection conditions.

Corrosion current density The corrosion rate expressed in A/m^2 (for iron 1 A/m^2 corresponds to 1.17 mm/year thickness loss).

Corrosion inhibitor A chemical substance that prevents corrosion or reduces the corrosion rate.

Corrosion fatigue Fatigue-type cracking of metal caused by repeated or fluctuating stresses in a corrosive environment.

Corrosion potential See Free Corrosion Potential.

Corrosion rate The rate at which corrosion proceeds, usually expressed as µm/year.

Corrosion resistance Ability of a material to withstand corrosion in a given system.

Corrosion resistant alloy (CRA) Alloy consisting of combination of metals such as iron, chromium, molybdenum, nickel, cobalt, titanium, with increased corrosion resistance with respect to carbon steel.

Corrosion tubercle Bulges on a metal surface due to the accumulation of corrosion products on small localized corrosion cavities.

Corrosiveness The tendency of an environment to cause corrosion.

Coulomb Unit of measurement of electric charge in the International System of Units (SI), measuring the quantity of charge transported by a 1 A current flowing for 1 s.

Counter electrode Electrode used to impose current to the working electrode. Also called auxiliary electrode.

Crevice corrosion Localized corrosion of an active-passive metal at a shielded surface from full exposure to the environment (typically in a gap).

Crevice critical gap size (CCGS) Minimum interstice that allows the aggressive environment to enter but impedes the diffusion of oxygen.

Critical crevice temperature (CCT) Maximum temperature without crevice attack for each gap size.

Critical pitting chloride concentration (CPCC) Threshold below which pitting does not initiate.

Critical pitting temperature (CPT) Maximum temperature at which stainless steel resists pitting attack, once fixed potential and environmental conditions.

Crystal grain Portion of metal consisting of a single crystal, originated during metal formation, in particular, during its solidification.

Cupronickel Copper alloy that contains nickel, plus iron and manganese as strengthening elements, used in seawater applications for its high corrosion resistance.

Current Flow of an electric charge, carried either by electrons in a conductive material or by ions in an electrolyte.

Current density The current to or from a unit area of an electrode surface, usually expressed in mA/m^2.

Depolarisation The reduction of overvoltage contributions.

Dezincification Selective corrosion of zinc in brass.

Differential aeration cell An electrochemical cell where the same metal is in contact with different oxygen concentration.

Diffusion Displacement of atoms and ions under the effect of a concentration gradient, governed by Fick laws.

Diffusion limiting current density The current density that corresponds to the maximum transport rate that a particular species can sustain because of the limitation of diffusion (often referred to as limiting current density).

Double layer The interface between a metal and an electrolyte where an electrical charge separation takes place (the simplest model is represented by a parallel plate condenser).

Drainage Drainage of electric current from a structure in contact with an electrolyte, by means of a metallic conductor (natural drainage) or an impressed current or galvanic anodes (forced drainage).

Drinking water See Potable Water.

Driving voltage The difference in potential which measures the metal's tendency to oxidise or more generally the energy available for a reaction to occur (also difference in potential between two metals).

Efficiency (of coating) Ratio of surface area of a metal covered by a coating layer to the bare metal surface.

Efficiency (of inhibitor) Percent decrease in corrosion rate caused by a corrosion inhibitor.

Electric field Zone of electrolyte where electric forces act, generated by the circulation of current.

Electrical continuity Ability of a structure composed of different metals to conduct electricity through electrical connections.

Electrical resistance Tendency of a conductor to oppose resistance to the passage of current when subjected to a voltage, measured in Ω. It causes the conductor heating.

Electro-osmosis The phenomenon of water diffusion through a coating or a porous media promoted by the passage of current.

Electrochemical cell A cell consisting of two electrodes, an anode and a cathode, immersed in an electrolyte; anode and cathode can be different metals or different areas on the surface of a single metal.

Electrochemical reaction A chemical reaction characterised by a gain or loss of electrons at electrode surfaces.

Electrical resistance probe (ER probe) Metal coupon of the same material of the monitored structure, which allows the calculation of corrosion rate by measuring its electrical resistance variation.

Electrochemistry Scientific discipline that studies the reactions taking place at the surface of a conductor exchanging current with an electrolyte, involving ionic and electronic species.

Electrode An electron conductor by means of which electrons are provided for, or removed from, an electrode reaction.

Electrode potential The potential of an electrode measured against a reference electrode.

Electrodeposition Electrochemical technique by which a metallic coating is obtained from the reduction of metal ions.

Electromotive force (EMF) Potential difference expressing the variation of electrical energy associated to a corrosion reaction, calculated from Gibbs free energy.

Electronic conductor Conductor in which the current is carried by electron migration.

Electrolysis The forced passage of electricity through a cell that produces chemical changes at the electrodes.

Electrolyte A solution containing ions that migrate in an electric field.

Enamel Smooth, glossy and durable coating made by a thin layer of glass or ceramic.

Equilibrium potential The electrode potential of a reversible electrode when it is not polarised. It is calculated by Nernst equation.

Erosion-corrosion A conjoint action involving corrosion and a mechanical action (erosion) in the presence of a moving corrosive fluid leading to accelerated loss of material.

Evans diagram (E-i) Diagram of potential–current density proposed by U.R. Evans (1889–1980) mostly used to represent corrosion processes.

Exfoliation Special type of intergranular corrosion typical of aluminium alloys which proceeds through preferential intergranular paths, usually parallel to the direction of extrusion or rolling, causing the metal to delaminate.

Exchange current density The rate of exchange of positive or negative charges between the metal-electrolyte interface of an electrode at its equilibrium potential.

Extreme value statistics It deals with the extreme deviations from the median of probability distributions. It aims to assess, from a given ordered sample of a given random variable, the probability of events that are more extreme than any previously observed.

Faraday A quantity of electric charge, equal to 96,485 C, required to oxidise or reduce one chemical equivalent.

Faraday law Fundamental law of electrochemistry stating that the mass formed or consumed in an electrochemical process is proportional to the circulated charge through the electrochemical equivalent.

Fatigue Mechanical phenomenon leading to failure under repeated and cyclic stresses lower than the material tensile strength.

Ferrite Body-centered cubic crystalline phase of iron-based alloys.

Ferritic steel A steel whose microstructure at room temperature consists predominantly of ferrite.

Finite element method (FEM) A numerical technique for solving field equations based on imposing the constrains on anodic and cathodic surface and the discretization of the domain.

Free corrosion potential The potential of a metal in an electrolyte under open-circuit conditions (also known as open-circuit potential or corrosion potential).

Fick laws Field equations which describe diffusion phenomenon.

Filiform corrosion Special type of crevice corrosion that occurs beneath paints or lacquers on a coated metal surface.

Flade potential Defined by Friedrich Flade (1880–1916). See Passivity Potential.

Fouling Deposits including accumulation and growth of marine organisms, beneath which biofilm can form.

Freshwater Water characterized by a low salinity level (< 2 g/L).

Fretting Wear damage induced under load in presence of repeated micrometric slips between two surfaces.

Fretting-corrosion A conjoint action involving corrosion and fretting between two metallic surfaces, triggering fatigue cracks.

Galvanic anode Metal or alloy used to obtain cathodic protection of more noble materials.

Galvanic cell See Electrochemical Cell.

Galvanic corrosion Accelerated corrosion of a metal because of an electrical contact with a more noble material (a metal or a non-metallic conductor) in a corrosive electrolyte (also called bimetallic corrosion).

Galvanic coupling A pair of dissimilar metals or conductive nonmetals in electrical contact in an electrolyte (also called bimetallic coupling).

Galvanic current The electric current in a galvanic cell.

Galvanic series Ranking of metals and alloys according to their free corrosion potential in a given electrolyte.

Galvanized steel Carbon steel coated with a protective zinc layer.

Galvanostat Instrument that feeds an electrochemical cell with constant current.

Galvanostatic Refers to a technique for maintaining a constant exchanged current on an electrode.

Gaussian distribution See Normal Distribution.

Graphitization Selective corrosion of grey cast iron, consisting in the etching of iron and consequent increase in the surface content of graphite.

Gumbel distribution It is one of most used Extreme Value Statistics, often used for localized corrosion analysis (for instance, maximum pit depth).

Hardness (of metal) Measure of the resistance to localized plastic deformation induced by either mechanical indentation or abrasion.

Hardness (of water) Water content of Ca^{2+} and Mg^{2+} ions, i.e. ions that contribute to the formation of limestone deposits.

Heat affected zone (HAZ) Portion of metal that did not melt during welding, but underwent microstructural modification.

High strength steel Steel subjected to heat treatments or mechanical treatments to reach high tensile resistance.

High strength low-alloy steel (HSLA) Low-alloy steel that is heat treated to reach high tensile resistance.

Huey test Test evaluating susceptibility to intergranular attack of stainless steel by immersion in boiling nitric acid.

Hydrogen embrittlement Loss of ductility of a metal resulting from the diffusion in the metal of atomic hydrogen.

Hydrogen induced cracking (HIC) Internal damage as small cracks (stepwise type) caused by atomic hydrogen recombining inside carbon steel with high sulphur content.

Hydrogen overvoltage Overvoltage associated with the reaction of hydrogen evolution.

Immunity A metal state that corresponds to an electrode potential more negative than the equilibrium potential.

Imperfect passivity A passivity condition of an active-passive metal in which pitting cannot start but can propagate, if already initiated.

Impressed current An electric current supplied by a device employing a power source that is external to the electrode system.

Impressed current anode An anode used in impressed cathodic protection systems. It stands an anodic reaction as metal dissolution, oxygen or chlorine evolution.

Initiation time Time needed for corrosion conditions to onset, and hence for corrosion to start propagating.

Intergranular corrosion Type of localized corrosion attack occurring at grain boundaries of sensitized alloys, typically stainless steels.

Ion An electrically charged atom or complex of atoms.

Ion activity The molar concentration of an ion multiplied by the ion activity coefficient.

Ionic conductor Conductor in which current is carried by ions migration.

IR drop See Ohmic Drop.

IR-free potential Potential measured without the voltage error caused by the ohmic drop due to the protection current or any other flowing current.

Knife-line attack Special intergranular corrosion attack occurring to welds, limited to a few crystalline grains parallel to the weld bead and penetrating steel thickness up to cut off the weld.

Langelier saturation index (LSI) Indicator of the degree of saturation of calcium carbonate in water. It is calculated as $pH-pH_S$ (actual pH minus saturation pH of calcium carbonate). Positive values indicate scaling tendency.

Linear polarisation resistance (LPR) Electrochemical technique used to measure the instantaneous corrosion rate of a metal in an electrolyte. It employs three electrodes: a working electrode, a reference electrode and an auxiliary electrode (or counter electrode).

Localized corrosion Corrosion attack occurring on a small portion of the exposed surface of metal.

Lognormal distribution A continuous distribution in which the logarithm of a variable has a normal distribution.

Low-alloy steel Steel containing intentional alloying element additions with a content less than 5%.

Low-carbon steel Steel with carbon content less than 0.3% and no other intentional alloying element.

Luggin probe A small tube or a capillary filled with electrolyte, terminating closely to the metal surface of an electrode to eliminate ohmic drop during potential measurements.

Macrocell Electrochemical cell setting up by the separation of anodic and cathodic surfaces.

Macrocouple Couple of two metals with different practical nobility.

Macrocouple current Current flowing from the anode to the cathode in a macrocouple.

Magnetite Iron oxide with chemical formula Fe_3O_4 showing electrical conductivity and magnetic properties.

Manganese-oxidizing-bacteria (MOB) Bacteria of Leptothrix strain type which ennobles the corrosion potential, then inducing pitting attack on low-grade stainless steels in fresh water.

Martensite Steel structure consisting of a carbon oversaturated solid solution with needle-like microstructure, high hardness and brittleness.

Martensitic stainless steel A stainless steel with martensitic microstructure at room temperature.

Mesa corrosion Corrosion occurring in presence of CO_2 at temperatures below 80 °C, typical of oil and gas facilities.

Microbiologically-induced corrosion (MIC) Corrosion attack promoted by the presence of specific bacteria, also known as microbial corrosion or biological corrosion.

Mobility Velocity of an ion in a given electrolyte divided by the electric field intensity, in $[m^2 \ (s \ V)^{-1}]$.

Monitoring (of corrosion) Inspection and control program followed to verify conditions of corrosion and protection of a structure.

Muntz alloy Brass with 60% copper 40% zinc and traces of iron used for corrosion resistant machine parts and boat hulls.

Nernst equation An equation that expresses the equilibrium potential of an electrochemical reaction.

Noble Expresses a low tendency to react or the positive direction of electrode potential.

Noble metal Metal with a low tendency to react (or corrode), such as gold or platinum. Its equilibrium potential is highly positive.

Noble potential A potential towards the positive end of a scale of electrode potentials.

Non-destructive testing (NDT) Techniques used to monitor the characteristics of components (e.g. material thickness, presence of cracks) without causing damage.

Normal distribution The most common distribution function for independent, randomly generated variables. Normal distribution is bell-shaped and symmetrical with respect to the mean value.

Off potential In cathodic protection, potential of a protected structure measured by switching the current off.

Ohm (Ω) Unit of measurement of electric resistance in the International System of Units (SI).

Ohmic drop Potential difference resulting from the passage of current inside a conductor. It depends on current intensity, electrode or electrolyte resistivity and path length.

Open-circuit potential The potential of an electrode in the absence of an external current flow. See also Free Corrosion Potential.

Overvoltage The change in potential of an electrode when current is exchanged with the electrolyte (see Polarisation).

Oxidation (1) An electrochemical reaction in which electrons are released. (2) Corrosion of a metal that is exposed to an oxidizing gas typically at elevated temperatures.

Oxygen limiting current density The current density that corresponds to the maximum transport rate of oxygen by diffusion to the metal surface (see diffusion).

Oxygen scavenger Substance used to reduce the oxygen content in a solution by a chemical reaction with dissolved oxygen. Most used compounds are sodium bi-sulphite and idrazine or its derivates.

Painting cycle Series of paint layers consisting of a primer in contact with the metal, an intermediate coat and a finishing coat in contact with the environment, each with different composition and function.

Passivation The thermodynamic condition (E, T, pH) of stability of metal oxides and hydroxides in Pourbaix diagrams.

Passive A state of passivity of a metal.

Passivity Kinetic condition of stability of metal oxides and hydroxide formed on a metal surface that leads to a practical halt of corrosion.

Passivity potential The lowest value of the passive range, also called Flade potential.

Pedeferri diagram (E-% Cl) A diagram potential–chloride content proposed by Pietro Pedeferri (1938–2008) to represent graphically the regions for occurrence or prevention of pitting of passive metals.

Perfect passivity A passivity condition of a metal in which localized corrosion (pitting and crevice) cannot initiate nor propagate.

pH A measure of acidity or alkalinity of an aqueous electrolyte. pH value is the negative logarithm of the hydrogen ion activity ($-\log[H^+]$ where $[H^+]$ = hydrogen ion activity).

Piontelli probe Device for IR-free potential measurements developed by Piontelli characterized by a capillary with a lateral channel with the aim to reduce the shielding effects (see Luggin capillary).

Pipeline Series of straight pipes welded together over a long distance, used to convey fluids (liquids and gases) from one location to another.

Pipeline inspection gauge (PIG) Pipeline inspection tool based on ultrasonic or electromagnetic principles to measure continuously surface thickness and check cracks or localized corrosion attacks.

Piping Complex network of pipes within the boundaries of a plant.

Pitting Localized corrosion of an active-passive metal that takes the form of cavities called pits. Surface area surrounding pits remains passive.

Pitting potential The lowest value of potential at which pits nucleate and grow.

Pitting resistance equivalent number (PREN) Index of stainless steels calculated from their composition, used to define the resistance to localized corrosion.

Poisson distribution Discrete distribution used to model the number of events occurring within a given time interval.

Polarisation The change of potential caused by a current flow across the electrode/electrolyte interface.

Polarisation curve A plot of current density versus electrode potential for a specific electrode/electrolyte combination.

Polarisation resistance The slope (dE/di) at the corrosion potential of a potential (E)-current density (i) curve. Corrosion rate is inversely proportional to the polarisation resistance (see Linear Polarisation Resistance).

Pole Extremity of a galvanic chain. The positive pole is the one with higher potential, the negative pole the other one.

Portland cement Cement obtained by grinding and heating a mixture of calcium carbonate, clay and sand (clinker) with small additions of gypsum.

Post-welding treatment Weld heating and cooling aimed at obtaining desired microstructure and mechanical properties of weld bead and heat affected zone.

Potable (or drinking) water Water with limited amount of specific salts and restrictions on bacteria and dangerous chemicals, according for example to the Council Directive 98/83/EC of the European Commission.

Potential Electrode potential with respect to a reference electrode.

Potentiodynamic It refers to a technique wherein the potential of an electrode with respect to a reference electrode is varied at a selected rate by application of an external current.

Potentiostat An instrument for automatically maintaining a constant electrode potential.

Potentiostatic It refers to a technique for maintaining a constant electrode potential.

Pourbaix diagram (E-pH) A graphical representation of thermodynamic stability of species for metal/electrolyte systems proposed by M. Pourbaix (1904–1998).

Practical nobility Free corrosion potential of a metal in a specific environment.

Precipitation hardening Heat treatment to increase mechanical resistance of a metallic alloy.

Primary passivation potential Minimum potential above which a stable oxide forms on a metal surface.

Protection current density The current density necessary to obtain cathodic protection of a metal in an environment. It equals the rate of cathodic processes taking place at the protection potential.

Protection potential Potential below which corrosion is controlled and pits do not propagate.

Quasi-immunity A state of metal when its potential is slightly more positive than the equilibrium potential so that corrosion rate is negligible.

Reduction An electrochemical reaction in which a species gains electrons.

Reference electrode An electrode with stable potential used as a reference to measure the potential of another electrode.

Reinforced concrete Composite material made of concrete and steel reinforcements.

Relative humidity The ratio, expressed as a percentage, of the amount of water vapor present in a given volume of air at a given temperature to the amount required to saturate the air at that temperature.

Resistivity Tendency of a material or of an electrolyte to oppose resistance to the passage of current, also called specific electrical resistance, measured in Ω m.

Rust Corrosion product of iron and steel consisting of various iron oxides and hydrated iron oxides.

Ryznar stability index Indicator of the degree of saturation of calcium carbonate in water. It is calculated as 2pHs–pH (two times the saturation pH of calcium carbonate minus actual pH). Values below 6 indicate scaling tendency.

Salinity Total content of salts dissolved in an aqueous solution.

Sampling Procedure to minimize bias errors in selecting elements from a population for testing. Types of sampling are: random, systematic, stratified, cluster.

Sandblasting Technique for preparation of the surface of a metal before the application of a paint.

Scaling (1) The formation of thick corrosion-product layers on a metal surface. (2) The deposition of water-insoluble constituents on a metal surface.

Seawater A brackish-type water with average salinity of 34–36 g/L.

Selective corrosion Corrosion affecting some metallic alloys in which one component is preferentially dissolved.

Sensitization (sensitized alloy) Sensitization of a stainless steel and nickel alloys is the presence of a chromium-depleted zone around the chromium carbides precipitated at grain boundaries.

Service life The time in which a structure (or equipment) can ensure all functions it was designed for, without extraordinary maintenance interventions.

Soil Complex agglomeration of solid particles, entrapping an aqueous solution, originated from the fragmentation of rock during environmentally induced physical, chemical and biological processes, mainly consisting of coarse particles, sand, clay and silt.

Solubilisation treatment Heat treatment of stainless steels and nickel alloys aiming to dissociate carbides and dissolve carbon in the metal matrix.

Solution Condensed phase with multiple constituents, whose properties can be varied by varying mass ratios of constituents.

Solid solution Atomic scale intermixing of more than one atomic species in the solid state.

Sour corrosion Corrosion in a production fluid of oil and gas industry containing H_2S.

Stabilized stainless steel Addition of stabilizers such as niobium or titanium avoids the chromium carbides precipitation during welding, then impeding sensitization.

Stainless steel (SS) Ferrous alloy with a minimum chromium content of 12% by mass to obtain a passive film on the surface.

Standard electrode potential The reversible potential for an electrode process when all products and reactions are at unitary activity on a scale in which the potential for the standard hydrogen reference electrode is zero.

Stent Endovascular prosthesis.

Strauss test Test evaluating susceptibility to intergranular attack of stainless steel by immersion in a boiling $Cu/CuSO_4$—sulphuric acid mixture.

Stray current External current interfering a metallic structure.

Stress corrosion cracking (SCC) Cracking of a material produced by the combined action of corrosion and tensile stress (residual or applied).

Stress intensity factor (K) Describes the intensification of the stress state at a crack tip, it is used to establish failure criteria in fracture mechanics.

Sulphate reducing bacteria (SRB) Anaerobic bacteria thriving in anoxic environment whilst not dying in the presence of oxygen. They catalyse the reduction of sulphate to sulphide then leading to corrosion conditions of iron.

Sulphide stress corrosion cracking (SSC) Form of Stress Corrosion Cracking (SCC) which proceeds by hydrogen embrittlement mechanism in the presence of hydrogen sulphide.

Surface preparation (before painting) Treatment of the metal surface before painting consisting of removal of dirt, oxides and any deposits. The most used technique is sandblasting.

Sweet corrosion Corrosion in production fluid of oil and gas industry containing CO_2 but not H_2S.

Tafel law Linear relationship between the overvoltage and the logarithm of current density.

Tafel slope The slope of the Tafel law in a potential—logarithm of current density plot.

Tap water Water in domestic piping; it generally meets the requirements of potable water.

Total dissolved solid (TDS) Amount of salts dissolved in a water.

Thermal treatment Heating and cooling of a metallic material aimed at obtaining desired mechanical properties.

Throwing power Distance at which macrocell current vanishes. In cathodic protection, it is the distance from the anode to the point of the cathodic surface still in protection.

Time of wetness Fraction of time in which an electrolyte wets a metal surface.

Transport number Ratio between current density due to the movement of one ion and current transported by all ions present in the electrolyte.

Ultimate tensile strength (UTS) Maximum tensile stress that can be borne by a material.

Ultrasonic testing (UT) High frequency sound waves introduced in the material to detect hidden surfaces or internal flaws.

Uniform corrosion Corrosion that is distributed more or less uniformly over the surface of a metal.

Valence Number of electrons released or gained by an atom or a molecule.

Volt (V) Unit of measurement of electric potential in the International System of Units (SI).

Voltmeter Instrument to measure electrical potential, in V, characterized by very high internal impedance (>10 MΩ).

Water cut Percentage of water phase in oil and gas production fluids.

Water hardness See Hardness.

Weathering steel Steel alloy containing small amounts of chromium, copper, phosphorus and nickel, which forms a protective patina of corrosion products after several years of exposure to the environment.

Weibull distribution It is one of most used Extreme Value Statistics, often used for localized corrosion analysis (for instance, lowest time-to-failure in SCC).

Weight loss Loss of mass of material caused by corrosion.

Weld decay Corrosion of welded zone occurring particularly on stainless steels.

White sandblasted surface Surface free of oxides, rust, rolling scale, corrosion products, or any other substance for at least 98% of the whole surface (see sandblasting).

Working electrode The test or specimen electrode in an electrochemical cell subjected to an exchange of current.

Yield strength Stress at which the material shows a clear deviation from the linear stress-strain relationship (for metals, equivalent to the stress causing a 0.2% residual deformation).

Index

Printed in the United States
By Bookmasters